NEUROMETHODS

Series Editor
Wolfgang Walz
University of Saskatchewan
Saskatoon, SK, Canada

For further volumes:
http://www.springer.com/series/7657

Microelectrode Biosensors

Edited by

Stéphane Marinesco

Université Claude Bernard Lyon 1, Lyon, France
and
Centre de Recherches en Neurosciences de Lyon (CRNL), Inserm

Nicholas Dale

School of Life Sciences, University of Warwick, Coventry, UK

Humana Press

Editors
Stéphane Marinesco
Université Claude Bernard Lyon 1
Lyon, France
and
Centre de Recherches en Neurosciences
de Lyon (CRNL), Inserm

Nicholas Dale
School of Life Sciences
University of Warwick
Coventry, UK

ISSN 0893-2336 ISSN 1940-6045 (electronic)
ISBN 978-1-4939-5901-3 ISBN 978-1-62703-370-1 (eBook)
DOI 10.1007/978-1-62703-370-1
Springer New York Heidelberg Dordrecht London

Printed on acid-free paper

Humana Press is a brand of Springer
Springer is part of Springer Science+Business Media (www.springer.com)

Preface to the Series

Under the guidance of its founders Alan Boulton and Glen Baker, the Neuromethods series by Humana Press has been very successful since the first volume appeared in 1985. In about 17 years, 37 volumes have been published. In 2006, Springer Science + Business Media made a renewed commitment to this series. The new program will focus on methods that either are unique to the nervous system and excitable cells or need special consideration to be applied to the neurosciences. The program will strike a balance between recent and exciting developments like those concerning new animal models of disease, imaging, in vivo methods, and more established techniques. These include immunocytochemistry and electrophysiological technologies. New trainees in neurosciences still need a sound footing in these older methods in order to apply a critical approach to their results. The careful application of methods is probably the most important step in the process of scientific inquiry. In the past, new methodologies led the way in developing new disciplines in the biological and medical sciences. For example, Physiology emerged out of Anatomy in the nineteenth century by harnessing new methods based on the newly discovered phenomenon of electricity. Nowadays, the relationships between disciplines and methods are more complex. Methods are now widely shared between disciplines and research areas. New developments in electronic publishing also make it possible for scientists to download chapters or protocols selectively within a very short time of encountering them. This new approach has been taken into account in the design of individual volumes and chapters in this series.

Wolfgang Walz

Preface

Chemical communication between brain cells lies at the heart of central nervous system functioning. Neurons and glial cells communicate by releasing neurotransmitters and respond to hormones carried by the bloodstream. They also rely on metabolites present in the interstitial fluid such as glucose and lactate for their energy supply.

To understand both chemical signaling and the flow of metabolites between cells, it is highly desirable to analyze the chemical composition of the extracellular medium in real time. We can distinguish at least three different extracellular compartments of significance to brain function: the cerebrospinal fluid, the bulk extracellular space in the brain parenchyma, and the synaptic cleft. An ideal method would allow us to sample from all three compartments in real time. While sampling cerebrospinal fluid is relatively easy for a number of methods, sampling the bulk extracellular space is harder and requires minimally invasive probes to reduce the disruption to the cellular networks in the brain as much as possible. The final compartment, the synaptic cleft, has been the traditional focus for analysis of chemical signaling in the brain through the use of well-established electrophysiological and pharmacological methods. As the dimensions of the synaptic cleft (50 nm in width) remain far smaller than any sensor probes yet developed, direct measurement of transmitter release in the synaptic cleft may arguably only be analyzed by selective expression of reporter molecules within the cleft itself.

Because extracellular concentrations of neurotransmitters can vary with very fast kinetics, it is essential that detection takes place with a temporal resolution in the order of seconds or less. Finally, a major problem is to achieve selective measurements in a complex medium like the brain extracellular fluid, containing probably many hundreds of chemical species.

It is probably fair to say that no one technique matches the ideal requirements of being able to sample from all three compartments with the necessary speed sensitivity and selectivity. Researchers have struggled for over 25 years against these difficulties to develop the efficient sampling techniques now available. Historically, the first method to be developed was to sample minute amounts of the extracellular fluid through microdialysis probes implanted in the brain parenchyma. The analysis of the chemical composition of the dialysate can then take place off-line using techniques such as high-performance liquid chromatography, capillary electrophoresis, or mass spectrometry. Microdialysis has many attractive attributes, such as the ability to sample many different analytes at the same time. Furthermore concentration of the relevant analytes can be measured directly. However in general the temporal resolution is limited, and the microdialysis probes are quite large and thus disrupt the extracellular space and can also provoke tissue reactions during long-term implantation.

A more recent avenue of research has been to develop microelectrodes that are sensitive to a specific neurotransmitter or metabolite through electrochemical detection. The first molecules to be detected in this fashion were directly oxidizable, like dopamine, norepinephrine, serotonin, or their metabolites. For molecules that are not directly electroactive, it is necessary to modify the microelectrode by grafting polymeric or enzymatic membranes capable of translating a local concentration into an electrical current: a concept referred to as biosensing. Biosensors fabricated on microelectrodes can be made much smaller than microdialysis probes and can have sub-second response times. The main target compartment for such microelectrode biosensors is almost certainly the bulk extracellular space. While the synaptic cleft remains out of direct reach, the bulk extracellular space is very important as it contacts all of the component cells of the brain and directly communicates with the synaptic cleft. Thus the transmitters and modulators that act in a more paracrine manner will have to diffuse from their sites of release or production through this extracellular space.

This book will cover the topic of electrochemical biosensor technology as applied to the neurosciences. The International Union of Pure and Applied Chemistry defines an electrochemical biosensor as "a self-contained integrated device, which is capable of providing specific quantitative or semi-quantitative analytical information using a biological recognition element (biochemical receptor) which is retained in direct spatial contact with an electrochemical transduction element." This book will not therefore cover the new developments in the field of dopamine or serotonin detection using unmodified carbon or platinum electrodes. The reader is invited to refer to other publications on this topic. Instead we will present the concept of microelectrode biosensing and demonstrate the progress that has been made in advancing the technology and how this has been applied to study the nervous system.

Part 1: Designs, Principles, and Methods

This first part addresses the specific issue of manufacturing biosensors that can be implanted in the central nervous system for neurotransmitter detection. It reviews the different technical options for manufacturing stable and sensitive enzymatic biosensors and covers the basic principles underlying electrochemical biosensors. This part also provides practical advice that will enable the reader to develop and prepare their own biosensors.

Part 2: Case Studies

This part provides examples of recent studies where biosensors have made a difference in bringing a new level of understanding of signaling mechanisms in neuroscience. The aim of this second part is not only to increase the readers' awareness about the potential importance of biosensors for their field of interest, but also to impart practical advice on how to use biosensors in integrated experimental programs, how to interpret the electrochemical signals and encourage awareness of the strengths, and potential pitfalls of this technique.

Part 3: The Future?

The book finishes by reviewing recent developments in biosensor technology that have not yet been applied to implantable microelectrodes, but may have great potential: the application of carbon nanotube based biosensors, impedance-based methods with antibody electrodes, aptamer and molecular imprinted polymer-based methods for selective analyte detection, and the potential of electrochemiluminescence techniques.

We hope that this collection of articles by leading practitioners in the field of biosensing will serve to inspire the wider neuroscience and physiology community to adopt these powerful methods in their own applications. We look forward to the widespread advances that will flow from the adoption of biosensing as part of the standard laboratory toolkit.

Lyon, France *Stéphane Marinesco*
Coventry, UK *Nicholas Dale*

Contents

Contributors

LOURDES AGÜÍ • *Department of Analytical Chemistry, Faculty of Chemistry, University Complutense of Madrid, Madrid, Spain*

SAMIR AKHTAR • *Cranfield Health, Cranfield University, Cranfield, Bedfordshire, UK*

JASON J. BURMEISTER • *Anatomy & Neurobiology, Morris K. Udall Parkinson's Disease Research Center of Excellence, Center for Microelectrode Technology, College of Medicine, University of Kentucky, Lexington, KY, USA*

RAYMOND CESPUGLIO • *Faculty of Medicine, Claude Bernard University of Lyon, Lyon, France*

THOMAS I.F.H. CREMERS • *Department of Pharmacy, University of Groningen, Groningen, The Netherlands; Brains On-Line BV, Groningen, The Netherlands*

NICHOLAS DALE • *School of Life Sciences, University of Warwick, Coventry, UK*

VERDA A. DAVIS • *Anatomy & Neurobiology, Morris K. Udall Parkinson's Disease Research Center of Excellence, Center for Microelectrode Technology, College of Medicine, University of Kentucky, Lexington, KY, USA*

MIRANDA (VAN DER ZEYDEN) EVERING • *Department of Pharmacy, University of Groningen, Groningen, The Netherlands; Brains On-Line BV, Groningen, The Netherlands*

DAN FEI • *Leicester School of Pharmacy, De Montfort University, Leicester, UK*

ROBERT J. FORSTER • *School of Chemical Sciences, Dublin City University, Dublin, Ireland*

OLIVIER FREY • *Bio Engineering Laboratory, Department of Biosystems Science and Engineering, Swiss Federal Institute of Technology (ETH) Zurich, Basel, Switzerland*

YI GE • *Cranfield Health, Cranfield University, Cranfield, Bedfordshire, UK*

GREG A. GERHARDT • *Anatomy & Neurobiology, Morris K. Udall Parkinson's Disease Research Center of Excellence, Center for Microelectrode Technology, College of Medicine, University of Kentucky, Lexington, KY, USA*

ALAIN GRATTON • *Department of Psychiatry, Douglas Mental Health University Institute, McGill University, Montréal, QC, Canada*

ERIN R. HASCUP • *Department of Psychiatry, Douglas Mental Health University Institute, McGill University, Montréal, QC, Canada*

KEVIN N. HASCUP • *Department of Psychiatry, Douglas Mental Health University Institute, McGill University, Montréal, QC, Canada*

JASON M. HINZMAN • *Anatomy & Neurobiology, Morris K. Udall Parkinson's Disease Research Center of Excellence, Center for Microelectrode Technology, College of Medicine, University of Kentucky, Lexington, KY, USA*

CASSIE HO • *Department of Chemistry, Simon Fraser University, Burnaby, BC, Canada*

PETER HUETTL • *Anatomy & Neurobiology, Morris K. Udall Parkinson's Disease Research Center of Excellence, Center for Microelectrode Technology, College of Medicine, University of Kentucky, Lexington, KY, USA*

NICOLE JAFFREZIC-RENAULT • *Laboratory of Analytical Sciences, Institute of Analytical Sciences, Claude Bernard University Lyon 1, University of Lyon, Villeurbanne, France*

ANDREA JAQUINS-GERSTL • *Department of Chemistry, University of Pittsburgh, Pittsburgh, PA, USA*

TIA E. KEYES • *School of Chemical Sciences, Dublin City University, Dublin, Ireland*
O. MEAGAN LITTRELL • *Anatomy & Neurobiology, Morris K. Udall Parkinson's Disease Research Center of Excellence, Center for Microelectrode Technology, College of Medicine, University of Kentucky, Lexington, KY, USA*
STÉPHANE MARINESCO • *FaUniversité Claude Bernard Lyon 1, Centre de Recherches en Neurosciences de Lyon (CRNL), Inserm U1028, CNRS UMR5292, Lyon, France*
CAROLINE MAUCLER • *INSERM U1028, CNRS UMR5292, Lyon Neuroscience Research Center, AniRA-Neurochem Technological Platform, Team WAKE, Lyon, France*
ADRIAN C. MICHAEL • *Department of Chemistry, University of Pittsburgh, Pittsburgh, PA, USA*
FARHAN MIRZA • *Cranfield Health, Cranfield University, Cranfield, Bedfordshire, UK*
LARISSA NETCHIPOROUK • *Faculty of Medicine, Claude Bernard University of Lyon, Lyon, France*
ROBERT D. O'NEILL • *UCD School of Chemistry and Chemical Biology, University College Dublin, Belfield, Dublin 4, Ireland*
VINAY PARIKH • *Department of Psychology and Neuroscience Program, Temple University, Philadelphia, PA, USA*
PIERRE PERNOT • *INSERM U1028, CNRS UMR5292, Lyon Neuroscience Research Center, AniRA-Neurochem Technological Platform, Team WAKE, Lyon, France*
SERGEY PILETSKY • *Cranfield Health, Cranfield University, Cranfield, Bedfordshire, UK*
JOSÉ M. PINGARRÓN • *Department of Analytical Chemistry, Faculty of Chemistry, University Complutense of Madrid, Madrid, Spain*
FRANCOIS POMERLEAU • *Anatomy & Neurobiology, Morris K. Udall Parkinson's Disease Research Center of Excellence, Center for Microelectrode Technology, College of Medicine, University of Kentucky, Lexington, KY, USA*
DAVID A. PRICE • *Anatomy & Neurobiology, Morris K. Udall Parkinson's Disease Research Center of Excellence, Center for Microelectrode Technology, College of Medicine, University of Kentucky, Lexington, KY, USA*
SI QIN • *Department of Pharmacy, University of Groningen, Groningen, The Netherlands; Brains On-Line BV, Groningen, The Netherlands*
JORGE E. QUINTERO • *Anatomy & Neurobiology, Morris K. Udall Parkinson's Disease Research Center of Excellence, Center for Microelectrode Technology, College of Medicine, University of Kentucky, Lexington, KY, USA*
MARTIN SARTER • *Department of Psychology and Neuroscience Program, Temple University of Michigan, Ann Arbor, MI, USA*
NATALIYA SHRAM • *Faculty of Medicine, Claude Bernard University of Lyon, Lyon, France*
INGRID STRÖMBERG • *Department of Integrative Medical Biology, Umeå University, Umeå, Sweden*
POOJA M. TALAULIKER • *Spinal Cord and Brain Injury Research Center, College of Medicine, University of Kentucky, Lexington, KY, USA*
FAMING TIAN • *Warwick Biosensor Group, School of Life Sciences, University of Warwick, Coventry, UK*
NATALIA VASYLIEVA • *INSERM U1028, CNRS UMR5292, Lyon Neuroscience Research Center, AniRA-Neurochem Technological Platform, Team WAKE, Lyon, France*

NGABI WAHONO • *Department of Pharmacy, University of Groningen, Groningen, The Netherlands; Brains On-Line BV, Groningen, The Netherlands*

SHENQI WANG • *College of Life Science and Technology & Advanced Biomaterials and Tissue Engineering Center, Huazhong University of Science and Technology, Wuhan, P. R. China*

CATHERINE E. WERNER • *Anatomy & Neurobiology, Morris K. Udall Parkinson's Disease Research Center of Excellence, Center for Microelectrode Technology, College of Medicine, University of Kentucky, Lexington, KY, USA*

BEN H.C. WESTERINK • *Department of Pharmacy, University of Groningen, Groningen, The Netherlands; Brains On-Line BV, Groningen, The Netherlands*

PALOMA YÁÑEZ-SEDEÑO • *Department of Analytical Chemistry, Faculty of Chemistry, University Complutense of Madrid, Madrid, Spain*

HUA-ZHONG YU • *Department of Chemistry, Simon Fraser University, Burnaby, BC, Canada*

Part I

Designs, Principles, and Methods

Chapter 1

Microelectrode Designs for Oxidase-Based Biosensors

Stéphane Marinesco and Olivier Frey

Abstract

The electrode is a vital element of a microelectrode biosensor, which is designed to transform the recognition of a biological molecule into an easily quantifiable electrical signal. The most widely used enzymatic biosensors rely on oxidases for detecting neurotransmitters and metabolites in the central nervous system. These enzymes oxidize their substrates while producing peroxide (H_2O_2) from molecular oxygen (O_2). H_2O_2 is then oxidized on the microelectrode to produce an oxidation current that is directly proportional to the analyte concentration and can readily be measured by a low-current potentiostat. Therefore, the microelectrode serves both as a substrate to immobilize the enzyme and as a material to catalyze H_2O_2 oxidation and pass electrical current. In this chapter, we review two major types of microelectrode design. For many years, neuroscientists have used handmade wire electrodes consisting of a platinum (or carbon) wire inserted into a pulled glass capillary. Within the last 10 years, exciting new methods relying on microsystem technologies have emerged. Using these processes, microprobes with multiple integrated microelectrodes can be generated on a semi-industrial scale allowing simultaneous detection of several molecules of interest in different brain areas. Wire electrodes, which can reach extremely small sizes, could remain useful for specific neuroscience applications requiring very high spatial resolution; however, microfabricated silicon needles will undoubtedly contribute to future developments in the field of microelectrode biosensors allowing multi-analyte and network-oriented neuronal monitoring.

Key words Microprobe, Microelectrode, Platinum, Silicon, BioMEMS, Peroxide

1 Introduction

In a microelectrode biosensor, the microelectrode itself is a very important component. In an electrochemical biosensor, defined as "a self-contained integrated device" composed of "a biological recognition element in contact with an electrochemical transduction element" (1), the microelectrode is the transducer designed to pass electrical current in response to neurotransmitter or metabolic molecule presence in the surrounding medium. The role of the microelectrode, however, is much more demanding than just passing current. Translating the recognition of a molecule into an electrical signal can be performed in at least two different ways.

Stéphane Marinesco and Nicholas Dale (eds.), *Microelectrode Biosensors*, Neuromethods, vol. 80,
DOI 10.1007/978-1-62703-370-1_1, © Springer Science+Business Media, LLC 2013

(1) In many cases, the molecule of interest can simply be bound to a biological receptor, and although this binding process does not generate an electrical current, it often perceptibly changes the impedance of the electrode. Detecting changes in electrode impedance is the focus of Chap. 14 of this book. Molecular recognition using molecularly imprinted polymers or aptamers is developed in detail in Chaps. 15 and 17, respectively. To our knowledge, however, impedance methods have not yet been translated into operating systems that can be used on a routine basis in neuroscience research. (2) Most microelectrode biosensors presently used in neuroscience are based on the recognition of molecules by an enzyme belonging to the class of oxidoreductases that is directly immobilized on the surface of the microelectrode. The basic principle of these biosensors relies on shuttling some of the electrons involved in the redox reaction to the electrode.

1.1 Oxidase-Based Biodetection

Oxidoreductases are enzymes that catalyze the transfer of electrons from one molecule to another. In such reactions, one of the molecules is the substrate and the other one is contained within the polypeptidic chain and is called the cofactor. Flavins (derivatives of vitamin B2) and nicotinamides (NAD^+/NADH or $NADP^+$/NADPH) are the most frequent cofactors (2). Because their activity involves a transfer of electrons, these enzymes are especially well suited for the design of electrochemical biosensors. Twenty-two different subtypes of oxidoreductases have been identified (3), but only oxidases have been used so far for biosensors in neuroscience. Oxidases use O_2 as an electron acceptor and catalyze oxidation reactions of the type:

$$\text{Substrate} + O_2 \rightarrow \text{Oxidized product} + H_2O_2 \text{ or } H_2O \quad (1)$$

The main interest of these enzymes is the production of hydrogen peroxide (H_2O_2) as a result of oxidative activity. Because H_2O_2 can be readily oxidized on an electrode, such oxidases are prime candidates for biosensor development. Generally, oxygen as co-substrate is available in non-limiting concentrations both in most biological media and in vivo, which is another advantage of using oxidases. There are, however, specific pathological situations in which PO_2 is low, such as hypoxia or ischemia; thus the enzymatic reaction can become O_2-limited, which might impact the performance of the biosensor. This is especially true if the concentration of the enzyme substrate is high, such as for glucose or lactate. Additionally, we should note that oxidase activity is pH-dependent: enzymatic activity typically displays a bell-shaped relationship with pH, with maximal activity at a pH value that may or may not be the physiological brain pH. For example the optimal pH range for glucose oxidase (*Aspergillus niger*) is 3.5–6.5 (4), while it is 6.0–8.5 for glutamate oxidase (*Streptomyces* sp. X-119-6) (5).

Moreover, enzymatic activity is also temperature-dependent and increases with temperature until the enzyme is denatured. Since enzymatic activity directly impacts biosensor performance, pH, temperature, and PO_2 can potentially limit the detection capabilities (see for example (6)).

Oxidases can use a wide variety of cofactors like iron or copper ions, iron-containing hemes, or flavins (7). In neuroscience research, most oxidases utilized for biosensors have flavin adenine dinucleotide (FAD) as their cofactor. For example, glucose oxidase, glutamate oxidase, choline oxidase, and d-amino acid oxidase are FAD-dependent enzymes (5, 8–10). Exceptions such as l-lactate oxidase, which uses flavin mononucleotide (FMN) (11), and xanthine oxidase, which contains two FAD molecules, two molybdenum atoms, and eight iron atoms (12), can also be used for biosensor fabrication. For FAD-dependent enzymes, the oxidation of the substrate is a two-step reaction:

$$\text{Substrate} + \text{enzyme} - \text{FAD} \rightarrow \text{Oxidized product } + \text{enzyme} - \text{FADH}_2 \qquad (2)$$

Followed by

$$\text{Enzyme} - \text{FADH}_2 + \text{O}_2 \rightarrow \text{Enzyme} - \text{FAD} + \text{H}_2\text{O}_2 \qquad (3)$$

Oxidase-based biosensors translate such enzymatic activity into an electrical current.

1.2 Measuring Oxidative Enzymatic Activity

The simplest and most classical way of converting oxidative enzymatic activity into an electrochemical signal is the reoxidation of the enzymatic reaction product H_2O_2 into molecular oxygen:

$$\text{H}_2\text{O}_2 \rightarrow 2\ \text{H}^+ + \text{O}_2 + 2\text{e}^- \qquad (4)$$

This reaction produces two electrons per molecule of H_2O_2 oxidized on the electrode surface. The oxidation current, whose magnitude is directly related to the substrate concentration, can be recorded by a dedicated acquisition system (potentiostat). Alternative detection methods are to monitor the consumption of oxygen or the production of other reaction products with specific sensors (e.g., NH_3 for glutamate oxidase).

Another conversion of oxidative enzymatic activity into an electrical current can be performed in principle by skipping the second step of the reaction scheme (reaction 1.3) and directly reoxidizing the FAD cofactor by applying a suitable potential. However, this process is much more complicated than it might appear: directly oxidizing $FADH_2$ implies that electrons can shuttle from FAD to the electrode (13). Such electrical wiring could in theory be achieved by single-wall carbon nanotubes (14), platinum nanoparticles (15), or chemical mediators like prussian blue or ruthenium purple (16). They have the further advantage that in most cases the applied potentials are significantly lower than H_2O_2

oxidation potential (for example, ±0 V) making the detection less susceptible to interferences. However, such electrochemical designs have not yet been applied to neuroscience research on a routine basis. A more detailed discussion on the possibilities of direct electrical wiring between the redox center of the enzyme and the electrode surface is developed in Chaps. 4 and 13.

In the rest of this chapter, we will focus on microelectrode designs that are best suited for H_2O_2 detection, since it is such a frequent reaction by-product.

2 Electrode Materials for H_2O_2 Detection

The classical electrode material for detecting H_2O_2 is platinum. H_2O_2 oxidation has been shown to be catalyzed by the presence of platinum oxides (Pt-$(OH)_2$) on the surface of the electrode (17):

$$H_2O_2 + Pt(OH)_2 \rightarrow Pt + 2\ H_2O + O_2 \tag{5}$$

$$Pt + 2H_2O \rightarrow Pt(OH)_2 + 2H^+ + 2e^- \tag{6}$$

A similar reaction can occur on other noble metals for example palladium, ruthenium, or rhodium (17), although not on gold, which yields very low H_2O_2 oxidation currents (18). For in vivo neuroscience applications, thin wires are interesting for fabricating implantable devices. Platinum wires are now commercially available in a wide variety of choices, but because platinum is a rather soft metal, wires smaller than 50 μm in diameter can be difficult to manipulate. For this reason, more rigid platinum/iridium alloys are currently employed for microelectrode fabrication, in particular those containing 90% platinum and 10% iridium (available, for example, at Goodfellow Corporation, Oakdale, PA, USA).

Another electrode material commonly used in neuroscience is carbon. Carbon can be purchased as fibers that can be very thin (down to 5 μm, Goodfellow, Oakdale, PA, USA), and those fibers can be further etched to reach even smaller diameters by passing current in a mixture of sulfuric acid and chromic acid (19, 20). Such carbon fiber microelectrodes have been widely used for the detection of biogenic amines in vivo, especially dopamine (21). However, carbon is not optimal for H_2O_2 detection, as Fig. 1 demonstrates. Cyclic voltammograms obtained with platinum wire and carbon fiber microelectrode display the different characters of the electrode materials. At platinum electrodes, H_2O_2 is reduced at potentials lower than 100–200 mV (cathodic current) and oxidized above 300 mV (anodic current). In most microelectrode biosensors recording schemes, the platinum microelectrode is held at a constant potential of 500–700 mV vs. Ag/AgCl to achieve continuous and fast-response H_2O_2 detection (constant potential amperometry).

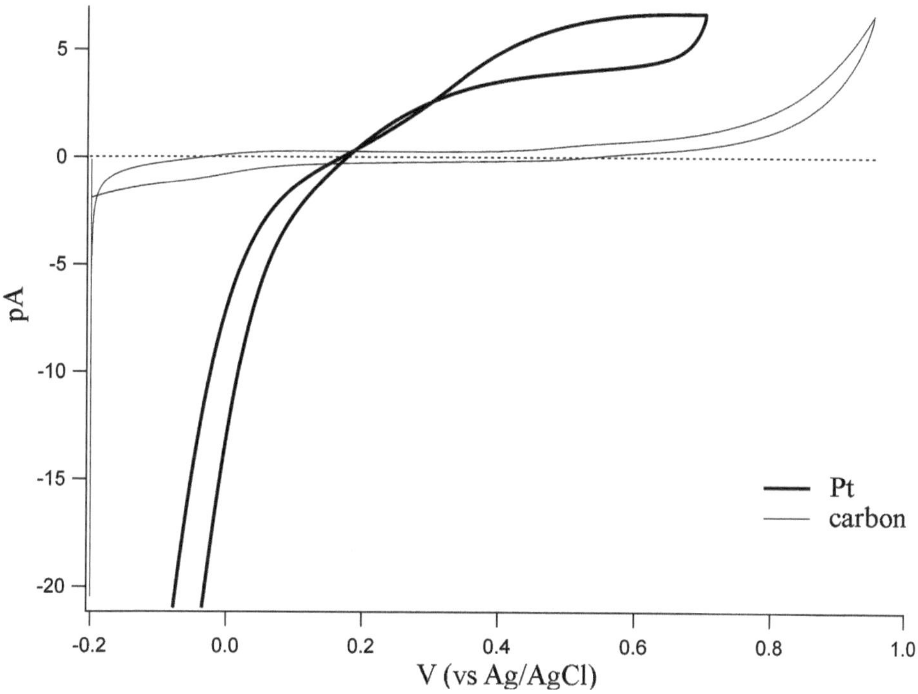

Fig. 1 H_2O_2 oxidation with Pt or Carbon. Example cyclic voltammograms obtained in 20 μM H_2O_2 with a Pt wire (*red*) or carbon fiber (*black*) electrode vs. Ag/AgCl. Note H_2O_2 reduction below, and oxidation above 0.2 V for Pt, whereas H_2O_2 is only oxidized at potentials higher than 0.7 V with carbon

With carbon, however, H_2O_2 detection is much less favorable. There is no detectable reduction, and oxidation starts at much higher potentials, around 750 mV. Therefore, although H_2O_2 detection is feasible in principle with carbon fiber microelectrodes, it is not commonly used for biosensor preparation.

An elegant compromise consisting of an electrochemical metallization of carbon electrodes has been proposed by O'Connell et al. (22). Metallization is performed by electrodepositing a thin layer of either rhodium or ruthenium. This procedure advantageously combines the valuable mechanical properties of carbon fibers with the optimal electrochemical properties of certain metals for H_2O_2 detection. It has been applied to microelectrode biosensors used for brain implantation (23) and could potentially have a broad interest in neuroscience.

3 Procedures for Manufacturing Platinum Wire Microelectrodes

Platinum wire microelectrodes can be readily prepared by hand in the laboratory. Platinum 90%/iridium 10% wires with a diameter of 25 μm (Goodfellow, Oakdale, PA, USA) provide a useful basis for small, thin microelectrodes compatible with brain implantation. Such wires are encased in a glass micropipette. A tight junction

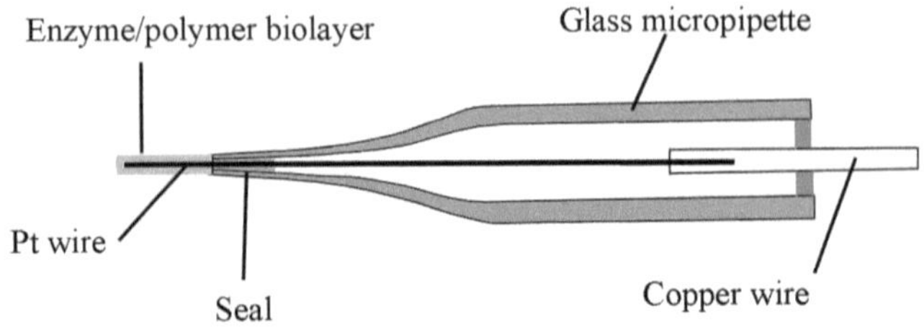

Fig. 2 Schematic representation of a handmade biosensor prepared with a Pt wire encased in a glass micropipette

between the wire and the glass at the tip of the micropipette is necessary to avoid leakage of the recording medium inside the electrode, causing high background currents. Placing the platinum wire inside the glass capillary and pulling the two at the same time usually result in a reliable tight junction. Alternatively, the platinum wire can be inserted into the already pulled glass micropipette. The tip of the micropipette is then broken manually with fine forceps to let the wire protrude out, and a tight seal is achieved either by melting the glass locally or by applying a small drop of epoxy resin (for example Araldite® 5 min). The platinum wire is then glued to a sturdier copper wire (or any metallic wire) with silver conductive paint (available for example from Electrolube, RS components, Beauvais, France), and the copper wire is glued to the back of the glass micropipette with epoxy resin. Finally, the part of the platinum wire extending out of the micropipette tip is cut to the desired length using fine scissors, to make a functional microelectrode capable of detecting H_2O_2 (Fig. 2).

Even though microwires are very small and suitable to reach deep brain structures, the main disadvantage of such handmade platinum microelectrodes is their lack of reproducibility. Variations in microelectrode sensitivity to H_2O_2 arise from differences in the surface of the platinum active part that is cut by hand, a problem that can be partly solved by precise polishing of the electrode tip. In addition, the sensitivity of the platinum wire can also vary depending on the lot purchased from the supplier. Another limitation of these microelectrodes is their inability to detect more than one analyte at a time, while in many cases researchers need to detect several neurotransmitters. Moreover, carefully controlling amperometric recordings requires the use of a second “blank” or “null” biosensor in which the enzyme is replaced by bovine serum albumin. Multi-electrode devices have been presented recently by assembling single wires into bundles to create tetrodes (24) or arrays (25, 26). Such elaborate fabrication relies heavily on the experience of the technician: it is very tedious work that suffers from limited accuracy and reproducibility of the relative electrode positioning. For all these reasons, several groups have investigated the possibility of manufacturing platinum microelectrodes on silicon or ceramic needles using microfabrication technologies.

4 Microprobe Arrays Fabricated Using Microfabrication Techniques

With increased use of photolithographic processes for microelectromechanical systems (MEMS) it seemed likely, as early as the late 1960s, that the same microfabrication techniques could also be used for the construction of microelectrodes. The main interest for such microfabrication methods is that microprobe layouts can be designed and reproduced at high resolution with spatial accuracy in the micrometer range. Moreover, the processes are executed in a parallel mode allowing mass production in a very reproducible way. As a result, the last 30 years have seen tremendous development of MEMS in the biomedical field, thereby creating a new branch known as BioMEMS.

With respect to in vivo neuronal recordings, microfabrication techniques allow the integration of multiple electrodes on a custom-designed single shank, which significantly increases the number of recording sites per microprobe cross section and reduces brain damage. Several shanks can be precisely assembled together in order to generate two- or even three-dimensional configurations of microelectrodes that are tailored for specific experiments and precisely reach the targeted brain areas. We will now present an overview on recent advances in microprobes fabricated using microtechnological processes.

4.1 General Principles in Microtechnology Processes

In general, micromachining fabrication starts with flat, polished substrates such as silicon or glass, on which a copy of a master pattern is transferred using photolithography (Fig. 3). For this purpose, a light-sensitive material called photoresist is applied onto the silicon substrate and either destabilized (positive photoresist) or stabilized (negative photoresist) by exposure to UV light. The destabilized photoresist can then be removed with a developer. This core technique is used to either structure the substrate itself using different etching steps (bulk micromachining) or to pattern specific layers that were previously deposited on the substrate (surface micromachining). In Fig. 3, for example, a positive resist is applied onto a substrate that was previously coated with a thin layer of a conductive or nonconductive material (Fig. 3a). It is then exposed with UV light through a mask containing transparent and opaque regions defining the pattern to be transferred into the resist (Fig. 3b). A subsequent development step only dissolves the exposed regions of the resist and provides a mask on the substrate surface (Fig. 3c) that allows etching of the underlying layer or substrate itself (Fig. 3d1) before finally removing the resist itself (Fig. 3e1). Alternatively, the developed photoresist can also be used to pattern metal layers (Fig. 3d2). In this case, the thin metal layer is deposited after the photolithographic step. The unwanted metal present on the resist is removed by dissolving the resist with

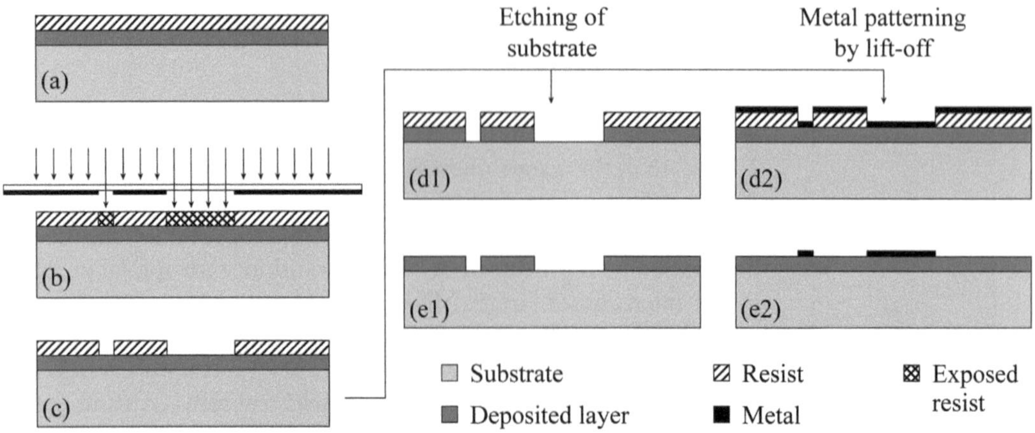

Fig. 3 Schematic cross section of a basic photolithographic process on a substrate with one added thin-film layer to either structure the underlying layer by subsequent etching using the photoresist as mask (**d1**–**e1**) or pattern a metal layer deposited after structuring the photoresist by a process called liftoff (**d2**–**e2**)

the developer and leaves back a structured metal layer that was deposited directly onto the substrate (lift-off, Fig. 3e2).

With progressive development of the fabrication techniques in the last decades, several structuring, modification, and deposition techniques have continually been added and improved to allow a higher design flexibility and improved accuracy and reliability of micromachining (27). This includes methods to deposit or grow different dielectric layers with good insulation properties as well as conductive layers of high metal quality. In fact, there are now an impressive number of variations on the theme of microfabrication and photolithography. In most cases, single-crystal silicon available from the semiconductor industry is used as substrate. It can be structured using different wet etching solutions as well as dry/plasma etching processes. There are also a variety of dielectric layers such as silicon oxide and silicon nitride that can be grown or deposited on top and structured afterwards. Furthermore, different metal layer depositions (evaporation, sputtering, electroplating) are available to create conductive lines and electrodes. Other organic materials such as polymers and plastics have also been successfully incorporated into microfabrication processes. All of them are executed on wafer-level or even on multiple wafers at the same time and therefore allow a high level of parallelization. Moreover a variety of different designs can be produced in a single fabrication run. The engineering part lies in an intelligent sequential combination of deposition and etching processes for creating complex structures. Silicon substrates also allow the integration of electric circuitry for signal acquisition and processing right in the microprobe. It is clearly beyond the scope of this chapter to cover the range of processes presently available in MEMS technology.

Instead, we will focus on the specific methods developed for microelectrode fabrication.

4.2 Multi-Electrode Microprobe Designs Using Layer-by-Layer Microfabrication

The microtechnological processes explained above are planar in nature. Patterns designed using conventional graphical software are transferred onto a flat surface in a serial manner. Therefore an essential characteristic of microelectrodes designed through layer-by-layer microfabrication is their two-dimensional organization that can be defined with a resolution of 1 μm in modern academic research cleanroom facilities.

When designing microprobes with multiple microelectrodes using such technology, we distinguish a general construction consisting of three major steps. First, the overall shape of the microprobe with the optimum length and thickness is defined by the substrate and acts as a mechanical backbone. Then active (sensing) metallic spots are patterned on the surface of the needle, each of which becomes a separate microelectrode that can later be functionalized with enzymes and screening polymer layers to form individual biosensors. Finally, the whole microprobe has to be packaged, i.e., each microelectrode has to be connected to the recording unit through convenient, easy-to-handle wires and connectors and protected from harsh environments by an appropriate encapsulation.

In practice, the microelectrodes are first generated on the whole wafer surface, then the microprobes are "cut out," and finally are electrically and mechanically connected to the appropriate recording device. We should note that once optimized to reach an acceptable yield, this full process is often difficult to modify. The masks used for photolithography are defined in the beginning, and changes in the fabrication process imply modifying the masks, executing new fabrication runs, and adapting again the parameters, which is often costly and time-consuming. It is therefore important to decide what applications are targeted and carefully choose the parameters of the microprobes. In this regard, layer-by-layer microfabrication is often less flexible than handmade fabrication procedures. This disadvantage is compensated by the possibility of mass-fabrication of microprobes that eventually saves time and resources.

4.2.1 Structuring of the Microprobe

The contour of the microprobes including shape, length, and width is designed with respect to future experiments. Several shanks can be arranged in parallel to create combs with a defined pitch (Fig. 4, left). In addition, the number, size, shape, and arrangement of several microelectrodes can be defined and integrated on the surface of every shank. Because all of these electrodes need to be connected to the rear end of the probe with metal tracks, and because each of these tracks needs to be at least 5 μm wide, the number of electrodes actually constrains the overall width of the microprobe (generally 100–250 μm). The thickness of the

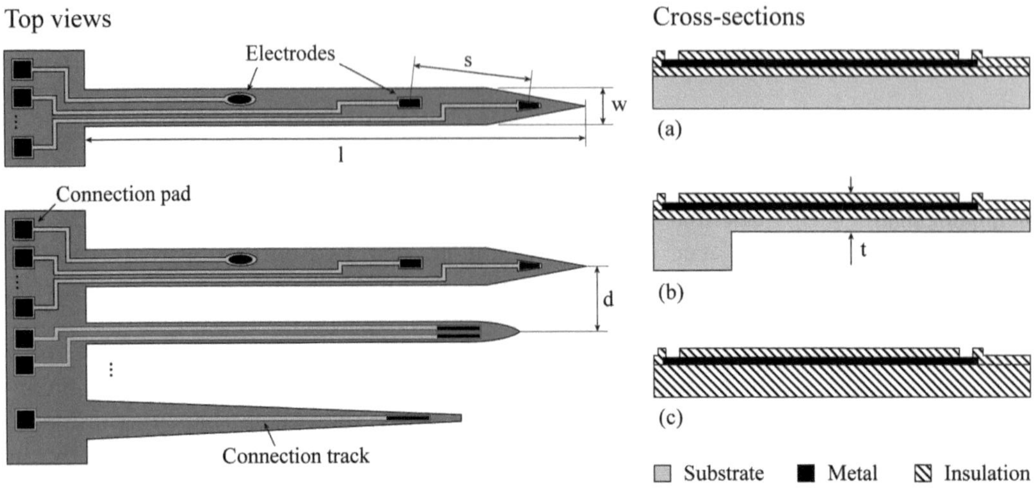

Fig. 4 (*left*) Schematic drawing of a single microprobe and comb illustrating design variations and important parameters. (*right*) Cross sections showing the general construction of a microfabricated probe consisting of the substrate and additional alternating layers of conducting and insulating materials on *top*

microprobe is usually defined by the thickness of the substrate (Fig. 4a,c) unless a controlled thinning from the backside is performed (Fig. 4b). If the backside surface stays flat, electrodes can also be placed there. Furthermore, if a nonconductive substrate is used, no insulation beneath the metal layer is required (Fig. 4c). Because many microprobes can be placed on a wafer substrate (normally 10–20 cm in diameter), different microprobe array variations can be produced in a single process run in parallel (parametric design).

Pioneering work in layer-by-layer microprobe fabrication was presented in the early 1970s by Wise and coworkers (28). They used integrated-circuit fabrication techniques to realize multi-electrode microprobes on silicon substrates with controlled inter-electrode spacing. Since then, such processes have been continuously improved and additional features such as microfluidic channels (29) and integrated circuits have been added in more recent years (30–36). Microelectrode devices based on this technology are now commercially available from NeuroNexus Technologies Inc., Ann Arbor, USA. To define the microprobe shape boron-doped regions are created in the silicon substrate. These regions act as an anisotropic wet etch stop, i.e., creating sharp, straight edges whereas undoped regions are selectively dissolved. Thus, a variety of microprobe shapes with a thickness of about 15 μm and parallel layouts to create microprobe combs can be fabricated. For this purpose, a metal layer is deposited on the substrate as a thin film, patterned using photolithography and sandwiched between two insulation layers—in most cases silicon dioxide and silicon nitride stacks. The top insulation layer is opened again to create microelectrodes along the shank and connection pads at the base, while metal tracks for interconnection remain covered (cf. Fig. 4). Similar procedures have been pursued by other groups with slight variations regarding microprobe structuring and metals used (37–40).

Considerable progress has been made recently to better control the lateral dimension of the microprobes and accurately transfer designed contours, by using dry etching techniques, especially anisotropic deep reactive ion etching (DRIE). First, a DRIE structuring of the probe shape is performed from the top side of the wafer, followed by a wet etching from the back side to thin down the shafts and release the probes (41–43). DRIE produces vertical sidewalls and therefore allows an exact translation of the mask onto the silicon substrate. To control the thickness of the microprobes, silicon-on-insulator (SOI) substrates consisting of a buried oxide layer (serving as etch stop) between bulk silicon can also be used (44–46). Alternatively, wafers can simply be etched through the whole thickness from one side. In this case, the microprobe thickness corresponds to the wafer thickness (44, 47, 48). With better control of the etch rate of DRIE, the etch stop is not required anymore and the thickness can be controlled by the etch process itself. With such control, the fabrication of the microprobe only requires two consecutive DRIE steps, one from the top and another from the back side of standard silicon substrates (49, 50). Such a DRIE process will be described in more detail in Sect. 4.3.

Due to their increased strength and electrical inertness, ceramic-based materials have been chosen as alternative substrate. Platinum electrodes are patterned on 125-μm-thick, 1-in. ceramic square wafers by means of photolithography (51, 52) and polyimide is deposited as a passivation layer. Single microprobes are detached using either a diamond blade or an industrial laser (for details, refer to Chap. 2). Additionally, glass substrates have been used as a structural backbone (53, 54).

Finally, photostructurable epoxy resins are now increasinglyt used as substrates to fabricate microprobes. For example, microprobes with a width below 100 μm and integrated gold or platinum electrodes have been recently produced based on SU-8, a chemically inert and biocompatible material (55, 56). Fabrication processes still have to be optimized, but direct patterning can lead to a rapid and cost-effective production.

The choice of the substrate material can considerably impact the biocompatibility of the implanted microprobes. For example, the mechanical mismatch between stiff microprobes and soft biological tissue can aggravate inflammation at the recording site, leading to tissue damage and signal loss in chronic and semi-chronic experiments (57–59). For these purposes, it is possible to fabricate flexible microprobes with considerable lower Young's moduli. Pickard et al. (60) used photofabrication techniques to produce a nickel–iron microelectrode array on a Kapton film specifically designed for biological implantation. Like SU-8, Kapton provides both the structural and passivation material (Fig. 4c). Alternatively, micromachined polyimide structures with a multilayer metallization have also been successfully prepared (61–63). Reactive ion etching can be used to precisely pattern the probe contour and open the access to the metal layers, even on these new organic materials.

A completely different approach—mentioned here due to its commercial availability from Blackrock Microsystems, Salt Lake City, USA—is the so-called Utah array. It consists of out-of-plane needles on a thin backbone and a single metal electrode on each tip (64).

4.2.2 Generating Biosensor Microelectrodes on the Microprobe Surface

Microelectrodes are created on the wafer surface by selectively etching the top insulation layer and exposing the underlying, patterned thin-film metal. Typically noble metals like Pt, Ir, or gold are employed. Carbon electrodes have also been described recently (65). For oxidase-based biosensors, however, Pt is by far the most frequently used material due to its excellent electrochemical properties for H_2O_2 detection (Sect. 2). The surface of the metal microelectrode must not be covered or contaminated in any way during deposition or subsequent fabrication steps, since this can reduce the electro-catalytic activity of the electrode and result in drastically lower sensitivity of the biosensor. Therefore, the surface of the electrode needs to remain electrochemically stable and clean. High-quality insulation layers are required between the electrodes and between the metal layer and the substrate in order to avoid cross talk and capacitive coupling while being immersed in different solutions for up to several months. Compared with planar microelectrodes used for electrophysiology (i.e., recording of local field potentials or single units) the microelectrodes designed for electrochemical biosensors are often slightly larger to increase the active sensor area, however, with a loss of spatial resolution. For this reason the *effective* surface area of the electrode is sometimes increased by a controlled roughening such as the deposition of platinum black. A good compromise for biosensor microelectrodes is a surface area in the range of 2×10^3–10×10^3 μm^2 guaranteeing an adequate sensitivity and detection limit with currently used enzyme immobilization methods. The shape and number of microelectrodes as well as their relative placement on the shank can be chosen in relation to the intended application in a way to precisely place them in several areas of the brain. An important value and benefit of such microprobes is certainly the large number of integrated electrodes per shank cross section, which is inversely proportional to the brain damage caused by the microprobe. However, analog to planar multi-electrodes arrays (MEAs), there are a couple of important points that have to be considered in the development phase when placing microelectrodes close to each other, especially when some of them are selectively coated with biosensitive layers. The inter-electrode distance is certainly one of the key parameters and has to be optimized. For a high spatial resolution or in the case of a differential measurement (e.g., enzymatic sensor and null sensor), where two electrodes should ideally be in the same detection area, the distance has to be as small as possible. A small inter-electrode spacing, however, increases the risk of chemical cross talk. This can occur (1) during the deposition of the enzymatic membrane when the employed method is not sufficiently spatially controlled and adjacent electrodes are partially coated or (2) during recording when H_2O_2 generated at one biosensor electrode is diffusing to a neighboring one where it is oxidized, leading to an nonspecific signal. This becomes a critical issue when electrode spacing is chosen below 100 μm (66, 67). For the first point, the electrodes can be designed in a way that they simplify and support the subsequent biolayer functionalization. An example is to place the electrodes in a small cavity, which defines a restricted volume for the enzyme membrane, protects it during implantation, and increases the diffusion length between the metal surfaces. Chemical cross talk in some cases can become a serious problem. It depends on a variety of factors including the measurement environment and related diffusion coefficients (blood flow, brain tissue), the enzyme immobilization method (membrane density, enzyme load), and especially the concentration levels of the target analytes. Glucose for example is present in substantially higher concentrations (5–15 mM) in the brain compared with acetylcholine and glutamate (<20 μM), for which chemical cross talk has not been observed using electrodes separated by more than 100 μm. It is clear, that depending on the application, this issue has to be considered during the design of the electrode layout.

4.2.3 Microprobe Packaging and Assembly

After fabrication, microprobes have to be "packaged," i.e., they have to be fixed on a larger device for handling and to provide electrical connection to the acquisition setup. All parts have to be protected both mechanically and electrically from the environment where they will be employed. Packaging is one of the main cost factors in the microchip industry that is often underestimated.

Packaging depends on the final application of the device (1) Microprobes designed for anesthetized animals can be placed on a stick-like printed circuit board (PCB), electrically connected using wirebonds, and encapsulated with liquid-resistant epoxy to protect the wirebonds. The PCB has to be designed in such a way that it can be easily and reliably fixed to a stereotactic frame. The choice of the correct epoxy for assembly is crucial since contact of the electrolyte solution with the wiring will instantaneously result in device failure. (2) In the case of freely moving animals space is much more limited, especially for rodents. The microprobe is often equipped with a small electric socket for reversible connection to the acquisition setup which allows the animal to be released between recordings. There are two main approaches: microprobes are packed in small monolithic devices that are directly fixed to the skull, or microprobes are assembled with highly flexible ribbon cables that are either fabricated directly together with the microprobe (68) or in a separate process and connected afterwards (69–71). The latter method allows implanting the microprobe into the brain tissue without contact with the skull or the connecting cable. Because brain and skull can display small relative movements, inflammation caused by the frictions between the microprobe and the tissue is often generated. Such a "floating" implantation device can help reduce this effect and improve biocompatibility, especially during chronic implantation.

Microprobe arrays can also be assembled in three-dimensional multi-shank combs. Such devices have been presented by Wise's laboratory (28, 30) and more recently by the NeuroProbes consortium (72–75). Multiple combs are stacked on a microfabricated base, thereby creating a 3D needle array with a potential resolution of several hundreds of electrodes with controlled spacing.

4.2.4 Functionalization of the Microprobe Arrays

Once packaged, microprobes have to undergo several post-fabrication procedures before they can be used. These include various cleaning steps of the electrodes and optionally the deposition of additional metal layers. Then, functionalization consisting of immobilization of biosensitive membranes is required for biosensor preparation; therefore, it is important to use an array-compatible deposition method that is spatially controlled, reproducible, and can be performed in parallel. Such procedures are required for reliable generation of multi-analyte arrays. Dip-coating methods in general coat the whole needle-shank and thus are not appropriate for generating different sensors on a single shank, but are clearly a method of choice for wire electrodes. Drop-coating procedures are serial and require high-precision spotters for reproducible coating. However, such procedures are optimized for planar devices and are not always appropriate for 3D arrays. Electrochemical-based deposition methods have the advantage of displaying inherent spatial control and allowing the deposition of functional layers on specific electrodes in parallel. The electrochemical entrapment of enzymes on an electrode has been achieved by numerous methods. Different polymers mostly based on polyphenylenediamine (76–78) and polypyrrole (79–81) were used over the last decades to immobilize the enzyme right out of the solution during polymerization. The methods were continuously adapted including layer-by-layer depositions and multienzyme membranes either to detect new analytes or to improve enzyme stability and biosensor characteristics such as sensitivity, selectivity, response time, and lifetime. Similarly, charged enzymes and proteins have been directly deposited on electrodes by electrochemically aided adsorption (EAA) (82, 83). Their stability could be improved by co-deposited collagen (84) or bovine serum albumin (85). To prevent leaching of the deposited enzyme molecules, chemical cross-linking using glutaraldehyde (49, 86), electrodeposition paints (EP) (87), as well as sol–gels (88) are employed as immobilization matrices. Alternatively, also microfluidic structures can be integrated on the microprobes using microfabrication techniques to provide precise drop deposition (89).

Depending on the electrochemical scheme, and on the applied working potential used for detection, unwanted oxidation of endogenous electroactive molecules like ascorbic acid, catecholamines, uric acid, or serotonin has to be considered. To prevent such interference, screening layers such as polyphenylenediamine (PPD) and/or Nafion can be added to the microelectrode design (6). Finally, the functionalized biosensor microprobes often need to be stored in a cold (~4°C) environment and obviously should not break during handling and transportation. These are further issues that have to be considered during packaging.

4.3 Example Procedure for Microprobe Fabrication

Within the European project NeuroProbes (IST-027017) we fabricated silicon microprobes that are designed for recordings in the medial prefrontal cortex (mPFC) of the rat brain. Therefore, the dimensions of the microprobe were chosen to be 6.5 mm in length with an allover width of 100 μm and allover thickness of 30–100 μm (adjustable with the process). Two 50 μm × 150 μm thin-film Pt microelectrodes are integrated in a 10-μm deep recess, separated by 200 μm and located 500 μm from the tip (Fig. 5a). Additionally, a third, non-recessed electrode with dimensions of 40 μm × 500 μm, is positioned further back to be used as a shank-integrated reference electrode. Additional electrodes can be integrated on the same microprobe at application-desired positions with minor widening of the shank. To access multiple areas in the brain simultaneously, 4-shank combs were designed consisting of four parallel-aligned microprobes of the same layout with pitches of 550 μm and 2 mm, respectively (Fig. 5b, c). All fabrication processes were performed in the facilities of the common microtechnology laboratory of SAMLAB and CSEM (COMLAB), Neuchatel, Switzerland.

The microprobes are fabricated starting with a 300-μm-thick, (100)-oriented, 4-in. silicon wafer. The process is illustrated schematically in Fig. 6. A first mask is transferred to a previously grown thermal oxide (200 nm) using photolithography and buffered hydrofluoric acid (BHF, Fig. 6a). 10 μm deep recesses are etched using KOH (Fig. 6b). A well-chosen design produces vertical side walls along the long side of the electrode and two

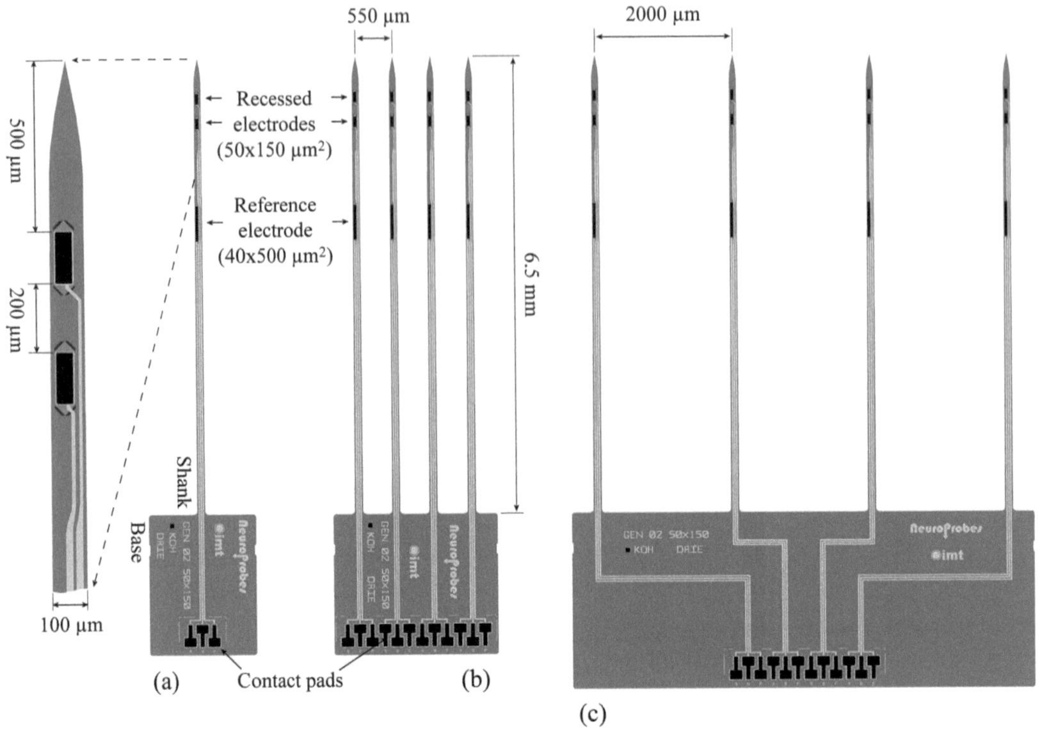

Fig. 5 Layout of the silicon microprobes: (**a**) Single microprobe with three electrodes of which the two close to the tip are recessed by a KOH process. (**b**) 4-shank comb consisting of four single microprobe shafts with a pitch of 550 µm. (**c**) 4-shank comb with a 2,000-µm pitch

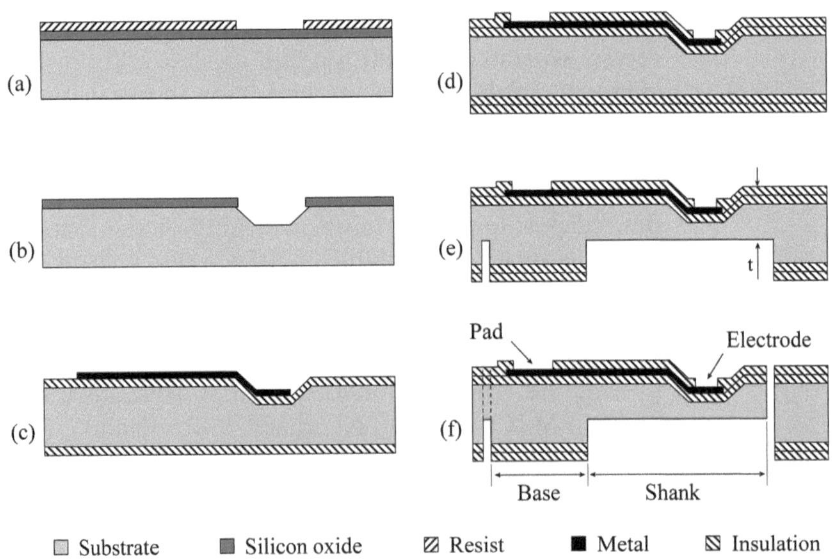

Fig. 6 Cross section of the complete fabrication process for silicon microprobe arrays (see text for details; layer thicknesses are not on scale)

inclined faces towards the front and the back of the electrode (short side). Then, the thermal oxide is completely removed with BHF. 200 nm of thermally grown dry oxide and 200 nm LPCVD silicon nitride layers are successively deposited as a bottom insulation layer. Both are pinhole-free and display excellent passivation properties. The metallization (20 nm of Ta and 130 nm of Pt, consecutively deposited by electron-beam evaporation) is patterned by a lift-off process (Fig. 6c). Here, a total resist thickness of about 8 μm is required to reliably cover the edges of the recesses and produce well-defined metal structures. The interconnection tracks are laid across the inclined side walls of the KOH recess where the electrode is placed. The recess anchors and protects the enzymatic membrane on the electrodes during tissue insertion and removal.

A second 200 nm LPCVD silicon nitride layer and a 200 nm PECVD oxide layer are used as top passivation and are opened at the recording sites and back bonding pads by reactive ion etching (Fig. 6d). The microelectrode structuring is completed with a H_3PO_4 dip (10 min at 160°C) to make sure that all nitride is removed from the electrode surface. The opening is slightly smaller than the metal structure to avoid exposure of the underlying Ta adhesion layer.

The microprobes are formed by a first thinning of the wafer from the backside using deep reactive ion etching (DRIE) to the intended needle thickness. This is performed under the future shaft and around the base (Fig. 6e). Because of well-controlled etch rates, the shaft thickness can be controlled to within 5 μm. A second DRIE etch from the front side shapes the needle contour and releases the microprobes. For both etching steps thick photoresist is used as mask (Fig. 6f).

In Fig. 7a a fully processed wafer including a detailed view of the single probes is shown. All silicon microprobes—single probes and combs—remain in the wafer attached to the remaining silicon grid by small silicon joints. A slight pressure on the base breaks them off and releases the chosen microprobe (array). In total over 250 single probes and arrays are fabricated in parallel on a single 4-in. wafer.

Figure 7b shows two SEM micrographs of the front part of a single silicon microprobe and a close-up view of the two KOH-recessed electrodes. The DRIE process creates a highly reproducible shape of the probe with respect to the designed masks (>2 μm accuracy). Very sharp tips that facilitate insertion into the brain can be achieved. An assessment of the quality of the Pt microelectrodes by cyclic voltammetry in 1 M H_2SO_4 solution shows good surface characteristics and a reproducible roughness factor of ~2, determined using the hydrogen desorption charge (210 μC/cm^2 (90), Fig. 7c).

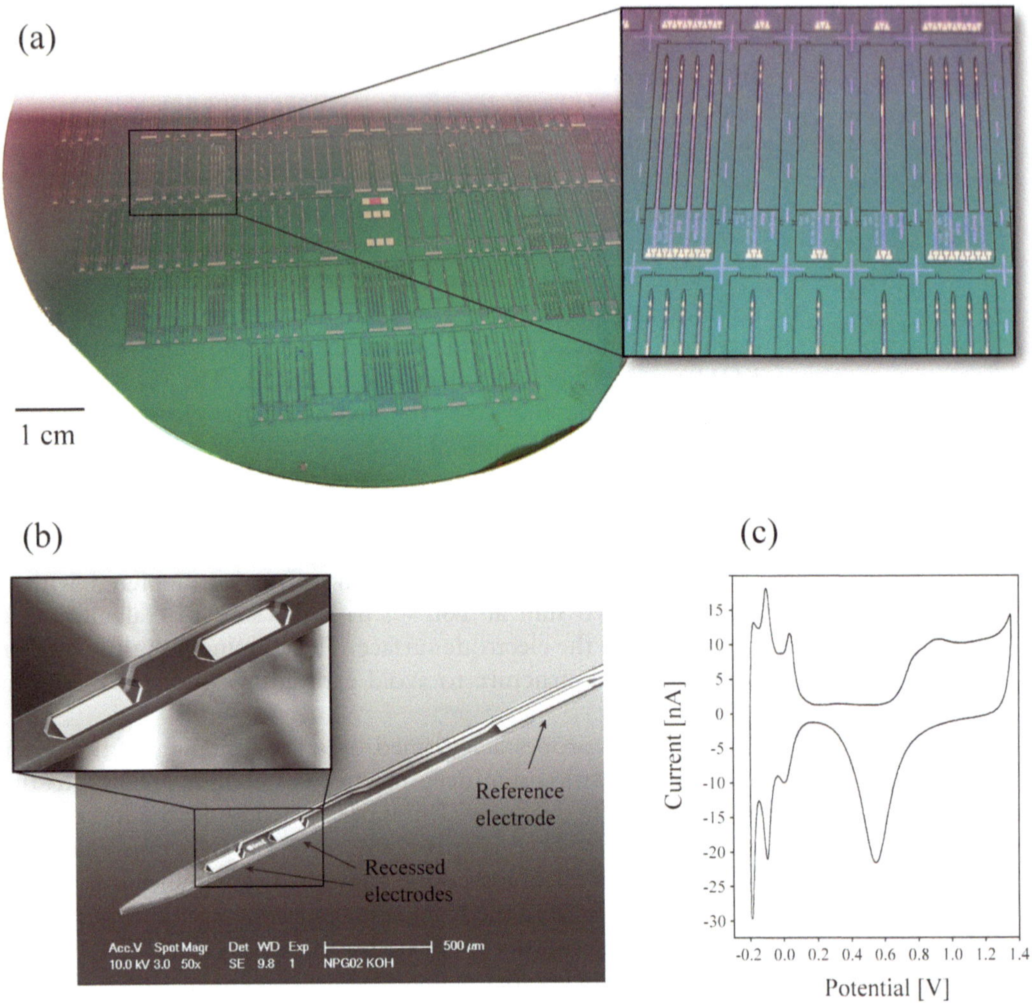

Fig. 7 (**a**) Processed wafer: the remaining wafer structure acts as a grid, to which the probes are attached until use. The *inset* shows a close-up view of single microprobes and combs. (**b**) Front part of the shank of the silicon microprobe with two recessed electrodes and the reference electrode. (**c**) Cyclic voltammogram (15th cycle) of a single Pt electrode (50 μm × 150 μm) in 1 M H_2SO_4 (100 mV/s)

5 Conclusion

Silicon is clearly a substrate of choice for the realization of microprobe arrays with integrated electrodes. It is precisely machinable, allows the direct integrations of circuitry (91–93) or microfluidic channels (29, 44, 94–96), and is inherently biocompatible, even with additional passivation layers. Using photolithographic processes multi-electrode arrays can be designed to record from multiple target areas simultaneously.

With the continued innovation in layer-by-layer microfabrication processes, silicon-based biosensors will abound in neuroscience laboratories in the future. Examples of future innovation and challenges to be considered are as follows:

1. Microprobe fabrication and the subsequent packaging are expensive and time-consuming steps. However, the life span of the final biosensor devices is mostly limited by the biosensitive layer containing denaturizing enzymes. Electrode cleaning to regenerate the initial metal electrode quality is difficult for silicon microprobe arrays and results in single-use devices. Therefore, innovative packaging solutions have to be provided that either are very simple and inexpensive or allow to reversibly exchange the life-limiting part, hence, the microprobe arrays themselves.
2. The high number of biosensor electrodes in a single implantable device has high demands on the connectivity and the handling of the acquired data. Instead of thick wire bundles leaving the animals head, new approaches such as integrated multi-potentiostats or multi-plexing circuits close to the array are required to reduce cable number and simplify connectivity. This has also advantages of reducing noise contributions.
3. To record in freely moving animals small setups are required that do not restrict the animal's behavior. Preferably these systems are equipped with a transponder for wireless data transmission.

An important drawback of silicon probes is their overall fragility especially when their cross section is thinned down to about of 20–30 μm. In fact, silicon is quite a brittle material that can break when microprobes are longer than 1 cm and less than 10–20 μm thickness (97, 98). Smaller handmade electrodes can be prepared using Pt wire or carbon fibers encased in a glass micropipette. Such cylindrical electrodes can reach diameters in the order of 5–25 μm. These types of microelectrodes are useful for neuroscience applications requiring extremely small probe size, such as implantation in minute hypothalamic or brainstem nuclei, or recordings in thin brain slices. In fact, the wide variety of in vivo or in vitro applications will probably require a multiplicity of microelectrode designs including silicon-based needles and handmade glass micropipettes. These microelectrode designs should be a decisive step towards the online monitoring of neurotransmitter release in the living brain, which is key to understanding chemical communication between brain cells and brain functioning in general.

Acknowledgements

SM was supported by Inserm U628, Université Claude Bernard Lyon I, and by grants from the Agence Nationale pour la Recherche (ANR-09-BLAN-0063 Neurosense). OF was supported by the

IMT of EPF Lausanne and the D-BSSE of ETH Zurich. Biosensor microprobes were developed at SAMLAB and the Microsystems Technology Division of the CSEM, Neuchâtel, Switzerland, within the NeuroProbes project (EU IP IST-027017). We are grateful to Marta K. Lewandowska for useful suggestions on an earlier version of this manuscript.

References

1. Thevenot DR, Toth K, Durst RA, Wilson GS (1999) Electrochemical biosensors. Recommended definitions and classification. Pure Appl Chem 71:2333–2348
2. Walsh C (1980) Flavin coenzymes: at the crossroads of biological redox chemistry. Acc Chem Res 13:148–155
3. Moss GP (2010) International union of biochemistry and molecular biology: recommendations on biochemical and organic nomenclature, symbols and terminology etc. (http://www.chem.qmul.ac.uk/iubmb/enzyme/EC1/)
4. Nakamura S, Fujiki S (1968) Comparative studies on the glucose oxidases of *Aspergillus niger* and *Penicillium amagasakiense*. J Biochem 63:51–58
5. Arima J, Tamura T, Kusakabe H, Ashiuchi M, Yagi T, Tanaka H, Inagaki K (2003) Recombinant expression, biochemical characterization and stabilization through proteolysis of an l-glutamate oxidase from *Streptomyces sp.* X-119-6. J Biochem 134:805–812
6. Pernot P, Mothet JP, Schuvailo O, Soldatkin A, Pollegioni L, Pilone M, Adeline MT, Cespuglio R, Marinesco S (2008) Characterization of a yeast d-amino acid oxidase microbiosensor for d-serine detection in the central nervous system. Anal Chem 80:9
7. Sedewitz B, Schleifer KH, Gotz F (1984) Purification and biochemical characterization of pyruvate oxidase from Lactobacillus plantarum. J Bacteriol 160:273–278
8. Keilin D, Hartree EF (1952) Specificity of glucose oxidase (notatin). Biochem J 50:331–341
9. Pollegioni L, Falbo A, Pilone MS (1992) Specificity and kinetics of *Rhodotorula gracilis* d-amino acid oxidase. Biochim Biophys Acta 1120:11–16
10. Ohishi N, Yagi K (1979) Covalently bound flavin as prosthetic group of choline oxidase. Biochem Biophys Res Commun 86:1084–1088
11. Yorita K, Janko K, Aki K, Ghisla S, Palfey BA, Massey V (1997) On the reaction mechanism of l-lactate oxidase: quantitative structure-activity analysis of the reaction with para-substituted l-mandelates. Proc Natl Acad Sci USA 94:9590–9595
12. Enroth C, Eger BT, Okamoto K, Nishino T, Nishino T, Pai EF (2000) Crystal structures of bovine milk xanthine dehydrogenase and xanthine oxidase: structure-based mechanism of conversion. Proc Natl Acad Sci USA 97:10723–10728
13. Willner I, Heleg-Shabtai V, Blonder R, Katz E, Tao GL, Buckmann AF, Heller A (1996) Electrical wiring of glucose oxidase by reconstitution of FAD-modified monolayers assembled onto Au-electrodes. J Am Chem Soc 118:10321–10322
14. Patolsky F, Weizmann Y, Willner I (2004) Long-range electrical contacting of redox enzymes by SWCNT connectors. Angew Chem 43:2113–2117
15. Bahshi L, Frasconi M, Tel-Vered R, Yehezkeli O, Willner I (2008) Following the biocatalytic activities of glucose oxidase by electrochemically cross-linked enzyme-Pt nanoparticles composite electrodes. Anal Chem 80:8253–8259
16. Tian F, Llaudet E, Dale N (2007) Ruthenium purple-mediated microelectrode biosensors based on sol–gel film. Anal Chem 79:6760–6766
17. Hall SB, Khudaish EA, Hart AL (1998) Electrochemical oxidation of hydrogen peroxide at platinum electrodes. Part 1: An adsorption controlled mechanism. Electrochim Acta 43:579–588
18. Gooding JJ, Praig VG, Hall EA (1998) Platinum-catalyzed enzyme electrodes immobilized on gold using self-assembled layers. Anal Chem 70:2396–2402
19. Armstrong-James M, Fox K, Millar J (1980) A method for etching the tips of carbon fibre microelectrodes. J Neurosci Methods 2:431–432
20. Kuras A, Gutmaniene N (2000) Technique for producing a carbon-fibre microelectrode with the fine recording tip. J Neurosci Methods 96:143–146
21. Wightman RM (2006) Detection technologies, probing cellular chemistry in biological systems with microelectrodes. Science 311:1570–1574
22. O'Connell PJ, O'Sullivan CK, Guilbault GG (1998) Electrochemical metallisation of carbon electrodes. Anal Chim Acta 373:261–270

23. Schuvailo OM, Soldatkin OO, Lefebvre A, Cespuglio R, Soldatkin AP (2006) Highly selective microbiosensors for in vivo measurement of glucose, lactate and glutamate. Anal Chim Acta 573–574:110–116
24. Gray CM, Maldonado PE, Wilson M, McNaughton B (1995) Tetrodes markedly improve the reliability and yield of multiple single-unit isolation from multi-unit recordings in cat striate cortex. J Neurosci Methods 63:43–54
25. Musallam S, Bak MJ, Troyk PR, Andersen RA (2007) A floating metal microelectrode array for chronic implantation. J Neurosci Methods 160:122–127
26. Ulbert I, Halgren E, Heit G, Karmos G (2001) Multiple microelectrode-recording system for human intracortical applications. J Neurosci Methods 106:69–79
27. Madou MJ (1997) Fundamentals of microfabrication. CRC, Boca Raton, FL
28. Wise KD, Angell JB, Starr A (1970) An integrated-circuit approach to extracellular microelectrodes. IEEE Trans Biomed Eng BM17(3):238–247
29. Chen JK, Wise KD, Hetke JF, Bledsoe SC (1997) A multichannel neural probe for selective chemical delivery at the cellular level. IEEE Trans Biomed Eng 44:760–769
30. Bai Q, Wise KD, Anderson DJ (2000) A high-yield microassembly structure for three-dimensional microelectrode arrays. IEEE Trans Biomed Eng 47:281–289
31. Hoogerwerf AC, Wise KD (1994) A 3-dimensional microelectrode array for chronic neural recording. IEEE Trans Biomed Eng 41:1136–1146
32. Najafi K, Wise KD, Mochizuki T (1985) A high-yield IC-compatible multichannel recording array. IEEE Trans Electron Devices 32:1206–1211
33. Perlin GE, Wise KD (2009) Ultra-compact integration for fully-implantable neural microsystems. In: IEEE 22nd international conference on micro electro mechanical systems, Sorrento, pp 228–231
34. Wise KD (2009) Wireless integrated microsystems: wearable and implantable devices for improved health care. In: International conference on solid-state sensors, actuators and microsystems (transducers), Denver, USA, pp 1–8
35. Wise KD, Anderson DJ, Hetke JF, Kipke DR, Najafi K (2004) Wireless implantable microsystems: high-density electronic interfaces to the nervous system. Proc IEEE 92:76–97
36. Yao Y, Gulari MN, Wiler JA, Wise KD (2007) A Microassembled low-profile three-dimensional microelectrode array for neural prosthesis applications. J Microelectromech Syst 16: 977–988
37. Ensell G, Banks DJ, Ewins DJ, Balachandran W, Richards PR (1996) Silicon-based microelectrodes for neurophysiology fabricated using a gold metallization nitride passivation system. J Microelectromech Syst 5:117–121
38. Leong KH, Ensell G, Wall P, Pickard RS (1990) Multichannel microelectrode probes machined in silicon. Biosens Bioelectron 5:303–310
39. Lin LW, Pisano AP (1999) Silicon-processed microneedles. J Microelectromech Syst 8: 78–84
40. Takahashi K, Matsuo T (1984) Integration of multi-microelectrode and interface circuits by silicon planar and 3-dimensional fabrication technology. Sens Actuat 5:89–99
41. Kewley DT, Hills MD, Borkholder DA, Opris IE, Maluf NI, Storment CW, Bower JM, Kovacs GTA (1997) Plasma-etched neural probes. Sens Actuat A Phys 58:27–35
42. Oh SJ, Song JK, Kim JW, Kim SJ (2006) A high-yield fabrication process for silicon neural probes. IEEE Trans Biomed Eng 53:351–354
43. Yoon TH, Hwang EJ, Shin DY, Park SI, Oh SJ, Jung SC, Shin HC, Kim SJ (2000) A micromachined silicon depth probe for multichannel neural recording. IEEE Trans Biomed Eng 47:1082–1087
44. Cheung KC, Djupsund K, Dan Y, Lee LP (2003) Implantable multichannel electrode array based on SOI technology. J Microelectromech Syst 12:179–184
45. Ensell G, Banks DJ, Richards PR, Balachandran W, Ewins DJ (2000) Silicon-based microelectrodes for neurophysiology, micromachined from silicon-on-insulator wafers. Med Biol Eng Comput 38:175–179
46. Norlin P, Kindlundh M, Mouroux A, Yoshida K, Hofmann UG (2002) A 32-site neural recording probe fabricated by DRIE of SOI substrates. J Micromech Microeng 12:414–419
47. Errachid A, Ivorra A, Aguilo J, Villa R, Zine N, Bausells J (2001) New technology for multi-sensor silicon needles for biomedical applications. Sens Actuat B Chem 78:279–284
48. Wassum KM, Tolosa VM, Wang JJ, Walker E, Monbouquette HG, Maidment NT (2008) Silicon wafer-based platinum microelectrode array biosensor for near real-time measurement of glutamate in vivo. Sensors 8:5023–5036
49. Frey O, Holtzman T, McNamara RM, Theobald DEH, van Der Wal PD, de Rooij NF, Dalley JW, Koudelka-Hep M (2010) Enzyme-based choline and L-glutamate biosensor electrodes on silicon microprobe arrays. Biosens Bioelectron 26:477–484
50. Herwik S, Kisban S, Aarts AAA, Seidl K, Girardeau G, Benchenane K, Zugaro MB, Wiener SI, Paul O, Neves HP, Ruther P (2009) Fabrication technology for silicon-based microprobe arrays used in acute and sub-chronic

neural recording. J Micromech Microeng 19: 074008
51. Burmeister JJ, Moxon K, Gerhardt GA (2000) Ceramic-based multisite microelectrodes for electrochemical recordings. Anal Chem 72:187–192
52. Burmeister JJ, Pomerleau F, Palmer M, Day BK, Huettl P, Gerhardt GA (2002) Improved ceramic-based multisite microelectrode for rapid measurements of L-glutamate in the CNS. J Neurosci Methods 119:163–171
53. Lin CW, Lee YT, Chang CW, Hsu WL, Chang YC, Fang W (2009) Novel glass microprobe arrays for neural recording. Biosens Bioelectron 25:475–481
54. Vrouwe E, Keldermann AJ, Bloom M (2008) Microfluidic glass needle arrays for drug dosing during neural recording. In: 12th International conference on miniaturized systems for chemistry and life science (microTAS), San Diego, CA, pp 378–380
55. Liu SY, Liu G, Tian YC, Chen YP, Yu HQ, Fang F (2007) An innovative microelectrode fabricated using photolithography for measuring dissolved oxygen distributions in aerobic granules. Environ Sci Technol 41:5447–5452
56. Tijero M, Gabriel G, Caro J, Altuna A, Hernández R, Villa R, Berganzo J, Blanco FJ, Salido R, Fernández LJ (2009) SU-8 microprobe with microelectrodes for monitoring electrical impedance in living tissues. Biosens Bioelectron 24:2410–2416
57. Seymour JP, Kipke DR (2007) Neural probe design for reduced tissue encapsulation in CNS. Biomaterials 28:3594–3607
58. Stieglitz T (2004) Considerations on surface and structural biocompatibility as prerequisite for long-term stability of neural prostheses. J Nanosci Nanotechnol 4:496–503
59. Subbaroyan J, Martin DC, Kipke DR (2005) A finite-element model of the mechanical effects of implantable microelectrodes in the cerebral cortex. J Neural Eng 2:103–113
60. Pickard RS, Joseph PL, Collins AJ, Hicks RCJ (1979) Flexible printed-circuit probe for electrophysiology. Med Biol Eng Comput 17: 261–267
61. Stieglitz T, Beutel HMS, Meyer JU (2000) Micromachined, polyimide-based devices for flexible neural interfaces. Biomed Microdevices 2:283–294
62. Takeuchi S, Suzuki T, Mabuchi K, Fujita H (2004) 3D flexible multichannel neural probe array. J Micromech Microeng 14:104–107
63. Mercanzini A, Cheung K, Buhl DL, Boers M, Maillard A, Colin P, Bensadoun JC, Bertsch A, Renaud P (2008) Demonstration of cortical recording using novel flexible polymer neural probes. Sens Actuat A Phys 143:90–96
64. Campbell PK, Jones KE, Huber RJ, Horch KW, Normann RA (1991) A silicon-based, three-dimensional neural interface: manufacturing processes for an intracortical electrode array. IEEE Trans Biomed Eng 38:758–768
65. Zachek MK, Park J, Takmakov P, Wightman RM, McCarty GS (2010) Microfabricated FSCV-compatible microelectrode array for real-time monitoring of heterogeneous dopamine release. Analyst 135:1556–1563
66. Strike DJ, Hengstenberg A, Quinto M, Kurzawa C, Koudelka-Hep M, Schuhmann W (1999) Localized visualization of chemical cross-talk in microsensor arrays by using scanning electrochemical microscopy. Mikrochim Acta 131:47–55
67. Yu PG, Wilson GS (2000) An independently addressable microbiosensor array: What are the limits of sensing element density? Faraday Discuss 116:305–317
68. Hetke JF, Lund JL, Najafi K, Wise KD, Anderson DJ (1994) Silicon ribbon cables for chronically implantable microelectrode arrays. IEEE Trans Biomed Eng 41:314–321
69. Kisban S, Herwik S, Seidl K, Rubehn B, Paul O, Ruther P, Stieglitz T, Jezzini A (2007) Microprobe array with low impedance electrodes and highly flexible polyimide cables for acute neural recording. In: 29th Annual international conference of the IEEE engineering in medicine and biology society, Lyon, pp 175–178
70. Kisban S, Janssen P, Herwik S, Stieglitz T, Paul O, Ruther P (2008) Hybrid microprobes for chronic implantation in the cerebral cortex. In: 30th Annual international conference of the IEEE engineering in medicine and biology society, Vancover, BC, pp 2016–2019
71. Ruther P, Herwik S, Kisban S, Seidl K, Paul O (2010) Recent progress in neural probes using silicon MEMS technology. IEEJ Trans Elect Electron Eng 5:505–515
72. Aarts AAA, Neves HP, Puers RP, Van Hoof C (2008) An interconnect for out-of-plane assembled biomedical probe arrays. J Micromech Microeng 18:064004
73. Neves HP, Orban GA, Koudelka-Hep M, Ruther P (2007) Development of multifunctional probe arrays for cerebral applications. In: 3rd International IEEE EMBS conference on neural engineering, Kohala coast, HI, pp 104–109
74. Neves HP, Ruther P (2007) The neuroprobes project. In: 29th Annual international conference of the IEEE engineering in medicine and biology society, Lyon, pp 6443–6445, 16 Nov 2007
75. Ruther P, Aarts AAA, Frey O, Herwik S, Kisban S, Seidl K, Spieth S, Schumacher A,

Kofudelka-Hep M, Paul O, Stieglitz T, Zengerle R, Neves HP (2008) The neuro-probes project - multifunctional probe arrays for neural recording and stimulation. In: 13th Annual conference of the IFESS, Freiburg, pp 238–240
76. O'Neill RD, Lowry JP, Rocchitta G, McMahon CP, Serra PA (2008) Designing sensitive and selective polymer/enzyme composite biosensors for brain monitoring in vivo. Trends Anal Chem 27:78–88
77. Lowry JP, Mcateer K, Elatrash SS, Duff A, O'Neill RD (1994) Characterization of glucose oxidase-modified poly(phenylenediamine)-coated electrodes in-vitro and in-vivo - homogeneous interference by ascorbic-acid in hydrogen-peroxide detection. Anal Chem 66:1754–1761
78. McMahon CP, O'Neill RD (2005) Polymer-enzyme composite biosensor with high glutamate sensitivity and low oxygen dependence. Anal Chem 77:1196–1199
79. Cosnier S, Innocent C, Allien L, Poitry S, Tsacopoulos M (1997) An electrochemical method for making enzyme microsensors: application to the detection of dopamine and glutamate. Anal Chem 69:968–971
80. Llaudet E, Botting NP, Crayston JA, Dale N (2003) A three-enzyme microelectrode sensor for detecting purine release from central nervous system. Biosens Bioelectron 18:43–52
81. Malitesta C, Palmisano F, Torsi L, Zambonin PG (1990) Glucose fast-response amperometric sensor based on glucose oxidase immobilized in an electropolymerized poly(*o*-phenylenediamine) film. Anal Chem 62:2735–2740
82. Johnson KW (1991) Reproducible electrodeposition of biomolecules for the fabrication of miniature electroenzymatic biosensors. Sens Actuat B Chem 5:85–89
83. Lee JS, Nakahama S, Hirao A (1991) A new glucose sensor using microporous enzyme membrane. Sens Actuat B Chem 3:215–219
84. Wang SS, Vieth WR (1973) Collagen-enzyme complex membranes and their performance in biocatalytic modules. Biotechnol Bioeng 15:93–115
85. Strike DJ, de Rooij NF, Koudelka-Hep M (1993) Electrodeposition of glucose-oxidase for the fabrication of miniature sensors. Sens Actuat B Chem 13:61–64
86. Strike DJ, De Rooij NF, Koudelka-Hep M (1995) Electrochemical techniques for the modification of microelectrodes. Biosens Bioelectron 10:61–66
87. Kurzawa C, Hengstenberg A, Schuhmann W (2002) Immobilization method for the preparation of biosensors based on pH shift-induced deposition of biomolecule-containing polymer films. Anal Chem 74:355–361
88. Llaudet E, Hatz S, Droniou M, Dale N (2005) Microelectrode biosensor for real-time measurement of ATP in biological tissue. Anal Chem 77:3267–3273
89. Frey O, van der Wal PD, de Rooij NF, Koudelka-Hep M (2009) Biosensor microprobe functionalisation using an on-chip dispensing approach. Procedia Chemistry 1:281–284
90. Bard AJ, Faulkner LR (2001) Electrochemical methods: fundamentals and applications, 2nd edn. Wiley, New York
91. Wise KD (2009) Wireless integrateg microsystems: wearable and implantable devices for improved health care. In: International solid-state sensors, actuators and microsystems conference (transducers), IEEE, Denver, CO, pp 1–8
92. Seidl K, Herwik S, Torfs T, Neves HP, Paul O, Ruther P (2011) CMOS-based high-density silicon microprobe arrays for electronic depth control in intracortical neural recording. J Microelectromech Syst 20:1439–1448
93. Torfs T, Aarts AAA, Erismis MA, Aslam J, Yazicioglu RF, Seidl K, Herwik S, Ulbert I, Dombovari B, Fiath R, Kerekes BP, Puers R, Paul O, Ruther P, Van Hoof C, Neves HP (2011) Two-dimensional multi-channel neural probes with electronic depth control. IEEE Trans Biomed Circ Syst 5:403–412
94. Frey O, van der Wal PD, Spieth S, Brett O, Seidl K, Paul O, Ruther P, Zengerle R, de Rooij NF (2011) Biosensor microprobes with integrated microfluidic channels for bi-directional neurochemical interaction. J Neural Eng 8(6):066001
95. SeidlK, Spieth S, Herwik S, Steigert J, Zengerle R, Paul O, Ruther P (2010) In-plane silicon probes for simultaneous neural recording and drug delivery. J Micromech Microeng 20:105006
96. Spieth S, Brett O, Seidl K, Aarts AAA, Erismis MA, Herwik S, Trenkle F, Tatzner S, Auber J, Daub M, Neves HP, Puers R, Paul O, Ruther P, Zengerle R (2011) A floating 3D silicon microprobe array for neural drug delivery compatible with electrical recording. J Micromech Microeng 21:125001–125017
97. Najafi K, Hetke JF (1990) Strength characterization of silicon microprobes in neurophysiological tissues. IEEE Trans Biomed Eng 37:474–481
98. Najafi K, Ji J, Wise KD (1990) Scaling limitations of silicon multichannel recording probes. IEEE Trans Biomed Eng 37:1–11

Chapter 2

Microelectrode Array Fabrication and Optimization for Selective Neurochemical Detection

Kevin N. Hascup, Erin R. Hascup, O. Meagan Littrell, Jason M. Hinzman, Catherine E. Werner, Verda A. Davis, Jason J. Burmeister, Francois Pomerleau, Jorge E. Quintero, Peter Huettl, and Greg A. Gerhardt

Abstract

A major goal of our research is to develop an implantable device for routine amperometric recordings of L-glutamate and other neurotransmitters in the mammalian central nervous system. Specifically, we wanted to develop a microelectrode that is (1) mass produced such that other laboratories can easily utilize the same recording technology, (2) designed to study multiple brain regions and neurotransmitters in various in vitro and in vivo systems, and (3) configured for "self-referencing" recordings, which allows for measurements of resting or tonic levels of neurotransmitters, cross-checking of the selectivity of the microelectrode measures, and improved signal-to-noise ratio by noise subtraction. The present chapter documents our current capabilities of measuring L-glutamate and several other neurotransmitters with rapid temporal resolution using mass-fabricated microelectrode arrays formed on ceramic. We have routinely demonstrated that these electrodes have fast temporal resolution (<1 s), excellent spatial resolution (microns), and low detection limits (≤200 nM) and cause minimal damage (50–100 μm) to surrounding brain tissue. While not a comprehensive assessment of the technology, this chapter contains a large amount of information regarding the fabrication, use, and potential pitfalls of this technology.

Key words Glutamate, Microelectrode array, Biosensor, Amperometry, In vivo electrochemistry, Photolithography, Neurotransmission

1 Introduction

Historically, the idea of glutamate as a neurotransmitter was met with much skepticism. While the quick action and high potency of glutamate on central nervous system (CNS) neurons were clearly demonstrated in the early 1960s, the high cytosolic concentrations and rapid removal without degradative enzymes went against what was expected of a neurotransmitter (1). Today, glutamate is regarded as the most prevalent excitatory neurotransmitter in the mammalian CNS. It has long-lasting effects on neuronal function,

Stéphane Marinesco and Nicholas Dale (eds.), *Microelectrode Biosensors*, Neuromethods, vol. 80,
DOI 10.1007/978-1-62703-370-1_2, © Springer Science+Business Media, LLC 2013

such as the establishment of long-term potentiation or depression of synaptic transmission (2–4), and on neuronal structure, such as the initiation of neuronal migration, axonal growth, and synapse formation during development (5–7). Furthermore, dysregulation of the glutamate system is involved in several neurodegenerative disorders including Parkinson's disease, Alzheimer's disease, Huntington's disease, and amyotrophic lateral sclerosis (8, 9) as well as neuropsychiatric disorders such as schizophrenia, addiction, anxiety, and depression (10–13).

Measures of extracellular glutamate signaling in the mammalian CNS have been dominated by microdialysis techniques over the past two decades. Microdialysis utilizes semipermeable membranes to control the diffusion of extracellular neurotransmitters along their concentration gradients (14, 15) allowing for measurements of multiple analytes at femtomolar concentrations. Microdialysis probes have a diameter of 150–400 μm with an average length of 1–4 mm (16). Dialysate samples are usually collected every 5–20 min. Although recent advances in the detection methods have allowed for sub-minute sampling rates (17–19), these approaches have not been practical for routine use by many laboratories.

Few microdialysis studies have found evidence that neurons contribute significantly to extracellular glutamate levels under resting (basal) conditions (for review, see (20)). The general consensus from microdialysis studies is that maintenance of extracellular glutamate levels does not follow a classical exocytotic release criterion; that is resting glutamate levels are minimally tetrodotoxin (TTX) or calcium dependent (20–22). This has led to the conclusion that the origin of resting glutamate is not neuronal, but rather astrocytic, presumably involving a reversal of the high-affinity glutamate transporters (23) or by the cystine–glutamate antiporter (22, 24).

More recently, our laboratory has been able to measure glutamate in vivo with enzyme-based microelectrode arrays (MEAs) coupled to constant-potential amperometric recording techniques (25–34). Our laboratory and others using similar microelectrodes have demonstrated that resting glutamate levels are highly TTX and Ca^{++} dependent, thus supporting a neuronal origin of at least 50% resting glutamate (30, 34–37). There are several key differences between the two techniques that may account for the inability of microdialysis to properly study resting glutamate levels. First, the temporal resolution of microdialysis is on the order of minutes, which is too slow to detect glutamatergic neural transmission that occurs over milliseconds to seconds (38). Our enzyme-based MEAs coupled with constant-potential amperometric recording techniques are capable of detecting the fast glutamate neurotransmission at less than 1 s recording intervals. Second, the larger size of the microdialysis probes makes it difficult to sample glutamate close to the synaptic cleft (39). Since glutamate has rapid clearance kinetics, being able to sample closer to the synaptic cleft may be necessary to study neuronal glutamate. Third, implantation of the microdialysis probe results in significant short and long-term trauma, which is evidenced by histological, physiological, biochemical, and neurochemical changes in CNS tissue that can occur in a 2.8 mm circumference surrounding the implant site (40–42). This damage also alters neurotransmitter release and uptake near the probe implant site (42, 43). Our MEAs have been shown to cause less damage to the surrounding CNS tissue as compared to microdialysis (33, 42).

The following chapter begins by explaining some of the theory and principles surrounding in vivo electrochemical techniques. We have placed particular emphasis on understanding the benefits and pitfalls of each technique and the importance of selectively determining analyte concentrations. The remainder of the chapter focuses on the development and optimization of MEAs selective for numerous neurochemicals such as acetylcholine, choline, lactate, glucose, and glutamate. Measurements utilizing these MEAs in freely moving animals are discussed in Chapter 8.

2 Principles of in Vivo Electrochemistry

In principle, in vivo electrochemistry is quite simple, yet it is a powerful means to continuously monitor overflow of neurotransmitters in the extracellular space (44, 45). A potentiostat controls an applied potential at the working electrode versus a Ag/AgCl reference electrode. The working and reference electrodes must be in ionic contact. The working electrode is usually an inert material, such as platinum (Pt), iridium, or carbon that can readily act as a source or sink for electrons. If the potential is sufficient, molecules are either oxidized or reduced directly at the working electrode surface depending upon their intrinsic electrochemical properties. The currents generated from Faradaic reactions are linear with respect to the concentration of the electroactive molecule(s) in the tissue surrounding the electrode (46). This basic principle allows for in vitro calibration methods for in vivo studies (46–48).

The technique has rapidly progressed since its introduction in the early 1970s. Several electrochemical techniques have been developed for faster waveforms, and, subsequently, enhanced data acquisition. These techniques have greatly improved our understanding of in vivo neurotransmitter release and reuptake. High-speed chronoamperometry (CA), fast-scan cyclic voltammetry (FSCV), and constant-potential amperometry are briefly discussed below along with their strengths and weaknesses. For a more thorough understanding of these electrochemical techniques, the reader is referred to (44–46, 48).

Predominantly used to measure biogenic amines, high-speed CA provides fast electrochemical monitoring with adequate analyte resolution (46). A square wave is applied to the working electrode at a typical frequency of 5–25 Hz. The applied potential is stepped to a level slightly higher than the oxidation potential of the analyte where it is held constant for a fixed period (20–100 ms) and then returned to the resting potential. The potential step results in the oxidation and reduction of the analyte and has two key advantages. First, it prevents the accumulation of oxidized analyte at the recording surface. Second, the reduction/oxidation (redox) ratio provides chemical resolution since this ratio is unique for several neurotransmitters including dopamine (DA), serotonin (5-HT), and norepinephrine. Third, the method is very sensitive because the background non-Faradaic current rapidly decreases allowing for recordings that are predominantly composed of the Faradaic signal.

FSCV uses a triangular waveform that is intermittently applied to the working electrode at rates usually up to 300 V/s allowing for fast temporal resolution. The advantage of the triangular waveform allows for the oxidation and reduction of analytes along a linear range of applied potentials. To achieve chemical resolution in FSCV, several scans before stimulation are averaged and then subtracted from each of the scans taken during the stimulated release. This results in background-subtracted cyclic voltammograms of the electroactive analyte, which are very characteristic of the biogenic amines, such as DA (45, 46). Similar to high-speed CA, the biogenic amines are easily identified (46). One disadvantage of this technique is that a large amount of background current is inherent to FSCV that can lead to differential currents due to changes in pH, particularly on carbon-fiber electrodes. A change in the hydronium ion concentration perturbs the oxidation and reduction potentials, but these are fairly distinct from biogenic amines (46).

One of the simplest and most widely used electrochemical techniques for monitoring rapid neurochemical events is constant-potential amperometry. This technique involves the measurement of current at a constant fixed potential. Since the current can be continuously monitored, neurochemical events can be measured as quickly as 1 ms (46, 48). Since the voltage is applied continuously to the working electrode and the current is directly proportional to the concentration at all times, the non-Faradaic background current recorded from the electrode is low, allowing for sensitive measurements of electrochemically active molecules (47, 48). Ultimately, the filtering, signal-to-noise ratio, sampling rate, and organic layer composition will determine the temporal resolution (46).

While constant-potential amperometry provides the best temporal resolution of any electrochemical technique, it also has the worst chemical resolution. It is difficult to distinguish multiple electroactive molecules that are present in vivo. To further complicate matters, several interferent compounds (ascorbic acid, AA; 3,4-dihydroxyphenylacetic acid, DOPAC; 5-hydroxyindoleacetic acid, 5-HIAA; and uric acid, UA) also exist and can be measured at oxidation potentials similar to those of biogenic amines. The in vivo concentrations of these interferents are often 100–1,000 times larger than the analyte of interest (45, 49). Thus, surface modifications are needed on the working electrode to improve its selectivity for the analyte of interest (46).

Our laboratory is well aware of the limitations of constant-potential amperometry. As such, we have developed a unique MEA that when coupled with specific surface modifications and self-referencing techniques allows for the selective monitoring of specific neurochemicals including, but not limited to, DA, glutamate, choline, acetylcholine, glucose, and lactate. These electrodes have fast temporal resolution (<1 s), excellent spatial resolution (microns), and low detection limits (≤200 nM) and cause minimal damage (50–100 μm) to surrounding brain tissue (42).

3 Self-Referencing

Historically, microelectrodes have been designed using a single carbon-fiber or Pt wire. With these types of microelectrodes, selectivity is achieved by using surface coatings to allow specific molecules access to the recording surface while blocking others. The molecules that pass through this coating are electrochemically detected on the working electrode surface (26). Our laboratory has taken this process one step further and developed a MEA consisting of multiple, uniform recording surfaces that are patterned in a precise geometrical configuration (Sect. 4). This precise patterning allows us to apply similar proteins (Sect. 5) on two separate, yet spatially adjacent, recording surfaces. By employing offline subtraction methods, we are able to eliminate unknown artifacts from the analyte signal and enhance the signal-to-noise. We have coined this procedure "self-referencing" and demonstrations of these principles have been shown elsewhere (26, 27, 50).

Application of similar coatings on two recording surfaces allows for recognition of possible interfering agents and their removal from the analyte signal. Essentially, a self-referencing MEA is the electrochemical equivalent of performing double-beam spectroscopy. Ideally, the only difference between the two sites is that one responds to the analyte of interest while the other does not. This is accomplished by coating an enzyme specific to an analyte of interest on one recording site while a chemically inactive protein is applied to the other recording site referred to as the control or sentinel site. The enzyme-coated site detects the analyte of interest in addition to everything that the control site detects (more thoroughly discussed in Sect. 5.3.1). Applying an inactive protein layer to the control site is necessary to minimize differences between the diffusional properties of molecules reaching the different Pt surfaces (26, 50). Without this coating, the control sites may respond faster to interferents than the analyte-detecting sites. Ensuring similar response times of the recording surface is imperative to remove the background and interfering signals from the analyte signal during offline subtraction.

Besides being able to remove interferents that contribute to an analyte signal, self-referencing recordings can remove periodic or random noise. This is an obvious advantage because smaller changes in current (lower detection limits) can be achieved (26). Burmeister and Gerhardt (26) have shown that the in vitro detection limit is greatly improved with our self-referencing technique. Higher in vitro detection limits most likely result from current induction of the magnetic stir plate during in vitro calibration (Sect. 6.1). In vivo results often show enhanced signal-to-noise ratio when the self-referencing subtraction technique is employed.

Additionally, self-referencing is useful during real-time (sub-second) monitoring of analytes in the brain. If the analyte-detecting site responds and there is no response on the control sites, the signal is considered to be due to analyte. For this situation, the control site is not used to quantify the analyte, rather it is used to determine if an interfering signal is present (26).

The remainder of this chapter details the necessary procedures for utilizing self-referencing MEAs. While our laboratory predominantly focuses on glutamate detection, similar procedures can be utilized for measuring other neurochemicals. The specific enzymes needed to measure these additional neurochemicals are listed in Table 1. Where applicable, these are also discussed in this chapter.

4 Microelectrode Arrays for Neurochemical Monitoring

Computer-controlled potentiostats in combination with microelectrodes designed by photolithographic procedures allow for simultaneous measures from multiple recording sites on a single biocompatible substrate. This has become an extremely important development considering electrodes only measure analytes from small foci (microns) of tissue directly surrounding the recording surface. This is useful for studying small or layered structures in the brain (51) or the spinal cord. Additionally, electrodes with multiple recordings sites can be geometrically arranged to measure analyte concentrations from two or more distinct brain regions. Pilot data from early MEA designs suggest a minimum of 30 μm of separation between recording sites to prevent diffusion of hydrogen peroxide to adjacent recording sites (cross talk).

Table 1
Enzymes

Enzyme	Substrate	Product	Use
Acetylcholinesterase	Acetylcholine	Choline, acetic acid	Measure choline with choline oxidase
Alcohol oxidase	Alcohol, O_2	H_2O_2, Aldehyde	Measure alcohols
D-Amino acid oxidase	D-amino acids, O_2	H_2O_2, NH_4^+, 2-oxo acid	Measure D-amino acids
L-Amino acid oxidase	L-amino acids, O_2	H_2O_2, NH_4^+, 2-oxo acid,	Measure L-amino acids
Ascorbate oxidase	Ascorbate, O_2	Dehydroascorbate	Remove ascorbate
Aspartate oxidase	Aspartate, O_2	H_2O_2, NH_4^+, Oxaloacetate	Measure aspartate
Catalase	H_2O_2	O_2	Remove H_2O_2
Cholesterol esterase	Cholesteryl oleate, or other cholesterol ester, Taurocholate	Cholesterol, oleic acid, or other corresponding ketone	Measure cholesterol ester with cholesterol oxidase
Cholesterol oxidase	Cholesterol, O_2	H_2O_2, 4-cholesten-3-one	Measure cholesterol
Choline oxidase	Choline, O_2	H_2O_2, Betaine	Measure choline
Galactose oxidase	D-galactose, O_2	H_2O_2, D-Galacto-hexodialose	Measure galactose
Glucose oxidase	Glucose, O_2	H_2O_2, Gluconic Acid	Measure glucose
L-glutamate oxidase	L-glutamate, O_2	H_2O_2, α-ketoglutarate	Measure L-glutamate
GABase	GABA, α-ketoglutarate	Succinic semialdehyde, L-glutamate, NADPH, H^+	Measure GABA with L-glutamate oxidase
Glutaminase	Glutamine	NH_4^+, L-glutamate	Measure glutamine with L-glutamate oxidase
Glycerol kinase	Glycerol, ATP	Glycerol-3-phosphate, ADP	Measure glycerol with glycerol-3-phosphate oxidase
Glycerol-3-phosphate oxidase	Glycerol-3-phosphate, O_2	H_2O_2, Dihydroxyacetone phosphate	Measure ATP with glycerol kinase
Hexokinase	ATP, glucose	Glucose-6-phosphate, ADP	Measure ATP with L-glucose oxidase
Horseradish peroxidase	H_2O_2	H_2O, O_2	Use as mediator for H_2O_2 detection
Lactate oxidase	Lactate, O_2	H_2O_2, Pyruvate	Measure lactate
Lysine oxidase	Lysine, O_2	H_2O_2, NH_4^+, 6-amino-2-oxohexanoic acid	Measure lysine
Pyruvate oxidase	Pyruvate, O_2, phosphate	H_2O_2, CO_2, acetyl phosphate	Measure pyruvate
Sarcosine Oxidase	Sarcosine, O_2	H_2O_2, Glycine, formaldehyde	Measure sarcosine
Xanthine oxidase	Xanthine, O_2	H_2O_2, Uric acid	Measure xanthine

4.1 Microelectrode Array Fabrication

The MEAs developed by our laboratory are mass fabricated using photolithographic methods (52–55). There are several major advantages of photolithography over handmade electrodes. First, the technique allows routine production of reproducible recording surfaces as small as 10 × 10 μm. Second, multiple microelectrodes are patterned onto a single fabrication substrate (usually 2.5 × 2.5 cm) allowing for increased fabrication number at a decreased cost. Third, photolithographic methods are used to manufacture numerous microelectrode designs with multiple recording sites in well-defined, highly reproducible geometrical configurations. This makes the method ideal for designing a microelectrode that can be easily, repetitively, and precisely positioned within a single or even multiple brain structures simultaneously with the aid of stereotaxic equipment.

Our multisite microelectrodes are currently constructed in conjunction with Hybrid Circuits (Sunnyvale, California). The fabrication process is also thoroughly explained in previous publications (25, 26, 47, 48, 50). Initially, microelectrode photographic masks are designed on a computer-aided design (CAD) program where arrays of 4–16 recording sites are arranged on templates. A 2.5 cm × 2.5 cm × 125 μm thick polished ceramic wafer (alumina, Al_2O_3, Coors Ceramic, Coors Superstrate 996) serves as a common substrate for the microelectrode arrays. Ceramic reduces the cross talk from adjacent connecting lines. Additionally, ceramic is strong and rigid, which aids in precise stereotaxic placement into tissues with minimal flexing or breaking. The ceramic substrate may be polished or lapped down to achieve microelectrodes as thin as 37.5 μm (50).

Following cleaning, photoresist is spun onto the ceramic wafer. Collimated light passing through the photomask exposes the photoresist, thus transferring the microelectrode images onto the wafer. Locations of the recording sites, connecting lines, and bonding pads are not exposed to light. Solvents are used to remove unexposed photoresist from the wafer. Next, an adhesion layer of titanium (Ti) is used to allow noble metals (Pt or Ir) to adhere to the ceramic substrate. Titanium (500 Å thick) is sputtered onto the developed photoresist-covered ceramic wafer. Following the adhesion layer, the active recording metal layer, Pt for most of our work, is sputtered onto the substrate (2,300 Å thick). Solvents are used to remove the developed photoresist and unwanted metals leaving the recording pads, connecting lines, and bonding pads.

Once the Pt recording sites, connecting lines, and bonding pads are in place, the connecting lines are coated with an insulator using another photolithographic step. The connecting lines act as wires to connect the bonding pads to the recording sites and must be insulated from aqueous environments. To accomplish this, the microelectrodes are again coated with photoresist. A second photomask is used to define the areas where polyimide is to be placed.

Polyimide acts as an insulator to define the recording site active area as well as the bonding pads. It also reduces cross talk between the connecting lines. After the photoresist is developed, polyimide is spun onto the wafer (2–4 μm thick). Once the insulating layer is applied, only the recording sites and bonding pads are exposed. The photoresist and excess polyimide are removed and the remaining polyimide is cured at 200°C. The complete fabrication process is depicted in Fig. 1. Many 1-cm-long microelectrodes (34–64) can be patterned onto the wafer simultaneously to facilitate production and increase the number of microelectrodes that are made from the ceramic substrate (Fig. 2). Patterning can also be performed on the reverse side of the ceramic wafer to increase recording site density (Sect. 7.1).

After the formation of the microelectrodes on the ceramic wafer, a diamond saw or laser is used to form or "cut out" the individual microelectrodes. A major advantage of the diamond saw is

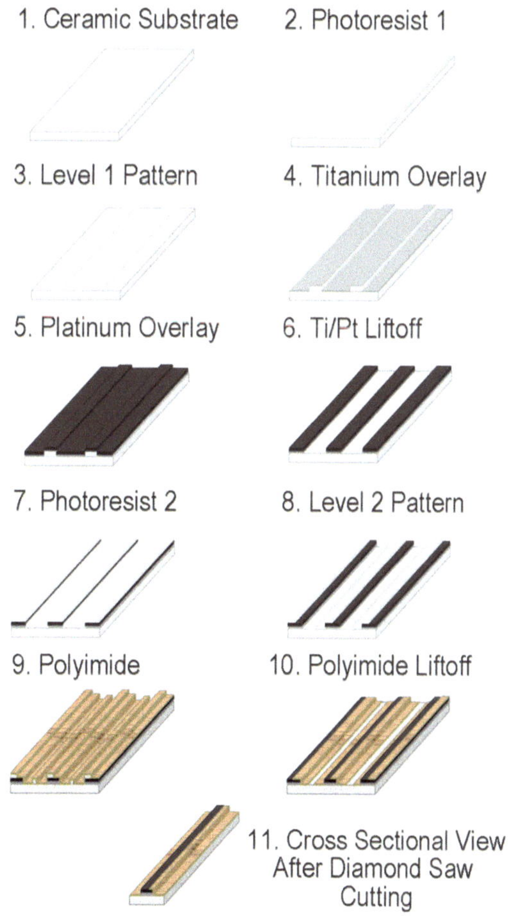

Fig. 1 Photolithographic process. Schematic representation of the photolithographic process used for MEA fabrication

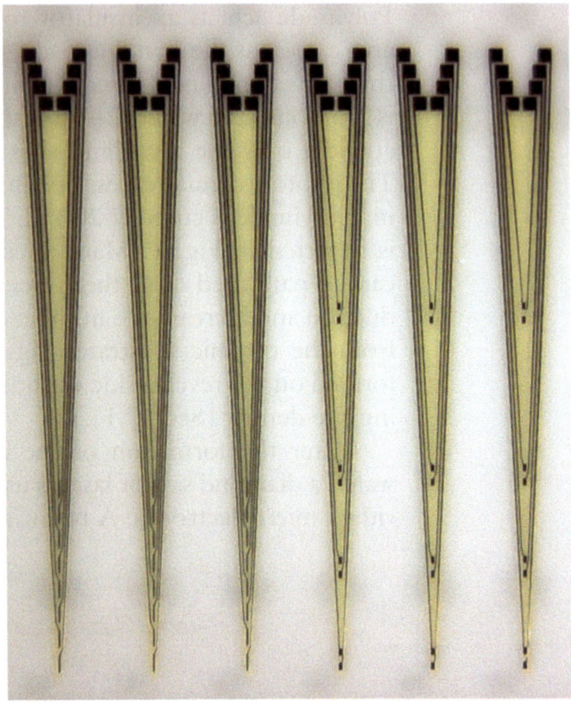

Fig. 2 Ceramic wafer. A ceramic wafer with two different recording site configurations patterned on its surface. Each electrode tip contains eight recording sites. Recording lines connect each recording site and run the length of the electrode tip for connection to the PCB. The length of each tip is approximately 1 cm

that it produces highly polished edges for reduced tissue damage during implantation, but it can lead to small variations in tip sizes. As previously mentioned, the wafer thickness is polished to 125 μm. Once cut, the microelectrode tip is ~9,000 μm long and ~1,000 μm at the widest point furthest from the Pt recording sites, where the tip is connected to the printed circuit board (see below). More importantly, the tapered tip of the microelectrode is considerably smaller and varies between 100 and175 μm. We have found that the tapered design of the microelectrode reduces tissue damage surrounding the recording sites.

The ceramic microelectrode tip with Pt recording sites is attached to a printed circuit board (PCB) holder for handling and connection to recording equipment. To connect the ceramic microelectrode to the PCB holder, each bonding pad is wire bonded to an individual Pt recording site on the ceramic microelectrode tip. The tips are epoxied onto the paddle for stability and to insulate the wire bonds. Cutting and assembly are currently performed in conjunction with Hybrid Circuits, Inc. (Sunnyvale, CA). The fully constructed MEAs for anesthetized and freely moving animal recordings are shown in Fig. 3a, b, respectively.

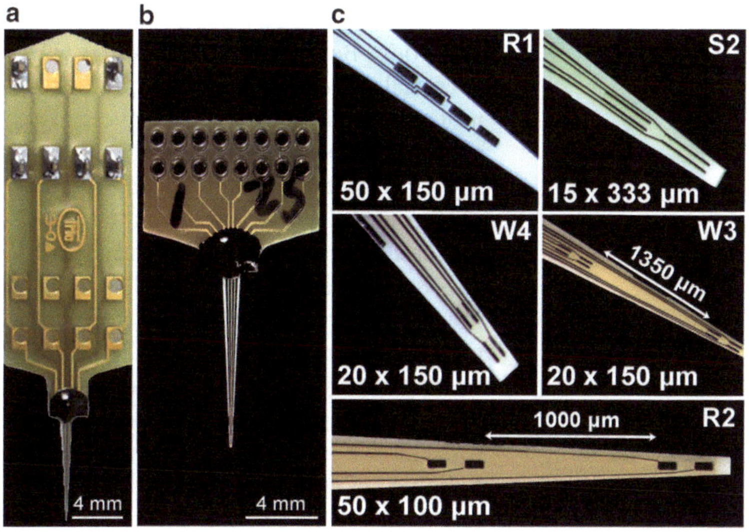

Fig. 3 Commonly used MEAs. Photographs of the fully fabricated anesthetized (**a**) and freely moving (**b**) MEAs showing the PCB and wire bonded tip with black epoxy. (**c**) Magnified images of tips with several Pt recording sites patterned in unique geometrical configurations. The name of each tip is shown in the *upper right*, while the size of the recording sites is shown in the *lower left*. Where applicable, distance between groups of recording sites is also labeled

4.2 Microelectrode Array Designs

As previously mentioned, the major advantage of photolithography is the ability to design microelectrodes with well-defined, highly reproducible geometrical configurations. Our most commonly used MEAs have four or eight Pt recording sites and are shown in Fig. 3c. MEAs are patterned to measure analytes within layered structures or from multiple brain structures. The R1 provide a larger recording distance that is useful for large brain regions or layered structures, while the S2 and W4 provide dual detection in smaller brain structures. The W3 was designed for simultaneous measurements in the arcuate and median sulci of the Non-Human Primate prefrontal cortex. One of the newest designs is the R2. The 1 mm spacing between pairs of sites makes this microelectrode ideal for measuring multiple brain structures such as the CA1 and CA3 regions of the hippocampus, the prelimbic and infralimbic regions of the prefrontal cortex, and the caudate/putamen and nucleus accumbens of the rat neostriatum. Additional designs are presented in (48).

5 Microelectrode Array Preparation

Prior to in vivo studies, MEAs are subjected to numerous cleaning and coating procedures to ensure successful detection of specific neurotransmitters. Few neurochemicals are electrochemically active on Pt at potentials of ≤+0.7 V versus a Ag/AgCl reference electrode. Besides biogenic amines (DA, 5-HT, NE), most neurochemicals that are oxidizable at this potential are often not the molecule of interest to the investigator. AA, DOPAC, 5-HIAA, and UA are present in the extracellular fluid in 100- to 1,000-fold higher concentrations than biogenic amines and amino acid transmitters (45, 49). To discriminate against these molecules, we apply protective films onto the surface of our MEA. Furthermore, enzymes are applied to the MEA for those interested in studying non-biogenic amine neurotransmitters. Oxidase enzymes are used to produce an electrochemically active reporter molecule, such as hydrogen peroxide (H_2O_2) that is readily oxidizable on Pt surfaces. The following sections thoroughly detail our coating procedure for selective, self-referencing measurements of L-glutamate. These same principles can be applied to numerous other neurotransmitters as shown in Table 1, but the proportions of enzymes vs. other materials are often specific to a given enzyme.

5.1 Cleaning Procedures

MEAs received from Hybrid Circuits undergo a chemical cleaning procedure to remove any particulate matter or residue from the Pt recording sites that may have been deposited during the manufacturing process. The recording surfaces of new microelectrodes are placed into separate, stirred solutions of Fisherbrand Citrisolv™ (Fisher Scientific, Catalog #04-355-121), isopropyl alcohol, and ddH_2O. The MEA tips are placed in each solution for 5 min. After each chemical cleaning procedure, a thin film can develop on the Pt recording sites, so a Kimwipe® (Kimberly–Clark Professional) or cotton swab is delicately swiped across the tip of the microelectrodes to remove this film or any excess ddH_2O. MEAs are dried in an oven for 5 min at 180°C prior to exclusion layer or enzyme coating.

Typically, enzyme-based microelectrodes for slice and anesthetized animal recordings are reused for several experiments once they have been cleaned and re-coated. MEAs used for studies in awake animals cannot be reused, at the present time. The cementing process for chronic implantations precludes the recovery of these microelectrodes. Over prolonged use, applied protein layers (described below) may degrade analyte detection. A slightly different cleaning procedure is employed for microelectrodes that have been previously coated with an enzyme layer. This enzyme layer must first be removed prior to reuse. The MEA tip is placed into a stirred solution of ddH_2O at 80°C for 15 min to soften the protein matrix on the Pt recording sites. Following this, the MEAs are cleaned as described in the preceding paragraph with one exception. MEAs are left in the stirred solution of isopropyl alcohol for 15 min. All MEAs are tested for response to H_2O_2 to ensure good sensitivity of the Pt recording surfaces. Also, the analyte of interest for the previously coated enzyme layer is tested to be certain that it is no longer present. A second round of chemical cleaning may be necessary to fully remove the enzyme layer.

5.2 Exclusion Layer Coatings

Exclusion layer coatings are applied after successful cleaning of the MEA and are used to optimize their recording properties. Exclusion layers block or minimize undesirable electrochemically active compounds (such as AA, DOPAC, UA, and 5-HIAA) found in high concentrations in the extracellular fluid of the CNS (45, 49). By blocking these undesired molecules from the Pt recording sites, the MEA has better selectivity for the analyte of interest. The planar geometry of the MEAs often affects how materials adhere to the recording surfaces. It should be noted that exclusion and enzyme layer coatings can inhibit compounds from diffusing to the microelectrode surface by creating a diffusion barrier, thus limiting the response time of the microelectrode. For this reason, coating parametrics are studied to determine optimal coating procedures for the exclusion layers to block interferents (26, 57–60). Selectivity is calculated as a ratio of the analyte slope to the interferent slope (25) and ratios for several different analytes are shown in Table 2.

5.2.1 Nafion®

Nafion® is an anionic Teflon® derivative and is one of the most widely used methods for improving the selectivity of electrochemical recordings in CNS tissue (45). Historically, Nafion® is used on the surface of carbon-fiber microelectrodes for the selective detection of biogenic amines (57–60). The negatively charged sulfonic acid groups substituted into the polymer matrix of Nafion® repel anions from the Pt recording surfaces. If these anions are not blocked, they will cause high background signals that interfere with reliable recordings. At the same time, the negatively charged sulfonic acid groups on Nafion® concentrate cationic species such as DA, NE, and 5-HT. In addition, the hydrophobic properties of a given species can compete with anionic charge rejection. Thus, Nafion® does not reject the 5-HIAA anion well; its hydrophobic nature allows 5-HIAA to penetrate despite its anionic charge. Nevertheless, Nafion® films provide very satisfactory anion exclusion and this protective barrier can be used on both untreated and electrochemically pretreated carbon-fiber and Pt recording surfaces (45). With the development of the multisite microelectrodes, Nafion® is a logical choice for an exclusion layer and is reliably used in our studies of L-glutamate especially in regard to measurements in awake, freely moving rats and mice over multiple days (33, 34).

Nafion® (5% in aliphatic alcohols, Sigma-Aldrich, Catalog #27,470-4) is aliquoted into a 2 mL brown glass vial. The MEA tip is then lowered half-way into the aliquot of Nafion® and then rotated in a circle five times lasting approximately one second per rotation. The tip is then pulled straight out of the Nafion® solution. Next, MEAs are placed into an oven at 180°C for four minutes to allow

Table 2
Selectivity ratios

Enzyme (final analyte concentration)	Exclusion layer	*N*	Selectivity ratios (XX:1)					Reference
			AA	DA	UA	DOPAC	NE	
L-glutamate oxidase (200 μM)[a]	Nafion	9	389 ± 153	0.15 ± 0.01	780	890 ± 110	0.43 ± 0.04	(26)
L-glutamate oxidase (60 μM)[a]	mPD	51	167 ± 19	(N/A)	(N/A)	(N/A)	(N/A)	(66)
Acetylcholinesterase (80 μM)[b]	mPD	16	254 ± 51	101 ± 26	(N/A)	(N/A)	(N/A)	(65)
Choline oxidase (40 μM)[b]	Nafion	12	885 ± 80	1000 ± 0	571 ± 132	441 ± 142	334 ± 142	(75)
Glucose oxidase (20 mM)[c]	Polyurethane	(N/A)	0.07 ± 0.26	0.19 ± 0.03	0.63 ± 0.24	0.47 ± 0.27	0.27 ± 0.03	(26)
Lactate oxidase (20 mM)[c]	Polyurethane	16	0.23 ± 0.04	0.04 ± 0.03	0.07 ± 0.06	−0.05 ± 0.07	0.22 ± 0.04	(68)

Data reported as mean ± standard error of the mean
[a]250 μM AA; 10 μM UA, DOPAC, DA and NE
[b]250 μM AA; 20 μM UA and DOPAC; 2 μM DA and NE
[c]250 μM AA and UA and 20 μM DA, NE, and DOPA

the Nafion® layer to cure on the tip. In general, lower temperature curing produces thicker films compared to higher temperature curing. Also, longer drying times at higher temperatures produce MEAs with lower responses to both AA and to analytes such as H_2O_2. Microelectrodes coated with excessive Nafion® show little or no detectable response to electrochemically active molecules (59, 60). MEAs coated with not enough Nafion® record high levels of background AA once inserted into the brain. After the curing procedure, MEAs are allowed to cool to room temperature for 30 min before enzyme coating (Sect. 5.3.1). An important concept to remember is that once a prepared Nafion®-coated MEA is placed into an aqueous solution (either in vitro or in vivo) the tip must remain wet or, at the very least, removed from solution for only a limited time. When Nafion® dries after soaking; it can crack allowing areas where interferents pass through to the Pt recording surface. If this occurs, the electrode is no longer selective for the analyte being measured.

5.2.2 1,3-Phenylenediamine

However, since Nafion® does slightly concentrate biogenic amines (45) to the recording surface, other exclusion layers are used in CNS regions with high monoamine tissue levels. The organic molecule 1,3-phenylenediamine (mPD) is another chemical used to create an exclusion layer for Pt recording electrodes. A potential is applied to a solution of mPD causing mPD to electropolymerize on Pt recording surfaces. Historically, mPD or 1,2-phenylenediamine (*o*-phenylenediamine) is used on carbon-fiber microelectrodes for selectivity and to prevent electrode fouling, a process where oxidizable molecules such as 5-HT can adhere to the carbon-fiber surface, thus preventing detection. The selectivity of electropolymerized mPD is likely achieved by forming a size exclusion layer that prevents larger molecules such as AA, DA, and DOPAC from reaching the recording surface. Smaller molecules, such as nitric oxide and H_2O_2, are still able to pass through the matrix (61). Since H_2O_2 is a reporter molecule for oxidase enzymes, this makes mPD an ideal exclusion layer for our enzyme-based multisite microelectrodes (61, 62).

The electropolymerizing procedure is completely different from Nafion® coating but has three distinct advantages. (1) The matrix formed by mPD blocks DA, NE, and 5-HT from reaching the Pt recording sites. This helps make the microelectrode more selective for measuring the analyte of interest. (2) The mPD is electropolymerized through the enzyme layer and onto the microelectrode after the electrode is coated with an enzyme layer. The selectivity over DA and AA is greatly diminished if enzymes were applied after the mPD was plated onto the electrode surface. The glutaraldehyde cross-linking process seems to be detrimental to the mPD layer function, maybe by increasing pore sizes. The best results are obtained when enzymes were applied before plating with mPD. The benefits of plating mPD after enzyme application include fast response, improved sensitivity due to H_2O_2 being generated close to the electrode surface, recycling of H_2O_2 back to oxygen, and improved stability of the polymerized layer (63, 64). The enzymatic activity is not affected by mPD electropolymerization. In addition, we have found that the most consistent results are obtained if the enzymes are allowed to cure on the recording sites for at least 72 h before plating with mPD. This ensures that the enzymes will not dissipate from the MEA surface when put into solution (65). (3) Because a potential must be applied, Pt recording sites can also be selectively coated with mPD. Removing the applied potential between a Pt recording site and the headstage prevents mPD from being electropolymerized onto that site. The procedure for electropolymerizing mPD to the Pt recording sites is outlined below.

Our laboratory has found that 1,3-phenylenediamine dihydrochloride 99% (Acros Organics, Catalog# 216650250) works best for electroplating and matrix formation on our Pt recording sites. First, a solution of 5 mM mPD (0.0905 g) is prepared in a sparged-reduced oxygen solution of 50 mM phosphate-buffered saline (PBS; 10.0 mM NaH_2PO_4, 40.0 mM Na_2HPO_4, 100.0 mM NaCl; pH 7.4). Sparging is accomplished by bubbling nitrogen gas through 125 mL of 50 mM PBS solution for 20 min to remove oxygen. This procedure is done to minimize oxidation of mPD by oxygen in the solution. The mPD is dissolved into the 50 mM PBS by slowly mixing the solution on a stir plate. Inversion or rapid mixing of the solution can reintroduce oxygen into the solution. Once the solution is prepared, we recommend using it immediately. Gently pour approximately 40 mL of the 5 mM mPD solution into a 50 mL beaker. Connect the MEA to the FAST16mkII (Fast Analytical Sensing Technology, Quanteon, LLC, Nicholasville, KY) and lower the tip half-way into the solution along with a glass Ag/AgCl reference electrode (Bioanalytical Systems, Inc. RE-5B, Catalog #MF-2079). Use the FAST16mkII custom software to apply a potential as a triangular wave with an offset of −0.5 V, peak-to-peak amplitude equal to ±0.25 V, at a frequency of 0.05 Hz, for a period of 20 min. Once electropolymerization is complete, MEAs are rinsed with ddH_2O to remove excess mPD. Allow the MEAs to cure for an additional 24 h in a low-humidity environment at room temperature before in vivo use (51, 66).

5.2.3 Polyurethane

An additional polyurethane layer is applied over top of other exclusion and enzyme layers for measuring glucose and lactate that exists in millimolar concentrations in the extracellular fluid. Polyurethane has been widely used in other implantable devices (67), but it has several added benefits when applied over the MEAs. First, it helps to improve the linearity of the MEA for high concentrations of glucose or lactate. Second, polyurethane does not interfere with the O_2 dependence of either glucose or lactate oxidase (26, 68). (The importance of O_2 dependence is described in Sect. 5.3.3.)

Third, polyurethane provides an additional exclusion layer that prevents the diffusion of biogenic amines, thereby making the MEA nearly interferent free (68). A 5% polyurethane (Fluka, Catalog #81367) solution dissolved in a mixture of tetrahydrofuran (98%) and dimethylformamide (2%) is prepared. Seventy-two hours after enzyme layer coating, the MEA tip is dipped into the solution and allowed to dry at room temperature for 24 h.

5.3 Enzymes

Enzymes provide a means to convert a molecule that is not inherently electroactive, and thus not measurable, into a reporter molecule such as H_2O_2 that is oxidized at the Pt recording surfaces of the MEA. The current measured from the oxidation of H_2O_2 generated during the enzymatic breakdown is directly proportional to the analyte concentration (47). Table 1 provides a list of available oxidase enzymes and their potential uses. Some compounds require multiple enzymes to convert them to a reporter molecule, such as acetylcholine, adenosine, and γ-aminobutyric acid (GABA). A chemical cross-linking procedure is used to immobilize the enzymes to the Pt recording surface. This helps to stabilize the enzymes and makes them active for longer periods of time. In the case of L-glutamate oxidase, we have shown reliable performance in awake, freely moving rodents for at least 7 days post-implantation (33, 34). Our laboratory and others have used several of these enzymes for in vivo neurochemical measurements including L-glutamate oxidase (25–34, 37, 50, 51, 66, 69–73), acetylcholinesterase (65, 74), choline oxidase (75–79), L-lactate oxidase (68), and L-glucose oxidase (26). An extensive description of the L-glutamate oxidase coating procedure is described in the following section.

5.3.1 L-Glutamate Oxidase Coating Procedures

Properly cleaned microelectrodes are ready for enzyme immobilization. We have carried out an extensive series of parametric studies to determine this optimized mixture for enzyme immobilization, as follows. A stock solution of L-glutamate oxidase, recombinant (EC 1.4.3.11) from US Biological (Swampscott, MA; Catalog #G4001-01), is prepared by adding 50 μL of ddH_2O to the lyophilized, purified enzyme (25 units) to make a final concentration of 0.5 U/μL. This stock solution is viable for ~6 months when stored at 4°C. All proteins and enzymes are brought to room temperature prior to coating. Bovine serum albumin (BSA; 0.010 g) Fraction V, 99% (Sigma-Aldrich, Catalog #A-3059), is added to a 1.5 mL microcentrifuge tube. A 985 μL aliquot of ddH_2O is added and the BSA is dissolved by manual agitation. Once dissolved, 5 μL of glutaraldehyde solution, Grade I, 25% (Sigma-Aldrich Catalog #G-6257), is added to the BSA mixture and manually mixed by inversion five times. This solution, referred to as the protein matrix, is then set aside for 5 minutes until it turns a faint yellow color. Glutaraldehyde cross-links proteins, which adhere to the microelectrode surface when cured. The BSA serves as a matrix to protect the oxidase enzyme activity during immobilization. Next, 4 μL of the BSA/glutaraldehyde mixture is removed and added to a 300 μL microcentrifuge tube and to this 1 μL of the L-glutamate oxidase stock solution (0.5 U/μL) is added and mixed by micropipette agitation. This 5 μL solution has a final concentration of 1% BSA, 0.125% glutaraldehyde, and approximately 1% L-glutamate oxidase. Once the solutions are prepared, they should be used immediately for MEA coating.

All enzymes are currently applied to the Pt recording sites by hand. Solutions are drawn up into a 10 μL Gastight® Hamilton microsyringe (Hamilton Co. Catalog #80065) and slowly dispensed to form a small droplet of solution at the tip of the microsyringe. Using a dissecting microscope, the droplet is lightly brushed on to the Pt recording sites. The solution quickly dries, leaving behind a thin, translucent layer of enzyme that is visible underneath a dissecting microscope. Two additional coats of enzyme are applied in the same manner with one minute dry times in between each coat. This procedure is complete once a visible film remains following coating. The advantage of using multisite microelectrodes is that a pair of recording sites is coated with the enzyme mixture and the adjacent sites are coated with the protein matrix (BSA and glutaraldehyde) in the same manner as applying the enzyme mixture.

Once the enzyme and/or inactive protein matrix is applied, microelectrodes are stored at room temperature. *Enzyme-coated microelectrodes must cure at room temperature for 72 h prior to calibration and experimentation.* The enzyme/protein layers on "uncured microelectrodes" can dissolve or lift off when put into solution. Complete curing increases enzyme layer adhesion to the microelectrode surface. This provides better sensitivity to L-glutamate as well as increase the shelf-life of the microelectrodes. Our laboratory recommends using the microelectrodes within three weeks after coating. Optimal curing times may vary depending on temperature, humidity, and the type of enzyme.

5.3.2 Other enzymes

In general, the preparation of the coating solution and enzyme application to the MEA surface is consistent for many of the enzymes we have tested. Glutaraldehyde and BSA are always used for enzyme cross-linking and protection during immobilizing. However, the relative concentrations of glutaraldehyde, BSA, and enzyme may vary. We refer the reader to previously published papers regarding the enzyme solution, coating parametrics, and curing times for glucose oxidase (26), choline oxidase (75), lactate oxidase (68), and acetylcholinesterase (65).

5.3.3 O_2 Dependence

As previously mentioned, the O_2 dependence of oxidase enzyme-coated MEAs is a concern considering air saturated O_2 levels (200 μM) are considerably more than concentrations in vivo

(50 μM) (80, 81). Oxygen is a required cofactor of oxidase enzymes for the production of the reporter molecule, H_2O_2. However, a portion of oxygen is returned to the tissue by the oxidation of H_2O_2 to O_2 and H^+ at the microelectrode surface. We have examined if low O_2 concentrations would affect MEA performance with several different enzymes (Table 3). To determine an enzyme's O_2 dependence, the MEA response was monitored as the O_2 concentration was varied over several analyte concentrations. The O_2 concentration was varied from air saturated levels (~200 μM) by purging the buffer with N_2 (99.98% pure). The O_2 concentration in solution was measured using a World Precision Instruments model ISO_2 dissolved oxygen meter with a resolution of 0.1%. As shown in Table 3, the enzymes tested have a low O_2 dependence particularly at in vivo concentrations (50 μM) (65, 66, 68, 75). But, certain animal models, such as those during recordings of stroke episodes, may require correction by use of an O_2 sensing microelectrode.

5.3.4 Response Times

We do not know the complete relationship between enzyme thickness and response time. Experience shows that microelectrode sensitivity is lower and response time is faster for very thin coatings. Sensitivity increases with enzyme thickness up to a point then levels off; however, response time becomes longer as a thicker diffusion barrier is applied to the microelectrode (65). Enzyme thicknesses for several enzymes were chosen because they yield acceptable sensitivity and response times.

MEA response time was measured using a custom flow stream system. The flow stream consisted of a syringe pump (Razel®, A-99), with a 60 mL syringe, an HPLC sample injector (Beckman, 210A) having a 500 μL sample loop, a polyethylene reservoir, and a polyethylene tubing (1.14 mm i.d.). The MEA tip was placed at the end of the tubing and a Ag/AgCl reference electrode was placed in the reservoir. The reservoir and approximately 15 cm of

Table 3
Oxygen dependence

	O_2 dependence % of air saturation (200 μM)				
Enzyme	**0 μM O_2**	**50 μM O_2**	**100 μM O_2**	**150 μM O_2**	**Reference**
L-glutamate oxidase	75–85	85	90	95	(66)
Acetylcholinesterase	65	78–93	90–100	100	(65)
Choline oxidase	70–77	85	90	95	(75)
Lactate oxidase	70–100	70–85	80–95	90–95	(68)

polyethylene tubing were immersed in a temperature-controlled water bath held at 37°C. Hydrogen peroxide (8.8 μM, 300 μL) or appropriate analyte from the sample loop was injected into the flow stream at a rate of 2.17 mL/min. A bubble was injected between the sample and flow stream to decrease mixing of the sample with the PBS carrier yielding a rapid step in concentration. The microelectrode response time was calculated as the time for the response to rise from 5% to 90% of the maximum amplitude for the concentration step (65).

Flow stream analysis was used to compare the response time of MEAs that were coated with mPD to high-temperature cured Nafion®. No significant decrease in response time for H_2O_2 detection was observed when the MEAs were coated with Nafion® or mPD compared with an uncoated MEA (65). The average 90% response time to 8.8 μM H_2O_2 for the mPD-coated microelectrode was 0.119 s ± 0.008 ($n = 12$) while the response time for Nafion®-coated microelectrodes was 0.117 s ± 0.006 ($n = 16$). An uncoated microelectrode yielded a 90% response in 0.111 s ± 0.007 ($n = 20$). Thus, neither Nafion® nor mPD significantly slowed the response time for H_2O_2 (65).

Additionally, we have tested the MEA response times for several different enzymes (65, 68, 75). These data indicate that enzyme kinetics were responsible for the majority of the slower response of the enzyme-coated microelectrodes. However, the enzyme layer itself does act as a diffusion barrier to H_2O_2 detection contributing to some of the delayed signal response. Our strategy in the future is to use thinner and fewer enzyme layers to minimize microelectrode response times due to increased diffusion time. The relationship between enzyme thickness and response time however needs further study. For this purpose an automated enzyme coater is being developed to precisely apply enzymes in a controlled thickness to improve MEA response times.

6 Microelectrode Array Calibration

We often observe that the microfabricated MEAs have highly reproducible Pt recording surfaces. However, manufacturing procedures only investigate the geometric surface area of the microelectrodes. Since each Pt recording site on a MEA can respond differently to H_2O_2 and current enzyme coating procedures can result in different layer thicknesses, each MEA must be calibrated in vitro prior to experimentation to determine standard curves. (The argument for pre- versus post-calibration is discussed in Sect. 6.3.) Thus, calibrations are used to equate a change in current from the oxidation of H_2O_2 to a proportional change in analyte concentration from the oxidase enzyme generating the H_2O_2. A known concentration of analyte is added to 40 mL of 50 mM PBS solution at 37°C to generate a current in picoamperes (pA) that is measured by the FAST16mkII system. The FAST16mkII system software records the current for each addition of analyte, creates a calibration curve for each Pt recording site, and stores the sensitivity to analyte for each site. Also, known interferents such as AA are added during the calibration to test the selectivity of the recording sites to the analyte of interest versus interferents. The calibration data are recalled and used to determine the concentration of analyte being measured from the change in current during in vivo experiments. The general procedure for calibrating an L-glutamate oxidase MEA is outlined below.

6.1 General Calibration Procedure

Coated MEA tips are soaked in a solution of 50 mM PBS at 37°C for at least 1 h prior to calibration. The soaking time allows for activation of the enzyme layer and better diffusion of molecules to the Pt recording sites. All stock solutions for L-glutamate calibrations are prepared in ddH_2O and listed below:

1. 20 mM AA (ascorbic acid powder, Fisher Scientific, Catalog #A62)
2. 20 mM L-glutamic acid sodium salt hydrate, minimum 99% (Sigma-Aldrich, Catalog #G1626)
3. 2 mM DA (3-hydroxytyramine hydrochloride, Sigma-Aldrich, Catalog #H-8502)
4. 8.8 mM H_2O_2 (30 wt.% in H_2O; Sigma-Aldrich, Catalog #216763)

Solutions are stored in a refrigerator at 4°C. AA and H_2O_2 are made fresh daily since they oxidize in solution. L-glutamate is made fresh on a weekly basis. DA is kept for a month provided the stock solution contains 0.01 M perchloric acid.

For calibrations, a FAST16mkII recording system in combination with a headstage current-to-voltage converter of 2 nA/V is used. Forty mL of 50 mM PBS is measured and added to a 50 mL beaker. This beaker is placed in a recirculating water bath (Quanteon, LLC) resting upon a battery operated, portable magnetic stir plate (Barnant Co., Model #700-0153). Because the enzyme is temperature and pH dependent, the PBS is maintained at physiological temperature (37°C) and pH (7.4). We use a battery operated stir plate to reduce potential AC (60 Hz) interference that can affect the recordings. A 10×3 mm^2 stir bar is added to the 50 mM PBS and the solution is slowly stirred to prevent vortex formation. A glass Ag/AgCl reference electrode (Bioanalytical Systems, Inc. RE-5B, Catalog #MF-2079) is placed in the buffer solution. Finally, the microelectrode tip is lowered half-way into the buffer solution. The FAST16mkII recording

Table 4
Calibration data

Enzyme	N	Range	Sensitivity (oxidation current) (pA/μM)	R^2	LOD	Pt surface (μm²)	Reference
L-glutamate oxidase	63	0–60 μM	16.7 ± 0.8	0.998 ± 0.000	524 ± 26 nM	7,500	(27)
Acetylcholinesterase	16	0–80 μM	4.7 ± 0.8	0.998 ± 0.001	120 ± 20 nM	4,995	(65)
Choline oxidase	12	0–40 μM	13.2 ± 1.7	0.999 ± 0.001	438 ± 66 nM	5,000	(75)
Glucose oxidase	(N/A)	0–20 mM	30.2 ± 6.5	0.988 ± 0.008	640 ± 140 μM	2,500	(26)
Lactate oxidase	12	0–20 mM	89.4 ± 13.4	0.997 ± 0.001	78 ± 13 μM	7,500	(68)

Data reported as mean ± standard error of the mean

system applies a potential of +0.7 V versus the glass Ag/AgCl reference to the Pt MEA recording surfaces. Once calibration has begun, the current is allowed to reach a stable baseline (for a minimum of 15 min, but may take longer). Once a stable baseline is reached, calibrations are performed using final buffer concentrations of 250 μM AA and 20, 40, and 60 μM L-glutamate, which is used to generate the standard curve. The concentration of L-glutamate chosen to generate the curve is based on the concentration we expect to observe during our experiments. Selectivity ratios for glutamate over AA are calculated in addition to the slope (sensitivity), limit of detection (LOD), and linearity (R^2) for glutamate for all MEAs. A sample of these is shown in Table 4.

The MEAs are also tested during calibration to determine the precision of the four Pt recording sites using DA (2 μM final concentration) and H_2O_2 (8.8 μM final concentration) as test substances. The test substances are used to check selectivity and can be used to normalize the enzyme-coated and protein matrix sites for self-referencing recordings. The additions of test substances do not factor into calculating the standard curve for the analyte of interest. MEAs are only used if the Pt recording sites had in vitro responses to DA and H_2O_2 that were within 10% of each other. After calibration, MEA tips are stored in 50 mM PBS until experimentation.

6.2 Other Enzymes

Many of the aspects for calibrating MEAs coated with L-glutamate oxidase are applied to other enzymes. All enzymes are calibrated at physiological temperature (37°C) and pH (7.4) in a solution of 50 mM PBS. Also, the type and concentrations of interferents and test substances remain consistent. The main differences are the type and concentration of analyte added to generate the standard curve. Obviously, the analyte used is dependent upon the type of enzyme coated to the MEA. However, the concentration of the analyte may vary depending upon relative in vivo concentration ranges. Again, we refer the reader to previously published papers regarding these differences for glucose oxidase (26), choline oxidase (75), lactate oxidase (68), and acetylcholinesterase (65). Calibration range, sensitivity, LOD, and linearity for these enzymes are shown in Table 4.

6.3 Pre- Vs. Post-Calibration

Unfortunately, a thoroughly satisfactory in vivo calibration technique has not been devised. All calibrations are done prior to in vivo recording. Post-calibration procedures are unreliable means for verifying electrode sensitivity and selectivity (27, 82). Adsorption of blood, tissue, and proteins to the recording surface or even damage during explants can affect the characteristics in a manner that is different from its properties when first implanted into brain tissue (62). It has been suggested that enzyme-based sensor pre-calibration does not compensate for ~50% loss of sensitivity following implantation, resulting in underestimation of the in vivo analyte result (82). However, Hinzman and colleagues (66) have demonstrated no difference in pre- vs. post-calibration performance for L-glutamate oxidase-coated MEAs (4,995 μm^2). They reported similar pre- vs. post-calibration results for limit of detection (1.1 ± 0.1 μM vs. 1.5 ± 0.4 μM; $p = 0.37$), selectivity (93 ± 21 vs 83 ± 26; $p = 0.77$), and sensitivity (-2.1 ± 0.2 pA/μM vs. -2.2 ± 0.1 pA/μM; $p = 0.68$) ($n = 24$).

7 Future Directions

7.1 Measuring Multiple Neurotransmitters Simultaneously

A major advantage of microdialysis is the ability to measure multiple neurotransmitters simultaneously (assuming good separation of peaks during offline detection by high-performance liquid chromatographic-based methods). While our laboratory has successfully demonstrated the ability to measure acetylcholine and choline simultaneously (65), this is an essential element of the enzymes involved. Ideally, the long-term goal is to design an MEA to measure multiple neurotransmitters from the same brain structure. As previously mentioned, photolithographic methods allow us to design MEAs with well-defined, highly reproducible geometric configurations that can be patterned onto the front and back of a ceramic wafer. By using both sides of the ceramic wafer we can greatly enhance the recording site density of an MEA. We have designed several 16-site MEAs, one of which is shown in Fig. 4. By increasing the recording site density we can selectively coat recording sites for self-referencing measurements of different analytes in the same brain region.

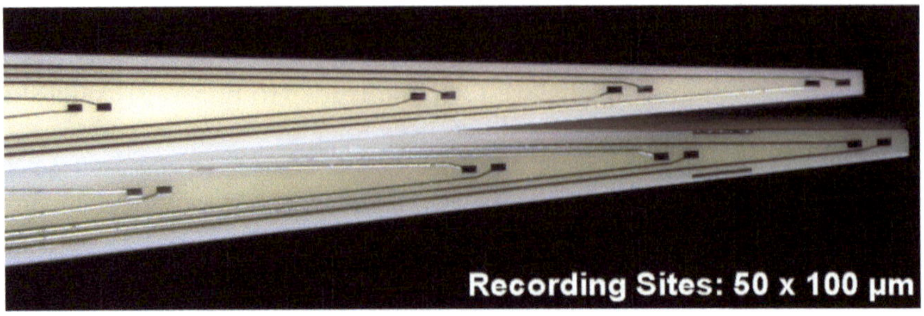

Fig. 4 16-Site MEA. The electrode tip of a 16-site MEA. Photolithographic techniques allows for patterning on the front and back of the ceramic wafer to increase surface recording site density. This photograph shows the Pt recording sites patterned on both sides by using a mirror to reflect the image from the underside of a microelectrode array

7.2 Simultaneous Electrophysiological and Electrochemical Recordings

The ability to simultaneously measure local field potentials (LFPs) and neurochemical dynamics would greatly enhance our understanding of neuromodulation. This would have profound implications for studying behavior or disorder related changes in the CNS. Until recently, few studies were capable of achieving simultaneous electrophysiological and electrochemical measures from the same recording preparation. And, those that did either switched between neuronal spike recording and voltammetric recordings (83–85) or required the use of two separate recording systems (86, 87). This made the experimental setups cumbersome and expensive while making data analysis and interpretation difficult. However, Zhang and colleagues (86, 87) are the first to demonstrate simultaneous LFPs and local neurochemical measurements using a single biosensor coupled with constant-potential amperometry. They successfully demonstrated that the high-frequency component of the amperometric signal resembled LFPs. Using a single setup and biosensor for the simultaneous acquisition of electrophysiological and neurochemical information can greatly enhance the field of in vivo electrochemistry (86, 87).

Acknowledgements

This work was supported by NSF EEC-0310723; NIH/NIDA DA017186; CEBRA, Phase II.

Disclosure

Greg A. Gerhardt, Ph.D. is the sole proprietor of Quanteon, LLC.

References

1. Krnjevic K (2005) From 'soup physiology' to normal brain science. J Physiol 569:1–2
2. Artola A, Singer W (1987) Long-term potentiation and NMDA receptors in rat visual cortex. Nature 330:649–652
3. Cotman CW, Monaghan DT, Ganong AH (1988) Excitatory amino acid neurotransmission: NMDA receptors and hebb-type synaptic plasticity. Annu Rev Neurosci 11:61–80
4. Ito M (1989) Long-term depression. Annu Rev Neurosci 12:85–102
5. Mattson MP, Dou P, Kater SB (1988) Outgrowth-regulating actions of glutamate in isolated hippocampal pyramidal neurons. J Neurosci 8:2087–2100
6. Brewer GJ, Cotman CW (1989) NMDA receptor regulation of neuronal morphology in cultured hippocampal neurons. Neurosci Lett 99:268–273
7. Komuro H, Rakic P (1993) Modulation of neuronal migration by NMDA receptors. Science 260:95–97
8. Doble A (1999) The role of excitotoxicity in neurodegenerative disease: implications for therapy. Pharmacol Ther 81:163–221
9. Danbolt NC (2001) Glutamate uptake. Prog Neurobiol 65:1–105
10. Mechri A, Saoud M, Khiari G et al (2001) Glutaminergic hypothesis of schizophrenia: clinical research studies with ketamine. Encéphale 27:53–59
11. Lopez-Moreno JA, Gonzalez-Cuevas G, Moreno G et al (2008) The pharmacology of the endocannabinoid system: functional and structural interactions with other neurotransmitter systems and their repercussions in behavioral addiction. Addict Biol 13:160–187
12. Mathew SJ, Price RB, Charney DS (2008) Recent advances in the neurobiology of anxiety disorders: implications for novel therapeutics. Am J Med Genet C Semin Med Genet 148C:89–98
13. McNally L, Bhagwagar Z, Hannestad J (2008) Inflammation, glutamate, and glia in depression: a literature review. CNS Spectr 13: 501–510
14. Ungerstedt U (1991) Microdialysis–principles and applications for studies in animals and man. J Intern Med 230:365–373
15. Di Chiara G, Tanda G, Carboni E (1996) Estimation of in-vivo neurotransmitter release by brain microdialysis: the issue of validity. Behav Pharmacol 7:640–657
16. Kennedy RT, Watson CJ, Haskins WE et al (2002) In vivo neurochemical monitoring by microdialysis and capillary separations. Curr Opin Chem Biol 6:659–665
17. Lada MW, Vickroy TW, Kennedy RT (1997) High temporal resolution monitoring of glutamate and aspartate in vivo using microdialysis on-line with capillary electrophoresis with laser-induced fluorescence detection. Anal Chem 69:4560–4565
18. Tucci S, Rada P, Sepulveda MJ et al (1997) Glutamate measured by 6-s resolution brain microdialysis: capillary electrophoretic and laser-induced fluorescence detection application. J Chromatogr B 694:343–349
19. Rossell S, Gonzalez LE, Hernandez L (2003) One-second time resolution brain microdialysis in fully awake rats - protocol for the collection, separation and sorting of nanoliter dialysate volumes. J Chromatogr B Analyt Technol Biomed Life Sci 784:385–393
20. Timmerman W, Westerink BHC (1997) Brain microdialysis of GABA and glutamate: what does it signify? Synapse 27:242–261
21. van der Zeyden M, Oldenziel WH, Rea K et al (2008) Microdialysis of GABA and glutamate: analysis, interpretation and comparison with microsensors. Pharmacol Biochem Behav 90:135–147
22. Kalivas PW (2009) The glutamate homeostasis hypothesis of addiction. Nat Rev Neurosci 10:561–572
23. Montiel T, Camacho A, Estrada-Sanchez AM et al (2005) Differential effects of the substrate inhibitor L-trans-pyrrolidine-2,4-dicarboxylate (PDC) and the non-substrate inhibitor DL-threo-beta-benzyloxyaspartate (DL-TBOA) of glutamate transporters on neuronal damage and extracellular amino acid levels in rat brain in vivo. Neuroscience 133:667–678
24. Baker DA, Xi ZX, Shen H et al (2002) The origin and neuronal function of in vivo nonsynaptic glutamate. J Neurosci 22:9134–9141
25. Burmeister JJ, Moxon K, Gerhardt G (2000) Ceramic-based multisite microelectrodes for electrochemical recordings. Anal Chem 72: 187–192
26. Burmeister JJ, Gerhardt G (2001) Self-referencing ceramic-based multisite microelectrodes for the detection and elimination of interferences from the measurement of L-glutamate and other analytes. Anal Chem 73:1037–1042
27. Burmeister J, Pomerleau F, Palmer M et al (2002) Improved ceramic-based multisite microelectrode for rapid measurements of L-glutamate in the CNS. J Neurosci Methods 119:163–171
28. Pomerleau F, Day BK, Huettl P et al (2003) Real time in vivo measures of L-glutamate in the rat central nervous system using ceramic-based

multisite microelectrode arrays. Ann NY Acad Sci 1003:454–457
29. Nickell J, Pomerleau F, Allen J et al (2005) Age-related changes in the dynamics of potassium-evoked L-glutamate release in the striatum of Fischer 344 rats. J Neural Transm 112:87–96
30. Day BK, Pomerleau F, Burmeister JJ et al (2006) Microelectrode array studies of basal and potassium-evoked release of L-glutamate in the anesthetized rat brain. J Neurochem 96:1626–1635
31. Nickell J, Salvatore MF, Pomerleau F et al (2007) Reduced plasma membrane surface expression of GLAST mediates decreased glutamate regulation in the aged striatum. Neurobiol Aging 28:1737–1748
32. Quintero JE, Day BK, Zhang Z et al (2007) Amperometric measures of age-related changes in glutamate regulation in the cortex of rhesus monkeys. Exp Neurol 208:238–246
33. Rutherford EC, Pomerleau F, Huettl P et al (2007) Chronic second-by-second measures of L-Glutamate in the central nervous system of freely moving rats. J Neurochem 102: 712–722
34. Hascup KN, Hascup ER, Pomerleau F et al (2008) Second-by-second measures of L-glutamate in the prefrontal cortex and striatum of freely moving mice. J Pharmacol Exp Ther 324:725–731
35. Kulagina NV, Shankar L, Michael AC (1999) Monitoring glutamate and ascorbate in the extracellular space of grain tissue with electrochemical microsensors. Anal Chem 71: 5093–5100
36. Oldenziel WH, Dijkstra G, Cremers TIFH et al (2006) In vivo monitoring of extracellular glutamate in the brain with a microsensor. Brain Res 1118:34–42
37. Hascup ER, Hascup KN, Stephens M et al (2010) Rapid microelectrode measurements and the origin and regulation of extracellular glutamate in rat prefrontal cortex. J Neurochem 115(6):1608–1620
38. Kinney GA, Overstreet LS, Slater NT (1997) Prolonged physiological entrapment of glutamate in the synaptic cleft of cerebellar unipolar brush cells. J Neurophysiol 78:1320–1333
39. Drew KL, Pehek EA, Rasley BT et al (2004) Sampling glutamate and GABA with microdialysis: suggestions on how to get the dialysis membrane closer to the synapse. J Neurosci Methods 140:127–131
40. Clapp-Lilly KL, Roberts RC, Duffy LK et al (1999) An ultrastructural analysis of tissue surrounding a microdialysis probe. J Neurosci Methods 90:129–142
41. Bungay PM, Newton-Vinson P, Isele W et al (2003) Microdialysis of dopamine interpreted with quantitative model incorporating probe implantation trauma. J Neurochem 86: 932–946
42. Hascup ER, af Bjerken S, Hascup KN et al (2009) Histological studies of the effects of chronic implantation of ceramic-based microelectrode arrays and microdialysis probes in rat prefrontal cortex. Brain Res 1291:12–20
43. Yang H, Peters JL, Michael AC (1998) Coupled effects of mass transfer and uptake kinetics on in vivo microdialysis of dopamine. J Neurochem 71:684–692
44. Marsden CA, Joseph MH, Kruk ZL et al (1988) In vivo voltammetry–present electrodes and methods. Neuroscience 25:389–400
45. Adams RN (1990) Invivo electrochemical measurements in the CNS. Prog Neurobiol 35:297–311
46. Michael DJ, Wightman RM (1999) Electrochemical monitoring of biogenic amine neurotransmission in real time. J Pharm Biomed Anal 19:33–46
47. Hascup KN, Rutherford EC, Quintero JE et al (2006) Second-by-second measures of L-glutamate and other neurotransmitters using enzyme-based microelectrode arrays. In: Michael AC, Borland LM (eds) Electrochemical methods for neuroscience. CRC, Boca Raton, FL
48. Gerhardt GA, Burmeister JJ (2000) Voltammetry in vivo for chemical analysis of the nervous system. In: Meyers RA (ed) Encyclopedia of analytical chemistry. Wiley, Chichester
49. Martin KF, Marsden CA (1987) Invivo electrochemistry - principles and applications. Life Sci 41:865–868
50. Burmeister JJ, Gerhardt GA (2003) Ceramic-based multisite microelectrode arrays for in vivo electrochemical recordings of glutamate and other neurochemicals. Trends Anal Chem 22:498–502
51. Stephens ML, Quintero JE, Pomerleau F et al (2009) Age-related changes in glutamate release in the CA3 and dentate gyrus of the rat hippocampus. Neurobiol Aging 32(5):811–820
52. Photolithography (2001) Texas engineering extension service (ed), Texas
53. Bargon J (ed) (1984) Methods and materials in microelectronic technology (IBM Research Symposia Series), Plenum, New York, NY
54. Wire Bond (1998) Texas Engineering Extension Service (ed), Texas
55. Gerhardt GA, Burmeister JJ (2006) Neurochemical arrays. In: Grimes CA (ed) Encyclopedia of sensors. American Scientific, Stevenson Ranch, CA
56. Borland LM, Shi GY, Yang H et al (2005) Voltammetric study of extracellular dopamine near microdialysis probes acutely implanted in the striatum of the anesthetized rat. J Neurosci Methods 146:149–158

57. Gerhardt GA, Oke AF, Nagy G et al (1984) Nafion-coated electrodes with high selectivity for CNS electrochemistry. Brain Res 290:390–395
58. van Horne C, Hoffer BJ, Stromberg I et al (1992) Clearance and diffusion of locally applied dopamine in normal and 6-hydroxydopamine-lesioned rat striatum. J Pharmacol Exp Ther 263:1285–1292
59. Hebert MA, van Horne CG, Hoffer BJ et al (1996) Functional effects of GDNF in normal rat striatum: presynaptic studies using in vivo electrochemistry and microdialysis. J Pharmacol Exp Ther 279:1181–1190
60. Hoffman AF, Lupica CR, Gerhardt GA (1998) Dopamine transporter activity in the substantia nigra and striatum assessed by high-speed chronoamperometric recordings in brain slices. J Pharmacol Exp Ther 287:487–496
61. Friedemann MN, Robinson SW, Gerhardt GA (1996) *o*-Phenylenediamine-modified carbon-fiber electrodes for the detection of nitric oxide. Anal Chem 68:2621–2628
62. Lowry JP, Miele M, O'Neill RD et al (1998) An amperometric glucose-oxidase/poly(o-Phenylenediamine) biosensor for monitoring brain extracellular glucose: in vivo characterisation in the striatum of freely-moving rats. J Neurosci Methods 79:65–74
63. Chen X, Matsumoto N, Hu Y et al (2002) Electrochemically mediated electrodeposition/electropolymerization to yield a glucose microbiosensor with improved characteristics. Anal Chem 74:368–372
64. Matsumoto N, Chen X, Wilson GS (2002) Fundamental studies of glucose oxidase deposition on a Pt electrode. Anal Chem 74:362–367
65. Burmeister JJ, Pomerleau F, Huettl P et al (2008) Ceramic-based multisite microelectrode arrays for simultaneous measures of choline and acetylcholine in CNS. Biosens Bioelectron 23:1382–1389
66. Hinzman JM, Thomas TC, Burmeister JJ et al (2010) Diffuse brain injury elevates tonic glutamate levels and potassium-evoked glutamate release in discrete brain regions at two days post-injury: an enzyme-based microelectrode array study. J Neurotrauma 27:889–899
67. Bindra DS, Zhang Y, Wilson GS et al (1991) Design and in vitro studies of a needle-type glucose sensor for subcutaneous monitoring. Anal Chem 63:1692–1696
68. Burmeister JJ, Palmer M, Gerhardt G (2005) L-Lactate measures in brain tissue with ceramic-based multisite microelectrodes. Biosens Bioelectron 20:1772–1779
69. Dash MB, Douglas CL, Vyazovskiy VV et al (2009) Long-term homeostasis of extracellular glutamate in the rat cerebral cortex across sleep and waking states. J Neurosci 29:620–629
70. Konradsson-Geuken A, Gash CR, Alexander K et al (2009) Second-by-second analysis of alpha 7 nicotine receptor regulation of glutamate release in the prefrontal cortex of awake rats. Synapse 63:1069–1082
71. Stephens ML, Pomerleau F, Huettl P et al (2010) Real-time glutamate measurements in the putamen of awake rhesus monkeys using an enzyme-based human microelectrode array prototype. J Neurosci Methods 185:264–272
72. Thomas TC, Grandy DK, Gerhardt GA et al (2009) Decreased dopamine D4 receptor expression increases extracellular glutamate and alters its regulation in mouse striatum. Neuropsychopharmacology 34:436–445
73. Parikh V, Ji J, Decker MW et al (2010) Prefrontal beta2 subunit-containing and alpha7 nicotinic acetylcholine receptors differentially control glutamatergic and cholinergic signaling. J Neurosci 30:3518–3530
74. Giuliano C, Parikh V, Ward JR et al (2008) Increases in cholinergic neurotransmission measured by using choline-sensitive microelectrodes: enhanced detection by hydrolysis of acetylcholine on recording sites? Neurochem Int 52:1343–1350
75. Burmeister J, Palmer M, Gerhardt G (2003) Ceramic-based multisite microelectrode array for rapid choline measures in brain tissue. Anal Chim Acta 481:65–74
76. Parikh V, Pomerleau F, Huettl P et al (2004) Rapid assessment of in vivo cholinergic transmission by amperometric detection of changes in extracellular choline levels. Eur J Neurosci 20:1545–1554
77. Parikh V, Sarter M (2006) Cortical choline transporter function measured in vivo using choline-sensitive microelectrodes: clearance of endogenous and exogenous choline and effects of removal of cholinergic terminals. J Neurochem 97:488–503
78. Parikh V, Kozak R, Martinez V et al (2007) Prefrontal acetylcholine release controls cue detection on multiple timescales. Neuron 56: 141–154
79. Parikh V, Sarter M (2008) Cholinergic mediation of attention: contributions of phasic and tonic increases in prefrontal cholinergic activity. Ann NY Acad Sci 1129:225–235
80. Hu Y, Mitchell KM, Albahadily FN et al (1994) Direct measurement of glutamate release in the brain using a dual enzyme-based electrochemical sensor. Brain Res 659:117–125
81. Dixon BM, Lowry JP, O'Neill RD (2002) Characterization in vitro and in vivo of the oxygen dependence of an enzyme/polymer biosensor for monitoring brain glucose. J Neurosci Methods 119:135–142
82. Wilson GS, Gifford R (2005) Biosensors for real-time in vivo measurements. Biosens Bioelectron 20:2388–2403

83. Williams GV, Millar J (1990) Concentration-dependent actions of stimulated dopamine release on neuronal activity in rat striatum. Neuroscience 39:1–16
84. Stamford JA, Palij P, Davidson C et al (1993) Simultaneous "Real-Time" electrochemical and electrophysiological recording in brain slices with a single carbon-fibre microelectrode. J Neurosci Methods 50:279–290
85. Cheer JF, Heien ML, Garris PA et al (2005) Simultaneous dopamine and single-unit recordings reveal accumbens gabaergic responses: implications for intracranial self-stimulation. Proc Natl Acad Sci USA 102:19150–19155
86. Zhang H, Lin SC, Nicolelis MA (2009) Acquiring local field potential information from amperometric neurochemical recordings. J Neurosci Methods 179:191–200
87. Zhang H, Lin SC, Nicolelis MA (2010) Spatiotemporal coupling between hippocampal acetycholine release and theta oscillations in vivo. J Neurosci 30:13431–13440
88. Bao X, Pal R, Hascup KN et al (2009) Transgenic expression of Glud1 (Glutamate Dehydrogenase 1) in neurons: in vivo model of enhanced glutamate release, altered synaptic plasticity, and selective neuronal vulnerability. J Neurosci 29:13929–13944

Chapter 3

The Advantage of Microelectrode Technologies for Measurement in Delicate Biological Environments Such as Brain Tissue

Andrea Jaquins-Gerstl and Adrian C. Michael

Abstract

Measuring the concentration of neurotransmitters in the extracellular space of living brain tissue is vital in our understanding of neurochemical events. Electrochemical microelectrodes and microsensors are commonly used for in vivo neurotransmitter monitoring; each makes a unique contribution to the study of neuroscience. Electrochemical methods are versatile techniques, due to the small size of microelectrodes and the high spatial and temporal resolution of microsensors. These methods are gaining widespread attention especially in the study of neurochemical environments and molecules. This chapter will discuss the advantages and applications of microdialysis, microelectrodes, and enzyme-based microsensors as they apply to neuroscience. Specific applications of dopamine measurements will be also addressed and a brief description of what the future holds for microsensors. Finally, this chapter will focus on the promising role of enzyme-based electrochemical microsensors.

Key words Microelectrodes, Enzyme-based microsensors, Microdialysis, Dopamine

1 Introduction

J.W.L. Thudichum's "*A Treatise on the Chemical Constitution of the Brain*," published in 1884 (1), is credited with being the first report on the role of chemistry in brain function. The concept that chemical signals pass information between brain cells was well established in the early part of the twentieth century, leading to the understanding that disruptions of neurotransmission processes lie at the heart of numerous brain pathologies that afflict large fractions of the human population. Today, our understanding of these pathologies remains incomplete. We can treat the symptoms of brain disorders, such as Parkinson's and Alzheimer's disease, attention deficit hyperactivity disorder, schizophrenia, and substance abuse, but we cannot necessarily prevent or cure them. Thus, the advancement of techniques to more deeply probe

Stéphane Marinesco and Nicholas Dale (eds.), *Microelectrode Biosensors*, Neuromethods, vol. 80,
DOI 10.1007/978-1-62703-370-1_3, © Springer Science+Business Media, LLC 2013

neurotransmission and its regulation remains a significant goal. In particular, techniques that allow real-time chemical monitoring in living animals, to allow correlations between brain chemistry and behavior, are considered to be of key importance. This chapter explores the emerging role of enzyme-based electrochemical microsensors in this effort.

2 Microelectrodes and Microdialysis in Neuroscience

The roots of present efforts to develop and apply enzyme-modified microsensors in neuroscience lie in the fields of in vivo voltammetry and microdialysis sampling. To place the enzyme-modified microsensors in context, we open this chapter by briefly describing the achievements and challenges associated with these related approaches.

2.1 In Vivo Voltammetry

Modern in vivo voltammetry descends directly from the pioneering polarographic research of Jaroslav Heyrovsky, who received the Nobel Prize in Chemistry in 1959. Ralph N. Adams pioneered the incorporation of solid electrodes, i.e., electrodes made of metals or carbon, into the techniques of electrochemistry (Adams' 1969 monograph "Electrochemistry at Solid Electrodes" is highly recommended reading for newcomers to this field). Building on his earlier investigations into the electrochemistry of phenolic compounds, Adams recognized the opportunity to apply electrochemistry to the in vivo detection of biogenic amine neurotransmitters, such as dopamine, norepinephrine, adrenaline, and serotonin, capitalizing on the presence of easily oxidizable catechol and indole moieties in their chemical structures. Hence, Adams implanted small carbon electrodes into a dopamine-rich region of the rat brain and monitored changes in the observed oxidation currents triggered by the administration of amphetamine, a drug believed to exert its psychoactive effects by increasing dopamine release from neuronal terminals and thereby driving up the dopamine concentration in the extracellular space (2–4). At first, the experiment appeared to be a wild success, because, as expected, amphetamine increased the measured oxidation current and this was interpreted to mean that the electrode had measured a change in dopamine concentration in vivo. As it turned out, this was a case of beginner's bad luck, because subsequent measurements showed that amphetamine also induces a large increase in the extracellular concentration of ascorbic acid (vitamin C), which is also easily oxidized at carbon electrodes at the same potentials as dopamine. Thus, these early missteps heightened awareness of the critical importance of clearly establishing the sensitivity, selectivity, and temporal response of in vivo electrochemical measurements.

A major milestone in the advancement of in vivo voltammetry was the invention of the carbon fiber microelectrode by Francois Gonon and colleagues, which is still in widespread use today for in vivo voltammetry and, to a lesser extent, as a substrate for enzyme-modified microsensors (5, 6). Gonon found that by electrochemically pretreating these electrodes in various ways it was possible to resolve in vivo oxidation currents into separate peaks at distinct potentials. The electrochemical pretreatments also produced highly sensitive electrodes (7, 8). Thus, this work was a major contribution to enhancing the sensitivity and selectivity of in vivo measurements. Nevertheless, the temporal response of the pretreated electrodes was too slow for some purposes, such as measuring neurotransmitter concentrations in the range of frequencies at which neurons generate action potentials. Julian Millar addressed this challenge by introducing fast-scan cyclic voltammetry (FSCV) in 1985 (9), a technique that is still widely used today. The capabilities of FSCV performed in vivo at carbon fiber microelectrodes have evolved into an enormously powerful technique that enables the detection of dopamine release associated with the action potentials occurring in the brain of awake, behaving rats (we will return to this topic a bit later in this chapter).

2.2 Microdialysis

Although FSCV is a powerful method, its applicability is limited to dopamine and perhaps norepinephrine and serotonin (although serotonin measurements have proven very challenging) (10–23). Since so many molecules play vital roles in brain function, great effort has been devoted to finding more widely applicable in vivo techniques. In addition, a priority was placed on techniques that would avoid the selectivity challenges brought to light by the early efforts to use electrochemistry in the brain. As a result, microdialysis sampling has become a technique of major importance in neurochemistry (24–33). Microdialysis recovers molecules from the brain extracellular space for subsequent analysis by any suitable technique, including HPLC, electrophoresis, mass spectrometry, and so on. Review articles provide lengthy lists of the molecules that have been quantified in brain dialysis samples (33–35). Microdialysis is easily performed in awake animals, as the technique does not involve noise-prone oxidation currents, so the literature contains innumerable publications that relate microdialysis results to animal behavior. In light of the existence of microdialysis, it is worthwhile considering whether the continued development of in vivo electrochemical techniques, including the development of enzyme-modified microsensors, is warranted.

Perhaps not surprisingly, microdialysis has limitations and some drawbacks. For example, a negative drawback of microdialysis is its limited spatiotemporal resolution. With conventional approaches, dialysis samples are analyzed on the time frame of several minutes while advanced approaches reach the sub-minute time frame.

However, interesting and significant neurotransmitter events also occur on the sub-second and millisecond timescale. A second drawback of microdialysis probes, not unrelated to the first, is the extent to which the probes damage the delicate and sensitive brain tissue from which the samples are recovered. Typical diameters of conventional microdialysis probes range from 200 to 300 μm much larger than the intercapillary distances in the brain. So, when a probe is inserted, it damages the very tissue from which it samples (36–44). The probe disturbs microcirculation, creates edema, changes the rate of glucose utilization, and opens the blood–brain barrier (44–49). The extent of injury is surprising, as far as 1.4 mm from the probe. Drew and Michael found that tissue within 100 μm of the probe track was significantly depleted of neuronal terminals (39, 50). The traumatized tissue presents a barrier to diffusion between the dialysis membrane and the synapse (51, 52). The sum of these processes may largely inflict the physiology of neurotransmission.

Together these findings demonstrate that microelectrodes are valuable tools for in vivo studies of dopamine; by using these very small devices to measure dopamine, it is now possible to examine dopamine through a new set of lenses. The use of a microdialysis probes would not have led to such great insight and novel findings. This is also true for spectroscopic methods such as positron emission tomography and functional magnetic imaging; real-time concentrations of neurotransmitters cannot be determined. In addition to limited temporal and spatial resolution, questions arise to the penetration trauma associated with microdialysis, how this affects the tissue surrounding the probe and its impact on neurochemical measurements. The ability to measure ultra-fast and ultra-small concentrations of dopamine gives motivation to examine other molecules in this same manner, questioning what other molecules and important advances are possibly being overlooked or missed. A question that many think the use of microsensors can answer.

3 Dopamine: The Gold Standard

Dopamine is considered by many to be the gold standard in neurotransmitter analysis. This neurochemical transmitter is implicated in a number of diseases states, including Parkinson's disease, schizophrenia, and drug abuse (53–55). Investigations into this ubiquitous neurotransmitter have been employed by microdialysis, microelectrodes, and microsensors. There are countless journals published on dopamine; over 110,000 research papers have been published in the last 60 years. This section will briefly highlight some new advances made in dopamine measurements and their pathological implications. All of these advancements use voltammetry techniques due to high temporal and spatial resolution requirements along with minimal damage at the implant site of the microelectrode.

Recently Wightman et al. used FSCV to investigate the varying effects of reinforce magnitude (intracranial stimulation parameters) on cue-evoked dopamine release in the nucleus accumbens (NAc) shell of rats performing intracranial self-stimulation (ICSS) (56). This electrochemical technique enables the detection of dopamine release with sub-second temporal resolution in "real time" during behavior. In this behavioral paradigm a rat depresses a lever to electrically stimulate select brain pathways. Often cue-evoked dopamine release in the NAc shell is associated with ICSS which develops with learning. Authors found that the concentration of the cue-associated dopamine transients increases significantly with increasing reinforcer magnitude. Also, increase in cue-evoked dopamine significantly correlated with shorter latencies to lever press. This study further establishes that cue-evoked dopamine in the NAc plays a role in learned, reward-seeking behavior. These transient time-locked increases in dopamine concentrations would not have been observed with microdialysis probes.

Similar research performed by Phillips and colleagues examined the role of dopamine signaling during stimulus-reward learning (57). The exact role that dopamine plays in reward-related processes remains controversial. Individuals make a variety of choices using a complex process that assigns value to the rewards and associated stimuli. However, often individuals vary in the degree to which cues bias choice and potentially promote maladaptive behavior. FSCV was used in conjunction with a carbon fiber microelectrode to characterize the pattern of phasic dopamine signaling in the NAc region during Pavlovian conditioning in rats. Food cues were used to probe the role of dopamine in stimulus-reward learning (Fig. 1). This study showed that intact dopamine transmission is not required for all forms of learning in which reward cues became effective predictors. Dopamine acts selectively in a form of stimulus-reward learning in which incentive salience is assigned to reward cues. This work offers great insight into a form of stimulus-reward learning which is present in increased susceptibility of impulse control across a broad range of psychiatric disorders.

Studies involving electrically evoked dopamine release by FSCV are also exploring the heterogeneity of dopamine release within the rat striatum. Previous studies attribute the heterogeneity of dopamine terminals to variations in the density of dopamine terminals at the microelectrode recording site. However, Michael et al. concluded that the heterogeneity of evoked dopamine release derives from variations in the extent to which dopamine terminals are autoinhibited (12). Authors demonstrate that low-amplitude, slow evoked dopamine responses occur even though recording electrodes are close to dopamine terminals, not by diffusion. Moreover,

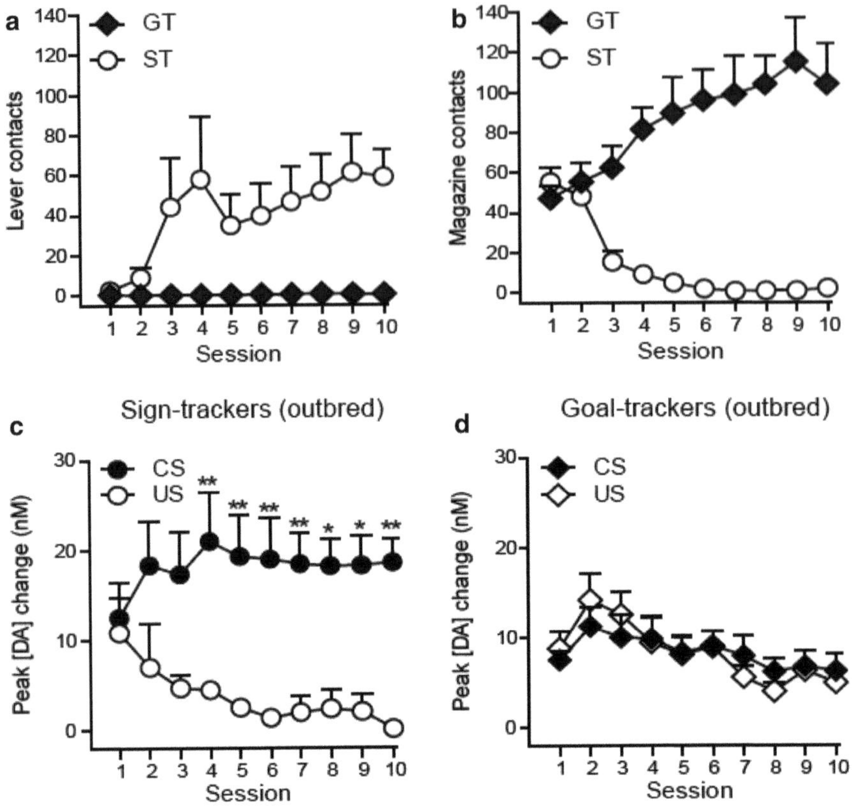

Fig. 1 Conditional responses and phasic dopamine release in outbred rats after extended training. Sign-tracking behavior ($n=4$) is shown in panel (**a**) and goal-tracking behavior ($n=4$) in panel (**b**) for rats that received ten training sessions. Mean + SEM (**a**) number of lever–CS contacts, (**b**) number of food–tray contacts during lever–CS presentation. Mean + SEM change in dopamine concentration in response to CS and US presentation for ten sessions of training in (**c**) sign-trackers and (**d**) goal-trackers. (Bonferroni post-hoc comparisons of CS- and US-evoked dopamine release $^*P < 0.05$, $^{**}P < 0.01$). Reproduction reprint permission of Nature © 2011 (57)

the D2 agonist and antagonist, quinpirole and raclopride, respectively, affect the slow responses in a manner consistent with the known functions of presynaptic D2 autoreceptors. This study demonstrates that the striatum contains dopamine terminals operating on multiple time courses: some fast, some slow, and others a hybrid response (Fig. 2). One might question whether these different signals ultimately are responsible for different functions that are associated with dopamine.

4 Microsensors in Neuroscience

Microsensors are gaining widespread attention and use. Our focus is on microsensors used in the brain; most of these sensors use oxidase enzymes with amperometric detection methods. Many examples of

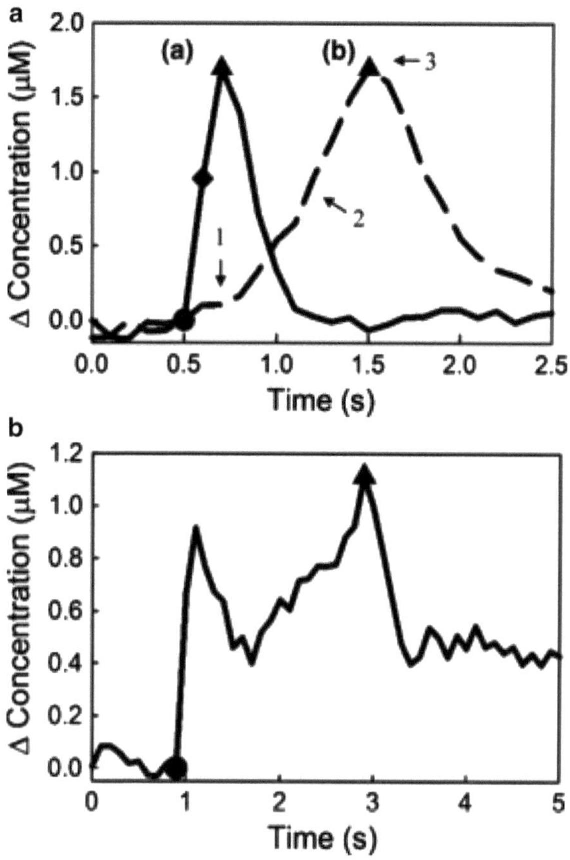

Fig. 2 (**a**) Kinetic heterogeneity of evoked DA release in the rat striatum. The symbols mark the start (*circles*) and finish (*triangles*) of the stimuli. The evoked change in extracellular DA was recorded by FSCV with 400-μm long carbon fiber microelectrodes. (**a**) A fast-type response: DA increases rapidly during the stimulus and falls rapidly after the stimulus. The amount of DA evoked after the first six stimulus pulses (*diamond*) is more than half of the amount of DA evoked after 12 pulses (*triangle*), indicating that the rate of DA release is decelerating. (**b**) A slow-type response: evoked DA release begins slowly (*arrow 1*) and accelerates as the stimulus continues (*arrow 2*). At the end of stimulation the signal returns toward baseline immediately (*arrow 3*), showing no signs of overshoot. (**b**) A carbon fiber electrode location exhibiting both fast-type and slow-type release. This hybrid-type response has two peaks. The slope of the ascending part of the first peak is steeper than the second peak counterpart. The rapid rise of the first peak demonstrates a fast-type response, while the second peak has a slow-type response. Adapted from (12). Reproduction reprint permission of Journal of Neurochemistry © 2009 (12)

enzyme-modified microsensors exist in the current literature, such as glucose oxidase (GOx), horseradish peroxidase (HRP), choline oxidase (ChOx), and acetylcholinesterase (AChE). These microsensors are often used in electrochemical detection and are often extended for use in the neurosciences. The most common microsensors used in the brain are glutamate, hydrogen peroxide, and choline (Ch).

The most well-known work in this field was demonstrated by Adam Heller; he pioneered the field of electrical connection ("wiring") of redox centers of enzymes to electrodes through electron-conducting redox hydrogels (58, 59). Unlike many other diffusion redox mediators, he used glucose–oxidase catalyzing redox hydrogels which have no leachable components. These redox hydrogels are cross-linked polymers which usually contain osmium polypyridine-based redox centers tethered to cross-linked water-soluble polymer backbones. The glucose is electro-oxidized through catalysis by glucose dehydrogenase and a "wiring" $Os^{2+/3+}$ redox mediator. The redox mediator is reduced by the glucose-reduced enzyme and is electro-oxidized on the carbon electrode. Glucose sensors are based on this technology and are in development for many commercial uses (60–63).

An important characteristic is the size of the microsensors; these small sensors can gain access to extracellular pools of neurotransmitters that are typically inaccessible to microdialysis probes (42, 64). One such example is extracellular glutamate, crucial in our understanding of glutamatergic systems in the central nervous system. Many analytical techniques do not fulfill necessary requirements. Microdialysis does not detect neuronally derived glutamate (65–67). A milestone in the development of glutamate sensors was the observation made by Kulagina et al. in 1999, who reported that hydrogel-coated glutamate microsensors were able to detect tetrodotoxin sensitive glutamate in the brain (64). It was not until 2006 that Day et al. reported similar findings (68). Many applications for the detection of glucose, choline, glutamate, and peroxide with a microsensor were further developed by the research group of Adrian C. Michael (69–74) and others.

Another glutamate microsensor that has shown excellent response time (1–2 s) and temporal stability (3-week half-life) is the work of Monbouquette (75). Amperometric detection was used to detect small amounts of glutamate (80 ± 10 nA μM^{-1} cm^{-2}) and selectively against ascorbate and dopamine in vivo. The microsensor consisted of Pt black (PtBlk)-coated electrode in conjunction with the enzyme l-GluOx. The Pt wire had a film of electropolymerized, overoxidized polypyrrole; this film permits passage of H_2O_2 to the electrode surface while rejecting common interferents.

Becoming more common, microsensors can measure more than just one analyte at a time, and can be configured for "self-referencing" (Fig. 3) (76). One of the most unique contributions in the field of microsensors is the eight-site microelectrode arrays by Greg Gerhardt and colleagues (77). Fabrication processes now allow microelectrodes to contain eight platinum (Pt) recording sites. Along with this new technology they have also began testing a more flexible multisite microelectrode in monkeys and humans for recording deep brain stimulation. These new eight Pt site microelectrodes are capable of simultaneously recording amperometric data from each site on a second-by-second basis. This is especially advantageous in that multiple analytes can be measured from a single microelectrode in the same brain region and recording from several

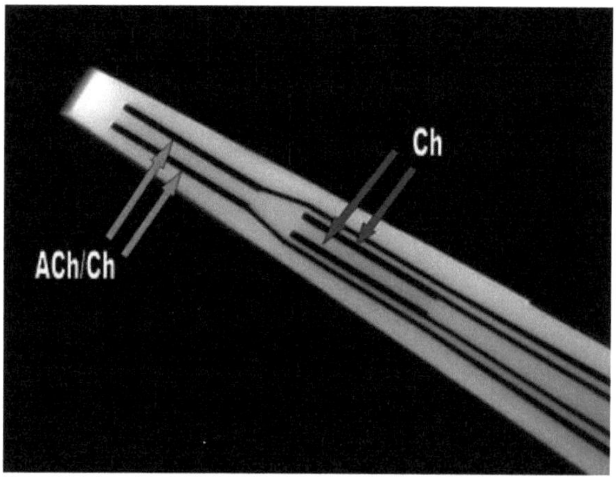

Fig. 3 High magnification photomicrograph of the recording sites of the ACh-sensitive microelectrode array (MEA) illustrating the two pairs of Pt channels. The first pair, located 100 μm from the tip of the MEA, was sensitive to both ACh- and Ch-based signals, whereas the second pair located 100 μm up the shaft of the MEA was sensitive to only Ch-based signals. Reproduction reprint permission of European Journal of Neuroscience © 2006 (78)

different anatomical brain regions can be achieved from multisite microelectrodes. Another novel development in this work is that the microsensors can be configured for "self-referencing" recordings (sentinel sites); this allows cross-checking of selectivity (previously mentioned). This technique is beneficial for qualitatively assessing the validity of analyte signaling during an experiment and the ability to subtract off interferents and noise. Future designs will incorporate 16 Pt recording sites on the same microelectrode, allowing for four or more analytes to be simultaneously measured.

In recent years considerable progress has been made in the application of enzyme-based microsensors, especially in the detection of acetylcholine (ACh) and glutamate (Fig. 4). The work of Bruno and Gerhardt has provided great insight into measurements of both ACh and glutamate (78). One such example is the regulation of cholinergic transmission; rapid measurements, second-by-second changes of ACh release, and clearance in anesthetized rats were observed using a multichannel microelectrode array. Changes in current selectively produced by the hydrolysis of ACh to Ch and the subsequent oxidation of Ch and hydrogen peroxidase (H_2O_2) were measured, and self-referencing techniques were used which isolated the current generated by the release of ACh.

The literature for microsensors is vast; however, work by Agnesi successfully coupled microsensors into a wireless platform (79). Authors developed the wireless instantaneously neurotransmitter concentration system (WINCS) which is coupled to digital telemetry using FSCV to measure dopamine concentrations. In this study WINCS was used with fixed potential amperometry (FPA) to

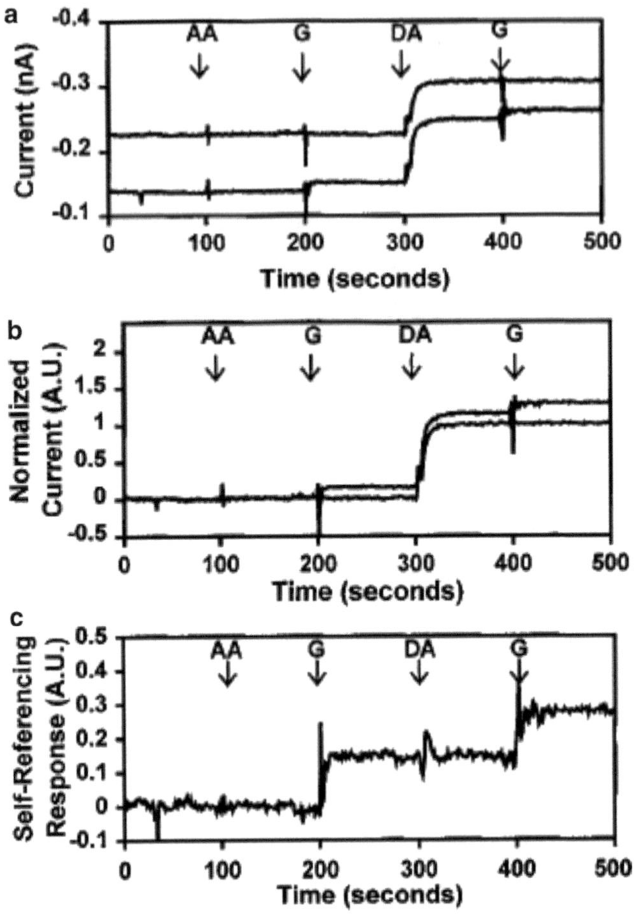

Fig. 4 Electrode response versus time plots illustrating the self-referencing capabilities of the glutamate sensor. The *arrows* indicate concentration steps of 250 μM ascorbic acid (AA), 10 μM glutamate (G), and 10 μM dopamine (DA). The *upper traces* in (**a**) and (**b**) are the control electrode responses and the *lower traces* are the glutamate electrode responses. (**a**) is the raw current response, (**b**) is the normalized, baseline-subtracted signal, and (**c**) is the subtracted signal resulting from glutamate (arbitrary units, A.U.) Reproduction reprint permission of Analytical Chemistry © 2001 (76)

measure extracellular concentrations of dopamine, glutamate, and adenosine. By combining enzyme-linked biosensors and FPA, non-electroactive analytes can be monitored in real time. Authors performed an extensive study using three different recording electrodes (1) a carbon fiber microelectrode to measure dopamine, (2) a multiple enzyme-linked electrode of adenosine deaminase, nucleoside phosphorylase, and xanthine oxidase which was used to measure adenosine, and (3) a glutamate oxidase enzyme-linked electrode to measure glutamate. In most of these studies anesthetized rats were used; however, in order to mimic intraoperative human neurosurgery, cortical glutamate release was monitored using a pig model. By incorporating a transimpedance amplifier, a microprocessor, a

Blue tooth transceiver, and a battery-powered multilayer printed circuit board, striatal dopamine, glutamate, and adenosine were successfully measured both in vitro and in vivo.

Although the field of enzyme-based microsensors has grown substantially over the last several decades, relatively few applications have so far involved real-time measurements in the brain. Properties of the sensors in vivo are difficult to assess and often contradict in vitro results. Sensitivity, linearity, stability, and interferences under continuous operation are all important consideration for in vivo measurements. Fouling of the electrode surface, fouling of the surface of the outer bio-layer, and loss of enzyme activity due to proteases can all have serious consequences for in vivo measurements. However, despite these limitations we emphasize that in many circumstances microsensors are extremely useful analytical tools. Novel measurements such as these far exceed conventional microdialysis capabilities. This growing body of evidence using enzyme-based microsensors lays important ground work for future advances made in the field of neuroscience.

5 Conclusion

In this chapter, we have discussed the advantages and limitations of microelectrodes, microsensors, and microdialysis for monitoring neurotransmitters in the CNS. Microelectrodes and microsensors have come a long way from the earliest days; when coupled to electrochemical techniques these devices make a unique contribution to the field of neurochemistry. It is becoming increasingly clear that microsensors provide a wealth of neurochemical information. Despite the complexity of living systems and the problems associated with in vivo measurements using microsensors, significant progress has been made in improving reliability, extending capabilities to higher sensitivity and higher selectivity, and faster response times. Furthermore, with the development of brain imaging methods it is now possible to study neurotransmitter dynamics in humans. The next generation holds great promise in advancing a number of fields, all of which should bring us closer to understanding the multifaceted system of communication between neurons in the living brain.

References

1. Thudichum JLW (ed) (1884) A treatise on the chemical constitution of the brain. Bailliere, London
2. Adams RN (1976) Probing brain chemistry with electroanalytical techniques. Anal Chem 48:1126A–1138A
3. Kissinger PT, Hart JB, Adams RN (1973) Voltammetry in brain tissue–a new neurophysiological measurement. Brain Res 55:209–213
4. Wightman RM, Strope E, Plotsky PM, Adams RN (1976) Monitoring of transmitter metabolites by voltammetry in cerebrospinal fluid following neural pathway stimulation. Nature 262:145–146
5. Gonon F, Buda M, Cespuglio R, Jouvet M, Pujol JF (1980) In vivo electrochemical detection of catechols in the neostriatum of anaesthetized rats: dopamine or DOPAC? Nature 286:902–904

6. Ponchon JL, Cespuglio R, Gonon F, Jouvet M, Pujol JF (1979) Normal pulse polarography with carbon fiber electrodes for in vitro and in vivo determination of catecholamines. Anal Chem 51:1483–1486
7. Gonon FG, Fombarlet CM, Buda MJ, Pujol JF (1981) Electrochemical treatment of pyrolytic carbon fiber electrodes. Anal Chem 53:1386–1389
8. Gonon FG, Navarre F, Buda MJ (1984) In vivo monitoring of dopamine release in the rat brain with differential normal pulse voltammetry. Anal Chem 56:573–575
9. Millar J, Stamford JA, Kruk ZL, Wightman RM (1985) Electrochemical, pharmacological and electrophysiological evidence of rapid dopamine release and removal in the rat caudate nucleus following electrical stimulation of the medial forebrain bundle. Eur J Pharmacol 3:341–348
10. Kozminski KD, Gutman DA, Davila V, Sulzer D, Ewing AG (1998) Voltammetric and pharmacological characterization of dopamine release from single exocytotic events at rat pheochromocytoma (PC12) cells. Anal Chem 70:3123
11. Limberger N, Trou SJ, Kruk ZL, Starke K (1991) "Real time" measurement of endogenous dopamine release during short trains of pulses in slices of rat neostriatum and nucleus accumbens: role of autoinhibition. Naunyn Schmiedebergs Arch Pharmacol 344:623–629
12. Moquin KF, Michael AC (2009) Tonic autoinhibition contributes to the heterogeneity of evoked dopamine release in the rat striatum. J Neurochem 110:1491–1501
13. Peters JL, Michael AC (2000) Changes in the kinetics of dopamine release and uptake have differential effects on the spatial distribution of extracellular dopamine concentration in rat striatum. J Neurochem 74:1563–1573
14. Phillips PEM, Stuber GD, Heien MLAV, Wightman RM, Carelli RM (2003) Subsecond dopamine release promotes cocaine seeking. Nature 422:614–618
15. Roitman MF, Stuber GD, Phillips PEM, Wightman RM, Carelli RM (2004) Dopamine operates as a subsecond modulator of food seeking. J Neurosci 24:1265–1271
16. Robinson DL, Phillips PE, Budygin EA, Trafton BJ, Garris PA, Wightman RM (2001) Sub-second changes in accumbal dopamine during sexual behavior in male rats. Neuroreport 12:2549–2552
17. Robinson DL, Venton BJ, Heien ML, Wightman RM (2003) Detecting subsecond dopamine release with fast-scan cyclic voltammetry in vivo. Clin Chem 49:1763–1773
18. Pihel K, Schroeder TJ, Wightman MA (1994) Rapid and selective cyclic voltammetric measurements of epinephrine and norepinephrine as a method to measure secretion from single bovine adrenal medullary cells. Anal Chem 66:4532–4537
19. Pihel K, Hsieh S, Wightman MA (1995) Electrochemical detection of histamine and 5-hydroxytryptamine at isolated mast cells. Anal Chem 67:4514–4521
20. Moquin KF, Michael AC (2011) An inverse correlation between the apparent rate of dopamine clearance and tonic autoinhibition in subdomains of the rat striatum: a possible role of transporter-mediated dopamine efflux. J Neurochem 117:133–142
21. Wang Y, Moquin KF, Michael AC (2010) Evidence for coupling between steady-state and dynamic extracellular dopamine concentrations in the rat striatum. J Neurochem 114:150–159
22. Dugast C, Brun P, Sotty F, Renaud B, Suaud-Chagny MF (1997) On the involvement of a tonic dopamine D2-autoinhibition in the regulation of pulse-to-pulse-evoked dopamine release in the rat striatum in vivo. Naunyn Schmiedebergs Arch Pharmacol 355:716–719
23. Venton BJ, Zhang H, Garris PA, Phillips PE, Sulzer D, Wightman RM (2003) Real-time decoding of dopamine concentration changes in the caudate-putamen during tonic and phasic firing. J Neurochem 87:1284–1295
24. Kennedy RT, Thompson JE, Vickroy TW (2002) In vivo monitoring of amino acids by direct sampling of brain extracellular fluid at ultralow flow rates and capillary electrophoresis. J Neurosci Methods 114:39–49
25. Kennedy RT, Watson CJ, Haskins WE, Powell DH, Strecker RE (2002) In vivo neurochemical monitoring by microdialysis and capillary separations. Curr Opin Chem Biol 6:659–665
26. Nordstrom CH (2009) Cerebral energy metabolism and microdialysis in neurocritical care. Childs Nerv Syst 26:465–472
27. Obrenovitch TP, Richards DA, Sarna G, Symon L (1993) Combined intracerebral microdialysis and electrophysiological recordings:methods and applications. J Neurosci Methods 47:139–145
28. Parsons LH, Justice JB Jr (1994) Quantitative approaches to in vivo brain microdialysis. Crit Rev Neurobiol 8:189–220
29. Ungerstedt U, Hallstrom A (1987) In vivo microdialysis–a new approach to the analysis of neurotransmitters in the brain. Life Sci 41:861–864
30. Ungerstedt U, Rostami E (2004) Microdialysis in neurointensive care. Curr Pharm Des 10:2145–2152

31. Werner C, Engelhard K (2007) Pathophysiology of traumatic brain injury. Br J Anaesth 99:4–9
32. Benveniste H, Hansen AJ, Ottosen NS (1989) Determination of brain interstitial concentrations by microdialysis. J Neurochem 52:1741–1750
33. Watson CJ, Venton BJ, Kennedy RT (2006) In vivo measurements of neurotransmitters by microdialysis sampling. Anal Chem 78:1391–1399
34. Perry M, Li Q, Kennedy RT (2009) Review of recent advances in analytical techniques for the determination of neurotransmitters. Anal Chim Acta 653:1–22
35. Shippenberg TS, Thompson AC (2001) Overview of microdialysis. Curr Protoc Neurosci (Chapter 7, Unit 7.1)
36. Benveniste H, Diemer NH (1987) Cellular reactions to implantation of a microdialysis tube in the rat hippocampus. Acta Neuropathol 74:234–238
37. Benveniste H, Drejer J, Schousboe A, Diemer NH (1987) Regional cerebral glucose phosphorylation and blood-flow after insertion of a microdialysis fiber through the dorsal hippocampus in the rat. J Neurochem 49:729–734
38. Borland LM, Shi G, Yang H, Michael AC (2005) Voltammetric study of extracellular dopamine near microdialysis probes acutely implanted in the striatum of the anesthetized rat. J Neurosci Methods 146:149–158
39. Clapp-Lilly KL, Roberts RC, Duffy LK, Irons KP, Hu Y, Drew KL (1999) An ultrastructural analysis of tissue surrounding a microdialysis probe. J Neurosci Methods 90:129–142
40. Lu Y, Peters JL, Michael AC (1998) Direct comparison of the response of voltammetry and microdialysis to electrically evoked release of striatal dopamine. J Neurochem 70:584–593
41. Peters JL, Michael AC (1998) Modeling voltammetry and microdialysis of striatal extracellular dopamine: the impact of dopamine uptake on extraction and recovery ratios. J Neurochem 70:594–603
42. Yang H, Peters JL, Michael AC (1998) Coupled effects of mass transfer and uptake kinetics on in vivo microdialysis of dopamine. J Neurochem 71:684–692
43. Zhou F, Braddock JF, Hu Y, Zhu X, Castellani RJ, Smith MA, Drew KL (2002) Microbial origin of glutamate, hibernation and tissue trauma: an in vivo microdialysis study. J Neurosci Methods 119:121–128
44. Zhou F, Zhu X, Castellani RJ, Stimmelmayr R, Perry G, Smith MA, Drew KL (2001) Hibernation, a model of neuroprotection. Am J Pathol 158:2145–2151
45. Fassbender K, Schneider T, Bertsch D, Schlueter M, Fatar M, Ragoschke A, Kühl S, Kischka U, Hennerici M (2000) Temporal profile of release of interleukin-1β in neurotrauma. Neurosci Let 284:135–138
46. Major O, Shdanova T, Duffek L, Nagy Z (1990) Continuous monitoring of blood-brain barrier opening to 51Cr-EDTA by microdialysis after probe injury. Acta Neurochir Suppl 51:46–48
47. Morgan ME, Singhal D, Anderson BD (1996) Quantitative assessment of blood-brain barrier damage during microdialysis. J Pharmacol Exp Ther 277:1167–1176
48. Planas AM, Justicia C, Sole S, Friguls B, Cervera A, Adell A, Chamorro A (2002) Certain forms of matrix metalloproteinase-9 accumulate in the extracellular space after microdialysis probe implantation and middle cerebral artery occulusion/reperfusion. J Cereb Blood Flow Metab 22:918–925
49. Woodroofe MN, Sarna GS, Wadhwa M, Hayes GM, Loughlin AJ, Tinker A, Cuzner ML (1991) Detection of interleukin-1 and interleukin-6 in adult rat brain, following mechanical injury, by in vivo microdialysis: evidence of a role for microglia in cytokine production. J Neuroimmunol 33:227–236
50. Peters JL, Miner LH, Michael AC, Sesack SR (2004) Ultrastructure at carbon fiber microelectrode implantation sites after acute voltammetric measurements in the striatum of anesthetized rats. J Neurosci Methods 137:9–23
51. Jaquins-Gerstl A, Michael AC (2009) Comparison of the brain penetration injury associated with microdialysis and voltammetry. J Neurosci Methods 183:127–135
52. Mitala CM, Wang Y, Borland LM, Jung M, Shand S, Watkins S, Weber SG, Michael AC (2008) Impact of microdialysis probes on vasculature and dopamine in the rat striatum: a combined fluorescence and voltammetric study. J Neurosci Methods 174:177–185
53. Koob GF, Bloom FE (1988) Cellular and molecular mechanisms of drug dependence. Science 242:715–723
54. Martinez-Leon NC (2006) Psychopathology of attention deficit/hyperactivity disorder (AD/HD). Int J Clin Health Psychol 6:379–399
55. Seeman P, Ulpain C, Bergeron C, Riederer P, Jellinger K, Gabriel E, Reynolds GP, Tourtellotte WW (1984) Bimodal distribution of dopamine receptor densities in brains of schizophrenics. Science 225:728–730
56. Beyene M, Carelli RM, Wightman RM (2010) Cue-evoked dopamine release in the nucleus accumens shell tracks reinforcer magnitude

during intracranial self-stimulation. Neuroscience 169:1682–1688
57. Flagel SB, Clark JJ, Robinson TE, Mayo L, Czuj A, Willuhn I, Akers CA, Clinton SM, Phillips PEM, Akil H (2011) A selective role for dopamine in stimulus-reward learning. Nature 469:53–57
58. Heller A (1992) Electrical connection of enzyme redox centers to electrodes. J Phys Chem 96:3579
59. Heller A, Gregg BA (1990) Cross-linked redox gels containing glucose oxidase for amperometric biosensor applications. Anal Chem 62:258
60. Wang J, Freiha B, Naser N, Gonzalez RE, Wollenberger U, Ozsoz M, Evans O (1995) Amperometric biosensing of organic peroxides with peroxidase modified electrodes. Anal Chem 67:1326–1331
61. Wagner JG, Schmidtke DW, Quinn CP, Fleming TF, Bernacky B, Heller A (1998) Continuous amperometric monitoring of glucose in a brittle diabetic chimpanzee with a miniature subcutaneous electrode. Proc Natl Acad Sci USA 95:6379–6382
62. Schmidtke DW, Freeland AC, Heller A, Bonnecaze RT (1998) Measurement and modeling of the transient difference between blood and subcutaneous glucose concentrations in the rat after injection of insulin. Proc Natl Acad Sci USA 95:294–299
63. Feldman B, Brazg R, Schwartz S, Weinstein R (2003) A continuous glucose sensor based on wired enzyme technology-results from a 3-day trial in patients with type I diabetes. Diabetes Technol Ther 5:769–779
64. Kulagina NV, Shankar L, Michael AC (1999) Monitoring glutamate and ascorbate in the extracellular space of brain tissue with electrochemical microsensors. Anal Chem 71:5093–5100
65. Herrera-Marschitz M, You ZB, Goiny M, Meana J, Silveria R, Goukhin OV, Chen Y, Esponza S, Petterson E, Loidl CF, Lubec G, Andersson K, Ntlander I, Terenius L, Ungerstedt U (1996) On the orgin of extracellular glutamate levels monitiored in the basal ganglia of the rat by in vivo microdialysis. J Neurochem 66:1726–1735
66. Semba J, Kito S, Toru M (1995) Characterzation of extracellular aminoacids in striatum of freely moving rats by in vivo microdialysis. J Neural Transm 100:3–52
67. Shiraishi L, Kamiyama Y, Huttemeir PC, Beneniste H (1997) Extracellular glutamate and dopamine measured by microdialysis in the rat striatium during blockade of synaptic transmission in anesthetized and awake rats. Brain Res 759:221–227
68. Day BK, Pomerleau F, Burmeister JJ, Huettl P, Gerhardt GA (2006) Microelectrode array studies of basal and potassium-evoked release of l-glutamate in the anesthetized rat brain. J Neurochem 96:1626–1635
69. Garguilo MG, Huynh N, Proctor A, Michael AC (1993) Amperometric sensors for peroxide, choline and acetylcholine based on electron transfer between HRP and a redox-polymer. Anal Chem 65:523–528
70. Garguilo MG, Michael AC (1994) Quantitation of choline in the extracellular fluid of brain tissue with amperometric microsensors. Anal Chem 66:2621–2629
71. Garguilo MG, Michael AC (1995) Optimization of amperometric microsensors for monitoring choline in the extracellular fluid of brain tissue. Anal Chim Acta 307:291–299
72. Garguilo MG, Michael AC (1995) Enzyme-modified electrodes for peroxide, choline and acetylcholine. Trends Anal Chem 14:164–169
73. Garguilo MG, Michael AC (1996) Amperometric microsensors of monitoring choline in the extracellular fluid of brain. J Neurosci Methods 70:73–82
74. Kulagina NV, Michael AC (2003) Monitoring hydrogen peroxide in the extracellular space of the brain with amperometric microsensors. Anal Chem 75:4875–4881
75. Hamdi N, Wang J, Walker E, Maidment NT, Monbouquette HG (2006) An electroenzymatic l-glutmate microsensor selective against dopamine. J Electroanal Chem 591:33–40
76. Burmeister J, Gerhardt GA (2001) Self-referencing ceramic-based multisite microelectrodes for detection and elimination of interferences from the measurement of l-glutamate and other analytes. Anal Chem 73:1037–1042
77. Burmeister JJ, Pomerleau F, Palmer M, Day BK, Huettl P, Gerhardt GA (2002) Improved ceramic-based multisite microelectrodes for rapid measurements of l-glutamate in the CNS. J Neurosci Methods 119:163–171
78. Bruno JP, Gash C, Martin B, Zmarowski A, Pomerleau F, Burmeister J, Huettl P, Gerhardt GA (2006) Second-by-second measurement of acetylcholine release in prefrontal cortex. Eur J Neurosci 24:2749–2757
79. Agnesi F, Tye SJ, Bledsoe JM, Griessenauer CJ, Kimble CJ, Bennet KE, Garris PA, Blaha CD, Lee KH (2009) Wireless instantaneous neurotransmitter concentration system-based amperometric detection of dopamine, adenosine, and glutamate for intraoperative monitoring. J Neurosurg 111:256–265

Chapter 4

Hexacyanoferrates as Mediators for Microelectrode Biosensors

Faming Tian and Nicholas Dale

Abstract

Hexacyanoferrates such as Prussian blue or Ruthenium Purple have great potential for use in microelectrode biosensors. They act as surface-bound mediators that greatly facilitate the detection of H_2O_2 at modest potentials, thereby increasing the selectivity of the biosensors versus electroactive interferences. We review the structure of this class of mediator, their electrochemical properties, and their actions as "artificial peroxidases." We give a brief outline of how hexacyanoferrate layers can be deposited onto microelectrodes and activated and the means by which they can be integrated into enzymatic biosensors. We discuss the performance of such biosensors and their potential limitations with regard to stability, ionic sensitivity, and speed of response.

Key words Microbiosensor, Prussian blue, Ruthenium purple, Metal hexacyanoferrate, Electron mediator, Hydrogen peroxide detection

1 Introduction

The polynuclear transition metal hexacyanometallates of the general formula $M^A_k[M^B(CN)_6]_l\ xH_2O$, where M^A and M^B are transition metal ions with different formal oxidation numbers and k, l, and x are stoichiometric numbers, represent an important class of insoluble mixed valence compounds. Besides their importance for basic aspects of solid-state chemistry, electrochemistry, and structural attributes, the study of polynuclear transition metal hexacyanometallates has been driven by the desire to apply them in electrocatalysis, electrochromism, ion-exchange, and ion sensing research areas.

Among the polynuclear transition metal hexacyanometallates, Prussian blue, the oldest coordination compound, accidentally discovered by the Berlin artist, Diesbach, in 1704, is the classical prototype transition metal hexacyanide. Following its discovery, Prussian blue was initially used only as an important pigment for color-related application such as paints, printing inks, laundry dyes,

Stéphane Marinesco and Nicholas Dale (eds.), *Microelectrode Biosensors*, Neuromethods, vol. 80,
DOI 10.1007/978-1-62703-370-1_4, © Springer Science+Business Media, LLC 2013

and lacquers due to the excellent stability and intensity of its color. But the study of Prussian blue (and its analogues—transition metal hexacyanoferrates) and fundamental investigation of its electrochemical and spectroelectrochemical properties were rapidly boosted by successful electrochemical deposition of thin adherent films of Prussian blue on various metal and semiconductor substrates, initially discovered by Neff in 1978 (1). Subsequently, PB-modified electrodes have been intensively investigated due to their chemical stability, electrochemical catalytic activity, electrochromic reactions, easy preparation, and low cost (2–4).

2 Hexacyanoferrates: Structure and Electrochemistry

The structure of PB has been extensively studied to provide insight into its electrochemical behavior and catalytic activity. The first structure of PB was discussed on the basis of powder diffraction patterns by Keggin and Miles (5), followed by a proposed structure based on X-ray diffraction and spectroscopic investigations by Ludi and coworkers (6–8). As a result the structure and nature of chemical bonding in PB are now reasonably well understood. The basic structure (Fig. 1) is a three-dimensional polymeric network consisting of alternating ferric and ferrous ions on a face-centered cubic lattice with a cell constant of 1.02 nm, indicating a very roomy crystal structure. The zeolitic nature of PB with channel diameters of about

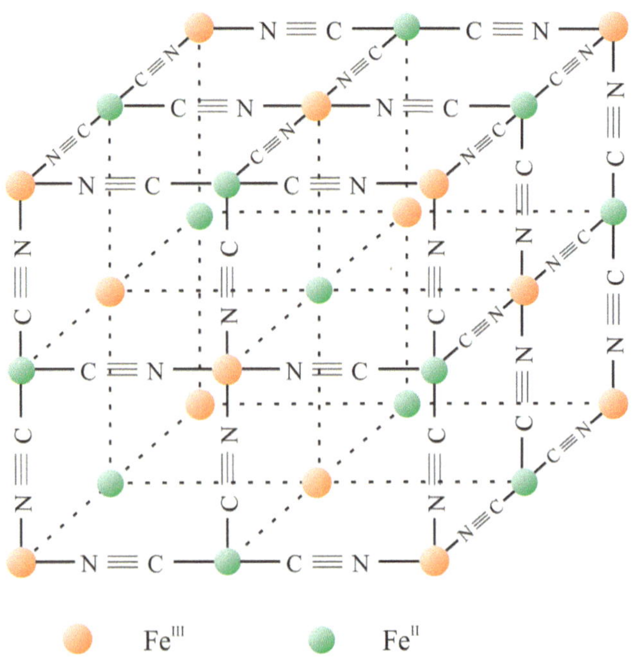

Fig. 1 Crystal structure of Prussian blue

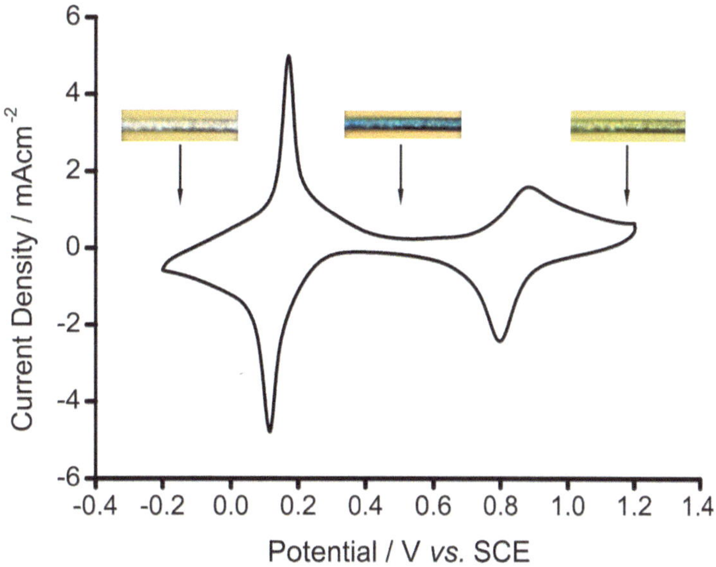

Fig. 2 A typical cyclic voltammogram of a PB-modified Pt electrode (length 0.5 mm, diameter 50 μm) at a scan rate of 50 mV s^{-1} in the absence of O_2. The supporting electrolyte was 1.0 M KCl (pH 4.0). *Inset pictures* show the colors of the PB film on the same electrode at different potential ranges (*silver* between −0.2 and 0 V; *blue* between 0.4 and 0.6 V; *yellow* between 1.0 and 1.2 V)

0.32 nm allows the diffusion of low-molecular-weight molecules (such as O_2 and H_2O_2) through the crystal. High-spin ferric ions [Fe(III)] are bridged octahedrally by nitrogen atoms of the cyanide ligands (–N≡C), and low-spin ferrous ions [Fe(II)] are surrounded by carbon atoms of cyanide ligands (–C≡N), respectively. The remaining charge is balanced either by potassium ions, as in so-called "soluble" PB [$KFe^{III}Fe^{II}(CN)_6$], or by ferric ions, as in "insoluble" PB [$Fe_4^{III}(Fe^{II}(CN)_6)_3$]. In some cases, coordinated water molecules and anions are also contained in structures of PB. It is worth noting that the term "soluble" was invented by dye makers and refers to the ease with which the potassium ion can be transported into the crystal during deposition rather than to the real solubility of PB. In fact, both "soluble" form and "insoluble" form of PB are highly insoluble with a solubility product constant K_{sp} of 10^{-40}.

PB films in both "soluble" and "insoluble" forms are electrochemically active. A typical cyclic voltammogram of PB films consists of two pair of reversible redox waves, as shown in Fig. 2. In between observed two pairs of waves the state of film is PB itself. The reversible redox waves with a mid-peak potential $E_{1/2}$ at 0.2 V and 0.9 V (vs. SCE) resepectively correspond to its reduction to Prussian white (PW, also called Everitt's salt) accompanied with the loss of color, and oxidation to Prussian yellow (PY, also called Berlin green). It is believed that redox waves at 0.2 V (vs. SCE) result from the redox reaction of the high-spin ferric ions $Fe^{3+/2+}$ in the PB film, while the other pair redox waves at 0.9 V (vs. SCE) are due to the reduction/oxidation of the low-spin ferrous ions $Fe^{II/III}(CN)_6^{4-/3-}$ in the PB film (9–11). Thus, the electron transfer mechanism for PB film at 0.2 V (vs. SCE) in the presence of potassium electrolyte has been described by two alternative reaction schemes:

$$\underset{\text{"soluble" PB}}{KFe^{III}Fe^{II}(CN)_6} + K^+ + e^- = \underset{\text{PW(Everitt's salt)}}{K_2Fe^{II}Fe^{II}(CN)_6} \qquad (1)$$

$$\underset{\text{"insoluble" PB}}{Fe_4{}^{III}[Fe^{II}(CN)_6]_3} + 4K^+ + 4e^- = \underset{\text{PW(Everitt's salt)}}{K_4Fe_4{}^{II}[Fe^{II}(CN)_6]_3} \qquad (2)$$

while the following alternative redox reactions have been proposed for waves at 0.9 V (vs. SCE):

$$\underset{\text{"soluble" PB}}{KFe^{III}Fe^{II}(CN)_6} = 2/3K^+ + 2/3e^- + \underset{\text{PY(Berlin green)}}{K_{1/3}(Fe^{III}(CN)_6)_{2/3}(Fe^{II}(CN)_6)_{1/3}}$$

$$\underset{\text{"insoluble" PB}}{Fe_4{}^{III}[Fe^{II}(CN)_6]_3} + 3A^- = \underset{\text{PY(Berlin green)}}{Fe_4{}^{III}[Fe^{III}(CN)_6A]_3} + 3e^-$$

where A^- represents the anions of the supporting electrolyte used. Apparently, the oxidation of PB film involves the loss of cations for the "soluble" form or the incorporation of electrolyte anions for "insoluble" form, while the reduction reactions are accompanied by the incorporation of cations into the film for both "soluble" and "insoluble" PB forms.

It is worth discussing further the redox reactions at 0.2 V (vs. SCE) as this pair of waves is adopted for most biosensors. A Nernst equation for redox reactions of the high-spin ferric ions involved, for example, in reaction (4.2) can be written as follows:

$$E = E^o + \frac{RT}{4F}\ln\frac{a(\text{PB})}{a(\text{PW})} + \frac{RT}{F}\ln a(\text{K}^+)$$

where a(PB) and a(PW) are the activities of Prussian blue and Prussian white, respectively, and $a(K^+)$ refers to the activity of potassium ions in the supporting electrolyte. According to this equation, the voltammetric potential would decrease with decreasing potassium ion concentration showing a Nernstian slope of 59 mV irrespective of whether the "soluble" or "insoluble" PB is involved. It is therefore evident that the amperometric response of a PB-based biosensor will change with alteration of potassium ions concentration in supporting electrolyte. Thus, it is very important to note that a constant concentration of potassium ions throughout experiment is crucial for holding an appropriate operational potential of a PB-based biosensor and ensuring that the amperometric response reflects changes in the target substrate rather than those caused by concentration changes in the supporting electrolyte.

It has been demonstrated that not all cations can be incorporated into PB and the redox reaction of PB/PW relies greatly on the species of cations in supporting electrolyte (12, 13). Cations including K^+, Rb^+, Cs^+, and NH_4^+ were found to be able to promote electrochemical activity of PB film. However, the redox waves around 0.2 V (vs. SCE) were rapidly lost after very few cycles in the supporting electrolyte with the presence of Na^+, Li^+, H^+, and all of the group II cations, leaving only one pair of redox waves around 0.9 V (vs. SCE) in the cyclic voltammogram of PB. The dependence of electrochemical activity on cations has been explained in terms of the hydrated ion radii and the channel radius of the PB lattice. As mentioned before, the zeolitic nature of PB film has a channel radius of about 0.16 nm which will easily allow a flux of hydrated cations such as K^+, Rb^+, Cs^+, and NH_4^+ whose radii are 0.125, 0.128, 0.119, and 0.125 nm, respectively, to pass through the channels and holes of PB crystal. During the reduction of the high-spin ferric ions, those cations with smaller hydrated radii will transport through the crystal of PB providing necessary charge compensation, while other cations with larger radii will block these redox processes. A more recent study (14) shows that the electrochemical behavior of PB films depends on the nature of the alkali cation present in the aqueous solution, and this dependence is related not only with the rate of transport of these species through the film but also with the ease of insertion–expulsion of cations during the electrochemical processes.

The redox waves for the PB film modified electrodes in oxygen-free solutions containing an excess of potassium chloride as a supporting electrolyte are not markedly affected by pH in the range from 1.5 to 7 except that the shape of the waves at 0.2 V (vs. SCE) is slightly changed by changing the pH of the solution. However, the PB film is gradually destroyed in the solutions with alkaline pH. For example after only a few scans at neutral pH, PB films can be disrupted, and a very low stability was observed when pH values are above 7. The marked instability of PB film at alkaline pH can be explained by the strong interaction between ferric ions and hydroxyl ions (OH^-) which forms $Fe(OH)_3$ at pH higher than 6.4, causing the destruction of the Fe(II)–CN–Fe(III) bond, finally solubilizing the PB polycrystal. This disintegration or leakage of PB film on surface of electrodes results in a decreased signal, especially for those PB films prepared with electrochemical synthesis procedures, producing a very low stability of fabricated PB-based biosensors. Consequently, the utility of PB film modified electrodes is limited to neutral or acidic solutions. This has been considered as a major disadvantage of PB film modified electrodes as it was impossible to utilize PB as electron mediator for catalysis of H_2O_2 produced by enzymes such as L-lysine oxidase, choline oxidase, and acetylcholinesterase whose optimum pH is in the basic range.

Efforts to improve the stability of PB film on the surface of electrodes have been intensive over the last two decades. Assuming that the instability of PB film at alkaline pH is only due to the reaction between hydroxyl ions and high-spin ferric ions, it is possible to reduce availability of the ferric ions for reaction with OH^- by slightly changing three-dimensional crystalline structures. Thus the hydrodynamic stability of the deposited PB film at alkaline pH can be firstly improved by small structural alternations introduced by different PB synthesis procedures. For instance, by growing more regular crystals in precursor solution with 0.1 M HCl during the electrochemical deposition of PB and applying additional post-treatment (such as electrochemical activation and heating treatment), excellent operational stability has been achieved on PB-modified glassy carbon electrode (15). A long operational lifetime and an increased stability of PB at alkaline pH (up to 9) were further obtained on carbon-based electrodes by adopting a chemical synthesis method for the modification of electrode surface with PB. Secondly, the stability of PB was found to be greatly related to matrix of electrode substrate modified by PB. Amorphous carbon-based materials such as glass carbon, graphite, and carbon nanotube provide a superior matrix for deposition of PB (16). In a chemical synthesis procedure, a large excess amount of PB is chemically adsorbed onto those carbon matrices, compared with amount of PB deposited in electrochemical synthesis procedure. This acts as a PB reservoir minimizing the effect of the leakage due to the hydrolysis of ferric ions at alkaline pH (17). Moreover, it is noteworthy that carbon-based electrodes with electrochemically deposited PB also show favorable pH stability in alkaline solution. This behavior suggests that the structure of PB crystals is protected in the carbon matrix and not influenced by either the proton or the hydroxyl ion concentration in the bulk solution.

3 Potential as Mediators and Concept of "Artificial Peroxidase"

It was first reported by Itaya et al. (18) that PB can act as a three-dimensional catalyst for detection of hydrogen peroxide. Hydrogen peroxide can diffuse into zeolitic structure of PB crystal and be electrochemically catalyzed by either high-spin ferric ions or low-spin ferrous ions. Firstly, hydrogen peroxide can be oxidized by the oxidized form of PB (Prussian yellow). The catalyzed oxidation of hydrogen peroxide commences at potential where the low-spin ferrous ions are oxidized (0.7–0.9 V vs. SCE) in acidic media:

$$H_2O_2 \rightarrow 2H^+ + 2e^- + O_2$$

This potential indicates that the trivalent low-spin ferrous ions act as catalyst for the oxidation of hydrogen peroxide. Secondly, hydrogen peroxide can also be electrochemically reduced by the reduced form of PB (Prussian white) during potential range from 0.2 to −0.2 V (vs. SCE). High-spin ferric ions are responsible for electrochemical reduction of hydrogen peroxide by comparing the electrode potential for hydrogen peroxide reduction and mid-peak potential for redox waves of PB/PW. Hydrogen peroxide is completely reduced to water before diffusion out of the crystal in acidic media:

$$H_2O_2 + 2e^- + 2H^+ \rightarrow 2H_2O$$

In neutral media the reduction of hydrogen peroxide catalyzed by PB has been found to be as follows (19):

$$H_2O_2 + 2e^- \rightarrow 2OH^-$$

For oxidase-based biosensors, in which hydrogen peroxide is measured on the electrode as a final electroactive product of enzymatic reactions, a low operational potential is always preferred for catalytic detection of hydrogen peroxide in order to obtain a high selectivity over potential interfering substances that commonly exist in physiological applications. However, it is noteworthy that Prussian white also has catalytic activity for the reduction of dissolved oxygen. However, the catalyzed reduction of dissolved oxygen and hydrogen peroxide commences at about 0.2 V and 0.5 V (vs. SCE) on a PB-modified glass carbon electrode, respectively (18). Furthermore, the rate of catalytic reduction of hydrogen peroxide to water is likely to be a lot faster than that of oxygen (18). Based on these facts, PB was first employed as electron mediator for hydrogen peroxide on a PB-modified carbon paste electrode by Boyer et al. (20). The selective catalytic activity for the reduction of hydrogen peroxide against dissolved oxygen was further demonstrated by Karyakin group. In the first instance, PB-modified glass carbon electrode was operated at +180 mV (vs. Ag/AgCl) for selective detection of hydrogen peroxide with a dependence of linear amperometric response on analyte concentration in the range 1 μM–5 mM (21, 22). Then by improving the electrochemical deposition method, successful growth of more regular PB crystals allowed PB-modified glass carbon electrodes to be operated at 0.0 V (vs. Ag/AgCl) for electrochemical reduction of hydrogen peroxide with a good stability of PB film. This low operational potential endows the PB-modified electrode with more power for reduction of hydrogen peroxide while remaining almost insensitive towards dissolved oxygen (23). The rate of oxygen reduction by specially deposited PB was found to be a hundred fold lower than the rate of hydrogen peroxide reduction, which confirms the previous prediction (19).

On a typical glass carbon electrode covered with 4–6 nmol cm^{-2} of PB film, the electrochemical rate constant is about 0.01 cm s^{-1}, which is a lot higher than that for hydrogen peroxide oxidation on platinum in neutral media which is less than 7×10^{-6} cm s^{-1}. This result further confirms that PB is an excellent electron mediator for electrochemical reduction of hydrogen peroxide. Further calculation shows that the bimolecular rate constant (k_{cat}) for the reduction of hydrogen peroxide by PB film deposited under optimized conditions is 3×10^3 M^{-1} s^{-1}, which is very similar to that measured for the peroxidase enzyme (2×10^4 M^{-1} s^{-1}). The high catalytic activity and selectivity for reduction of hydrogen peroxide motivated researchers to denote PB as "artificial peroxidase" (24).

Apart from Prussian blue, many other transition metal hexacyanoferrates (MHCFs) are also found to be able to act as electron mediators to shuttle electrons between hydrogen peroxide and substrate electrodes, such as copper (25–28), cobalt (29), nickel (30), manganese, palladium (31), and chromium hexacyanoferrates (32). Thus they have been employed on modified electrodes for hydrogen peroxide detection. For example, copper-, cobalt-, and nickel hexacyanoferrate-modified carbon paste electrodes were shown to electrocatalyze the cathodic reduction of hydrogen peroxide (33). Compared with PB, all those transition metal hexacyanoferrates exhibit excellent anti-interference ability towards dissolved oxygen and other easily oxidized compounds such as ascorbic acid, uric acid, and acetaminophen. Though showing a similar or lower ability to catalyze electro-reduction of hydrogen peroxide as PB, they are much more stable over a wide range of pH, especially in neutral and alkaline solutions. This improvement on pH stability easily achieved in those transition metal hexacyanoferrates compared with PB allows modified electrodes to be operated at physiological pH levels well tolerated by immobilized enzymes, which might be one of the main reasons that drove the research on these alternative transition metal hexacyanoferrates for biosensing applications. In order to improve sensitivity for hydrogen peroxide detection on MHCF-modified electrodes, hybrid (or mixed) hexacyanoferrates of two different metals have been developed such as CuCoHCF (34), FeCoHCF (35), FeRuHCF (36), and CuPdHCF (37). These hybrid hexacyanoferrates not only keep the usual properties of single component, but also show improved performance including enhanced stability in alkaline solution, unique cation permeability, as well as improved sensitivity for hydrogen peroxide measurement in comparison to either single component. Modified electrodes based on electroactive rare earth metal hexacyanoferrate, such as neodymium, gallium, and cerium hexacyanoferrates, also present good electrocatalytic activity towards the cathodic reduction of hydrogen peroxide (38–40).

4 Hexacyanoferrate-Modified Electrodes: Deposition and Activation

In general, PB films can be prepared by two methods: chemical synthesis and electrochemical synthesis. Post-deposition treatments are usually taken after first step including PB activation process and stability improvement treatment with heating.

With the chemical synthesis method, firstly, on some metal and semiconductor substrates, such as Pt Au, Al, or graphite, the PB films can be deposited simply by inserting a pretreated electrode (cathodically polarized in HCl for about 1 h) in ferric ferricyanide solution. After several minutes, a blue film will be formed on the surface of electrode (1, 12). Secondly, PB-modified carbon paste electrodes can be prepared by mixing pretreated graphite powder with PB precipitate chemically obtained from a mixture either of hexacyanoferrate (II) and Fe(III) solution (41, 42) or of hexacyanoferrate(III) and Fe(II) solution (30, 33). And thirdly, PB can be adsorbed onto graphite powder (17), screen-printed graphite (16), carbon nanotube (43), or graphene (44) simply from a clear mixture of hexacyanoferrate(III) and Fe(III) solution (ferric–ferricyanide). The solution of ferric–ferricyanide in acidic media is a very strong oxidant, showing a highly positive open circuit potential (0.98 V vs. SCE for a Pt disk electrode). During chemical synthesis of PB, an auto-reduction or a catalytic reduction by carbon nanotube or graphene of the highly reactive ferric–ferricyanide complex, which presumably is initially formed, seems to occur.

Itaya et al. introduced an electrochemical deposition method involving the reduction of ferric ferricyanide—an equimolar mixture of hexacyanoferrate(III) and Fe(III) ions in acidic solution (9, 10). As this mixture is a clear stable solution where no precipitate is formed for several hours, a stable and reproducible PB film can be prepared from the mixture by both electrochemical synthesis method and chemical synthesis method. Chronopotentiometric investigation shows that galvanostatic transient behavior of a Pt electrode in equimolar ferric–ferricyanide mixtures has two basic plateaus at 0.7 V and 0.4 V (vs. SCE), respectively. At first plateau around 0.7 V, PB is deposited according to reduction of ferric–ferricyanide complex ($Fe^{III}Fe^{III}(CN)_6$). And at second plateau around 0.4 V, Fe^{3+} is firstly reduced to Fe^{2+} ions in the solution and then reacts with $Fe^{III}(CN)_6^{3-}$ producing the bulk precipitation of PB.

Electrodeposition of PB is mainly carried out with three different electrochemical techniques: galvanostatic, potentiostatic, and potentiodynamic cycling. For instance, electrodes can be cathodically polarized in mixed ferric–ferricyanide solution under galvanostatic conditions with a current density less than 50 $\mu A\ cm^{-2}$ for 30–120 s, where the electrochemical reduction is set at first plateau around 0.7 V (9–11, 45). With more driving force for electrochemical reduction of mixed ferric–ferricyanide solution, PB can also be electrodeposited under potentiostatic conditions by applying a constant potential of 0.4 V for a similar period on working electrode (23). Under these conditions, it was found that more regular PB crystals were obtained during the deposition from the ferric–ferricyanide solution contained with 0.1 M HCl. With additional posttreatment, excellent operational stability of PB-modified electrode has been achieved (15, 17, 46, 47). Potentiodynamic cycling methods such as cyclic voltammetry have also been widely adopted for electrochemical deposition of PB due to the fact that more uniform films are usually achieved with the cyclic voltammetric method (22, 48–51). Though the solutions for the PB preparation mentioned above are all composed of ferric ferricyanide, PB can also be deposited from a single or mixed ferro/ferricyanide using a cyclic voltammetric method (52–56).

After deposition of PB by either chemical or electrochemical synthesis, the electrode requires additional post-deposition treatments for stabilization of PB. These post-deposition treatments normally include an electrochemical activation procedure, removal of bonded water molecules in PB crystal by heating, and further electrochemical conditioning. Firstly, PB films are activated in the same supporting electrolyte solution as that used for PB deposition, which normally contains 0.1 M KCl and 0.1 M HCl, for example, by applying a dynamic cycling potential in the range between −0.05 and 0.35 V at a sweep rate of 50 mV s^{-1} until a reproducible cyclic voltammogram is obtained. Then, the PB-modified electrode is dried at 100° or 120° for 1–2 h to remove bound water molecules within PB crystals. The conductivity of PB is greatly affected by up to 14 bound water molecules per unit $Fe_4^{III}[Fe^{II}(CN)_6]_3$ (57). This bound water will affect the stability of PB-modified electrodes. An increased amount of electroactive PB was found on the electrode surface after the modified electrode was exposed at 100 °C for 1 h, compared with the amounts observed on electrodes treated at 20 °C (58). The PB-modified electrode can also be cathodically conditioned at −50 mV followed by electrochemical cycling in supporting solution for further improvement of PB stability (47).

PB analogues, other transition metal MHCFs, can be deposited using equivalent methods to PB deposition. With potentiostatic or galvanostatic methods, for instance, a two-step procedure is required for deposition of CuHCF film. A thin layer of copper is first deposited onto the surface of electrode at a constant cathodic current or potential in a solution with cupric ions, which is also called electroplating process. Then the electrode is polarized anodically in a ferricyanide solution with supporting electrolyte to form CuHCF film (25, 59). However, by potentiodynamically cycling electrode potential a mixture of potassium ferricyanide and a single transition metal salt such as Co^{2+} (29, 60–63), Zn^{2+} (64),

Cu^{2+} (26, 29, 65, 66), Pd^{2+} (31), Ni^{2+} (31, 67–70), In^{3+} (71), or $ZrOCl_2$ (72), the corresponding MHCF can be simply electrodeposited onto surface of electrode with aid of the respective supporting electrolyte. By employing a solution containing potassium ferricyanide and mixed transition metal salts, two salts in most cases, hybrid MHCFs such as FeMnHCF (73), NiFeHCF (73), NiCoHCF (74), NiPdHCF (75), FeCoHCF (35), FeRuHCF (36, 76), CuPdHCF (37), and CuCoHCF (34) can be electrodeposited with the same procedure. Just like PB, both single and hybrid transition metal MHCFs require electro-activation in same supporting electrolyte used for MHCF deposition. It has also been demonstrated that the stability of some transition metal MHCFs can be strongly enhanced by a simple and effective voltammetric conditioning method: repetitive potentiodynamic cycling of the freshly deposited MHCF film in a Ru(III) containing solution at pH 2 (67, 77).

Rare earth metal MHCFs can be popularly formed from the mixture of ferrocyanide and their inorganic salts $NdCl_3$ (78), $DyCl_3$ (79), $Ga(NO_3)_3$ (40), and $Ce(NO_3)_3$ (38) using a chemical synthesis method.

5 Preparation of PB/MHCF-Based Biosensors

Miniaturized biosensors based on enzymes have potential in clinical practice for real-time analysis of important body analytes. For oxidase-based biosensors, the amperometric signal was mostly obtained through electrochemical monitoring of the formation of hydrogen peroxide, which is one of bio-reaction products produced by biological recognition layer. Though hydrogen peroxide can be oxidized on surface of Pt electrode starting around +0.2 V (vs. Ag/AgCl), traditionally, a relatively high working potential ranging 0.5–0.7 V (vs. Ag/AgCl) is required to get a sufficient sensitivity for a biosensor. However, many substances usually present in biological samples (such as ascorbate and urate) can also be electrochemically oxidized at such a potential causing a high level of interference. Thus, the selectivity of the biosensor in such a detection mode is simply inadequate for such applications. In order to improve the selectivity of oxidase-based biosensor, permselective membranes such as Nafion and polyphenylenediamines were widely adopted in construction of biosensors. Those polymer membranes serve as screening barrier for electrochemical active species and protect biosensors from biofouling. *Please refer to Chaps. 2 and 6 for more details.*

Electron mediators have been widely employed as electrochemical transducer to circumvent the problem of high working potential, which subsequently provided the other protocol for eliminating electrochemical interferences on biosensor. There are

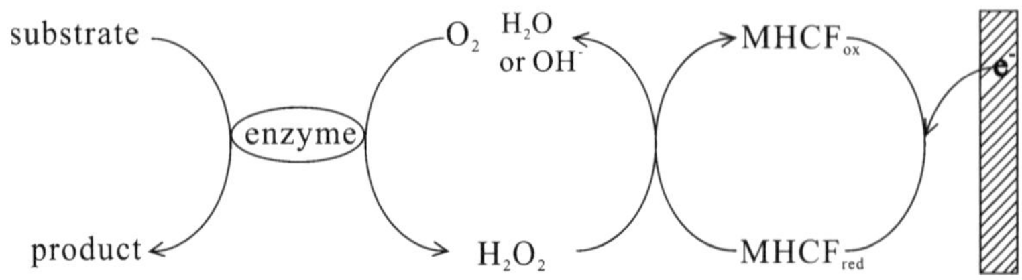

Fig. 3 Electrochemical reduction mechanism of hexacyanoferrate-mediated biosensors

many different methods for utilizing mediators with biosensors, the best and most convenient being so-called reagentless designs in which the mediator is permanently entrapped in the enzymatic biolayer (see (80) for a review). One widely used method for a reagentless designs involves "redox polymers" in which an electroactive mediator such as ferrocene or an osmium bipyridine complex is covalently attached on a polymeric backbone that acts as a hydrogel to both entrap the enzymes and provide electron transfer from the oxidase to the electrode surface (81–83). These biosensor designs are very selective but suffer from two potential drawbacks. Firstly the mediators have differing compatibilities with oxidases. Two solutions have been employed to counter this: firstly, to link the mediator to a peroxidase and electroreduce the peroxide produced by the oxidase rather than wiring the oxidase to the mediator directly (e.g., (81, 84, 85)) and, secondly, to tune the design of the mediator to the specific enzyme (83). Nevertheless both of these approaches require added complexity in biosensor design and fabrication.

By contrast, Prussian blue and other transition metal hexacyanoferrates offer a degree of simplicity. As they have strong and preferential electrocatalytic action towards the enzymatically produced hydrogen peroxide while displaying lower response to coexisting oxidizable substances owing to the low operational potential, they are potentially very advantageous as mediators in enzymatic biosensors and in effect provide an inorganic alternative to the mediated peroxidase detection of peroxide used by some investigators (e.g., (81)). The mechanism of hexacyanoferrate-mediated biosensors at low operational potential is shown in Fig. 3.

Enzyme immobilization is crucial in biosensor fabrication as it fundamentally determines performance of desired biosensor. Initially, a one-step procedure was attempted to construct PB-based amperometric biosensor, in which glucose oxidase or D-amino acid oxidase was dissolved in PB deposition solution followed by a simultaneous incorporation of enzyme into electrochemically deposited PB film on the surface of basal pyrolytic graphite electrode (86). However, the pH of the co-deposition solution does not favor both stability of PB film and enzyme activity. Specifically, the low pH of precursor solution required for formation of stable PB and other MHCFs is detrimental to the activity of most enzymes. Thus a further and separate enzyme immobilization step following MHCF formation is essential for construction of MHCF-mediated biosensor to avoid this harsh chemical condition. For example, different enzymes were directly doped into graphite power modified with chemically synthesized CuHCF and PB to prepare glucose biosensor (17, 33) and choline biosensor (17), respectively.

Other than simultaneously doping enzyme and mediators into the matrix of electrode, surface modification with enzyme on a mediated electrode is the most popular enzyme immobilization method. The enzyme can be immobilized onto surface of mediated electrode by different protocols including physical adsorption, entrapment within a polymer membrane, and encapsulation with the sol–gel technique.

With physical adsorption method, enzymes are incorporated onto electrode by casting or dip-coating one layer membrane (or multilayers) from a mixture of enzyme and all kinds of polymers such as self-gelatinizable copolymer 4-vinylpyridine grafted poly(vinyl alcohol) (45), hydrogel (31), poly(allylamine) hydrochloride (87), chitosan (88, 89), Tosflex (90), and Nafion (15, 21–23, 25, 46, 47, 91–93), respectively. These polymer membranes improve stability of biosensors by preventing both mechanical destruction of deposited MHCFs and escape of enzyme molecules from biocatalytic layer. Moreover, some polymer membranes such as perfluorinated cation exchange Nafion and cellulose structure-like chitosan also act as interference screening layer improving the selectivity of the biosensors. However, due to weak strength of the nonspecific adsorption of enzyme molecules on the surface of electrode, a continuous loss of enzyme molecules from the biosensor generally occurs during its application, which correspondingly decreases sensitivity of the biosensor. In order to circumvent the problem, cross-linking agents are widely employed to attach enzymes onto surface of electrode. The MHCF-mediated electrodes have been treated with carbodiimide (94) or glutaraldehyde (68, 95, 96) either before or after covering with enzyme solution. As a strong bifunctional cross-linking reagent, glutaraldehyde can drastically modify the enzyme, which inevitably leads to conformational changes and loss of enzymatic activity. Excessive intramolecular cross-linkages within the enzyme can be avoided by the addition of another inert protein such as bovine serum albumin (BSA) in cross-linking solution. And the enhancement of the intermolecular linkages between the enzyme and lysine residues of inert proteins can help to minimize deleterious effect of cross-linking process and thus diminish enzyme deactivation. All kinds of biosensors used for glucose (26, 63, 90, 97, 98), glutamate (99), choline, and acetylcholine (16) measurement have been prepared using the adsorption method combined with improvement of stability and sensitivity via the use of BSA and glutaraldehyde.

Electropolymerization is another attractive method for enzyme entrapment (100). This method shows many advantages including fast entrapment, controllable thickness, and porosity as well as tunable enzyme loading within the polymer film. For most polymer films, enzyme can be entrapped under mild chemical conditions without destroying enzyme's activity. The enzyme entrapment within a polymer membrane can be accomplished non-covalently and covalently. From a mixture of enzyme and monomers, enzyme can be non-covalently co-deposited onto the surface of MHCF-mediated electrode during electropolymerization process. Alternatively, with the help of cross-linking reagents, enzymes can also be covalently attached onto a polymer membrane with exposed functional groups such as carboxylic acids or amines. A variety of polymers were successfully employed to immobilize enzyme for construction of MHCF-mediated biosensors, of which polypyrrole is the most popular one due to its excellent water solubility and mild polymerization conditions. For instance, glucose oxidase (50, 101) and cholesterol oxidase (102) were entrapped within polypyrrole membrane on a PB-modified electrode to prepare glucose and cholesterol biosensor, respectively. Some other polymers such as polyaniline (103) and poly(*o*-phenylenediamine) (36, 101, 104) function not only as matrix for enzyme immobilization but also as a anti-fouling layer.

Sol–gel techniques are of particular interest to immobilize enzyme by surface- or bulk-modified electrodes since there are many attractive properties of sol–gel materials including low-temperature encapsulation, tunable porosity, negligible swelling, chemical inertness, physical rigidity, and favorable biocompatibility. Sol–gel film coated biosensors have attracted tremendous attention because of their fast response and easiness of preparation. In this type of biosensors, desired sol–gel film was traditionally prepared by dip-coating or spin-coating hydrolyzed sol on the surface of a suitable substrate. Glucose biosensors (27, 62) and lactate biosensors (105) have been prepared by dipping or casting the MHCF-modified electrode with a mixture of enzyme and hydrolyzed silanes. Though coating method is simple and fast, the prepared sol–gel film suffers from uncontrollable thickness, low porosity, and low adhesion to electrode surface. A more controllable and even faster sol–gel coating method was recently developed utilizing electrochemical deposition (106–110). In this method, at a sufficient cathodic potential, the pH on the surface of electrode was increased to act as a catalyst to enhance the condensation and polymerization of hydrolyzed sol solution forming a uniform gel layer in several seconds. The thickness of film coating can be controlled easily by manipulating current density during electrodeposition process. This film formation procedure provides a new methodology for sol–gel film formed potentially on any shape of conductive substrate. Enzymes can be rapidly encapsulated into a porous and rigid gel layer from a sol solution in the presence of enzymes. The activity of the immobilized enzyme is well retained within gel layer. This is mainly attributed to the enzyme structure being untouched by any covalent bond and benign conditions provided by gel layer.

6 Potential Limitations of Hexacyanoferrate-Based Biosensor and Solution–Speed, the Problem of Na^+ Ions

Temporal resolution is one of important characters of a biosensor for detecting a transient biological event occurred in tissue or slice. A fast response time is highly desirable to enable enzyme-based biosensors to measure changes in analyte concentration at physiological timescales (ms to s). The response time of an amperometric biosensor is often determined by measuring the time required to arise from 10 to 90 % of the maximum signal after applying a concentration step of substrate within the operating range of the biosensor. The response time of a biosensor greatly relies on the kinetics of the associated enzymatic reaction, multilayer structures, and the geometry dimensions of biosensors. Most MHCF-mediated biosensors are fabricated with a multilayer structure, where enzyme is immobilized on top of MHCF-modified electrode. Moreover, polymers are often employed for the purpose of enzyme immobilization and interferences screening. This inherent multilayer structure of most MHCF-mediated biosensors has a significant impact on their temporal resolution. For instance, the response time on a typical PB-mediated glucose biosensor based on a glass carbon disk electrode (1.5 mm diameter) is about 2–3 min in the micromolar glucose concentration range (22). The slow response time of most MHCF-mediated biosensors is one drawback restricting their application for real-time measurement. High temporal resolution can in principle be achieved by shrinking the thickness of multiple layers. Fast charge transport and close contact with analyte in nanostructured materials have given relatively fast response times from 10 to 20 s with other aspects of improved analytical performance (88, 93, 97). The biosensor dimensions also have a significant impact on the temporal resolution. Improved performance will result on ultramicroelectrodes due to fast mass transport rate of the analyte.

Inherently, the PB-based biosensor is restricted in its utilization for practical use by its cation-dependent properties. Na^+-based buffers are commonly used as supporting solution in clinical or physiological measurements. The electrochemistry of Prussian blue is highly affected by cations present in buffer solution. Those cations (such as K^+, NH_4^+, Rb^+, and Cs^+) whose hydrated ions fit the Prussian blue lattice (1.019 nm) can promote its electrochemistry. By contrast, all other cations including Na^+, H^+, Ca^{2+}, and Mg^{2+} are considered as blocking ones. As a result, potassium-based phosphate buffer solution is the most popular supporting electrolyte for those biosensors adopting Prussian blue as electron mediator.

To improve the electrochemical performance of PB film in potassium-free solutions, a hybrid thin film of PB was deposited on a glass carbon electrode in the presence of a cationic surfactant cetyltrimethylammonium bromide (CTAB) (111). A larger lattice channel size of the PB films was obtained with the help of CTAB during film formation (112). Compared to an ordinary PB-modified electrode, this CTAB-promoted PB-modified electrode displayed enhanced stability and interaction with Na^+ due to the larger lattice channels than those in PB-modified films. However, these CTAB-promoted films did not provide sharp current responses in the presence of Na^+, indicating that these increased channel sizes were still insufficient to enable the entry of this ion. The CTAB-promoted PB method was further developed by electrodepositing PB and Au nanoparticles in the presence of CTAB on the surface of a glass carbon electrode (113). This triple component nanocomposite film allows Na^+ transport within PB film by showing two pairs of well-defined redox peaks in a supporting electrolyte containing both K^+ and Na^+. The hybrid PB film retains well-defined and reversible voltammetric responses in Na^+-only-based electrolytes. The electrochemical reduction of H_2O_2 on the hybrid PB film modified electrode in the presence of physiological levels of Na^+ was superior to that of a PB-modified electrode with a low detection limit (0.1 μM) and a wide linear range of concentrations (0.882–195 μM). This hybrid PB film was successfully employed as an electron mediator for fabrication of A glucose biosensor used in a sodium-based physiological buffer.

Use of other metal hexacyanoferrates as electron mediators for construction of biosensors offers an alternative way to overcome the problem of Na^+-based buffers, as many PB analogues show good electrochemical behavior in these physiological buffers. For instance, well-defined and reversible cyclic voltammetric responses have been obtained on CoHCF (62, 114), NiHCF (68, 88), NiPdHCF (75), and FeRuHCF (76) modified electrode in sodium supporting electrolyte, respectively. Some of them have been successfully employed as mediators for preparation of biosensors for detection of glucose and oxalate.

Other polynuclear transition metal hexacyanometallates also show flexible electrochemistry in all kinds of supporting electrolyte containing different cations, such as ruthenium purple (RP) (115), which is known as ferric ruthenocyanide. RP film is generally deposited on the surface of electrode by a dynamic potential scanning in a defined potential range in a suitable aqueous solution containing Fe^{3+} and $Ru(CN)_6^{4-}$. An EQCM study (116) suggests that the Fe^{3+} and $Ru(CN)_6^{4-}$ species reacted to yield a soluble

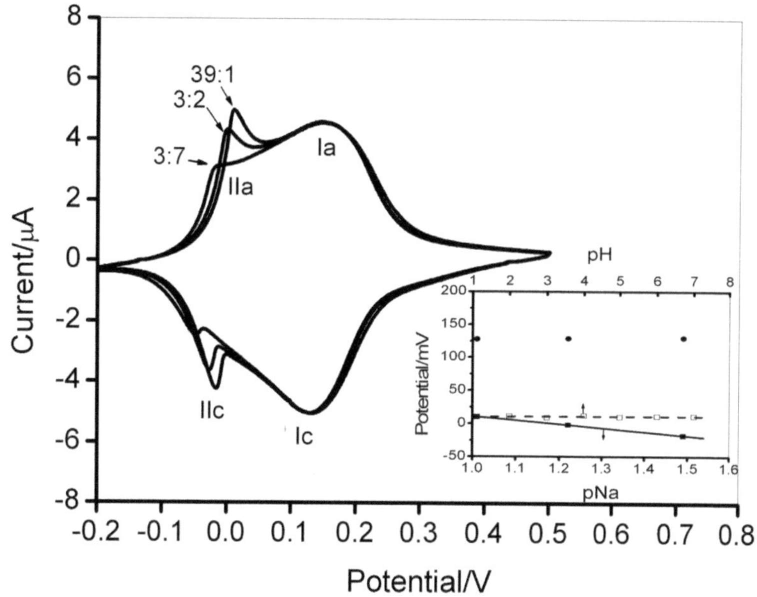

Fig. 4 Cyclic voltammograms of a gel-coated RP-modified microelectrode in phosphate buffer with different ratios of Na^+:K^+ (total monovalent concentration of 100 mM), pH 2.0. *Inset* shows dependence of peak potential (EIIa: *filled square* and *open square*, EIc: *filled circle*,) on pNa (*solid line*) and pH (*dashed line*). Reproduced with permission from (106). Copyright 2007 American Chemical Society

form of Fe^{III}–$Ru^{II}(CN)_6$ species, and then the Fe^{III}–$Ru^{II}(CN)_6$ species was reduced to Fe^{II}–$Ru^{II}(CN)_6$, which was subsequently deposited on the electrode surface to form the RP film. Like other MHCFs, RP shows significant sensitivity to hydrogen peroxide (117). Thus, it has been utilized as electrocatalytic mediator in glucose biosensors based on RP containing zeolite-modified conductive SnO_2 glass and on RP/industrial waste cinder hybrid-modified carbon paste electrode (118–120). Though RP has similar structure with a lattice constant of 1.042 nm to Prussian blue, its electrochemical property is slightly different from that of Prussian blue. It was found that electrochemistry of RP can survive in sodium-based phosphate buffer and two pairs of interesting redox peaks were observed (121). What is more, RP can even be directly synthesized in aqueous Na^+ electrolytic solution by an electrochemical deposition process, which further confirms the excellent Na^+ transport property within the RP film. Thus it is possible that RP can be used as a potential electrochemical mediator in fabrication of in vitro and/or in vivo biosensors. Recently, a stable and reliable RP film was successfully electrodeposited onto the surface of gold microelectrodes from aqueous solution (106). Figure 4 displays typical cyclic voltammograms at RP-modified sol–gel film coated gold electrode in an electrolyte solution containing NaCl and KCl with different ratios between them in a total

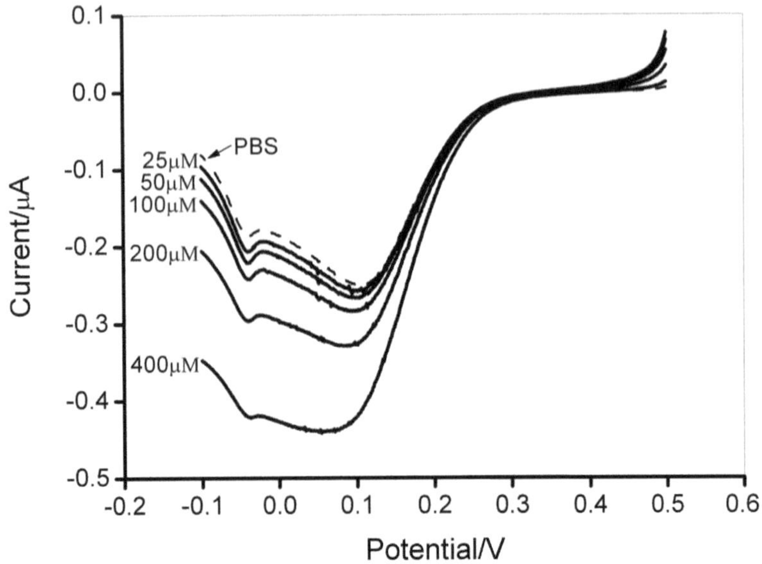

Fig. 5 Linear sweep voltammograms of a gel-coated RP-modified gold electrode in phosphate buffer (pH 7.4, NaCl 0.1 μM) with and without hydrogen peroxide at concentrations of 25, 50, 100, 200, and 400 μM. Scan rate: 10 mV s^{-1}. Reproduced with permission from (106). Copyright 2007 American Chemical Society

concentration of 100 mM. The electrochemistry of redox peak I was not altered with different concentrations of K^+ and Na^+ in these solutions, which indicates a non-monovalent-cation-dependent four electron process:

$$Fe_4^{III}[Ru(CN)_6]_3 + 4e^- = Fe_4^{II}[Ru(CN)_6]_3^{4-}$$

However, for redox peak II, both cathodic and anodic peak currents increased with the increment of Na^+ concentration. At the same time, the redox peak potentials were shifted positively with a separation of ~27.3 mV. The oxidation potential of peak II showed a pNa (pNa = −log(Na^+]) dependence of −58.4 mV/pNa, and it was also independent of the pH of supporting electrolyte (Fig. 4, inset). These results indicate that Na^+ takes part in the electrochemical process of redox peak II:

$$K_2Fe_2^{III}Fe_2^{II}[Ru(CN)_6]_3 + 2Na^+ + 2e^- = Na_2K_2Fe_4^{II}[Ru(CN)_6$$

The immobilized RP displayed dramatic electrocatalytic behavior to hydrogen peroxide in a sodium-based phosphate buffer (Fig. 5). In linear sweep voltammetry experiments, the sensing response of hydrogen peroxide on RP-modified electrode increased correspondingly with each increment of its concentration in phosphate buffer. The cathodic response to hydrogen peroxide occurred between potentials of +150 mV and −100 mV. This potentially allows highly selective operation of oxidase-based biosensors as small operational potentials can be selected to avoid amperometric

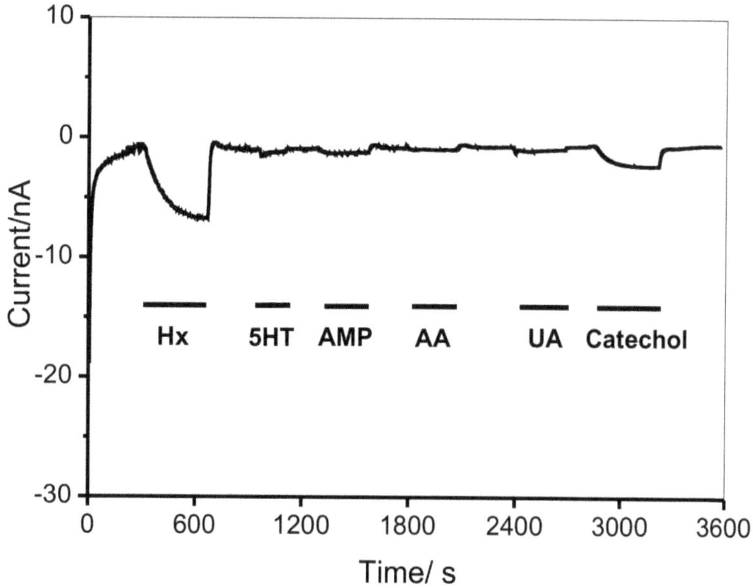

Fig. 6 Response of a hypoxanthine (Hx) biosensor to Hx (10 μM) and different interferences: serotonin (5HT, 100 μM), 4-acetaminophen (AMP, 1 mM), ascorbic acid (AA, 100 μM), uric acid (UA, 1 mM), and catechol (1 mM). Experiment was performed in phosphate buffer (pH 7.4, NaCl 100 mM) at −50 mV. Reproduced with permission from (106). Copyright 2007 American Chemical Society

responses to coexisting interferences. Sol–gel-derived ATP and hypoxanthine biosensors were further fabricated based on RP-modified gold microelectrodes using sol–gel electrochemical deposition method. The resulting biosensors had highly attractive performance characteristics including high sensitivity, wide linear calibration range, and excellent selectivity versus interferences (106). For instance, typical RP-mediated ATP biosensors showed high sensitivity to ATP of 126 nA μM^{-1} cm^{-2} and a linear range from 0.3 to 80 μM with a 10–90 % response rise time of 33 ± 2 s ($n = 8$). This is fairly similar to biosensor designs involving Os-bypridine mediators (81, 84, 85, 122). As expected, amperometric responses to 5HT (100 μM), ascorbic acid (100 μM), acetaminophen (1 mM), and uric acid (1 mM) were dramatically reduced on a RP-mediated hypoxanthine biosensor (Fig. 6). Though a response of 1.58 nA for 1 mM catechol was observed, this highly reactive chemical is unlikely to be present in physiological media. Interestingly no effect of interferences was found on the RP mediator (106), which contrasts with the observation that ascorbate and urate reduced the sensitivity of Os-mediated glutamate biosensors probably through direct interaction with the Os-bipyridine mediator (81).

The development of RP film deposition methods for microelectrodes combined with the ability to reliably deposit enzymatic layers on top of the RP film now makes RP a highly advantageous electron mediator in fabrication of oxidase-based microelectrode biosensors. The RP mediator, as it reduces peroxide at favorable potentials, is compatible with a wide range of oxidases and provides a highly selective transducer for detection of analytes in physiological fluids and in vivo.

References

1. Neff VD (1978) Electrochemical oxidation and reduction of thin films of prussian blue. J Electrochem Soc 125:886–887
2. Itaya K, Uchida I, Neff VD (1986) Electrochemistry of polynuclear transition metal cyanides: Prussian blue and its analogues. Acc Chem Res 19:162–168
3. Karyakin AA (2001) Prussian blue and its analogues: electrochemistry and analytical applications. Electroanalysis 13:813–819
4. Ricci F, Palleschi G (2005) Sensor and biosensor preparation, optimisation and applications of Prussian blue modified electrodes. Biosens Bioelectron 21:389–407
5. Keggin JF, Miles FD (1936) Structures and fomula of the Prussian blue and related compounds. Nature 137:577–578
6. Ludi A, Güdel H (1973) Structural chemistry of polynuclear transition metal cyanides. Inorg Chem Struct Bond 14:1–21
7. Buser HJ, Schwarzenbach D, Petter W, Ludi A (1977) The crystal structure of Prussian blue: Fe4[Fe(CN)6]3.xH2O. Inorg Chem 16:2704–2710
8. Herren F, Fischer P, Ludi A, Haelg W (1980) Neutron diffraction study of Prussian blue, Fe4[Fe(CN)6]3.xH2O. Location of water molecules and long-range magnetic order. Inorg Chem 19:956–959
9. Itaya K, Akahoshi H, Toshima S (1982) Electrochemistry of Prussian blue modified electrodes: an electrochemical preparation method. J Electrochem Soc 129:1498–1500
10. Itaya K, Ataka T, Toshima S (1982) Spectroelectrochemistry and electrochemical preparation method of Prussian blue modified electrodes. J Am Chem Soc 104:4767–4772
11. Itaya K, Shibayama K, Akahoshi H, Toshima S (1982) Prussian-blue-modified electrodes: an application for a stable electrochromic display device. J Appl Phys 53:804–805
12. Ellis D, Eckhoff M, Neff VD (1981) Electrochromism in the mixed-valence hexacyanides. 1. Voltammetric and spectral studies of the oxidation and reduction of thin films of Prussian blue. J Phys Chem 85:1225–1231
13. Rajan KP, Neff VD (1982) Electrochromism in the mixed-valence hexacyanides. 2. Kinetics of the reduction of ruthenium purple and Prussian blue. J Phys Chem 86:4361–4368
14. García-Jareño JJ, Giménez-Romero D, Vicente F, Gabrielli C, Keddam M, Perrot H (2003) EIS and Ac-electrogravimetry study of PB Films in KCl, NaCl, and CsCl aqueous solutions. J Phys Chem B 107: 11321–11330
15. Karyakin AA, Karyakina EE (1999) Prussian blue-based 'artificial peroxidase' as a transducer for hydrogen peroxide detection. Application to biosensors. Sens Actuators B Chem 57:268–273
16. Ricci F, Amine A, Palleschi G, Moscone D (2003) Prussian Blue based screen printed biosensors with improved characteristics of long-term lifetime and pH stability. Biosens Bioelectron 18:165–174
17. Moscone D, D'Ottavi D, Compagnone D, Palleschi G, Amine A (2001) Construction and analytical characterization of Prussian blue-based carbon paste electrodes and their assembly as oxidase enzyme sensors. Anal Chem 73:2529–2535
18. Itaya K, Shoji N, Uchida I (1984) Catalysis of the reduction of molecular oxygen to water at Prussian blue modified electrodes. J Am Chem Soc 106:3423–3429
19. Karyakin AA, Karyakina EE, Gorton L (1999) On the mechanism of H2O2 reduction at Prussian blue modified electrodes. Electrochem Commun 1:78–82
20. Boyer A, Kalcher K, Pietsch R (1990) Voltammetric behavior of perborate on Prussian-blue-modified carbon paste electrodes. Electroanalysis 2:155–161
21. Karyakin AA, Gitelmacher OV, Karyakina EE (1994) A high-sensitive glucose amperometric biosensor based on Prussian blue modified electrodes. Anal Lett 27:2861–2869
22. Karyakin AA, Gitelmacher OV, Karyakina EE (1995) Prussian blue-based first-generation biosensor. A sensitive amperometric electrode for glucose. Anal Chem 67:2419–2423

23. Karyakin AA, Karyakina EE, Gorton L (1996) Prussian-blue-based amperometric biosensors in flow-injection analysis. Talanta 43:1597–1606
24. Karyakin AA, Karyakina EE, Gorton L (1998) The electrocatalytic activity of Prussian blue in hydrogen peroxide reduction studied using a wall-jet electrode with continuous flow. J Electroanal Chem 456:97–104
25. de Mattos IL, Gorton L, Laurell T, Malinauskas A, Karyakin AA (2000) Development of biosensors based on hexacyanoferrates. Talanta 52:791–799
26. Fiorito PA, Brett CMA, Córdoba de Torresi SI (2006) Polypyrrole/copper hexacyanoferrate hybrid as redox mediator for glucose biosensors. Talanta 69:403–408
27. Pauliukaite R, Chiorcea Paquim AM, Oliveira Brett AM, Brett CMA (2006) Electrochemical, EIS and AFM characterisation of biosensors: trioxysilane sol-gel encapsulated glucose oxidase with two different redox mediators. Electrochim Acta 52:1–8
28. Baioni AP, Vidotti M, Fiorito PA, Córdoba de Torresi SI (2008) Copper hexacyanoferrate nanoparticles modified electrodes: a versatile tool for biosensors. J Electroanal Chem 622:219–224
29. Pauliukaite R, Florescu M, Brett CMA (2005) Characterization of cobalt- and copper hexacyanoferrate-modified carbon film electrodes for redox-mediated biosensors. J Solid State Electrochem 9:354–362
30. Garjonyte R, Malinauskas A (1998) Electrocatalytic reactions of hydrogen peroxide at carbon paste electrodes modified by some metal hexacyanoferrates. Sens Actuators B Chem 46:236–241
31. Ivekovic D, Milardovic S, Grabaric BS (2004) Palladium hexacyanoferrate hydrogel as a novel and simple enzyme immobilization matrix for amperometric biosensors. Biosens Bioelectron 20:872–878
32. Shan Lin M, Feng Tseng T (1998) Chromium(III) hexacyanoferrate(II)-based chemical sensor for the cathodic determination of hydrogen peroxide. Analyst 123:159–163
33. Wang J, Zhang X, Prakash M (1999) Glucose microsensors based on carbon paste enzyme electrodes modified with cupric hexacyanoferrate. Anal Chim Acta 395:11–16
34. Cui X, Hong L, Lin X (2002) Electrochemical preparation, characterization and application of electrodes modified with hybrid hexacyanoferrates of copper and cobalt. J Electroanal Chem 526:115–124
35. Tao W, Pan D, Liu Y, Nie L, Yao S (2005) An amperometric hydrogen peroxide sensor based on immobilization of hemoglobin in poly(o-aminophenol) film at iron-cobalt hexacyanoferrate-modified gold electrode. Anal Biochem 338:332–340
36. Mauko L, Ogorevc B, Pihlar B (2009) Response behavior of amperometric glucose biosensors based on different carbon substrate transducers coated with enzyme-active layer: a comparative study. Electroanalysis 21:2535–2541
37. Guadagnini L, Giorgetti M, Tarterini F, Tonelli D (2010) Electrocatalytic performances of pure and mixed hexacyanoferrates of Cu and Pd for the reduction of hydrogen peroxide. Electroanalysis 22:1695–1701
38. Bian XZ, Luo HQ, Li NB (2010) A new material based on nanostructured Prussian blue analogue film doped with Ce(III) for development of hydrogen peroxide sensor. Electroanalysis 22:1364–1368
39. Sheng QL, Yu H, Zheng JB (2007) Sol-gel derived carbon ceramic electrode for the investigation of the electrochemical behavior and electrocatalytic activity of neodymium hexacyanoferrate. Electrochim Acta 52:4506–4512
40. Yu H, Sheng QL, Zheng JB (2007) Preparation, electrochemical behavior and performance of gallium hexacyanoferrate as electrocatalyst of H_2O_2. Electrochim Acta 52:4403–4410
41. Dostal A, Meyer B, Scholz F, Schroeder U, Bond AM, Marken F, Shaw SJ (1995) Electrochemical study of microcrystalline solid Prussian blue particles mechanically attached to graphite and gold electrodes: electrochemically induced lattice reconstruction. J Phys Chem 99:2096–2103
42. Zakharchuk NF, Meyer B, Henning H, Scholz F, Jaworksi A, Stojek Z (1995) A comparative study of Prussian-blue-modified graphite paste electrodes and solid graphite electrodes with mechanically immobilized Prussian blue. J Electroanal Chem 398:23–35
43. Zhai J, Zhai Y, Wen D, Dong S (2009) Prussian blue/multiwalled carbon nanotube hybrids: synthesis, assembly and electrochemical behavior. Electroanalysis 21:2207–2212
44. Jin E, Lu X, Cui L, Chao D, Wang C (2010) Fabrication of graphene/prussian blue composite nanosheets and their electrocatalytic reduction of H_2O_2. Electrochim Acta 55:7230–7234
45. Deng Q, Li B, Dong S (1998) Self-gelatinizable copolymer immobilized glucose biosensor based on Prussian blue modified graphite electrode[dagger]. Analyst 123:1995–1999
46. Karyakin AA, Karyakina EE, Gorton L (2000) Amperometric biosensor for glutamate using

Prussian blue-based "artificial peroxidase" as a transducer for hydrogen peroxide. Anal Chem 72:1720–1723
47. Ricci F, Palleschi G, Yigzaw Y, Gorton L, Ruzgas T, Karyakin A (2003) Investigation of the effect of different glassy carbon materials on the performance of Prussian blue based sensors for hydrogen peroxide. Electroanalysis 15:175–182
48. Kulesza PJ, Doblhofer K (1989) The membrane properties of Prussian blue films on electrodes. J Electroanal Chem Interf Electrochem 274:95–105
49. Karyakin AA, Chaplin MF (1994) Polypyrrole–Prussian blue films with controlled level of doping: codeposition of polypyrrole and Prussian blue. J Electroanal Chem 370:301–303
50. Derwinska K, Miecznikowski K, Koncki R, Kulesza PJ, Glab S, Malik MA (2003) Application of Prussian blue based composite film with functionalized organic polymer to construction of enzymatic glucose biosensor. Electroanalysis 15:1843–1849
51. Salazar P, Martín M, Roche R, González-Mora JL, O'Neill RD (2010) Microbiosensors for glucose based on Prussian blue modified carbon fiber electrodes for in vivo monitoring in the central nervous system. Biosens Bioelectron 26:748–753
52. Gomathi H, Prabhakara Rao G (1990) Simple electrochemical immobilization of the ferro/ferricyanide redox couple on carbon electrodes. J Appl Electrochem 20:454–456
53. Pharr CM, Griffiths PR (1997) Infrared spectroelectrochemical analysis of adsorbed hexacyanoferrate species formed during potential cycling in the ferrocyanide/ferricyanide redox couple. Anal Chem 69:4673–4679
54. Yang R, Qian Z, Deng J (1998) Electrochemical deposition of Prussian blue from a single ferricyanide solution. J Electrochem Soc 145:2231–2236
55. Abbaspour A, Kamyabi MA (2005) Electrochemical formation of Prussian blue films with a single ferricyanide solution on gold electrode. J Electroanal Chem 584: 117–123
56. Kumar SS, Joseph J, Phani KL (2007) Novel method for deposition of Gold–Prussian blue nanocomposite films induced by electrochemically formed gold nanoparticles: characterization and application to electrocatalysis. Chem Mater 19:4722–4730
57. Inoue H, Nakazawa T, Mitsuhashi T, Shirai T, Fluck E (1989) Characterization of Prussian blue and its thermal decomposition products. Hyperfine Interact 46:723–731
58. de Mattos IL, Gorton L, Ruzgas T, Karyakin AA (2000) Sensor for hydrogen peroxide based on Prussian blue modified electrode: improvement of the operational stability. Anal Sci 16:795–798
59. Garjonyte R, Malinauskas A (1999) Operational stability of amperometric hydrogen peroxide sensors, based on ferrous and copper hexacyanoferrates. Sens Actuators B Chem 56:93–97
60. Joseph J, Gomathi H, Rao GP (1991) Electrodes modified with cobalt hexacyanoferrate. J Electroanal Chem Interf Electrochem 304:263–269
61. Tao W, Pan D, Liu Y, Nie L, Yao S (2004) Characterization and electrocatalytic properties of cobalt hexacyanoferrate films immobilized on Au-colloid modified gold electrodes. J Electroanal Chem 572:109–117
62. Florescu M, Barsan M, Pauliukaite R, Brett CMA (2007) Development and application of oxysilane sol–gel electrochemical glucose biosensors based on cobalt hexacyanoferrate modified carbon film electrodes. Electroanalysis 19:220–226
63. Crespilho FN, Ghica ME, Gouveia-Caridade C, Oliveira ON Jr, Brett CMA (2008) Enzyme immobilisation on electroactive nanostructured membranes (ENM): optimised architectures for biosensing. Talanta 76:922–928
64. Joseph J, Gomathi H, Rao GP (1997) Modification of carbon electrodes with zinc hexacyanoferrate. J Electroanal Chem 431: 231–235
65. Chen S-M, Chan C-M (2003) Preparation, characterization, and electrocatalytic properties of copper hexacyanoferrate film and bilayer film modified electrodes. J Electroanal Chem 543:161–173
66. Chen S-M, Li S-H, Thiagarajan S (2007) Preparation of bilayer platinum and copper hexacyanoferrate hybrid film modified electrode and its electrocatalytic properties. J Electrochem Soc 154:E123–E130
67. Cataldi TRI, De Benedetto GE (1998) On the ability of ruthenium to stabilize polynuclear hexacyanometallate film electrodes. J Electroanal Chem 458:149–154
68. Fiorito PA, Córdoba de Torresi SI (2005) Hybrid nickel hexacyanoferrate/polypyrrole composite as mediator for hydrogen peroxide detection and its application in oxidase-based biosensors. J Electroanal Chem 581:31–37
69. Lin Y, Cui X (2006) Electrosynthesis, characterization, and application of novel hybrid materials based on carbon nanotube-polyaniline-nickel hexacyanoferrate nanocomposites. J Mater Chem 16:585–592

70. Wang X, Zhang Y, Banks CE, Chen Q, Ji X (2010) Non-enzymatic amperometric glucose biosensor based on nickel hexacyanoferrate nanoparticle film modified electrodes. Colloids Surf B Biointerfaces 78:363–366
71. Dong S, Jin Z (1989) Electrochemistry of indium hexacyanoferrate film modified electrodes. Electrochim Acta 34:963–968
72. Liu S-Q, Chen Y, Chen H-Y (2002) Electrochemical studies on a zirconium hexacyanoferrate modified electrode and utility for the determination of rubidium. Electroanalysis 14:116–121
73. Bharathi S, Joseph J, Jeyakumar D, Rao GP (1991) Modified electrodes with mixed metal hexacyanoferrates. J Electroanal Chem Interf Electrochem 319:341–345
74. Kulesza PJ, Malik MA, Skorek J, Miecznikowski K, Zamponi S, Berrettoni M, Giorgetti M, Marassi R (1999) Hybrid metal cyanometallates electrochemical charging and spectrochemical identity of heteronuclear nickel/cobalt hexacyanoferrate. J Electrochem Soc 146:3757–3761
75. Kulesza PJ, Malik MA, Schmidt R, Smolinska A, Miecznikowski K, Zamponi S, Czerwinski A, Berrettoni M, Marassi R (2000) Electrochemical preparation and characterization of electrodes modified with mixed hexacyanoferrates of nickel and palladium. J Electroanal Chem 487:57–65
76. Pauliukaite R, Hočevar SB, Hutton EA, Ogorevc B (2008) Novel electrochemical microsensor for hydrogen peroxide based on iron-ruthenium hexacyanoferrate modified carbon fiber electrode. Electroanalysis 20:47–53
77. De Benedetto GE, Guascito MR, Ciriello R, Cataldi TRI (2000) Analysis by X-ray photoelectron spectroscopy of ruthenium stabilised polynuclear hexacyanometallate film electrodes. Anal Chim Acta 410:143–152
78. Sheng Q, Luo K, Zheng J, Zhang H (2008) Enzymatically induced formation of neodymium hexacyanoferrate nanoparticles on the glucose oxidase/chitosan modified glass carbon electrode for the detection of glucose. Biosens Bioelectron 24:429–434
79. Wu P, Cai C (2005) The solid state electrochemistry of dysprosium(III) hexacyanoferrate(II). Electroanalysis 17: 1583–1588
80. Chaubey A, Malhotra BD (2002) Mediated biosensors. Biosens Bioelectron 17:441–456
81. Oldenziel WH, Dijkstra G, Cremers TI, Westerink BH (2006) Evaluation of hydrogel-coated glutamate microsensors. Anal Chem 78:3366–3378
82. Mao F, Mano N, Heller A (2003) Long tethers binding redox centers to polymer backbones enhance electron transport in enzyme "Wiring" hydrogels. J Am Chem Soc 125: 4951–4957
83. Guschin DA, Castillo J, Dimcheva N, Schuhmann W (2010) Redox electrodeposition polymers: adaptation of the redox potential of polymer-bound Os complexes for bioanalytical applications. Anal Bioanal Chem 398:1661–1673
84. Garguilo MG, Michael AC (1996) Amperometric microsensors for monitoring choline in the extracellular fluid of brain. J Neurosci Methods 70:73–82
85. Kulagina NV, Shankar L, Michael AC (1999) Monitoring glutamate and ascorbate in the extracellular space of grain tissue with electrochemical microsensors. Anal Chem 71:5093–5100
86. Chi Q, Dong S (1995) Amperometric biosensors based on the immobilization of oxidases in a Prussian blue film by electrochemical codeposition. Anal Chim Acta 310:429–436
87. Ferreira M, Fiorito PA, Oliveira ON, Córdoba de Torresi SI (2004) Enzyme-mediated amperometric biosensors prepared with the Layer-by-Layer (LbL) adsorption technique. Biosens Bioelectron 19:1611–1615
88. Yang M, Jiang J, Lu Y, He Y, Shen G, Yu R (2007) Functional histidine/nickel hexacyanoferrate nanotube assembly for biosensor applications. Biomaterials 28:3408–3417
89. Wang X, Gu H, Yin F, Tu Y (2009) A glucose biosensor based on Prussian blue/chitosan hybrid film. Biosens Bioelectron 24: 1527–1530
90. Zen J-M, Kumar AS, Chung C-R (2003) A glucose biosensor employing a stable artificial peroxidase based on ruthenium purple anchored cinder. Anal Chem 75:2703–2709
91. Karyakin AA, Kotel'nikova EA, Lukachova LV, Karyakina EE, Wang J (2002) Optimal environment for glucose oxidase in perfluorosulfonated ionomer membranes: improvement of first-generation biosensors. Anal Chem 74:1597–1603
92. de Mattos IL, da Cunha Areias MC (2005) Automated determination of glucose in soluble coffee using Prussian blue-glucose oxidase-Nafion® modified electrode. Talanta 66:1281–1286
93. Li J, Qiu JD, Xu JJ, Chen HY, Xia XH (2007) The synergistic effect of Prussian-blue-grafted carbon nanotube/poly(4-vinylpyridine) composites for amperometric sensing. Adv Funct Mater 17:1574–1580

94. Jaffari SA, Turner APF (1997) Novel hexacyanoferrate(III) modified graphite disc electrodes and their application in enzyme electrodes—part I. Biosens Bioelectron 12:1–9
95. Wang Y et al (2003) Chitosan/Prussian blue-based biosensors. Meas Sci Technol 14:831
96. Fiorito PA, Córdoba de Torresi SI (2004) Optimized multilayer oxalate biosensor. Talanta 62:649–654
97. Wang S, Lu L, Yang M, Lei Y, Shen G, Yu R (2009) A novel cobalt hexacyanoferrate nanocomposite on CNT scaffold by seed medium and application for biosensor. Anal Chim Acta 651:220–226
98. Salazar P, Martín M, Roche R, O'Neill RD, González-Mora JL (2010) Prussian blue-modified microelectrodes for selective transduction in enzyme-based amperometric microbiosensors for in vivo neurochemical monitoring. Electrochim Acta 55:6476–6484
99. Varma S, Yigzaw Y, Gorton L (2006) Prussian blue-glutamate oxidase modified glassy carbon electrode: a sensitive l-glutamate and [beta]-N-oxalyl-[alpha],[beta]-diaminopropionic acid ([beta]-ODAP) sensor. Anal Chim Acta 556:319–325
100. McQuade DT, Pullen AE, Swager TM (2000) Conjugated polymer-based chemical sensors. Chem Rev 100:2537–2574
101. Garjonyte R, Malinauskas A (2000) Glucose biosensor based on glucose oxidase immobilized in electropolymerized polypyrrole and poly(o-phenylenediamine) films on a Prussian blue-modified electrode. Sens Actuators B Chem 63:122–128
102. Vidal J-C, Espuelas J, Garcia-Ruiz E, Castillo J-R (2004) Amperometric cholesterol biosensors based on the electropolymerization of pyrrole and the electrocatalytic effect of Prussian-blue layers helped with self-assembled monolayers. Talanta 64:655–664
103. Garjonyte R, Malinauskas A (2000) Amperometric glucose biosensors based on Prussian blue- and polyaniline-glucose oxidase modified electrodes. Biosens Bioelectron 15:445–451
104. Lukachova LV, Kotel'nikova EA, D'Ottavi D, Shkerin EA, Karyakina EE, Moscone D, Palleschi G, Curulli A, Karyakin AA (2003) Nonconducting polymers on Prussian blue modified electrodes: improvement of selectivity and stability of the advanced H_2O_2 transducer. IEEE Sens J 3:326–332
105. Yashina EI, Borisova AV, Karyakina EE, Shchegolikhina OI, Vagin MY, Sakharov DA, Tonevitsky AG, Karyakin AA (2010) Sol-gel immobilization of lactate oxidase from organic solvent: toward the advanced lactate biosensor. Anal Chem 82:1601–1604
106. Tian F, Llaudet E, Dale N (2007) Ruthenium purple-mediated microelectrode biosensors based on sol-gel film. Anal Chem 79: 6760–6766
107. Llaudet E, Hatz S, Droniou M, Dale N (2005) Microelectrode biosensor for real-time measurement of ATP in biological tissue. Anal Chem 77:3267–3273
108. Frenguelli BG, Wigmore G, Llaudet E, Dale N (2007) Temporal and mechanistic dissociation of ATP and adenosine release during ischaemia in the mammalian hippocampus. J Neurochem 101:1400–1413
109. Gourine AV, Dale N, Korsak A, Llaudet E, Tian F, Huckstepp R, Spyer KM (2008) Release of ATP and glutamate in the nucleus tractus solitarii mediate pulmonary stretch receptor (Breuer-Hering) reflex pathway. J Physiol 586:3963–3978
110. Tian F, Gourine AV, Huckstepp RTR, Dale N (2009) A microelectrode biosensor for real time monitoring of l-glutamate release. Anal Chim Acta 645:86–91
111. Vittal R, Kim K-J, Gomathi H, Yegnaraman V (2008) CTAB-promoted Prussian blue-modified electrode and its cation transport characteristics for K^+, Na^+, Li^+, and NH_4^+ ions. J Phys Chem B 112:1149–1156
112. Vittal R, Gomathi H (2002) Beneficial effects of cetyltrimethylammonium bromide in the modification of electrodes with cobalt hexacyanoferrate surface films. J Phys Chem B 106:10135–10143
113. Du J, Wang Y, Zhou X, Xue Z, Liu X, Sun K, Lu X (2010) Improved sensing in physiological buffers by controlling the nanostructure of Prussian blue films. J Phys Chem C 114:14786–14793
114. Arduini F, Cassisi A, Amine A, Ricci F, Moscone D, Palleschi G (2009) Electrocatalytic oxidation of thiocholine at chemically modified cobalt hexacyanoferrate screen-printed electrodes. J Electroanal Chem 626:66–74
115. Kingo Itaya TA, Toshima S (1982) Electrochemical preparation of a Prussian blue analogue: iron-ruthenium cyanide. J Am Chem Soc 104:3751–3752
116. Shen-Ming Chen S-HH (2003) Iron hexacyanoruthenate films and their electrocatalytic properties with nitrite and dopamine. J Electrochem Soc 150:D175–D183
117. Shyu S-C, Wang CM (1997) Characterization of iron-containing clay modified electrodes

and their applications for the detection of hydrogen peroxide and ascorbic acid. J Electrochem Soc 144:3419–3425

118. Suh-Ching Shyu CMW (1998) Characterizations of iron-containing clay modified electrodes and their applications for glucose sensing. J Electrochem Soc 145: 154–158

119. Chien Fu Chen CMW (1999) Ruthenium purple-containing zeolite modified electrodes and their application for the detection of glucose. J Electroanal Chem 466: 82–89

120. Jyh-Myng Zen ASK, Chung C-R (2003) A glucose biosensor employing a stable artificial peroxidase based on ruthenium purple anchored cinder. Anal Chem 75:2703–2709

121. Abe T, Toda G, Tajiri A, Kaneko M (2001) Electrochemistry of ferric ruthenocyanide (Ruthenium Purple), and its electrocatalysis for proton reduction. J Electroanal Chem 510:35–42

122. Belay A, Collins A, Ruzgas T, Kissinger PT, Gorton L, Csoregi E (1999) Redox hydrogel based bienzyme electrode for L-glutamate monitoring. J Pharm Biomed Anal 19:93–105

Chapter 5

Enzyme Immobilization on Microelectrode Biosensors

Natalia Vasylieva and Stéphane Marinesco

Abstract

Protein immobilization is a key step in biosensor preparation that impacts on the overall performance. Among the wide variety of immobilization techniques currently available, only a few have been reliably applied in vivo. In this chapter, we discuss five different enzyme immobilization procedures, namely covalent immobilization by cross-linking using glutaraldehyde, poly(ethylene glycol) diglycidyl ether (PEGDE), or a hydrogel matrix and enzyme entrapment in a sol–gel or polypyrrole-derived matrices. We describe in detail experimental protocols for each method as applied to glucose oxidase and glutamate oxidase-based biosensors. Enzymatic microelectrodes prepared using these five different procedures were then compared in terms of sensitivity, response time, linear range, apparent Michaelis–Menten constant, and stability. An optimal method would ideally be nontoxic, simple, and cheap and display high sensitivity, a large linear range, and relatively fast response time. Although there is no current consensus on an ideal immobilization method, PEGDE and sol–gel techniques are potentially promising procedures for in vivo laboratory studies.

Key words Enzyme electrode, Glucose, Glutamate, Immobilization techniques

1 Introduction

Enzymatic biosensors are now widely used for monitoring concentrations of molecules in vivo (1). For instance, they have been used in medical practice to monitor glycemia in diabetic patients (2, 3). They are also used in animals to detect neurotransmitters or metabolites with excellent temporal resolution in the central nervous system (4).

Such biosensors usually rely on the same basic design of an enzymatic layer immobilized on a platinum electrode. In fact, enzyme immobilization on the electrode surface is a critical step in assembling amperometric biosensors. The immobilized enzyme should be stable over time, especially in vivo at 37°C, and retain significant activity to provide optimal sensitivity. Enzyme immobilization can also impact the final operating parameters of the biosensor, such as response time, linear range, or selectivity. Finally, in the context of in vivo implantation, the biocompatibility and

Stéphane Marinesco and Nicholas Dale (eds.), *Microelectrode Biosensors*, Neuromethods, vol. 80,
DOI 10.1007/978-1-62703-370-1_5, © Springer Science+Business Media, LLC 2013

overall toxicity of the biosensor can be affected by the presence of chemicals used during immobilization.

A wide variety of immobilization techniques are now available. They can be divided into three major groups: cross-linking (5–9), entrapment in polymer matrices (10–13), and adsorption through physical or electrostatic interactions, e.g., the Langmuir–Blodgett film technique (14, 15). For biological applications, cross-linking and enzyme entrapment are the most widely used methods, because of their reliability and stability. Adsorption techniques often result in less stable immobilization and have not been validated in vivo to our knowledge.

In this chapter, we describe the influence of the immobilization method used for biosensor construction on its final operating parameters. We have chosen several methods that have been previously used in neuroscience research, namely cross-linking by glutaraldehyde or PEGDE, chemical binding to a polymer with or without redox centers, and entrapment in a sol–gel or in an electrochemically generated polypyrrole-derived matrix. Based on data from the literature and on own experience, we compare these methods in terms of simplicity, ease-of-use, toxicity, and overall final biosensor performance.

2 Immobilization Methods and Protocols

2.1 Cross-Linking

Cross-linking is the process of covalently binding two or more molecules. Proteins are natural polymers with a variety of functional groups and can be readily bound together by many cross-linking agents. This property has been widely applied to biosensor fabrication, in which stable binding of proteins to the electrode surface is required. The two main chemical functions in a protein that can react with a cross-linker are the amine ($-NH_2$) and carboxyl (–COOH) groups. All proteins possess at least one amino group at their N-terminal end and one on every lysine residue. Similarly, carboxyl groups are present at the C-terminal end and on glutamic acid and aspartic acid residues. Because they are charged at physiological pH, amino and carboxyl groups tend to be exposed on the outer surface of the protein, which makes them primary candidates for reacting with a cross-linking agent.

When choosing an appropriate cross-linker, an important parameter is the length of the spacer arm, i.e., the distance between the reactive groups. This length will determine, in part, the flexibility of the protein matrix and the degrees of freedom that the proteins retain once immobilized. The cross-linker that has been used for the longest time for biosensor preparation is glutaraldehyde, and this is still the reference. In our laboratory, we are now using poly(ethylene glycol) diglycidyl ether (PEGDE) with very positive results.

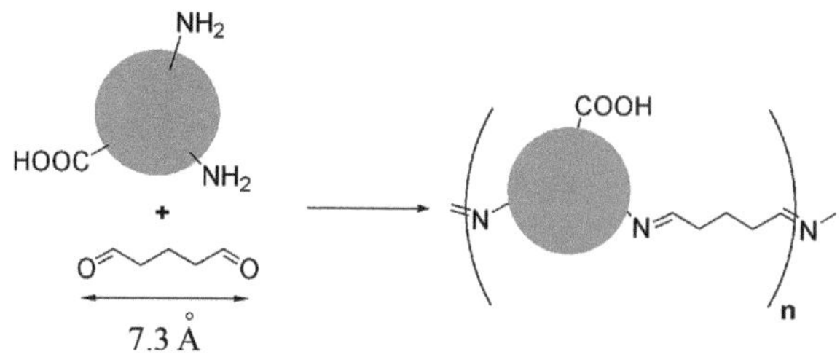

Fig. 1 *Glutaraldehyde method.* Reaction scheme for protein cross-linking by glutaraldehyde. Aldehyde groups react preferentially with amino groups present within the protein to produce covalent bonds between proteins

2.1.1 Glutaraldehyde

Glutaraldehyde (pentane-1,5-dial) is a small molecule (MW 100) bearing two aldehyde groups that can react with proteins (Fig. 1). Each free aldehyde group reacts with a different protein molecule, thus creating a ramified network (Fig. 1). Unbound aldehyde groups usually oxidize over time under normal storage conditions.

This method requires minimal equipment, and the reaction can be accomplished in a few minutes. Its simplicity has made it the reference method. It should be noted that glutaraldehyde is very toxic and should be manipulated with caution (fume hood, gloves, and goggles). Thus, even though glutaraldehyde is still widely used in scientific groups worldwide (9, 16, 17), its possible clinical applications are limited.

Glutaraldehyde cross-linking can be performed in at least two ways, by adding glutaraldehyde to the enzyme mixture or by exposing the enzyme to saturated vapor. Below, we present different procedures for immobilizing glucose oxidase and glutamate oxidase on a biosensor using glutaraldehyde. Unless specified otherwise, the glucose oxidase (EC 1.4.3.11) used was from *Aspergillus niger* and was purchased from Sigma-Aldrich (100–250 U/mg, Saint Quentin Fallavier, France), while the l-glutamate oxidase (EC 1.4.3.11) used was from *Streptomyces* sp. X119-6 and was purchased from Yamasa Corporation (Choshi, Chiba, Japan, also distributed by Seikagaku Inc. and Associated of Cape Cod Inc.).

Fixation in Saturated Vapor

To prepare a glucose biosensor, the working solution contains 60 mg/mL of glucose oxidase, 30 mg/mL of bovine serum albumin (BSA, Sigma), and 1% glycerol in 0.01 M phosphate-buffered saline (PBS), pH 7.4, while that for a glutamate biosensor contains 125 mg/mL of the enzyme, 60 mg/mL of BSA, and 1% glycerol in 0.01 M PBS, pH 7.4. The enzyme layer is deposited on the electrode surface by dipping the electrode in the enzyme solution. As the solution evaporates, it becomes thicker and an enzyme layer is deposited on the electrode. At this stage, however, the enzyme would quickly dissolve if the electrode was dipped into an aqueous solution. For immobilization, 5 mL of 25% glutaraldehyde (Sigma-Aldrich) is placed at the bottom of a closed dish. After ~2 min, the chamber is filled with near-saturating glutaraldehyde vapor; then the electrodes with their enzyme layer are placed in the vapor for about 3 min. The biosensors are then thoroughly washed and kept at 4°C for short-term storage and at −20°C for long-term storage (18).

Fixation in Liquid Phase

Enzyme immobilization using glutaraldehyde in solution has been thoroughly described by Greg A. Gerhardt's group (19) and is described in detail in Chapter 2. For glutamate biosensors, glutamate oxidase (~0.2 mg) is dissolved in 10 μL of a solution containing 2% BSA and 0.13% glutaraldehyde; then one drop (~0.5 μL) of this mixture is deposited onto one of the microelectrode sites on a silicon microprobe and allowed to dry. The microprobe is stored at −5°C before use and in 0.1 M PBS between measurements.

2.1.2 PEGDE Matrix

Another bifunctional molecule used for enzyme binding is PEGDE, a nontoxic chemical reagent that contains two epoxy groups able to react with amino-, hydroxyl-, and carboxyl groups. Under mild conditions of temperature, PEGDE reacts preferentially with amine functions present in proteins and creates a matrix for enzyme immobilization on the electrode surface very similar to those obtained using glutaraldehyde (Fig. 2). However, in contrast to glutaraldehyde, the kinetics of the reaction between proteins and epoxy groups at room temperature is rather slow (about 48 h), but can be considerably accelerated by a temperature increase (20). In the specific case of enzymatic biosensor preparation, this reaction

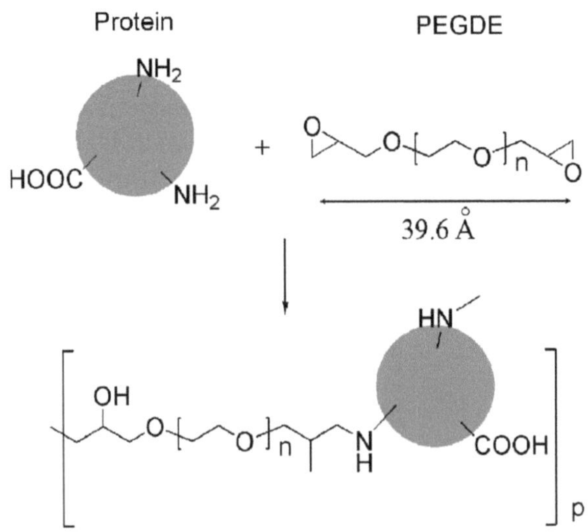

Fig. 2 *Poly(ethylene glycol) diglycidyl ether (PEGDE) method.* General reaction scheme of a protein with PEGDE. Amino groups present within proteins can react with the two epoxy rings of the PEGDE molecule, resulting in covalent immobilization

must occur under conditions that preserve enzymatic activity. PEGDE is a nontoxic, barely irritant molecule and can be used by non-qualified staff.

This method is very sensitive to heating conditions. If the curing temperature is too low, reliable enzyme immobilization does not occur, whereas too high a temperature can cause enzyme denaturation or inactivation. The amount of PEGDE added to the enzyme solution is also critical, as low amounts of PEGDE do not result in stable immobilization, whereas excess PEGDE leads to enzyme inactivation and low biosensor sensitivity. We now present two examples of the preparation of glucose and glutamate biosensors (21).

Enzyme Immobilization

To prepare a glucose biosensor, the working solution contains 60 mg/mL of glucose oxidase, 30 mg/mL of BSA (Sigma), 1% glycerol, and 20 mg/mL of PEGDE (average MW 526, Sigma-Aldrich) in 0.01 M PBS, pH 7.4, while, for the glutamate biosensor, the working solution contains 57 mg/mL of L*-glutamate oxidase, 80 mg/mL of BSA, 1% glycerol, and 80 mg/mL of PEGDE in 0.01 M PBS, pH 7.4. To prepare control microbiosensors, the oxidase is replaced by BSA (Sigma) and the working solution consists of 400 mg/mL BSA, 10 mg/mL of PEGDE, and 1% glycerol in 0.01 M PBS, pH 7.4. In all cases, the solution containing all components except PEGDE is stored as frozen aliquots, and the PEGDE is added just before biosensor preparation. Like for glutaraldehyde fixation, the enzyme layer is deposited by dipping the electrode in a thick enzyme solution, which results in a membrane that is still unstable and soluble in water. Immobilization occurs when the electrode is placed in a 55°C oven for 2 h. Biosensors are then kept at 4°C for short-term storage and at −20°C for long-term storage.*

For the two oxidases, the enzymatic membrane has a transparent light yellow appearance, is insoluble in water, and swells as it increased its water content.

2.1.3 Redox Hydrogel

Redox hydrogels are electron-conducting matrices created by cross-linking of water-soluble redox polymers that provide an electrical connection between the redox centers of enzyme molecules and the electrode. In such hydrogels, the polymer is usually a polyvinylpyridine, and the redox functions, tethered to the polymer backbones, are metal–organic complexes of $Os^{2+/3+}$ with substituted and unsubstituted pyridine (22). PEGDE is used as a cross-linking agent for binding enzymes to complexes of redox polymer, thereby producing efficient electron diffusion between the enzymatic redox centers and the electrode (20, 23, 24). The enzyme immobilization reaction is illustrated in Fig. 3a.

An electron mediator is immobilized in the matrix, together with the enzyme, and lowers the potential at which peroxide, released by the enzyme, is detected. Figure 3b shows the electron transfer pathway of the redox hydrogel-based biosensor (25, 26).

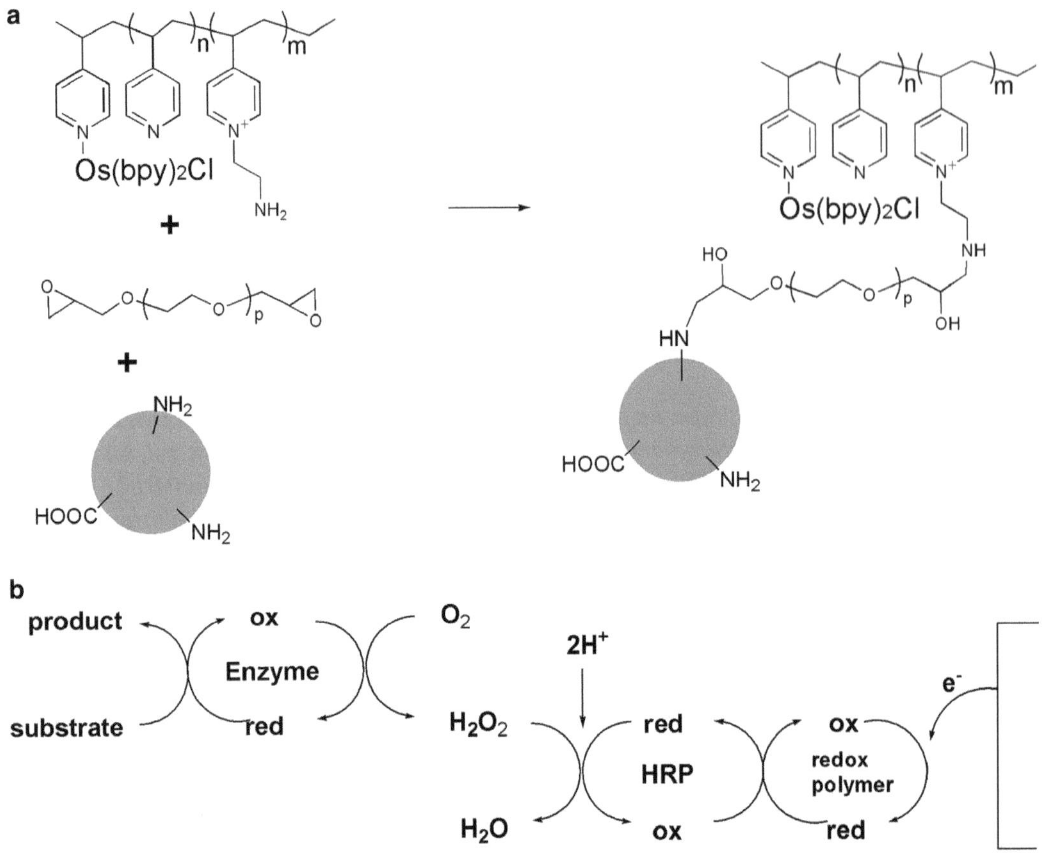

Fig. 3 *Redox hydrogel matrix method.* (**a**) Schematic illustration of enzyme immobilization in a hydrogel phase. The PEGDE cross-linker attaches the enzyme by its amine groups to the redox polymer backbone (polyvinylpyridine grafted with $Os^{2+/3+}$ complexes). (**b**) Electron transfer between the enzyme and the electrode by polymer redox centers

Both the redox polymer and PEGDE are highly water soluble, but the membrane resulting from the reaction is insoluble and has a swollen gel-like polymer structure. When the electrode is placed in a solution, the membrane increases its water content, which minimizes the diffusion constraints and provides easy permeation of the film by enzyme substrates and product molecules.

Several enzymes, including glucose oxidase or glutamate oxidase, have been successfully immobilized on biosensors using this technique (23, 25, 27). Moreover, glucose biosensors based on this or similar principles have been validated for glucose monitoring in diabetic patients (2, 3). However, the components of such redox hydrogels, when available commercially, are expensive or require several steps of organic synthesis. This method is therefore more difficult to implement in a nonexpert laboratory than those using only glutaraldehyde or PEGDE. Below, we present an experimental protocol designed by Adam Heller's laboratory and validated in neuroscience research by Adrian C. Michael's and Ben Westerink's groups (20, 27–29).

Redox Polymer Synthesis

0.494 g (0.864 mmol) of cis-bis(2,2′-bipyridine-N,N′)dichloroosmium (II) (Colonial Metals, Inc. or Johnson Matthey) and 0.430 g (4.09 mequiv) of poly(4-vinylpyridine) are heated under nitrogen in 18 mL of ethylene glycol for 2 h. After the solution is cooled to room temperature, 30 mL of dimethyl formamide and 1.5 g (7.3 mmol) of 2-bromoethylamine hydrobromide are added, and the solution stirred at 45°C overnight; then the crude polymer is precipitated by pouring the solution into rapidly stirred acetone. The hydroscopic precipitate is collected, dissolved in H_2O, filtered, and precipitated as a PF_6^- salt by addition of a solution of NH_4PF_6. The dry salt (0.49 g) is dissolved in 20 mL of acetonitrile, diluted with 50 mL of H_2O, and stirred for 2 h over 5.2 g of anion exchange beads (Bio-Rad AG1-X4, chloride form, TRISKEM); then the mixture is filtered and the filtrate evaporated under vacuum to ca. 10 mL. Concentrated HCl is then added to bring the pH to 2, and the solution is dripped into rapidly stirred acetonitrile. The precipitate formed is then filtered and dried in a vacuum desiccator (20).

Biosensor Construction

Unless stated otherwise, all reagents are dissolved in HEPES buffer, pH 8.

Glucose Sensor

The aqueous mixture of redox polymer and enzyme is composed of 20 μL of redox polymer (1 mg/mL in water), 4 μL of PEGDE (3 mg/mL in water), 20 μL of ascorbate oxidase (AAOx) (EC 1.10.3.3; from Curcurbita sp.) (10 mg/mL), 10 μL of horseradish peroxidase (2 mg/mL), and 10 μL of glucose oxidase (50 μg/mL). A 10 μL aliquot of the mixture is suspended on the end of a pipette and painted onto an electrode for 10 min. The sensor is then dried at 37°C for 1 h, soaked in water for 15 min, and dried in ambient air for 1 h (29).

Glutamate Biosensor

The aqueous mixture of redox polymer and enzyme is composed of 20 μL of redox polymer (1 mg/mL), 4 μL of PEGDE (3 mg/mL), 10 μL of HRP (0.474 U/μL, 3 mg/mL), 10 μL of AAOx (1.4 U/μL, 10 mg/mL), and 10 μL of glutamate oxidase (0.0242 U/μL, 2 mg/mL). Coating is performed by a 10 min manual dipping procedure. After coating, the sensors are cured for 1 h at 37°C, followed by a 10 min rinse in pure water, and drying in ambient air for 2 h (27).

In our hands, omitting the electron mediator from the hydrogel yields response times that are about ten times faster than those using $(Os(bpy)_2Cl)^+$ (28), a critical factor for the in vivo detection of neurotransmitters, such as glutamate. We have therefore modified the synthesis of redox hydrogels to obtain hydrogels without metal–organic complexes, as described below.

Hydrogel Without Metal–Organic Complexes

Polymer Synthesis

A solution containing 0.430 g (4.09 mequiv) of poly(4-vinylpyridine), 30 mL of DMF, and 1.5 g (7.3 mmol) of 2-bromoetylamine hydrobromide is prepared and stirred at 45°C overnight; then the crude polymer is precipitated by pouring the solution into rapidly stirred acetone. The hydroscopic precipitate is collected and dissolved in a minimal volume of H_2O *and the solution filtered and the filtrate again poured into rapidly stirred acetone, the precipitate collected and dissolved in a minimal volume of* H_2O*, and the solution filtered and evaporated under vacuum.*

Biosensor Construction

All reagents are dissolved in 0.01 M PBS, pH 7.4.

For glucose biosensors, 5 μL of the polymer (6 mg/mL), 10 μL of 6 mM SDS, 4 μL of PEGDE (80 mg/mL), and 5 μL of a mixture of 60 mg/mL (12,600 U/mL) of glucose oxidase and 30 mg/mL of BSA are mixed thoroughly. The enzyme layer is then deposited by dipping the tip of the electrode in the enzyme solution and the sensor dried at 45°C for 1 h and soaked in water.

For glutamate biosensors, the aqueous mixture of polymer and enzyme is prepared using 25 μL of polymer (15 mg/ mL), 5 μL of PEGDE (45 mg/mL), and 7 μL of glutamate oxidase (0.64 U/μL, 100 mg/mL). The resulting solution is rather liquid and difficult to deposit on a microelectrode, and it is usually necessary to concentrate it under vacuum at room temperature until viscous. The enzyme layer is deposited by dipping the tip of the electrode in the enzyme solution; then the sensor is dried at 45°C for 2 h and soaked in water.

Several alternatives to enzyme immobilization by cross-linking procedures are now available. A very promising approach involves enzyme entrapment in a matrix.

2.2 Enzyme Entrapment

The basic principle of enzyme entrapment consists in creating a polymer layer on the surface of the electrode. The enzyme to be immobilized on the biosensor is added to the monomer solution, and during synthesis of the polymer, enzyme molecules are mechanically or electrostatically entrapped in the matrix created by the polymer or adsorbed in the layer. Because there are no strong chemical bonds between the enzyme and the polymer (mainly physical constraints), it is presumed that such procedures result in better preservation of enzyme activity. In particular, biosensors prepared using these methods could, in principle, display enhanced sensitivity and selectivity. We will now describe the two main methods for enzyme entrapment, sol–gel matrices, and polypyrrole (and polyphenylenediamine) electropolymerization.

2.2.1 Sol–Gel Matrix

The sol–gel method is widely used in materials science, as well as in the biosensor domain, and creates a porous matrix containing molecules or substrates of interest on the electrode surface. The method consists of the hydrolysis of precursors (Reaction (5.1)), usually alkoxysilanes ($Si(OR)_4$), followed by polymerization of the products

(Reactions (2a) and (2b)). The process can be described by the following reactions (30, 31):

$$Si(OR)_4 + 4H_2O \rightarrow Si(OH)_4 + 4ROH \quad (1)$$

$$2Si(OH)_4 \rightarrow (HO)_3Si - O - Si(OH)_3 + H_2O \quad (2a)$$

$$Si(OH)_4 + Si(OR)_4 \rightarrow (HO)_3 - Si - O - Si(OR)_3 + ROH \quad (2b)$$

The rate of the first reaction is strongly dependent on the pH of the solution. It is very slow at pH 7 and is increased in strong alkaline or acidic media. In contrast, the rates of (Reactions (2a) and (2b)) are highest at a slightly alkaline pH (31, 32). To prepare a silica matrix containing an enzyme on the surface of an electrode surface, electrochemically induced pH changes are often used. For example, the electrochemical reduction of nitrate and H^+ ions added to the precursor solution causes a localized increase in pH through (Reactions (3) and (4)):

$$H^+ + e^- \rightarrow \frac{1}{2}H_2 \quad (3)$$

$$NO_{3^-} + 6H_2O + 8e^- \rightarrow NH_3 + OH^- \quad (4)$$

A gel is generated only where a negative potential is applied and the local pH increased (11, 32, 33) (Fig. 4). This procedure is especially interesting when using micro-fabricated probes with multiple electrodes etched on the surface; in this case, a given enzyme can be specifically immobilized on selected sensing spots by controlling the applied potential. Moreover, by adjusting the time during which the potential is applied to the electrode, the thickness of the matrix can

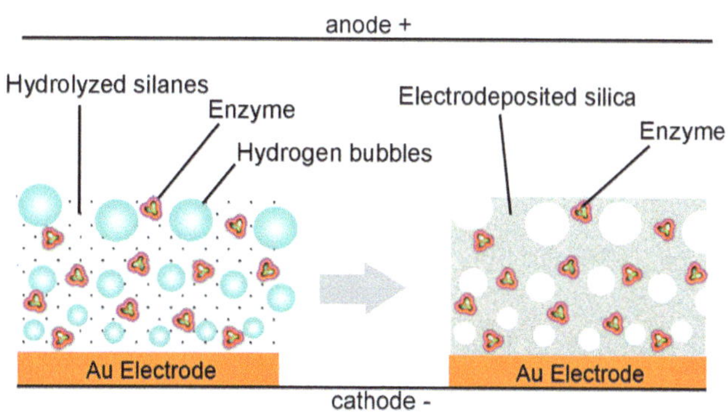

Fig. 4 *Sol–gel method.* Illustration of electrochemical deposition of a porous silica matrix on the electrode surface. The process of matrix (gel) formation is triggered by an electrochemically induced change in the pH in the initial solution (sol). Such pH changes are accompanied by the release of hydrogen bubbles that are important for the formation of the porous matrix structure, *adapted from* (35)

be controlled. Additionally, sol–gel-derived silica-based materials are well-suited for enzyme entrapment, because they are chemically inert and have a three-dimensional structure that allows some level of mobility for encapsulated biomolecules, thus preserving optimal catalytic activity (34). However, sol–gel matrices also have some disadvantages. For example, the precursors for sol–gel formation are toxic, which can limit potential clinical applications; in particular, tetramethoxysilane (TMOS) is toxic by inhalation. Other types of silanes can probably be chosen for the sol–gel procedure in order to decrease toxicity. In our hands, the sol–gel approach proved to be delicate to set up in the lab, especially because many different sol–gel protocols have been presented in the literature (e.g., (11, 34, 35)). Moreover, biosensors based on sol–gel procedures tend to display a significant initial decrease in sensitivity, probably due to a loss of biological material (34).

We propose two protocols for sol–gel matrix deposition on the surface of the electrode.

Method 1: Example of oxidase enzyme entrapment (36)

Two acidic solutions of hydrolyzed silanes are prepared containing, respectively, TMOS (Sigma-Aldrich) or (3-aminopropyl) trimethoxysilane (APTEOS, Sigma-Aldrich). The acidified TMOS solution is prepared from 7.39 mL of TMOS, 1.69 mL of distilled H_2O, and 0.11 mL of 0.04 M HCl, and it is sonicated on ice for 20 min and then can be stored on ice for several days. The APTEOS solution is prepared from 5.55 mL of APTEOS and 5 mL of H_2O; then the pH is adjusted to 3 with concentrated HCl and the solution sonicated on ice for 20 min prior to storing on ice. The sol solution is then prepared for electrochemical deposition. The mixture contains 1 part acidified TMOS solution, 1 part APTEOS solution, 2 parts 1 M glycerol, 1 part PEG 400 average MW, 4 parts 50 mM Tris pH 6.3, and 2 parts 0.4 M KNO_3. To prepare glutamate or ATP sensors, respectively, 1 U of L-glutamate oxidase (11) or 5 U of glycerol kinase plus 5 U of glycerol-3-phosphate oxidase (37) are dissolved in 10 μL of sol mix just before use and transferred into a small capillary. The working and reference Ag/AgCl electrodes are introduced from opposite ends and a potential between −0.9 and −1.2 V is applied for 10–40 s and a transparent, smooth, and robust gel layer is formed. The PEG and glycerol are added to stabilize the mixture of sol–gel and enzyme, while the nitrate is added to allow the formation of hydroxide ions and the subsequent pH increase on application of a negative potential. The biosensors produced are stored in PBS, pH 7.4.

Method 2: Example of hemoglobin entrapment (34)

A sol is prepared by mixing 2.125 g of TEOS with 2 mL of water and 2.5 mL of 0.01 M HCl for 12 h on a magnetic stirrer; then 0.07 M phosphate buffer, pH 6.0, is added to the silica sol to neutralize it and a one-tenth volume of 0.5 mM hemoglobin solution (Hb, MW 64000, Sigma) is added to the hydrolyzed sol. The electrode is then dipped into this mixture and a constant potential of −1.2 V is applied for 7 s, triggering electro-assisted deposition of the sol–gel layer. The electrode is then washed, dried for 60 min in air, and stored at 4°C. The stability of the deposited layer and the performance of the bioelectrode are improved by addition of 0.02 M cetyltrimethylammonium bromide, a cationic surfactant, to the starting solution.

For good reproducibility and a high quality silica layer, the working electrode must be thoroughly cleaned and dried for at least 1 h at 90°C before electrodeposition. Moreover, the authors noted that the addition of Au nanoparticles can improve biosensor performance (34).

An alternative entrapment method utilizing a biomimetic silification method has also been presented recently. Briefly, a silica gel layer is grown on the surface of the electrode after functionalization by poly-l-lysine. Like the sol–gel procedure, this method uses silane precursors for matrix formation and enzyme entrapment. Various enzymes can be immobilized under mild conditions using this procedure (38).

2.2.2 Polypyrrole-Derived Matrix

Electrochemical polymerization is a simple and attractive method for biosensor construction. Enzymes are entrapped either during the growth of a polymer generated by electrooxidation of a monomer contained in the enzyme solution (39) or in an adsorbed enzyme/monomer layer (40). Pyrrole and pyrrole-derived monomers are the most widely used components, because they are one of the few families of conductive polymers that are stable at ambient temperature and that can be easily used with different electrolytes, including aqueous solutions. During polymerization, the applied potential generates reactive pyrrole-cation radicals that react with neighboring monomer molecules, thereby producing a polymer layer on the electrode surface (39) (Fig. 5b). As in sol–gel procedures, it is easy to control the thickness or the position of the layer on specific sensing spots by adjusting the electrooxidation time or applying the voltage to specific places on a multielectrode microprobe. Additionally, the relatively low potential (+700 mV relative to the Ag/AgCl) required for electropolymerization provides sufficiently mild conditions to avoid enzyme denaturation (12). However, in our hands, polypyrrole entrapment proved to be one of the most difficult methods to implement in the laboratory. In fact, optimal results are obtained not with pyrrole itself, but with modified monomers in which a carbon chain is added to the pyrrole moiety. Such monomers are not commercially available and require multiphase organic synthesis (41), with even simplified synthetic schemes consisting of at least five steps (13). The pyrrole-derived monomer used in our studies was kindly supplied by Serge Cosnier's laboratory. We have tested the following method for electropolymerization of an absorbed pyrrole-enzyme layer (12, 40).

Enzyme Entrapment

The glucose oxidase solution used for biosensor preparation contains 60 mg/mL of glucose oxidase, 30 mg/mL of BSA (Sigma), and 1%

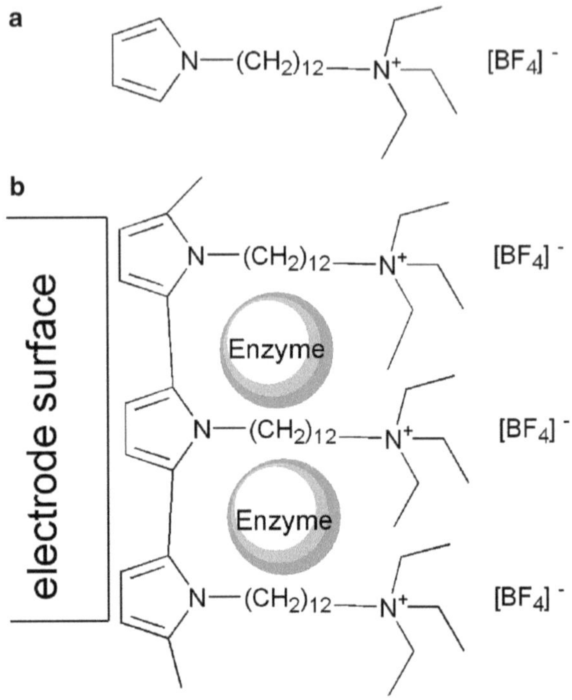

Fig. 5 *Polypyrrole method.* (**a**) Chemical structure of the monomer of [12-(pyrrol-1-yl) dodecyl] triethylammonium tetrafluoroborate (the inventor's version). (**b**) Schematic illustration of enzyme entrapment in the electrodeposited matrix

glycerol in 0.01 M PBS, pH 7.4. An aliquot of this solution (5 μL) is dried in vacuum at room temperature, and the pyrrole-derived monomer (300 mM), previously sonicated in water for 15 min, is added and carefully mixed to give a final concentration of 8.3 mM of monomer and 2.5 mg/mL of protein (42). The enzyme layer is deposited on the electrode surface by dip coating and left in vacuum for dehydration.

Polymerization is carried out in a two-electrode configuration system. The adsorbed enzyme-pyrrole-derived layer is polymerized in PBS at 800 mV constant potential versus an Ag/AgCl reference electrode for 10 min.

Data from the literature suggest that biosensors prepared using this procedure have a rather short lifetime. Electrodes stored at 4°C for 50 days lose about 50% of their sensitivity (43), whereas implantation in a biological preparation often causes partial or complete loss of sensitivity (13, 42).

In our list of possible enzyme entrapment methods, we should also mention polyphenylenediamine (PPD) polymers. Unlike polypyrrole, PPD is not electrically conductive, so the polymer layer cannot grow into such thick layers as those obtained with sol–gel or polypyrrole procedures. We have not tested this method in our laboratory. Instead, we invite the reader to refer to Sect. 6, which is entirely devoted to this procedure.

3 Quantitative Comparison of Different Immobilization Methods

We tested the different immobilization procedures described above on glucose and glutamate biosensors. When testing hydrogels, we focused exclusively on the method without redox centers, which results in more rapid response times and appears more suitable for monitoring fast changes in neurotransmitter levels. The sol–gel procedure used was method 1 (p. 104). Using polypyrrole, biosensors were prepared successfully using glucose oxidase, but the method proved very unreliable with other enzymes; those prepared with choline oxidase have a very low sensitivity and those prepared with D-amino acid oxidase could not be immobilized at all. For this reason, we abandoned this method and did not perform further tests with glutamate oxidase. To compare the influence of the immobilization method on the major biosensor parameters, we determined their sensitivity, response time, linear range, the apparent Michaelis–Menten constant, and stability.

3.1 Apparent Michaelis–Menten Constant and Linear Range

Enzymatic biosensors convert a specific substrate to its product while producing electrochemically detectable peroxide. The activity of most enzymes can be described by the Michaelis–Menten equation:

$$V(C) = \frac{V_{\mathrm{max}} \times C}{K_{\mathrm{m}} + C} \tag{6}$$

In turn, the current–concentration relationship that characterizes a biosensor can be approximated by the modified equation:

$$I(C) = \frac{I_{\mathrm{max}} * C}{K_{\mathrm{m}}^{\mathrm{App}} + C} \tag{7}$$

where I_{max} is the maximum current recorded by the biosensor and $K_{\mathrm{m}}^{\mathrm{App}}$ is the apparent Michaelis–Menten constant for the biosensor. $K_{\mathrm{m}}^{\mathrm{App}}$ not only is related to the K_{m} of the enzyme immobilized on the electrode, but also depends on other parameters such as the diffusion rates of the substrate, O_2, and peroxide within the enzyme layer. Because of this, the $K_{\mathrm{m}}^{\mathrm{App}}$ of a biosensor is often very different from the K_{m} of its free enzyme.

To determine the relationship between the current detected by the biosensor and the substrate concentration experimentally, we measured the current steps produced by addition of increasing amounts of enzyme substrate to the recording medium. Plotting the value of the oxidation current against the substrate concentration gives a curve that can be fitted to the modified Michaelis–Menten equation to yield the approximate value of $K_{\mathrm{m}}^{\mathrm{App}}$.

The $K_{\mathrm{m}}^{\mathrm{App}}$ values of the glucose biosensors were very variable, ranging from 1.1 ± 0.2 mM with glutaraldehyde fixation to

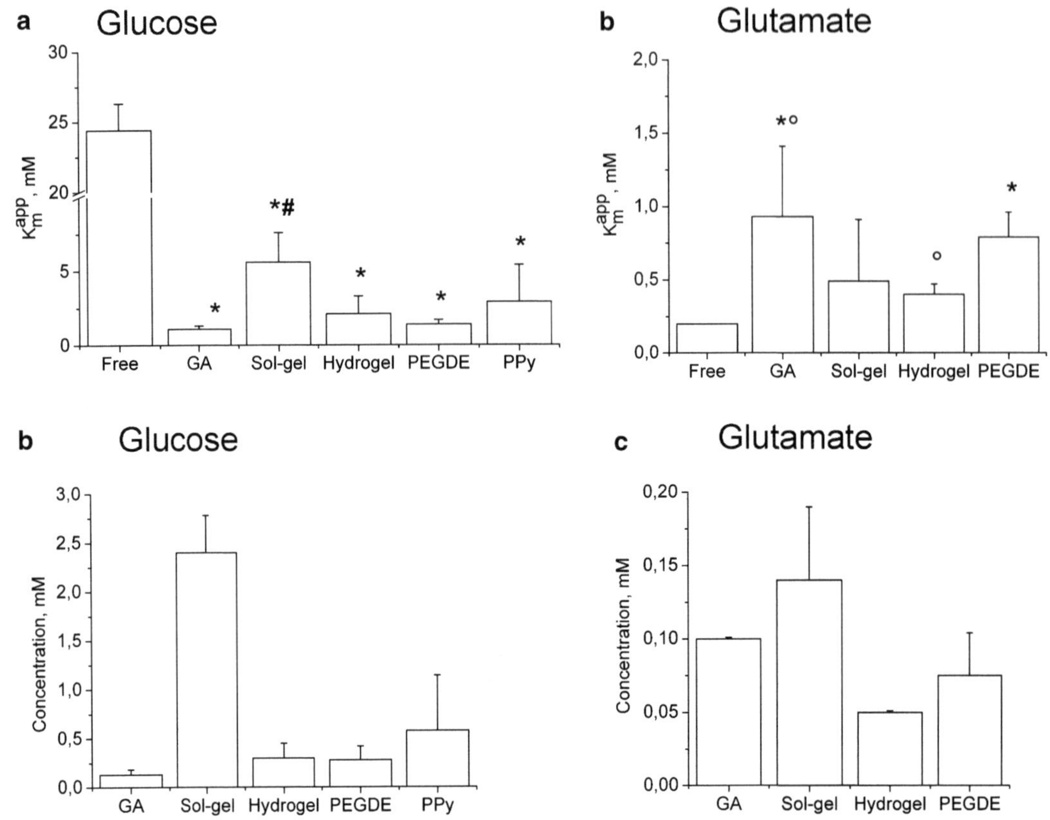

Fig. 6 *Apparent Michaelis–Menten constant and linear range.* Influence of the enzyme immobilization method on the apparent Michaelis–Menten constant of biosensors with immobilized glucose oxidase (**a**) or glutamate oxidase (**b**). Biosensor linear response range for detection of glucose (**c**) or glutamate (**d**) ($R^2 = 0.998$). The data are in good agreement with theoretical Michaelis–Menten kinetics, predicting linearity up to concentrations around $K_m^{App}/10$. Only biosensors prepared with the sol–gel method displayed larger linear ranges than those predicted by Michealis–Menten kinetics. The data are presented as the mean ± standard deviation (SD). The experiment sample size, n, was the number of biosensors constructed using the same protocol (4–6 electrodes). *Significantly different from the free enzyme, # significantly different from all other immobilization methods, ° significant difference between the two indicated methods. Statistical comparisons were performed using ANOVA, followed by a LSD post-hoc test ($p < 0.05$). *GA* glutaraldehyde; *Ppy* polypyrrole

5.6 ± 2 mM with sol–gel entrapment (Fig. 6a). Thus, the K_m^{App} was substantially lower than the K_m of the free enzyme, which is between 26 mM (44, 45) and 30 mM (46, 47). The K_m^{App} of glutamate biosensors were also very variable, ranging from 0.4 ± 0.1 mM for the hydrogel method to 0.9 ± 0.5 mM for glutaraldehyde fixation (Fig. 6b). In contrast, the glutamate biosensors had a K_m^{App} higher than the K_m of the free enzyme (0.2 mM, (48, 49)). These effects could result, in part, from substrate pre-concentration or exclusion from the membrane layer. In fact, substrate pre-concentration would decrease the K_m^{App}, whereas substrate exclusion from the membrane would increase it. Such apparent concentration effects have been described in the literature (18, 50).

To determine the linear range of our biosensors, we plotted the experimental current values obtained using the lowest concentrations that we tested and performed a linear fit. The linear range was defined as the highest concentration for which a successful fit at $R^2=0.998$ was still possible. For glucose and glutamate biosensors prepared with cross-linking methods (glutaraldehyde, hydrogel, or PEGDE) or with polypyrrole, the linear range was between 0 and 100–300 μM glucose ($n=6$) and between 0 and 75 μM glutamate ($n=4$–6). These results are in agreement with theoretical calculations predicting that the linear range should extend up to concentrations around $K_m^{App}/10$. Only biosensors prepared with a sol–gel matrix did not follow this rule, as their linear range extended up to 2.5 mM glucose ($n=6$, $R^2=0.998$) and 140 μM glutamate ($n=4$, $R^2=0.998$). These results suggest that biosensors with enzyme entrapped in a sol–gel matrix may not follow the theoretical kinetics of the Michaelis–Menten model.

3.2 Biosensor Sensitivity

Because biosensors are mostly used in their linear (or dynamic) range, the sensitivity of the sensor is defined as the initial slope of the current–concentration curve (the biosensor equivalent of K_m/V_{max} for the free enzyme) and is expressed in pA/μM of substrate or, more accurately, in μA/mM/cm² of electrode surface (in our studies, the microelectrode surface was about 8,300 μm²).

For glucose biosensors, the sensitivity varied from 12.2 ± 3.7 pA/μM ($n=35$) for a PEGDE matrix to 1 ± 0.1 pA/μM ($n=7$) for a sol–gel matrix (Fig. 7a). For glutamate biosensors, the glutaraldehyde, PEGDE, and hydrogel methods gave sensitivities between 21.6 ± 3.1 and 14.4 ± 8.2 pA/μM, with only sol–gel entrapment giving a significant lower value of 7.8 ± 3.5 pA/μM (Fig. 7b). It should be noted that sol–gel-based biosensors display excellent sensitivity with fresh-prepared electrodes; then the sensitivity rapidly decreases during the first hour of use, finally reaching a stable value.

Overall, our tests with glucose oxidase and glutamate oxidase indicate that covalent immobilization methods using protein cross-linking (glutaraldehyde, PEGDE, and hydrogel) yield more sensitive biosensors than methods based on enzyme entrapment (sol–gel and polypyrrole).

3.3 Response Time

Microelectrode biosensors designed for neuroscience research should detect neurotransmitters or metabolites with a high temporal resolution. Unlike the direct oxidation of biogenic amines on carbon-fiber electrodes, which allows the subsecond detection of dopamine or serotonin (51), enzymatic biosensors are limited by the kinetics of

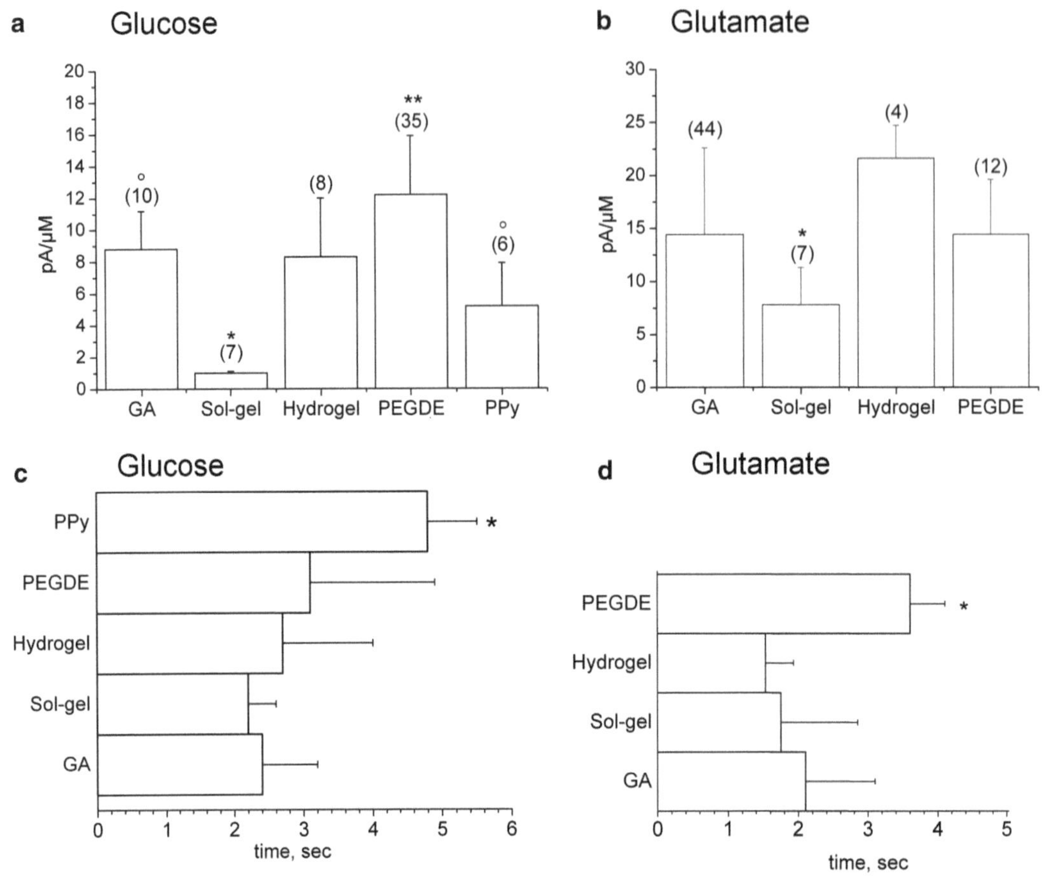

Fig. 7 *Biosensor sensitivity and response time.* (**a**) and (**b**): Dependence of biosensor sensitivity to glucose (**a**) or glutamate (**b**) on the method used for enzyme immobilization. Biosensors with enzyme entrapped in a sol–gel matrix showed the lowest sensitivity. *and ** reflect, respectively, a significantly lower or higher sensitivity compared to all other methods, °significant difference between the indicated methods. (**c**) and (**d**): Response time of glucose (**c**) and glutamate (**d**) biosensors as a function of the immobilization method. *Significant difference from all other methods, $n = 4$–6

the enzymatic reaction and usually display response times of the order of a few seconds. To determine the response time, biosensors are rapidly transferred from PBS to a solution containing a known concentration of the enzyme substrate, which generates a step of oxidation current at the biosensor. The response time is typically defined as the time required to reach 90% of the plateau response.

When testing the different types of biosensors, we found that the immobilization method significantly influenced the response time of the electrode (Fig. 7). For glucose biosensors, the response time was in the range of 2–3 s, except for a significantly slower response of polypyrrole-based biosensors (4.8 ± 0.7 s, $n = 6$ Fig. 7c). For glutamate biosensors, the response time was in the range of 1.5–2.1 s ($n = 4$), except for the significantly slower

PEGDE-based biosensors (3.6 ± 0.5 s, $n = 12$, Fig. 7d). Thus, both PEGDE and polypyrrole tended to yield biosensors with slightly slower response times.

3.4 Stability

For reliable in vivo experiments, in which biosensors are implanted in the brain for several hours or even days, the stability of the enzyme layer is a crucial parameter. An important aspect of biosensor stability is its interaction with the biological milieu and its susceptibility to biofouling processes taking place in the living brain. In this regard, a minimal tip size is advisable to decrease inflammation and/or hemorrhage during implantation. The biocompatibility of the materials used for electrode construction and the method of enzyme immobilization can have a significant impact on the lifetime of the biosensor in vivo. If the enzyme matrix is poorly compatible with brain tissue, membrane biofouling will lead to enzyme inactivation (52). For example, bringing biosensors with a enzyme-containing polypyrrole-derived or sol–gel matrix into contact with biological tissues frequently causes partial or complete loss of sensitivity (13, 42, 53).

We first tested the stability of the biosensors when left in PBS at 37°C. Biosensors formed by cross-linking by glutaraldehyde, hydrogel, or PEGDE and by polypyrrole-derived matrix methods showed remarkable stability when left for 2–3 h at 37°C. Biosensors prepared using PEGDE immobilization were stable for up to 3d at 37°C (21). In contrast, the sensitivity of sol–gel electrodes decreased within the first 2–3 h until a constant value was reached and then remained stable for several days or weeks.

In a second set of experiments on stability, we implanted biosensors in vivo. We only tried the three cross-linking methods using glutaraldehyde, hydrogel, or PEGDE. We were able to detect D-serine for up to 5 h in the frontal cortex of anesthetized rats using D-amino acid oxidase biosensors and compared their sensitivity before and after implantation and found no significant change. Sol–gel based biosensors have also been used in vivo for purine detection (54).

Thus, the methods that we chose to compare are indeed well-suited for in vivo experiments, except possibly in the case of some polypyrrole-based biosensors, which might be better adapted to in vitro (slice) experiments.

4 Conclusions

In reviewing these five major immobilization procedures available in the field of implantable microelectrode biosensors, one of our goals was to provide detailed experimental protocols that would allow the reader to set up these methods in his/her laboratory. Another goal was to identify the most efficient methods and to select them for future studies. In this regard, of the methods tested, PEGDE displayed promising properties:

- PEGDE is nontoxic, can be manipulated by nonexperts, and potential clinical applications can be considered.
- The PEGDE method is simple and cheap: the reagent is commercially available and enzyme immobilization involves only one reagent, without any requirement for custom chemical synthesis.
- Enzyme immobilization using PEGDE results in a very physically and enzymatically stable product (2–3 months at 4°C or 3 days at 37°C).

These advantages should make the PEGDE approach a valuable method for future biosensor applications. However, considering the operational parameters of all the biosensors that we tested, it is difficult to identify a single specific immobilization procedure with optimal overall biosensor performance. For example, PEGDE tends to give higher sensitivities that are mitigated by slightly slower response times. Additionally, PEGDE lacks the electrochemical control over membrane growth that is available with sol–gel or polypyrrole methods. In this regard, sol–gel procedures and their latest developments (10, 37) should attract great interest in the future.

We believe that more efforts should be made to come up with a consensual method for enzyme immobilization that would satisfy all the criteria of low toxicity, simplicity, stability, sensitivity, fast kinetic parameters, and fine control of membrane deposition. As biosensor technologies continue to develop and to spread to different laboratories, we have no doubt that such a consensus will emerge in the near future.

Acknowledgments

This study was supported by Inserm U628 and Université Claude Bernard Lyon I and by grants from the Agence Nationale pour la Recherche (ANR-09-BLAN-0063 Neurosense) and Institut Fédératif des Neurosciences de Lyon (Projet structurant IFNL 2008). NV received a PhD fellowship from the Ministère de la Recherche. Estelle Enriquez, Anne Meiller, and Caroline Maucler provided valuable technical assistance. We are grateful to Thomas Barkas for helpful comments on an earlier version of the manuscript.

References

1. Vaddiraju S, Tomazos I, Burgess DJ, Jain FC, Papadimitrakopoulos F (2010) Emerging synergy between nanotechnology and implantable biosensors: a review. Biosens Bioelectron 25:1553–1565
2. Heller A, Feldman B (2008) Electrochemical glucose sensors and their applications in diabetes management. Chem Rev 108:2482–2505

3. McGarraugh G (2009) The chemistry of commercial continuous glucose monitors. Diabetes Technol Ther 11(Suppl 1):S17–S24
4. Dale N, Hatz S, Tian FM, Llaudet E (2005) Listening to the brain: microelectrode biosensors for neurochemicals. Trends Biotechnol 23:420–428
5. Wilson R, Turner APF (1992) Glucose-oxidase—an ideal enzyme. Biosens Bioelectron 7:165–185
6. Burmeister JJ, Moxon K, Gerhardt GA (2000) Ceramic-based multisite microelectrodes for electrochemical recordings. Anal Chem 72:187–192
7. Netchiporouk LI, Shul'ga AA, Jaffrezic-Renault N, Olier R, Cespuglio R (1995) Properties of carbon fibre microelectrodes as a basis biosensors. Anal Chim Acta 303:275–283
8. Strike DJ, Derooij NF, Koudelkahep M (1993) Electrodeposition of glucose-oxidase for the fabrication of miniature sensors. Sens Actuators B Chem 13:61–64
9. Frey O, Holtzman T, McNamara RM, Theobald DE, van der Wal PD, de Rooij NF, Dalley JW, Koudelka-Hep M (2010) Enzyme-based choline and L-glutamate biosensor electrodes on silicon microprobe arrays. Biosens Bioelectron 26:477–484
10. McMahon CP, Killoran SJ, O'Neill RD (2005) Design variations of a polymer–enzyme composite biosensor for glucose: enhanced analyte sensitivity without increased oxygen dependence. J Electroanal Chem 580:193–202
11. Tian F, Gourine AV, Huckstepp RT, Dale N (2009) A microelectrode biosensor for real time monitoring of L-glutamate release. Anal Chim Acta 645:86–91
12. Coche-Guerente L, Cosnier S, Innocent C, Mailley P (1995) Development of amperometric biosensors based on the immobilization of enzymes in polymer films electrogenerated from a series of amphiphilic pyrrole derivatives. Anal Chim Acta 311:23–30
13. Llaudet E, Botting NP, Crayston JA, Dale N (2003) A three-enzyme microelectrode sensor for detecting purine release from central nervous system. Biosens Bioelectron 18:43–52
14. Okahata Y, Tsuruta T, Ijiro K, Ariga K (1989) Preparations of Langmuir-Blodgett films of enzyme lipid complexes—a glucose sensor membrane. Thin Solid Films 180:65–72
15. Singhal R, Takashima W, Kaneto K, Samanta SB, Annapoorni S, Malhotra BD (2002) Langmuir-Blodgett films of poly(3-dodecyl thiophene) for application to glucose biosensor. Sens Actuators B Chem 86:42–48
16. Stephens ML, Pomerleau F, Huettl P, Gerhardt GA, Zhang ZM (2010) Real-time glutamate measurements in the putamen of awake rhesus monkeys using an enzyme-based human microelectrode array prototype. J Neurosci Methods 185:264–272
17. Hnaien M, Hassen WM, Abdelghani A, Cotte S, Leonard D, Bessueille F, Jaffrezic-Renault N (2009) A conductometric biosensor for the estimation of the number of cleaving sites in peptides and proteins. Electrochem Commun 11:165–168
18. Pernot P, Mothet JP, Schuvailo O, Soldatkin A, Pollegioni L, Pilone M, Adeline MT, Cespuglio R, Marinesco S (2008) Characterization of a yeast D-amino acid oxidase microbiosensor for D-serine detection in the central nervous system. Anal Chem 80:1589–1597
19. Hu YB, Mitchell KM, Albahadily FN, Michaelis EK, Wilson GS (1994) Direct measurement of glutamate release in the brain using a dual enzyme-based electrochemical sensor. Brain Res 659:117–125
20. Gregg BA, Heller A (1991) Redox polymer films containing enzyme. 1. A redox-conducting epoxy cement: synthesis, characterization, and electrocatalytic oxidation of hydroquinone. J Phys Chem 95:5970–5975
21. Vasylieva N, Barnych B, Meiller A, Maucler C, Pollegioni L, Lin J-S, Barbier D, Marinesco S (2011) Covalent enzyme immobilization by poly(ethylene glycol) diglycidyl ether (PEGDE) for microeletrode biosensor preparation. Biosens Bioelectron 26:3993–4000
22. Heller A (2006) Electron-conducting redox hydrogels: design, characteristics and synthesis. Curr Opin Chem Biol 10:664–672
23. Gregg BA, Heller A (1991) Redox polymer films containing enzyme. 2. Glucose oxidase containing enzyme electrodes. J Phys Chem 95:5976–5980
24. Ohara TJ, Rajagopalan R, Heller A (1993) Glucose electrodes based on cross-linked (Os(bpy)2Cl)+/2+ complexed poly(1-vinylimidazole) films. Anal Chem 65:3512–3517
25. Kulagina NV, Shankar L, Michael AC (1999) Monitoring glutamate and ascorbate in the extracellular space of brain tissue with electrochemical microsensors. Anal Chem 71:5093–5100
26. Mikeladze E, Collins A, Sukhacheva M, Netrusov A, Csoregi E (2002) Characterization of a glutamate biosensor based on a novel gutamate oxidase integrated into a redox hydrogel. Electroanalysis 14:1052–1059
27. Oldenziel WH, Beukema W, Westerink BHC (2004) Improving the reproducibility of hydrogel-coated glutamate microsensors by using an automated dipcoater. J Neurosci Methods 140:117–126
28. Oldenziel WH, Westerink BHC (2005) Improving glutamate microsensors by opti-

mizing the composition of the redox hydrogel. Anal Chem 77:5520–5528
29. Cui J, Kulagina NV, Michael AC (2001) Pharmacological evidence for the selectivity of in vivo signals obtained with enzyme-based electrochemical sensors. J Neurosci Methods 104:183–189
30. Walcarius A (2001) Electroanalysis with pure, chemically modified, and sol-gel-derived silica-based materials. Electroanalysis 13:701–718
31. Ciriminna R, Sciortino M, Alonzo G, Schrijver AD, Pagliaro M (2010) From molecules to systems: sol-gel microencapsulation in silica-based materials. Chem Rev 111:765–789
32. Sayen S, Walcarius A (2003) Electro-assisted generation of functionalized silica films on gold. Electrochem Commun 5:241–348
33. Shacham R, Avnir D, Mandler D (1999) Electrodeposition of methylated sol-gel films on conducting surfaces. Adv Mater 11:384–388
34. Rozhanchuk T, Tananaiko O, Mazurenko I, Etienne M, Walcarius A, Zaitsev V (2009) Electroanalytical properties of haemoglobin in silica-nanocomposite films electrogenerated on pyrolitic graphite electrode. J Electroanal Chem 625:33–39
35. Yang S, Jia WZ, Qian QY, Zhou YG, Xia XH (2009) Simple approach for efficient encapsulation of enzyme in silica matrix with retained bioactivity. Anal Chem 81:3478–3484
36. Dale, N., Llaudet, E., and Droniou, M. (2004) COATINGS. WO2004/048603 A2, 10.06.2004
37. Llaudet E, Hatz S, Droniou M, Dale N (2005) Microelectrode biosensor for real-time measurement of ATP in biological tissue. Anal Chem 77:3267–3273
38. Tian FM, Wu WJ, Broderick M, Vamvakaki V, Chaniotakis N, Dale N (2010) Novel microbiosensors prepared utilizing biomimetic silicification method. Biosens Bioelectron 25:2408–2413
39. Foulds NC, Lowe CR (1986) Enzyme entrapment in electrically conducting polymers—immobilization of glucose-oxidase in polypyrrole and its application in amperometric glucose sensors. J Chem Soc Faraday Trans 1 82:1259–1264
40. Cosnier S, Innocent C (1992) A novel biosensor elaboration by electropolymerization of an adsorbed amphiphilic pyrrole-tyrosinase enzyme layer. Electroanal Chem 328:361–366
41. Coche-Guerente L, Deronzier A, Galland B, Labbe P, Moutet JC, Reverdy G (1991) Immobilization of redox anions in poly(amphiphilic pyrrolylalkylammonium) using a simple and monomer-saving one-step procedure in pure water-electrolyte. J Chem Soc Chem Commun 386–388
42. Poitry S, Poitry-Yamate C, Innocent C, Cosnier S, Tsacopoulos M (1997) Detection of glutamate released by neurons with an enzyme-based microelectrode. Electrochim Acta 42:3217–3223
43. Cosnier S, Fombon J-J, Labbea P, Limosin D (1999) Development of a PPO-polyamphiphilic pyrrole/electrode for on site monitoring of phenol in aqueous effluents. Sens Actuators B 59:134–139
44. Hayashi S, Nakamura S (1976) Comparison of fungal glucose oxidases. Chemical, physicochemical and immunological studies. Biochim Biophys Acta 438:37–48
45. Nakamura S, Hayashi S, Koga K (1976) Effect of periodate oxidation on the structure and properties of glucose oxidase. Biochim Biophys Acta 445:294–308
46. Szajani B, Molnar A, Klamar G, Kalman M (1987) Preparation, characterization, and potential application of an immobilized glucose oxidase. Appl Biochem Biotechnol 14:37–47
47. Kalisz HM, Hecht HJ, Schomburg D, Schmid RD (1990) Crystallization and preliminary X-ray diffraction studies of a deglycosylated glucose oxidase from Aspergillus niger. J Mol Biol 213:207–209
48. Arima J, Tamura T, Kusakabe H, Ashiuchi M, Yagi T, Tanaka H, Inagaki K (2003) Recombinant expression, biochemical characterization and stabilization through proteolysis of an L-glutamate oxidase from Streptomyces sp. X-119-6. J Biochem 134:805–812
49. Kusakabe H, Midorikawa Y, Kuninaka A, Yoshino H (1983) Occurrence of a new enzyme, L-glutamate oxidase in a wheat bran culture extract of Streptomyces sp. X-1 19-6. Agric Biol Chem 47:179–182
50. Demers N, Agostinelli E, Averill-Bates DA, Fortier G (2001) Immobilization of native and poly(ethylene glycol)-treated ('PEGylated') bovine serum amine oxidase into a biocompatible hydrogel. Biotechnol Appl Biochem 33:201–207
51. Robinson DL, Hermans A, Seipel AT, Wightman RM (2008) Monitoring rapid chemical communication in the brain. Chem Rev 108:2554–2584
52. Wisniewski N, Reichert M (2000) Methods for reducing biosensor membrane biofouling. Colloids Surf B Biointerfaces 18:197–219
53. Gourine AV, Dale N, Korsak A, Llaudet E, Tian F, Huckstepp R, Spyer KM (2008) Release of ATP and glutamate in the nucleus tractus solitarii mediate pulmonary stretch receptor (Breuer-Hering) reflex pathway. J Physiol 586:3963–3978
54. Dale N, Gourine AV, Llaudet E, Bulmer D, Thomas T, Spyer KM (2002) Rapid adenosine release in the nucleus tractus solitarii during defence response in rats: real-time measurement in vivo. J Physiol 544:149–160

Chapter 6

Characterization of Polymer–Enzyme Composite Biosensors for Brain Monitoring In Vivo

Robert D. O'Neill

Abstract

The application of biosensors for near real-time monitoring of key energy and signaling molecules in the intact brain poses major challenges at a number of levels. In this context, amperometric biosensor functionality is impacted by a variety of physicochemical and biological phenomena, including enzyme substrate and co-substrate sensitivity, interference from endogenous electroactive species, fabrication and response consequences of miniaturization for tissue implantation, and biocompatibility issues. Polymer–enzyme composite (PEC) implantable microbiosensors, incorporating a poly-*o*-phenylenediamine (P*o*PD) ultrathin permselective barrier, possess a variety of characteristics which make them suitable for long-term *in-vivo* electrochemical (LIVE) monitoring. This chapter reviews the use of P*o*PD in this context, and describes a battery of PEC sensitivity and selectivity parameters which allow development of the basic design in a systematic way in order to understand and improve their performance, and to diversify the analyte range of these novel probes of brain function.

Key words Long-term in-vivo electrochemistry (LIVE), Polymer-modified electrodes, Poly (phenylenediamine), Electropolymerization, Enzyme-based biosensors, H_2O_2 detection, Ascorbic acid interference, Glucose, Glutamate

1 Introduction

Understanding the mechanisms underlying the functioning of the human central nervous system stands at the forefront of modern science. Behavior, feelings, thoughts, and even consciousness itself may reflect the interplay of its electrical and chemical pathways. Although the measurement of real-time chemical events in the living brain is a supreme technical challenge, even in animal models, a growing number of analytical methodologies are being developed to study neurochemical dynamics in the intact brain. One such set of techniques involves the *in-situ* detection of substances in brain extracellular fluid (ECF), using implanted biosensors (1–4).

Stéphane Marinesco and Nicholas Dale (eds.), *Microelectrode Biosensors*, Neuromethods, vol. 80,
DOI 10.1007/978-1-62703-370-1_6, © Springer Science+Business Media, LLC 2013

The application of biosensors as analytical tools is a growing research topic in areas such as environmental surveillance, batch food analysis, and clinical monitoring, and is beginning to impact on quality-of-life issues (5–9). The choice of biosensor design for a particular application should be governed by diverse factors, including the chemical nature of the analytical medium (*e.g.*, lipophilic *vs.* hydrophilic), the sample size (*e.g.*, intracellular and extracellular monitoring *vs.* batch analysis), the time resolution and recording duration required, and the concentration of the target analyte relative to the corresponding interference compounds for the chosen technique (electrochemical, optical, gravimetric, tonometric, thermal, magnetoelastic, *etc.*) (10, 11). For *in-vivo* monitoring in the brain of behaving subjects, implantable biosensors showing good biocompatibility, sensitivity, selectivity, and stability in this strongly lipophilic environment are needed.

The chemical specificity of these biosensors can be very high, even to the level of stereoselectivity—for example, the use of L-glutamate oxidase (GluOx) in the design of amperometric biosensors to detect neurotransmitter L-glutamate (12, 13). However, the remarkable specificity of enzyme-based amperometric biosensors can be seriously undermined by interference from electroactive species present in the target medium, compromising the selectivity of the device. This problem is particularly pronounced for biosensors implanted in biological tissues for real-time monitoring (14–17), because separation techniques cannot be exploited to eliminate the interference, as is the case for online microdialysis approaches to *in-vivo* monitoring (18–20). Despite this drawback, considerable efforts have been made over the past two decades to overcome *in-vivo* biosensor sensitivity, selectivity, and stability issues, mainly because of the significant benefits of biosensor monitoring: small probe size, minimizing tissue damage (21–23), and high time resolution, allowing real-time correlation with animal behavior (24–27). As a result of these efforts, a range of amperometric enzyme-based biosensor designs, incorporating a permselective polymer layer to minimize interference, have already been applied successfully in many neurochemical studies (1, 2, 15, 17, 26, 28–31). Irrespective of the mechanism of interference rejection exploited (permselective layer, fine-tuning the applied potential, *etc.*), blocking the response from ascorbic acid (AA) should be the main focus of selectivity issues in the design of biosensors for brain monitoring applications. In addition to its very low redox potential (ubiquitous biological antioxidant), AA levels in the ECF are orders of magnitude higher than other interference species, such as dopamine, DOPAC, uric acid, *etc.* (32).

2 Designs of Biosensors for Brain Implantation

The design of an electrochemical biosensor is predicated firstly on the target analyte and the choice of appropriate biological element. For example, L-glutamate detection has been achieved using either glutamate receptor ion channels (33), L-glutamate oxidase (34), or L-glutamate dehydrogenase (35), and the corresponding electrochemical steps can be very different (33). However, the vast majority of enzyme-based amperometric biosensors exploit the biocatalytic oxidation of analyte, using oxidase enzymes containing the prosthetic group, flavin adenine dinucleotide (FAD). Despite this common foundation, a wide range of oxidase-based devices have been designed which make use of different signal transduction mechanisms and diverse selectivity strategies (4, 15, 16, 33, 35–38).

Another key factor which shapes biosensor design is the nature of the intended application matrix (16, 39). For example, devices for monitoring analyte concentration dynamics *in vivo* must be biocompatible over the time course of the implantation, both in terms of tissue effects on sensor functionality, and physiological reaction to the probe (23, 40, 41). In addition to possessing the appropriate level of biocompatibility, implantable oxidase-based biosensors must also fulfill the following minimum criteria for reliable analyte monitoring: appropriate size and geometry (24, 42), good sensitivity to the enzyme substrate (43), effective rejection of electroactive interference (44, 45), and low sensitivity to changes in pO_2 over the range of substrate and oxygen concentrations relevant to the intended application (12, 46). Even different tissues present distinct challenges. For example, biosensors designed to monitor glucose in blood or subcutaneous fat *in vivo* need a significantly greater oxygen tolerance compared to devices for monitoring brain ECF (47), because glucose levels are an order of magnitude higher in these peripheral tissues than in brain ECF (48), which increases the oxygen demand of the biosensor (see Sect. 4.1).

Within these restrictions, a number of biosensor designs have been developed and applied to monitor concentration dynamics of specific brain analytes *in vivo*. Among the first to be reported were those based on the organic conducting salt, tetrathiafulvalene tetracyanoquinodimethane (TTF^+–$TCNQ^-$) (49, 50). Although initially thought to involve direct electron transfer between the FAD center and this novel conducting surface, there is now evidence that signal generation is mediated by the $TCNQ^o$/$TCNQ^-$ redox couple available from the electrode material (51, 52). It is interesting to note that one of these seminal studies applied the concept of a "blank" sensor (one containing all the design elements of the biosensor, but lacking the enzyme) implanted beside the biosensor in order to minimize electroactive interference, using subtraction (49). This strategy was necessary because interference by AA can occur even at the low anodic applied potentials used for this class of biosensor (49, 53). The use of blank sensors, coupled with biosensors in "difference mode," is now seen as a vital component of reliable brain monitoring (24, 30, 44, 54–58), especially when the concentration of the target analyte is low relative to interference levels, although the use of a blank sensor does not mitigate the problem of oxygen dependence of oxidase-generated signals.

A more common design of biosensor used for analysis of brain systems (including brain slices *in vitro* and intact tissue *in vivo*) is that incorporating redox hydrogels, such as osmium poly(vinylpyridine) redox polymer (59–61), and this biosensor design has also been implanted in peripheral tissues (62). The Os–gel–HRP provides mediated electron transfer between the horseradish peroxidase, which reacts with the H_2O_2 generated according to Reactions (1) and (2), and the electrode operating at mild cathodic applied potentials (44). These devices may also include ascorbic acid oxidase and Nafion to reduce further interference by AA. Neurochemical studies *in vivo* with this class of biosensor have been limited to acute implantation in anesthetized preparations, and it is unclear at present whether they are stable enough to allow long-term brain monitoring in freely behaving animals. The presence of anesthetic is an important issue for *in-vivo* monitoring studies because different, and sometimes opposite, effects on neurochemical dynamics have been observed in the presence of anesthesia (16). Thus, for example, it is unwise to test the glutamate/ascorbate heteroexchange hypothesis in chloral-hydrate anesthetized rats, using this class of biosensor (58), when it has been demonstrated explicitly that chloral hydrate blocks this exchange (63).

A design, which has recently been implanted successfully for chronic recording in the brain of unanesthetized animals, exploits the anion-rejecting properties of Nafion as the main strategy for blocking electroactive interference at a ceramic-based biosensor (64, 65), in addition to a blank or "sentinel" electrode (see above) to increase further the selectivity of the detection system (24, 30). The biosensor signal is generated by the direct oxidation of H_2O_2 on the electrode surface (Reactions (1)–(3)) at a relatively high applied anodic potential. Studies carried out over 7 days showed that there was little loss of the sensitivity of the implanted differential device (24, 66), which indicates a promising future for this approach. Because of the intrinsic involvement of a blank sensor in the design, key neurotransmitters glutamate (65), and acetylcholine (67), both present in brain ECF at low concentrations, can be targeted with this approach. An issue which needs to be addressed, however, before widespread applications are undertaken is the oxygen dependence of these biosensors which has yet to be reported, although this is not likely to be problematic when the target analyte is at the low concentrations characteristic of neurotransmitters (12, 47). A variation on this design, which incorporates additional or replacement polymer layers deposited over a platinum substrate, has been applied for acute recording of brain glucose (68), lactate (69), and glutamate (57, 70).

2.1 Polyphenylenediamines as the Permselective Component of Implantable Biosensors

The use of a polyphenylenediamine (PPD) membrane, electrosynthesized from one of the three isomeric forms of phenylenediamine (PD), as a permselective film in the fabrication of implantable sensors and biosensors has been reported by a number of laboratories (26–29, 71–75). Poly-*ortho*-phenylenediamine (P*o*PD) is the most widely used form of PPD in the construction of polymer–enzyme composite (PEC) biosensors for a range of analytes, including glucose (26, 47, 76–85), lactate (86–88), L-glutamate (12, 28, 88–96), acetylcholine (97), D-serine (29, 98), and hydrogen peroxide (85, 99). The characteristics of specific enzymes, *e.g.*, their K_M value, determine to some extent the best method for their incorporation into the polymer film. For example, co-immobilization of glucose oxidase during electrosynthesis of P*o*PD appears to work best for this enzyme (see Sect. 6). For glutamate oxidase, however, the cost of the enzyme mitigates against a one-pot approach (90). Fortunately, the affinity of this enzyme is so high for its substrate (low K_M value; see Sect. 4) that alternative approaches, such as immobilization by a dip-evaporation method before or after electro-deposition of the polymer, yield biosensors with glutamate sensitivity close to optimum theoretical values—provided a polycation is incorporated into the PEC layer to mitigate the electrostatic repulsion between the surface enzyme and its anionic substrate (12, 100).

P*o*PD can be deposited electrochemically from *o*-phenylenediamine (*o*PD) monomer solutions at neutral pH to produce a thin self-sealing insulating polymer on the electrode surface (76). Film thickness estimations for P*o*PD generated under these conditions are typically in the region of 10–30 nm (76, 101–103). As well as possessing excellent interference rejection characteristics, P*o*PD also serves as a method of immobilizing enzymes and protects the electrode surface from biofouling (48, 77), making it ideal for biosensor applications *in vivo* (16). Polymers electrosynthesized from *meta*- and *para*-phenylenediamines (*m*PD and *p*PD) have been reported to possess similar *general* properties to P*o*PD (103–107), and they have been used, although not to the same extent, in biosensor construction (29, 105, 106, 108–114). Literature on the synthesis, structure, and properties of aromatic diamines, including the three PD isomers, has been thoroughly reviewed (115).

Recent reports have included within-study comparisons of the permselectivity properties of P*o*PD, P*m*PD, and P*p*PD films formed on palladium disks (104), ruthenium-coated carbon fibers (88),

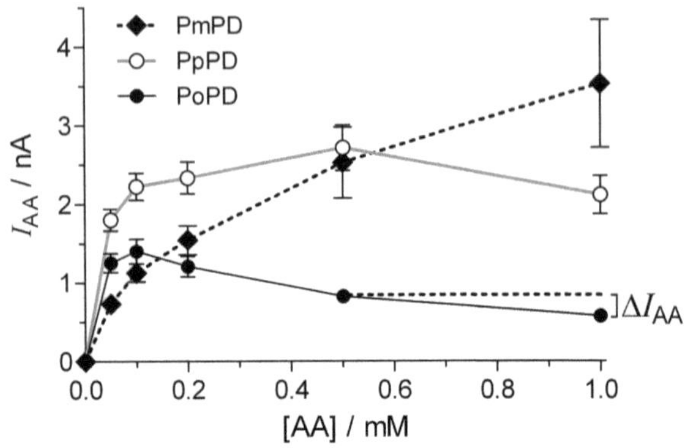

Fig. 1 A within-study comparison of AA calibration profiles shown as the mean ± SEM of the steady-state amperometric response for Pt–Ir wire cylinder electrodes modified with either P*o*PD ($n = 36$), P*m*PD ($n = 8$) or P*p*PD ($n = 8$). In addition to the significance of the absolute size of the AA response, the change in current associated with changes in AA concentration is of key concern when using these devices for *in-vivo* applications. In neurochemical monitoring, for example, where baseline AA levels are of the order of 0.5 mM in many brain regions, the change in sensor output caused by a doubling of AA concentration (ΔI_{AA}) can be very small for P*o*PD-based devices. Adapted with permission from (118)

and platinized glassy carbon electrodes (116), and concluded that P*m*PD was the most suitable for biosensor applications. These conclusions, however, are at odds with other studies which indicate that P*o*PD is superior to P*m*PD in this respect (103, 112). There are a number of possible explanations for this dichotomy. For example, the material used for the working electrode may affect the characteristics of the polymer deposited on it (92), and electrode size and/or geometry may also affect the apparent permeability of PPD layers (95, 117).

Therefore, a comparison of the three PPD films formed on Pt–Ir wire electrodes was carried out recently to determine the best PD isomer for use in the design of permselective layers for these small, implantable, low edge-density (95) cylinder-based biosensors (118); see Fig. 1. The findings for PPD-coated Pt–Ir electrodes were in general agreement with the Pd disk study (104) in identifying P*m*PD as the coating with the greatest permselectivity for H_2O_2 over AA at low concentrations of AA. The Pt–Ir study, however, included an investigation of the concentration dependence of the AA interference, and found that for AA levels greater than 200 μM, P*o*PD was superior (Fig. 1). Furthermore, stability studies indicated that the permselectivity of P*m*PD degraded more rapidly, supporting *o*PD as the choice of monomer for permselective membranes for use in long-term biological tissue implantation of metal-based biosensors where AA levels are high, such as in brain monitoring (73, 74, 78, 119, 120). Reversal of the P*o*PD and P*m*PD permselectivity ranking for low and high concentrations of AA emphasizes the value of studying the concentration dependence of interference effects, which is rarely reported in the literature, as well as the importance of specifying the phenylenediamine isomer used to generate the PPD layer (28).

The electrochemical signal in most PPD-based designs is generated by the direct oxidation of H_2O_2 on the electrode surface (Reactions (1)–(3)) at relatively high applied anodic potentials. A main advantage of PPD-based glucose biosensors is their stability, even during amperometric recording over days in the same freely behaving animal (48, 78, 119, 120); see Sect. 7. In addition, the pO_2 dependence of both glucose- (47, 119) and glutamate- (12, 42, 93) detecting forms of these biosensors has been quantified in detail (see Sect. 4.1), and shown to be adequate for brain monitoring. Thus, although each design of biosensor for brain monitoring has a key advantage, such as the small cross section of Os/gel/HRP-based carbon fiber electrodes (61, 121) and the self-referencing capability of the ceramic-based devices (30), the overall long-term stability and oxygen tolerance of PPD-based biosensors make them an attractive choice for further development (84, 85, 98).

3 PEC Biosensor Notation

To allow functionality in the electrochemically hostile environment of living brain tissue, implantable biosensors often contain a variety of components to enhance their sensitivity, selectivity, and stability. These agents are laid down in an order which depends on both their intended role and their physicochemical properties; therefore, in defining a systematic biosensor notation this order must be clear. The nomenclature employed here uses a slash (/) to indicate consecutive layering from left to right, while components deposited simultaneously are separated by a dash (–). For example, Pt/GOx/P*o*PD represents glucose oxidase (GOx) deposited on Pt (normally by dip evaporation (90, 93)) followed by electropolymerization of P*o*PD, Pt/P*o*PD/GOx indicates GOx deposited over electropolymerized P*o*PD, and Pt/P*o*PD–GOx signifies the co-deposition of enzyme and polymer by including GOx in the monomer solution (see Fig. 2). This notation is quite flexible and, where necessary, allows for details such as the geometry of the electrode, *e.g.*, disks (D) or cylinders (C) which can significantly affect biosensor performance (12, 93, 100) (see Figs. 3 and 4); the number of dip coatings applied (five in the following example); and the stock solution concentration of the agent deposited (*e.g.*, 2,000 U mL^{-1}): Pt_D/GOx(2 kU)$_5$/P*o*PD.

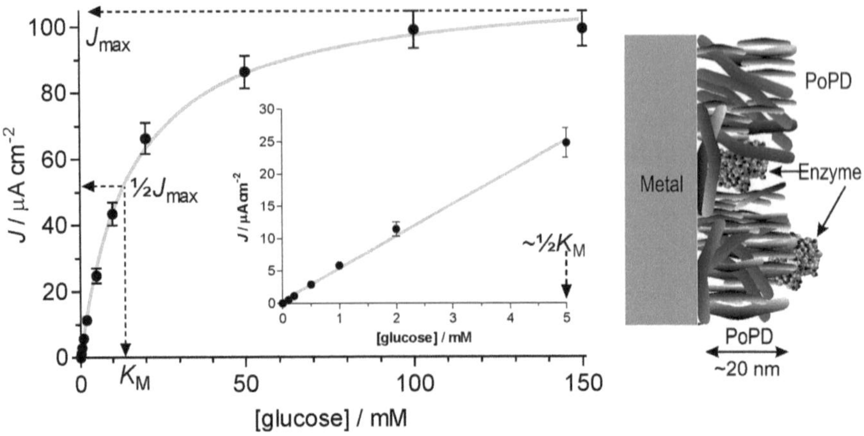

Fig. 2 Sample steady-state calibration data and nonlinear regression analysis for the biosensor design, Pt_C/P*o*PD–GOx (Eq. (7a), $R^2 = 0.998$, $n = 8$), illustrating the graphical significance of the Michaelis–Menten constants, J_{max} and K_M. The linear region slope (LRS) was obtained using linear regression up to 5 mM (~½K_M) glucose ($R^2 = 0.996$, $n = 8$; *inset*) and represents the most suitable measure of analytical sensitivity to enzyme substrate for each biosensor design. *Schematic*: representation of a PEC configuration for the Pt_C/P*o*PD–GOx design, illustrating trapped GOx (~8 nm diameter) in the P*o*PD layer deposited by the precipitation of insoluble chains formed during the electropolymerization of monomer solution containing the enzyme. Adapted from (84)

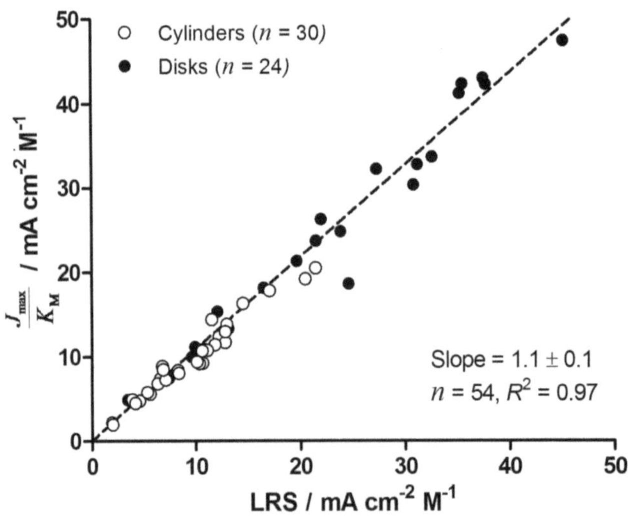

Fig. 3 A linear regression test of Eq. (10) for the Pt/GluOx/P*o*PD configuration of both Pt–Ir disk and cylinder geometries by determining the two Michaelis–Menten parameters (see Fig. 2) for each glutamate biosensor, calculating their ratio as J_{max}/K_M, and plotting this ratio against the experimentally measured linear region slope (LRS) for the corresponding electrode; Eq. (10) predicts a slope of unity. There was no significant difference between the regression slope for cylinder-based Pt_C/GluOx/P*o*PD (0.95 ± 0.04, $R^2 = 0.96$, $n = 30$) and that for disk-based Pt_D/GluOx/P*o*PD (1.10 ± 0.05, $R^2 = 0.96$, $n = 24$). These two populations were therefore pooled, and the resultant slope (1.1 ± 0.1, $R^2 = 0.97$, $n = 54$) confirms the validity of this relationship between J_{max}, K_M, and the LRS for biosensors obeying Michaelis–Menten kinetics. Reprinted with permission from Anal Chem 2006;78:2352–9 (93). Copyright 2006 American Chemical Society

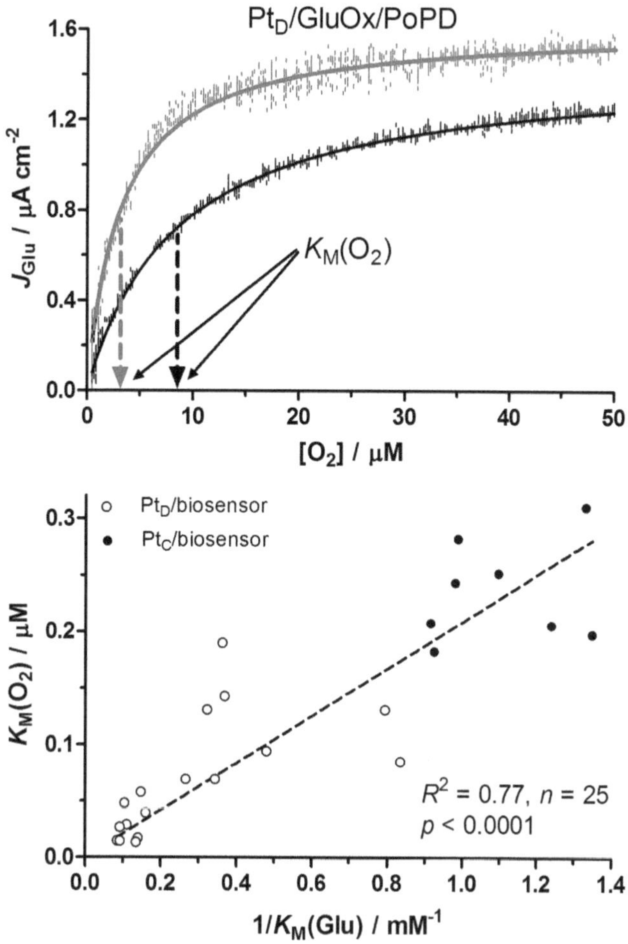

Fig. 4 *Top*: Examples of raw data for 50 μM glutamate (Glu), recorded amperometrically (+0.7 V *vs.* SCE) at 10 Hz with two Pt_D/GluOx/P*o*PD biosensors showing different Glu sensitivity (LRS values), plotted against oxygen concentration recorded simultaneously using an oxygen electrode. The curve in each case represents the nonlinear regression analysis, using Eq. (7b). Surprisingly, even for biosensors of the same geometry, size, and design, the device with the higher Glu LRS (*gray*) showed a lower oxygen dependence, $K_M(O_2)$. *Bottom*: Scatter plot and linear regression analysis for biosensor oxygen dependence, $K_M(O_2)$, *vs.* $1/K_M$(Glu), used as a measure of enzyme affinity for substrate, k_1 in Eq. (8) [see Eq. (9)]. There was no difference in the behavior of P*o*PD/GluOx *vs.* GluOx/P*o*PD polymer–enzyme configurations, and therefore single populations were formed for each of the geometrical designs: Pt_C/biosensor and Pt_D/biosensor. Although dispersed at either end of the distributions, the cylinder and disk forms of the biosensor displayed a continuity of behavior, suggesting enzyme loading was the main difference between these designs. The *Y*-axis has been normalized to correspond to values recorded for 1 μM Glu. Reprinted with permission from Anal Chem 2006;78:2352–9 (93). Copyright 2006 American Chemical Society

4 Enzyme-Related Biosensor Parameters

A number of sophisticated mathematical models of the behavior of enzymes in membranes have been described (122, 123). These complex analyses are often needed to understand and optimize the behavior of thick and/or conducting layers (124). However, a recent study has shown that substrate diffusion is not limiting for PPD layers incorporating an enzyme (82), due to their ultrathin nature (10–30 nm (76, 101, 103)). For example, the similar glutamate sensitivity of biosensors fabricated by entrapment of GluOx, using either pre-adsorption before P*o*PD deposition or a co-deposition protocol, led to the conclusion that P*o*PD thickness was similar to the diameter of the enzyme (125). The basic Michaelis–Menten enzyme parameters used here, therefore, provide more readily accessible insights into factors affecting the responsiveness of biosensors fabricated from the insulating form of PPD, and avoids the use of more complex analyses, such as those involving the Thiele modulus (126). The basis for this analysis has been reviewed previously (1), but is re-presented with further detail and clarification because it is key to the systematic development and characterization of biosensors advocated here. The parameters and reactions discussed in this chapter are summarized in Table 1.

For an oxidase enzyme (E) with FAD as prosthetic group, the catalytic conversion of substrate (S) and co-substrate (O_2) to products (P) and H_2O_2 can be written as Reactions (1)–(3):

$$\mathrm{S} + \mathrm{E/FAD} \rightarrow \mathrm{P} + \mathrm{E/FADH_2} \tag{1}$$

$$\mathrm{E/FADH_2} + \mathrm{O_2} \rightarrow \mathrm{E/FAD} + \mathrm{H_2O_2} \tag{2}$$

$$\mathrm{H_2O_2} \rightarrow \mathrm{O_2} + 2\mathrm{H^+} + 2\mathrm{e^-} \tag{3}$$

The two-substrate form of the Michaelis–Menten equation for the overall rate of reaction, v, is given by Eq. (4) (126, 127):

$$v = \frac{v'_{\mathrm{max}}}{1 + K'_{\mathrm{M}}(\mathrm{S})/[S] + K'_{\mathrm{M}}(\mathrm{O_2})/[O_2]} \tag{4}$$

If the concentration of the co-substrate is large and constant, Eq. (4) reduces to

$$v = \frac{v_{\mathrm{max}}}{1 + K_{\mathrm{M}}(\mathrm{S})/[S]} \tag{5}$$

Equation (5) represents the rate of product formation, which in this case reflects the rate of H_2O_2 generation. The H_2O_2 generated in the enzyme layer is the species detected at the electrode surface, but its bulk concentration is zero; it is the flux of H_2O_2

Table 1
Summary of equations and reactions discussed in this chapter

Reaction (1) (*enzyme–substrate reaction*): Generalized reaction of enzyme substrate (S) with the oxidized form of an oxidase enzyme (E) containing the FAD prosthetic group
Reaction (2) (*enzyme–co-substrate reaction*): Generalized reaction of enzyme co-substrate (O_2) with the reduced form of the oxidase enzyme in Reaction (1) to regenerate its oxidized state
Reaction (3) (*electro-oxidation of hydrogen peroxide*): The signal-generating step in many electrochemical oxidase-based biosensors is the electro-oxidation of H_2O_2, usually amperometrically at a relatively high fixed applied potential
Equation (4) (*two-substrate Michaelis–Menten equation*): Reflecting the involvement of Reactions (1 and 2) in the enzyme kinetics of FAD-containing oxidases, the two-substrate form of the Michaelis–Menten equation for the overall rate of reaction contains terms for the concentration of both substrate ($[S]$) and co-substrate ($[O_2]$), as well as ***composite*** rate constants, $K_M'(\mathrm{S})$ and $K_M'(\mathrm{O_2})$, respectively, for these two reactions (see Eq. 9)
Equation (5) (*one-substrate Michaelis–Menten equation*): To simplify Eq. (4), if the concentration of the co-substrate is kept constant, the one-substrate form of the Michaelis–Menten equation can be used to describe how the rate of the overall reaction depends on the concentration of substrate. The one remaining K_M term (the ***apparent*** Michaelis constant) is now phenomenologically the substrate concentration that generates half the maximum (enzyme saturated) overall reaction rate (see Fig. 2) but depends, for example, on the concentration of O_2 in the system
Equation (6) (*current converted Michaelis–Menten equation*): The overall rate of enzyme reaction defined by Eq. (5) determines the rate of H_2O_2 production. For electrochemical oxidase-based biosensors operated under well-defined conditions, the rate of H_2O_2 electro-oxidation [*i.e.*, the biosensor signal, Reaction (3)] is proportional to this rate of H_2O_2 production. Thus, Eq. (5) can be scaled linearly (by the $A\alpha zF$ term; see text) to express how the biosensor current varies with substrate concentration
Equation (7a) (*current-density converted Michaelis–Menten equation*): To compare the performance of biosensors with different electrode areas, Eq. (6) can be normalized by dividing the biosensor current by electrode area, calculated either geometrically or by electrochemical calibration. Just two parameters determine the biosensor calibration response over the entire range of substrate concentration (see Fig. 2): J_{max}, the asymptotal maximum current density observed at high substrate levels when the enzyme is saturated; and K_M, the substrate concentration that generates half the J_{max} value
Equation (7b) (*current-density converted Michaelis–Menten equation for co-substrate*): By analogy with Eqs. (5)–(7a), if the concentration of the substrate is constant, this form of the Michaelis–Menten equation can be used to describe how the biosensor current density depends on the concentration of co-substrate
Reaction (8) (*kinetic steps of a simple enzyme reaction*): This illustrates the three elementary reactions in a simple one-substrate enzyme process, and their corresponding rate constants
Equation (9) (*components of the Michaelis constant*): The "true" Michaelis constant, K_M, defined by this equation, reflects the balance of the process that forms enzyme–substrate complex, ES (k_1), and those that cause its decomposition ($k_{-1} + k_2$). Variations across biosensor populations in the "apparent" Michaelis constant (see Eq. 5) normally extracted from experimental calibrations (Fig. 2) have been used to study trends in enzyme–substrate binding (k_1), but can also be influenced by co-substrate factors (see text)

(continued)

Table 1 (continued)

Equation (10) (*relationship between J_{max}, K_M, and linear region slope*): For low substrate concentrations, the response of biosensors obeying Michaelis–Menten kinetics is nearly linear (Fig. 2). In the absolute limit ($[S] \rightarrow 0$), the slope is the tangent at $[S] = 0$, which has a value of J_{max}/K_M. Empirically, biosensors are effectively linear up to a substrate concentration of ½K_M (see Fig. 2). This linear region slope (LRS) has been shown to equate to the ratio of $J_{max}:K_M$ for various populations of biosensors (see Fig. 3)
Equation (11a) (*biosensor plateau response normalized with respect to H_2O_2 sensitivity*): The size of the plateau response of a biosensor (Fig. 2) is determined by three main factors: the density of active enzyme on the electrode surface, $[E]_{act}$; the value of k_2 [Reaction (8)]; and the sensitivity of the biosensor to H_2O_2 [Reaction (3)]. Variations in $[E]'_{act}$ often arise from changes in $[E]_{act}$ and so can be used to compare enzyme loading across populations of related biosensor designs
Equation (11b) ("*biosensor efficiency" parameter*): It is possible to focus in on the enzyme factors [J_{max} and K_M; see Eq. (10)] affecting biosensor LRS values by normalization with respect to H_2O_2 sensitivity
Reaction (12) (*disproportionation reaction of H_2O_2*): This reaction, which can take place both in solution and on surfaces, can lower the efficiency of biosensor signal generation [see Reaction (3)]
Equation (13) (*apparent hydrogen peroxide permeability*): Because the H_2O_2 calibration response is linear on both bare and PPD-modified Pt electrodes, *P*(HP)% can be defined as the ratio of H_2O_2 calibration slopes recorded with the same electrode before and after deposition of the polymer-enzyme composite (PEC) layer
Equation (14) (*apparent ascorbic acid permeability*): Because the AA calibration response is nonlinear on PPD-modified Pt electrodes, *P*(AA)% is defined as the ratio of the calibration current for 1 mM AA (physiologically relevant level), recorded with the same electrode before and after deposition of the PEC layer
Equation (15) (*permselectivity parameter*): The ratio of their respective apparent permeability values is one measure of the balance between interference rejection and signal transmission by the polymer layer
Equation (16) (*equimolar selectivity parameter*): This ratio of the currents for 1 mM AA and H_2O_2 is a more useful and convenient measure of the balance between interference rejection and signal transmission by the polymer layer
Equation (17) (*enzyme–substrate selectivity parameter*): This is an effective equimolar measure of the balance between interference rejection and signal generation/transmission by the PEC layer when the LRS is expressed in current/mM
Equation (18) (*concentration normalized enzyme–substrate selectivity parameter*): When enzyme–substrate selectivity [see Eq. (17)] is normalized with respect to estimated baseline concentrations of the main electroactive interference species (AA) and the enzyme substrate in brain extracellular fluid, the parameter generated indicates the percentage of the baseline current, recorded with an implanted biosensor in that brain region, which is due to baseline levels of endogenous AA
Equation (19) (*enzyme–substrate selectivity parameter for doubling ECF concentrations*): This parameter quantifies the effect of doubling the ECF levels of both AA and enzyme substrate on the response of an implanted biosensor

Many of the assumptions involved, *e.g.*, in generating limiting forms of certain equations, can be modeled under controlled calibration conditions *in vitro*. However, reliable use of biosensors in real samples, such as biological tissues *in vivo*, depends on how appropriate these assumptions are to the conditions of the specific application from the immobilized enzyme to the conducting surface that determines the current. Converting the rate (Eq. (5)) to current [Eq. (6)] is achieved by multiplying both sides by $A\alpha zF$, where α is the fraction of the flux that is oxidized on the electrode of geometric area, A (because some H_2O_2 is lost to the bulk (78)), z is the number of electrons involved in the reaction [$z=2$ for H_2O_2; see Reaction (3)], and F is the Faraday constant. Consistent with this analysis, many biosensor designs respond hyperbolically to the bulk concentration of enzyme substrate (see Fig. 2), and v_{max} [Eq. (5)] is related to I_{max} [Eq. (6)] through the $A\alpha zF$ term. The value of $K_M(S)$, usually written more simply as K_M (the concentration of substrate that yields ½v_{max}), does not change with this conversion. When comparing biosensors of different sizes, the response (I_S) can be normalized to current density [J_S, Eq. (7a)] by dividing by electrode surface area, A. Nonlinear regression analysis of J_S *vs.* [S] provides the basic response parameters: J_{max}, $K_M(S)$, and the coefficient of determination (goodness of fit), R^2 (see Fig. 2). Alternatively, if the concentration of S is fixed and O_2 levels are changed, then Eq. (7b) can be used to analyze the oxygen dependence of the signal (42, 119), where J'_{max} is the maximum (plateau) response for a particular concentration of S, and $K_M(O_2)$ is the apparent Michaelis constant for oxygen (see Sect. 4.1 and Fig. 4).

$$I_S = \frac{I_{max}}{1 + K_M(S)/[S]} \tag{6}$$

$$J_S = \frac{J_{max}}{1 + K_M(S)/[S]} \tag{7a}$$

$$J_S = \frac{J'_{max}}{1 + K_M(O_2)/[O_2]} \tag{7b}$$

The current density for the biosensor response to S, J_S, is a measure of the overall rate of the enzyme reaction, and J_{max} is the J_S value at enzyme saturation. Different values of J_{max}, determined under the same conditions, reflect differences in the reaction rate for saturated active enzyme on the surface ($k_2[E]_{act}$; see Reaction (8)) (47), provided the sensitivity of the electrode to H_2O_2 [Reaction (3)] does not vary much, as is the case for many of the PPD-modified Pt cylinders and disks used in brain monitoring (16, 78, 119); see Sects. 4.2 and 5 for discussion of biosensor H_2O_2 sensitivity and permeability. When the same enzyme preparation is being deposited under the same conditions of pH, temperature, *etc.*, using a variety of different surface components, it is unlikely that k_2 would vary greatly when the enzyme active site is deep within the globular protein structure, as is the case for FAD-based oxidases such as GOx (128) and GluOx (129). Therefore, variations in $k_2[E]_{act}$ (*i.e.*, J_{max}) between different biosensor designs are a good indication of variations in active enzyme loading ($[E]_{act}$). Thus, the main variables in determining the ***linear*** response (see Fig. 2) of a biosensor within a family of related ultrathin PEC-based designs (*e.g.*, Pt_D/PPD/GOx *vs.* Pt_C/PPD/GOx) are the amount of active enzyme trapped on the surface and the accessibility of substrate to that enzyme [k_1, Reaction (8)]. It should be noted here that, although Reaction (8) indicates that the first (k_1) step is an elementary reaction, it is not for substrate binding to enzyme immobilized on a biosensor surface. In practice, the k_1 reaction is a multistep process involving transport of substrate through the immobilization matrix to the enzyme's active site, followed by the binding event itself. The manipulation of enzyme kinetics for biosensor applications is often achieved by modification of the transport phase of the overall k_1 reaction (see, for example, Sect. 4.1).

The "true" Michaelis constant, K_M, is defined in terms of the rate constants for the generalized reactions [Reaction (8)] describing the conversion of substrate to products, catalyzed by the enzyme; see Eq. (9). When Eq. (7a) is used to approximate the two-substrate case, however, the K_M is more complex, containing co-substrate terms. Then K_M is the ***apparent*** Michaelis constant, and phenomenologically defines the concentration of substrate that gives half the J_{max} response (see Fig. 2). Thus changes in K_M often reflect changes in the access/binding constant, k_1, and variations in K_M have been interpreted in terms of barriers to substrate/enzyme binding (77, 130), as well as changes in oxygen demand (12, 47, 93, 131). Because the reactions involving k_{-1} and k_2 take place deep within the enzyme protein structure (see above), these are normally less affected by the immobilization matrix than is k_1. When comparing biosensor parameters based on different enzymes, *e.g.*, GOx and GluOx (L-glutamate oxidase), it may be necessary to specify the enzyme substrate, *e.g.*, K_M(G) *vs.* K_M(Glu). When analyzing the influence of co-substate (O_2) on the biosensor response, $K_M(O_2)$ and K_M(S) must be distinguished (12, 93).

$$E + S \underset{k_{-1}}{\overset{k_1}{\rightleftharpoons}} ES \xrightarrow{k_2} E + P \tag{8}$$

$$K_M = \frac{k_{-1} + k_2}{k_1} \tag{9}$$

The apparent Michaelis constant for S is also useful for defining the slope in the linear response region of the substrate calibration plot (LRS, linear region slope). The slope of the tangent of a hyperbola, as the independent variable approaches zero, is given by J_{max}/K_M; see Eq. (10). Of course, the response of a biosensor following Michaelis–Menten kinetics is never truly linear. However, for low concentrations of substrate, the response is "practically" linear (see Fig. 2 ***inset***), and all that remains is to define the substrate concentration range for which the linearity of the response is analytically acceptable. In practice, the range of the linear response to S is usually considered to extend up to ~½K_M (Fig. 2).

$$\underset{[S]\to 0}{\mathrm{Lt}}\ J_S = \frac{J_{max}}{1 + K_M / [S]} = \frac{J_{max}[S]}{[S] + K_M} \approx \frac{J_{max}}{K_M}[S] \approx \mathrm{LRS}\cdot[S] \qquad (10)$$

The involvement of K_M in defining the LRS is sometimes met with surprise, maybe because biosensor responses are far from linear at $[S] = K_M$. Nevertheless, the relationship, LRS $\approx J_{max}/K_M$ [Eq. (10)], is a very valuable one when designing biosensors for specific applications. For example, increasing enzyme loading is a common strategy in the quest for increased LRS. However, the experimentally measured LRS did not increase with J_{max} for Pt_D/GluOx/PPD biosensors because there was a corresponding increase in K_M(Glu) for this design due to electrostatic repulsion between the anionic glutamate analyte and the high polyanionic enzyme density on these disk surfaces (100). Electrostatic neutralization of this repulsion by incorporating polycationic polyethyleneimine (PEI) into the PEC matrix restored this valuable correlation [LRS $\propto J_{max}$ if K_M is constant; see Eq. (10)] by restoring K_M(Glu) values, allowing significant increases in LRS values (12). Furthermore, the quantitative validity of Eq. (10) was demonstrated by plotting individual values of the ratio J_{max}/K_M against the experimentally measured corresponding value of LRS for each member of a group of over fifty P*o*PD-based glutamate biosensors (93); Fig. 3 shows good agreement between Eq. (10) and experiment over an order of magnitude range of LRS.

4.1 Oxygen Dependence

In designing biosensors incorporating oxygen-dependent enzymes, the influence of pO_2 fluctuations on the biosensor response is a critical issue [see Reaction (2)]. The problem is quite complex because pO_2 varies both spatially and temporally in biological tissues. For example, there are pO_2 gradients with respect to distance from brain blood capillaries, values farthest from these being as low as 5 μM (132), although many biosensor designs, depending on their size, sample the average tissue O_2 concentration, estimated at ~50 μM (133, 134). pO_2 levels at a specific location also change with neuronal activity and drug administration (68, 119, 120, 135, 136). It is clear, therefore, that the O_2 dependence of biosensors designed to monitor analytes *in vivo* should be as low as possible. Turnover of O_2 in the enzyme layer [Reaction (2)] depends in part on the rate at which substrate binds to the enzyme [Reaction (1)], *i.e.*, on enzyme loading, substrate concentration, and the value of k_1 [Reaction (8)]. Both enzyme loading and the value of k_1 are tuneable to some extent in biosensor design (see below), and the concentration of substrate in the target medium must be considered in setting the limits of oxygen dependence.

Examples of the oxygen dependence of two Pt_D/GluOx/P*o*PD biosensors showing different glutamate sensitivity (here manifest as the response plateaux) are shown in Fig. 4 (top). The curve in each case represents the nonlinear regression analysis, using Eq. (7b). The smaller the value of $K_M(O_2)$, the lower the oxygen dependence because higher oxygen affinity leads to oxygen saturation at lower pO_2, thereby reducing biosensor dependency at higher pO_2 levels. Surprisingly, even for biosensors of the same geometry, size, and design, the device with the higher glutamate sensitivity, due to higher active enzyme loading (93), showed a lower oxygen dependence, $K_M(O_2)$, which appears to contradict the arguments above. To understand this phenomenon, $K_M(O_2)$ was plotted against $1/K_M$(Glu) (which served as an index of changes in k_1 (12, 47, 93)); see Fig. 4 (bottom). Firstly, regression analysis gave a reasonably good linear correlation, with an intercept on the $K_M(O_2)$ axis of effectively zero, consistent with the expectation that biosensor oxygen dependence should vanish as k_1 approaches zero. The positive correlation between $K_M(O_2)$ and $1/K_M$(Glu) is also expected because it indicates higher oxygen dependence with increasing k_1 values. The apparently anomalous behavior of the GluOx system, exemplified by the data shown in Fig. 4 (top) compared with Fig. 4 (bottom), stems from a phenomenon, discovered recently (93), of a major increase in K_M(Glu) with J_{max} and its consequence for the oxygen dependence of these biosensors. As GluOx loading was increased by adding additional enzyme dropwise, K_M(Glu) unexpectedly increased, due to a reduction in k_1 caused by the repulsion of the anionic glutamate substrate with the crowded polyanionic GluOx on the surface (12). This reduction in k_1 had the effect of reducing oxygen demand on a molecular (enzyme) level. Therefore, increased GluOx loading in these P*o*PD-based biosensors had dual beneficial effects of a modest increase in Glu sensitivity and a reduction in oxygen dependency in the LRS range of Glu concentrations (12, 93).

Others have used additional polymeric layers, such as polyurethane and Nafion deposited on the top of the enzyme-containing membrane (glucose oxidase, lactate oxidase, or glutamate oxidase), to hinder substrate binding in order to increase its K_M and extend the dynamic range of biosensors ($\sim\frac{1}{2}K_M$) up to 4 mM for glucose, 1.75 mM for lactate, and 0.25 mM for glutamate in one study (88). A Nafion film can thus reduce O_2 demand by the sensor, as well as increasing O_2 supply by acting as an O_2 reservoir (137), so a Nafion-coated glucose biosensor has been observed to be O_2 independent across a concentration range of 10–200 μM O_2 (138). The dependence of biosensor response on oxygen has also been limited by applying a combination of cellulose acetate and Nafion layers, which substantially reduced the glucose flux while maintaining a high diffusivity for oxygen. The ratio of substrate to oxygen was therefore reduced to such a level that the supply of oxygen to the enzyme layer is abundant, and the sensor worked reliably at concentration levels of ~15 μM O_2 (68).

Glucose is an example of an analyte with relatively high concentrations in brain ECF (compared to neurotransmitters, for example), and therefore is a useful model system for quantifying PEC sensitivity to changes in ECF pO_2. As predicted, the O_2 dependence of Pt/P*o*PD-based glucose biosensors became more acute at higher glucose concentrations (47, 119). For 10 mM glucose, a level not much greater than the normoglycemic value in human blood (139) and less than that observed under hyperglycemic diabetic conditions (140), the O_2 dependence was close to maximal (near-linear biosensor response to O_2 concentration) with a $K_M(O_2)$ value of ~150 μM (47). The Pt/P*o*PD-based glucose biosensor design would therefore not be suitable for monitoring blood glucose *in vivo*. In contrast, the glucose-response *vs.* O_2-concentration calibration plot for 0.5 mM glucose was hyperbolic, consistent with Eq. (7b), and displayed values of $K_M(O_2)$ in the range 5–10 μM (47). The 0.5 mM glucose current, therefore, showed a low sensitivity to changes in O_2 levels at average brain ECF levels of O_2 (~50 μM O_2). These *in-vitro* results suggest that Pt/PPD ~ GOx electrodes might be suitable for brain monitoring, where the estimated glucose level is much lower than in blood. Indeed, ECF glucose has been determined to be an order of magnitude lower than blood levels by both quantitative microdialysis (350 ± 16 μM (48)) and biosensor studies (300–400 μM (141)) in the striatum of the awake rat. The experimental oxygen dependence of PEC-based biosensors *in vivo* is discussed in Sect. 7 below.

4.2 The Importance of H_2O_2 Sensitivity

When the biosensor incorporates a component that significantly affects the sensitivity of the surface to H_2O_2, *e.g.*, Nafion or the lipid, phosphatidylethanolamine (PEA), then variations in J_{max} across sensor populations may be due to changes in either enzyme activity or H_2O_2 sensitivity. The activity of surface enzyme itself depends on two factors: the density of active enzyme molecules on the surface (active enzyme loading, $[E]_{act}$) and the value of the catalytic rate constant, k_2 [Reaction (8)]. Thus, enzyme activity is $k_2[E]_{act}$, and these two terms are rarely separable, especially for implantable microelectrodes where dip evaporation is used to deposit enzyme on the biosensor surface (*i.e.*, $[E]_{act}$ is not known independently). A useful way of monitoring changes in the effective activity of surface enzyme (across a population of related biosensors) is the parameter $[E]'_{act}$, defined in Eq. (11a), where slope(HP) is the calibration slope for hydrogen peroxide (HP) at the same biosensor. Variations in $[E]'_{act}$, determined under the same conditions, will reflect changes in $k_2[E]_{act}$, which in turn often arise from changes in $[E]_{act}$ (100).

$$[E]'_{act} = \frac{J_{max}}{\text{slope(HP)}} \quad (11a)$$

$$\text{BE\%} = \frac{\text{slope(S)}}{\text{slope(HP)}} \times 100\% \quad (11b)$$

An analogous, and possibly more useful, parameter which also normalizes the biosensor response with respect to H_2O_2 sensitivity is the biosensor efficiency (BE%); see Eq. (11b), where slope(S) is the LRS for enzyme substrate and slope(HP) is the calibration slope for H_2O_2 at the same biosensor. In theory, the absolute maximum value of BE% should be 100 % if there is enough active enzyme on the surface to generate sufficient H_2O_2 to mimic the flux of H_2O_2 to the surface in the H_2O_2 calibration, all the H_2O_2 produced by the enzyme is electro-oxidized on the surface, and the diffusion coefficients of S (D_S) and H_2O_2 were equal. Since D_S will almost invariably be less than that for H_2O_2 (~1.0×10^{-5} cm^2 s^{-1}) and some H_2O_2 is lost to the bulk even in quiescent solution (78), then $(\text{BE\%})_{max} < 100$ %. In other words, even a concentration of enzyme analyte in the linear response region will always produce a smaller biosensor signal than the same concentration of H_2O_2 for these H_2O_2-detecting devices [Reactions (1)–(3)]. An empirical maximum to date for BE% is close to 60 % (100), which is consistent with these factors.

4.3 Response Time and Limit of Detection

Two additional parameters related to biosensor response to enzyme substrate, S, are useful. Response times can be recorded in forced-convection environments, *e.g.*, a smoothly stirred solution, using a sufficiently high data-acquisition rate (42). To probe the response dynamics of the biosensor surface to enzyme substrate, the response time of the substrate should be compared to that of H_2O_2 injected in the same way (rate and location) into the electrochemical cell. A $t_{90\%}$ parameter has been defined as the time taken for the analyte response to reach 90 % of its maximum value from the start (<10 %) of the current upswing, and is similar to definitions used elsewhere (58, 64, 142). The limit of detection (LOD) of S is usually determined using the widely applied criterion of three times the standard deviation of the baseline current fluctuations. Response time and LOD are particularly crucial when designing biosensors to monitor fast transients in brain neurotransmitters whose ECF levels are invariably low, such as L-glutamate. PEC-based devices, such as Pt/GluOx/P*o*PD, show good behavior in these respects, with $t_{90\%}$ values of ~3 s (just 1 s slower than the response time to bulk H_2O_2) and an LOD of <0.5 μM glutamate (42, 90).

5 Permeability-Related Biosensor Parameters

The main analytical parameters of the surface polymer in PEC-based biosensors relate to (1) the sensitivity of the sensor to the target analyte and the enzyme-transducer molecule [H_2O_2; Reactions (1)–(3)] and (2) its ability to reject electroactive interference species. The starting point for understanding these properties is determining the sensitivity of H_2O_2 and ascorbic acid (AA, the archetypal interference molecule in brain ECF) at the bare metal. Firstly, the current *density* responses of AA at bare microdisks and cylinders are the same indicating that subtle differences in the geometry of diffusion profiles are not significant in this context; the same is true of H_2O_2 responses (117, 143). A direct comparison of the amperometric sensitivity of pure Pt and Pt–Ir has also been carried out (143, 144). A small loss of H_2O_2 sensitivity for the alloy compared with the pure Pt was observed, presumably due to the involvement of Pt oxides in H_2O_2 electro-oxidation (145). The greater sensitivity of both bare metal electrode types for H_2O_2 compared with AA is in line with diffusion coefficient values for the two analytes and diffusion-limited kinetics at the operational applied potential (0.7 V *vs.* SCE) (143).

$$2H_2O_2 \rightarrow 2H_2O + O_2 \tag{12}$$

Using a large population of wire electrodes, the hypothesis that ultrathin electro-deposited P*o*PD does not decrease the sensitivity of Pt–Ir to H_2O_2 has been tested (143). Surprisingly, the large sample size in that study revealed that P*o*PD-modified Pt–Ir showed a small, but statistically significant, *increase* in the H_2O_2 calibration slope (189 ± 10 mA cm^{-2} M^{-1}, $n = 51$) compared with the bare metal (171 ± 6 mA cm^{-2} M^{-1}, $n = 51$, $p = 0.05$ for a paired *t*-test). One possible explanation for the unexpected small increase in the efficiency of the electrochemical H_2O_2 oxidation on the coated metal is that the P*o*PD inhibits the H_2O_2 disproportionation reaction [Reaction (12)] at the electrode surface, the extent of which varies with electrode condition. It is clear, however, that the basic Pt–Ir/P*o*PD design displays near-optimal H_2O_2 sensitivity, so that the only effective way to improve selectivity significantly is to increase interference blocking by the polymer layer. The mechanically more robust Pt–Ir working electrodes are therefore used generally for implantable biosensors *in vivo*, and normally designated as Pt for simplicity of notation.

The transport of analytes from solution, through an insulating polymer coating on an electrode, to the underlying metal substrate is a complex process (146, 147). Parameters involved in membrane transport include the permeant's partition and diffusion coefficients in the polymer, as well as polymer thickness (148, 149). Because the thickness of the nonconducting form of P*o*PD is not known accurately, the "true permeability" of an analyte cannot be determined in a straightforward manner (148). Instead, the "apparent permeability" of an analyte has been defined [Eqs. (13) and (14)] as the ratio of the analyte currents recorded with the same electrode before and after deposition of the polymer. The apparent permeability is therefore a relative, normalized measure of the analyte flux to the metal surface when the polymer coating is present, and this parameter has proved useful in comparing the properties of permselective polymers for biosensor applications (1, 118). In line with the H_2O_2 calibration slope values discussed above, the apparent permeability of the PEC layer to H_2O_2 [Eq. (13)] is normally in the region of 100 ± 10 % for the basic Pt/Ox/P*o*PD designs (112, 117, 143).

$$P(\text{HP})\% = \frac{\text{slope(HP)at Pt/PEC}}{\text{slope(HP)at bare Pt}} \times 100\% \qquad (13)$$

The physicochemical basis of the efficient transport of H_2O_2 through P*o*PD coatings on electrode surfaces ($P(\text{HP})\% \approx 100$ %) has not been established, although this property is shared by P*m*PD and P*p*PD (104, 118), but not by other aromatic diamines (150). In contrast, for example, the equivalent $P(\text{HP})\%$ value for polyphenol has been estimated at <10 % (151). Where *P*-values do not vary far from 100 % (in contrast to $P(\text{AA})\%$ which can be as low as 0.1 %; see below), $P(\text{HP})\%$ may be not as much a permeability parameter as an index of the chemistry of this labile molecule at the electrode surface (152), *e.g.*, Reaction (12).

An unusual property of PPD-based PEC layers is their ability to reject AA more efficiently at higher, compared with lower, AA concentrations (Fig. 1). Indeed at extremely low levels of AA, the response of Pt/PPD electrodes is similar to bare Pt; however, even at 10 μM AA, the AA sensitivity (I_{AA}/[AA]) is considerably lower than at bare metal (112). Thus AA, and/or its oxidation products, trapped in the polymer matrix appears to cooperate efficiently with PPDs in a form of "self blocking" (103, 112), so that by concentrations in the hundred micromolar region the AA response is essentially flat, especially for P*o*PD and P*p*PD (Fig. 1). Because of this nonlinear behavior of AA at PPD-based Pt/PEC electrodes, the 1 mM value is often used in the corresponding definition of apparent AA permeability, $P(\text{AA})\%$ (Eq. (14)].

$$P(\text{AA})\% = \frac{I_{AA}(1\text{ mM})\text{ at Pt/PEC}}{I_{AA}(1\text{ mM})\text{ at bare Pt}} \times 100\% \qquad (14)$$

The main focus in using $P(\text{HP})\%$ for developing biosensor designs is to maintain its value as close to 100 % as possible (see above). The finding that values of $P(\text{HP})\%$ somewhat greater than 100 % have been observed (see Table 2) is a further indication that this is not a true permeability (84), possibly due to complications

Table 2
Permeability and selectivity values for both PEC-coated and enzyme-free PoPD-modified Pt_C electrodes

Design	*n*-value	*P*(HP)%	*P*(AA)%	LRS	*S*%	S_G%
Pt_C/P*o*PD	19	90 ± 5	0.11 ± 0.02	N/A[a]	0.09 ± 0.01	N/A[a]
Pt_C/GOx/P*o*PD	6	97 ± 7	0.34 ± 0.05	1.2 ± 0.1	0.20 ± 0.03	139 ± 36
Pt_C/P*o*PD/GOx	4	98 ± 2	0.51 ± 0.07	2.7 ± 0.1	0.29 ± 0.03	29 ± 8
Pt_C/GOx/ P*o*PD/GOx	4	127 ± 8	1.0 ± 0.1	1.7 ± 0.4	0.55 ± 0.07	108 ± 32
Pt_C/P*o*PD–GOx	8	120 ± 7	0.24 ± 0.04	5.0 ± 0.4	0.14 ± 0.02	7 ± 1

[a]Not applicable, because enzyme-free designs do not respond to glucose

Mean values ± SEM (*n* = number of electrodes) for the two apparent permeabilities, *P*(HP)% [Eq. (13)] and *P*(AA)% [Eq. (14)], the linear region slope (LRS in mA cm^{-2} M^{-1}) for glucose (see Fig. 1), and the two selectivity parameters, *S*% [Eq. (16)] and S_G% [Eq. (17)]. The electropolymerization solution contained 300 mM *o*PD in PBS; 5 mg mL^{-1} GOx was included for the co-immobilization design (Pt_C/P*o*PD–GOx). Adapted from (84)

such as Reaction (12). However, *P*(AA)% is very small (see Table 2) and probably reflects the poor ability of AA to permeate the PEC layer. The problems with interpreting *P*(HP)% also indicate that the permselectivity, S_P%, defined by Eq. (15), is likely not the most useful parameter for quantifying PEC selectivity (118).

$$S_P = \frac{P(\text{AA})\%}{P(\text{HP})\%} \times 100\% \qquad (15)$$

Instead, the equimolar selectivity, *S*%, defined by Eq. (16), is a good index of the relative sensitivities of the PEC-coated metal to the two analytes. Because *S*% represents a percentage interference, the smaller the value the more suitable the polymer is for biosensor applications, and values as low as ~0.2 % have been reported for P*o*PD-based devices (103, 143). The use of equimolar concentrations of HP and AA in Eq. (16) allows *S*% to be interpreted as the permselectivity of two analytes with an equal number of electrons transferred per molecule, as is the case here ($z = 2$). Although originally used to define the preferential permeation of ionic species through ion-exchange membranes (a phenomenon we might now call "ionic permselectivity"), the generic term "permselectivity" has recently been applied more generally to quantify the ratio of permeabilities of both charged and uncharged species through any membrane structure.

$$S\% = \frac{I_{\text{AA}}(1\text{mM}) \text{ at Pt/PEC}}{I_{\text{HP}}(1\text{mM}) \text{ at Pt/PEC}} \times 100\% \qquad (16)$$

However, the polymer composite can also affect the behavior of the trapped enzyme (47, 93). Thus, although *S*% is a good parameter for monitoring the effects of the polymer layer on relative H_2O_2 and AA sensitivities, and a good (low) value of *S*% is a prerequisite for a successful biosensor design, it is not sufficient. Having established a good *S*% value, it is necessary to quantify the effects of the polymer composite on enabling the production of H_2O_2 by the enzyme. The biosensor LRS for substrate is the best measure of the biosensor's analytical signal strength [see Eq. (10)], incorporating both its ability to generate H_2O_2 enzymatically and its sensitivity to this electroactive molecule [Reaction (3)]. Thus, we might define the equimolar substrate selectivity, S_S%, in terms of the LRS at 1 mM substrate by Eq. (17), which will be intrinsically larger (worse) than *S*% due to the factors discussed above for BE% [Eq. (11b)].

$$S_S\% = \frac{I_{AA}(1\,\text{mM})\ \text{at Pt/PEC}}{I_S(1\,\text{mM})\ \text{at Pt/PEC}} \times 100\% \qquad (17)$$

Finally, the substrate concentration in the ECF of the brain region of interest needs to be taken into account, and S_S(ECF)% defined as Eq. (18), where I_S(ECF) is the current for the estimated baseline concentration of target analyte, S, which is obtained by scaling the LRS. S_S(ECF)% is a measure of the relative contributions of AA and S to the baseline biosensor response recorded in the ECF. The value of I_{AA}(ECF) is close to that of I_{AA}(1 mM) used in other selectivity parameters above because of the high concentration of AA in the ECF of most brain areas and the self-limiting shape of the AA calibration profile (see Fig. 1).

$$S_S(\text{ECF})\% = \frac{I_{AA}(\text{ECF})\ \text{at Pt/PEC}}{I_S(\text{ECF})\ \text{at Pt/PEC}} \times 100\% \qquad (18)$$

A useful variation on S_S(ECF)% is $S_S(\text{ECF}_\Delta)$%, which is the substrate selectivity parameter based on doubling baseline ECF levels of both S and AA [Eq. (19)]. The significance of ΔI_{AA} is shown in Fig. 1, and I_S(ECF) has the same value as for Eq. (18) [*i.e.*, $I_S(\text{ECF}) = \Delta I_S(\text{ECF})$] because of the linearity of substrate response in this concentration range (see Fig. 2, inset). This latter parameter provides a measure of the percentage interference by AA in *changes* in the biosensor signal recorded in brain ECF, and therefore represents a key selectivity parameter for *in-vivo* applications.

$$S_S(\text{ECF}_\Delta)\% = \frac{\Delta I_{AA}\ \text{at Pt/PEC}}{I_S(\text{ECF})\ \text{at Pt/PEC}} \times 100\% \qquad (19)$$

It is not so important that some of these later selectivity coefficients generate larger values than the more common standard *S*% parameter. More critical than their absolute value is the evolution of the value of the selectivity parameter of choice with systematic fine tuning of the biosensor design.

6 Systematic Biosensor Design, Using Permeability and Selectivity Parameters

A large body of work on GOx-based biosensors incorporating an electrosynthesized PPD permselective layer for glucose detection has been published over the past two decades, and a limited selection is cited here (1, 26, 76–78, 81, 82, 88, 101, 137, 153–155). A broad range of variables are involved in the fabrication of these biosensors, including which of the three phenylenediamine monomers is used, the concentration of monomer, the background electrolyte and pH of the electropolymerization medium, the choice of cyclic voltammetry or fixed applied potential (and the value of applied potential) in the PPD electrosynthesis step, the size and shape of the electrode substrate, the mode of enzyme immobilization and its concentration, and the variety and concentration range of the interference species studied. Therefore, this system offers a good illustration of the usefulness of the sensitivity and selectivity parameters described here for systematic biosensor characterization in order to better understand the nature of the PEC layer (see Fig. 2) and to improve the performance of these biosensor devices.

The values for key parameters described above for three methods of GOx immobilization onto Pt–Ir cylinder electrodes are shown in Table 2 (84): dip evaporation before polymer electrosynthesis (Pt_C/GOx/P*o*PD), dip evaporation over the polymer using glutaraldehyde as a cross-linker (Pt_C/P*o*PD/GOx), and co-polymerization from the monomer solution containing enzyme (Pt_C/P*o*PD–GOx). Glucose calibrations for biosensors of all designs followed Michaelis–Menten hyperbolic behavior (Eq. (7a); see Fig. 2) and the enzyme-kinetic parameters have been discussed previously (84). The apparent HP permeability, *P*(HP)%, defined by Eq. (13), was similar for all designs, with an average value of 106 ± 8 % ($n = 41$) which was not significantly different from the ideal value of 100 % ($p > 0.42$). Therefore differences in the selectivity parameter, *S*% [Eq. (16)], across the designs should be influenced mainly by polymer interference rejection properties. Similarly, differences in the substrate selectivity parameter, S_S% [Eq. (17)], should be due to a combination of polymer interference rejection and the ability of the PEC layer to generate HP.

The apparent AA permeability, *P*(AA)%, calculated using Eq. (14), showed that all P*o*PD-modified designs blocked the 1 mM AA flux by ≥99 % compared to the bare metal, with *P*(AA)% ≤1 % (Table 2). The best blocking characteristics were displayed by Pt_C/P*o*PD. *i.e.*, by the enzyme-free P*o*PD, with *P*(AA)% = 0.11 ± 0.02 %, a value similar to that reported for P*o*PD deposited under the same conditions on low edge-density, pure Pt microfiber electrodes (95). The finding that *P*(AA)% for Pt_C/P*o*PD/GOx (0.34 ± 0.05 %) was significantly greater than that for the native polymer (Pt_C/P*o*PD, $p < 0.001$) suggests that the enzyme does not simply sit on top of the P*o*PD, but imbeds in the polymer, opening its structure and undermining its interference blocking to a small degree. Surprisingly, the GOx deposited before the P*o*PD had a similar effect on *P*(AA)% compared with the P*o*PD/GOx configuration. It is interesting to note that *P*(AA)% for the GOx/P*o*PD and P*o*PD/GOx configurations was additive when compared with Pt_C/GOx/P*o*PD/GOx, the worst AA-rejecting configuration (1.0 ± 0.1 %). This indicates that successive fabrication steps (in this case, GOx deposited both before and after the P*o*PD) perturb the polymer, compounding the slight undermining of its interference blocking properties. Finally, *P*(AA)% for the co-immobilized enzyme (GOx–P*o*PD, 0.24 ± 0.04 %), which had the best glucose LRS value (84), was second only to pure P*o*PD in terms of AA rejection (99.76 %) for these basic designs.

As expected from the corresponding definitions and the relatively constant value of *P*(HP)% across the biosensor designs described here (Table 2), the trend in the permselectivity parameter (*S*%), calculated using Eq. (16), paralleled that of *P*(AA)%. It is the equimolar substrate selectivity (S_G%, which is S_S% for glucose defined by Eq. 17), which most clearly reveals the superiority of the co-immobilized configuration among these basic designs (Table 2). The mean S_G% value for Pt_C/P*o*PD–GOx (7 ± 1 %, $n = 8$) was between 4 and 50 times smaller (better) than that for the other designs. Thus, in media containing equal concentrations of glucose and AA at about the 1-mM level, as is the case in brain extracellular fluid (48, 156), the baseline signal of the Pt_C/P*o*PD–GOx biosensor would have a ~7 % interference contribution from AA. However, because *changes* in the biosensor signal are far more important than their absolute output, interference in monitoring glucose changes would be considerably less than this value due to the self-limiting shape of the AA response (see ΔI_{AA} in Fig. 1).

6.1 Fabrication Protocol for the Pt_C/PoPD–GOx Design

Preparation of the co-immobilized configuration of cylinder biosensors involved the following steps. An ~5-cm length of Teflon-coated Pt–Ir wire (90:10, 125 μm diameter, Advent Research Materials Ltd., Eynsham, England) was stripped of 1 mm Teflon from both ends to expose the bare metal. After soldering one end of the wire to a gold contact, electropolymerization was carried out at +0.7 V *vs.* SCE for 15 min in phosphate-buffered saline (PBS, pH 7.4) containing 300 mM *o*PD and 5 mg mL^{-1} (~3 kU mL^{-1}) GOx. The Pt_C/P*o*PD–GOx biosensors thus formed were rinsed in deionized water to remove any unreacted *o*PD which has been classed as a possible carcinogen. (Extreme care should be exercised when working with *o*PD, whose other risk factors include irritation to eyes, skin, and lungs; controlled disposal procedures for *o*PD waste should also be in place.) The biosensors were then placed in fresh PBS background electrolyte, reconnected to the potentiostat (using a different SCE to avoid contamination with *o*PD), and settled at +0.7 V for several hours (usually overnight to provide very low, stable background currents, and also to model brain implantation where the biosensors are implanted the day before recording commences).

7 Brain Monitoring with PEC-Based Biosensors *In Vivo*

The high baseline levels of glucose in brain ECF (~0.5 mM (48)), coupled with the excellent permselectivity of P*o*PD-based biosensors, have made it possible to monitor physiologically and pharmacologically related concentration dynamics of this energy metabolite *in vivo* (48, 78, 88, 111, 119, 120). Many of these studies used devices of the type Pt/P*o*PD–GOx because these can be fabricated in a one-step procedure by incorporating enzyme in the monomer solution, in contrast to the time-consuming multiple dip-evaporation protocol for Pt/GOx/P*o*PD sensors, and because their permselectivity is superior (see Table 2).

As outlined in Sect. 2, a main advantage of P*o*PD-based glucose biosensors is their stability, even during amperometric recording over several days in freely behaving animals (48, 78, 119, 120), allowing long-term *in-vivo* electrochemical (LIVE) monitoring. Figure 5 shows the response of a Pt/P*o*PD–GOx biosensor, which had been implanted in rat brain striatum 12 days previously, to the systemic administration of insulin. The inset (Fig. 5) shows that this hormone produced a sustained depression in brain ECF glucose levels which could be reversed by the ingestion of food (access to food had been restricted for the previous 12 h). The effects of insulin on brain ECF glucose have been confirmed recently using a PPD-based biosensor incorporating a ruthenium layer (88), although the baseline currents observed *in vivo* with these devices were not as stable as those recorded with the Ru-free PEC design (Fig. 5).

In addition to the PEC layer maintaining its sensitivity to brain glucose, the permselective membrane was also capable of rejecting AA interference after long periods implanted in the tissue. Figure 5 also shows that systemic administration of AA, at a dose which doubled brain ECF AA levels (detected with a co-implanted carbon paste electrode, CPE), had no significant effect on the Pt/P*o*PD–GOx biosensor signal. A similar combination of bilaterally implanted biosensor and CPE has been used to confirm the *in-vitro* findings that, for physiologically relevant ECF concentrations of both glucose and oxygen, oxygen demand does not limit biosensor functionality (47, 119).

Taken together, the literature reports on Pt/P*o*PD–GOx indicate that these biosensors possess sufficient glucose sensitivity, selectivity, and stability to monitor this energy metabolite reliably in brain ECF over extended periods. The challenge remains, however,

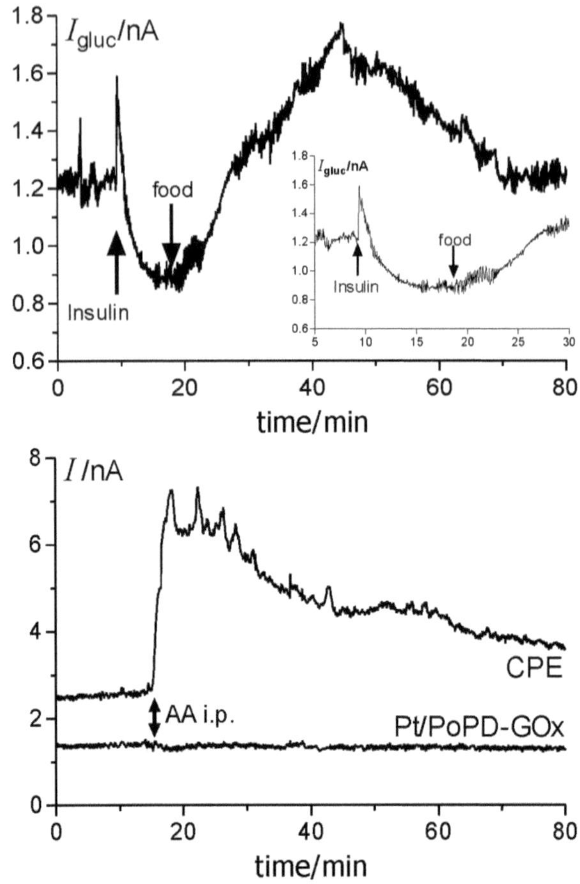

Fig. 5 *Top*: The response of a Pt/P*o*PD–GOx biosensor implanted in the striatum of a freely moving fasting rat recorded at 0.7 V *vs*. Ag-wire reference electrode over an 80-min period, 12 days after sensor implantation. Injection of insulin (15 U/kg, i.p.) caused a decrease in the signal which was reversed by the ingestion of food (see *Inset*). *Bottom*: Effect of ascorbate injection (2 g/kg, i.p.) on the amperometric response of a carbon paste electrode (CPE, held at 0.2 V) and a Pt/P*o*PD–GOx biosensor (held at 0.7 V) 8 days after bilateral implantation in rat dorsal striatum. Reprinted with permission from Anal Chem 1994;66:1754–61 (78). Copyright 1994 American Chemical Society

as to how to boost PEC performance even further to allow the analogous Pt/P*o*PD~GluOx devices to monitor brain ECF glutamate dynamics, at levels two orders of magnitude lower than those of glucose. The incorporation of the polycation PEI into the PEC design has provided an important step in that direction by enhancing the glutamate sensitivity significantly (12, 95, 100). Preliminary results *in vivo* with these PEI-modified biosensors indicate that long-term glutamate monitoring may soon be possible with additional fine turning of this design (54).

Further significant steps towards LIVE monitoring of neurotransmitter dynamics, using PPD-based biosensors, have been reported recently. Miniaturization of cylindrical electrodes by an order of magnitude (~10 μm diameter) has been achieved using both carbon fiber (71, 85, 88) and Pt wire (29, 95) substrates. In addition to reduced tissue damage, these microelectrodes also provide improved spatial and temporal resolution which is critical for neurotransmitter monitoring (96). Lately, the use of Prussian Blue to reduce H_2O_2 electrocatalytically at ~0 V *vs.* SCE has been combined with the permselective properties of P*o*PD to enhance further the interference rejection ability of the PEC layer (71, 85, 96). An additional major development in the application of LIVE technologies to investigate the relationship between neurochemical dynamics and overt behavior is the design of telemetric electronics (83, 157–159), which allow signals from implanted biosensors in completely untethered animals to be monitored during normal movement (26).

A major effort is also under way to exploit the advantages of biosensor devices for clinical applications (160–162). Clinical diagnosis routinely requires the assessment of species determined by chemical analysis, and biosensor technology offers several benefits over conventional diagnostic analysis, including simplicity of use, speedy results, capability for continuous monitoring, and coupling to low-cost, portable instrumentation (161, 163). However, because of the ethical issues associated with the implantation of electrochemical sensors into human tissue, as well as possible long-term effects of some of the components in biosensor devices, most of these studies have focused on *ex-vivo* analysis of clinical samples. One interesting example of this approach is the use of P*o*PD-based biosensors for analysis of brain microdialysates, which has been demonstrated for online monitoring of glucose and lactate in neurointensive care patients (164). Indeed, biosensor and microdialysis sampling technologies have been combined in a number of formats (25, 60, 165–168), including insertion of the biosensor into the microdialysis membrane (165, 166, 169–171), and P*o*PD has found a use as a permselective barrier in many of these approaches (142, 166, 172–175). This latter form has the advantage that it improves the biocompatibility of the combined probe by protecting the target tissue from the biosensor components and reactions, as well as providing protection to the biosensor from tissue reactions triggered by the presence of a foreign implant (40, 43, 176–181). The downside to this approach is that some of the intrinsic benefits of directly implantable biosensors are lost, including miniaturization to micron levels and sub-second time resolution of concentration dynamics in the extracellular space.

Acknowledgements

This work was supported by funding from Science Foundation Ireland (SFI), the Irish Research Council for Science, Engineering and Technology (IRCSET), and by University College Dublin (UCD).

References

1. O'Neill RD, Lowry JP, Rocchitta G, McMahon CP, Serra PA (2008) Designing sensitive and selective polymer/enzyme composite biosensors for brain monitoring *in vivo*. Trends Anal Chem 27:78–88
2. Wang J (2008) Electrochemical glucose biosensors. Chem Rev 108:814–825
3. Wilson GS, Johnson MA (2008) In-vivo electrochemistry: what can we learn about living systems? Chem Rev 108:2462–2481
4. Dale N, Hatz S, Tian FM, Llaudet E (2005) Listening to the brain: microelectrode biosensors for neurochemicals. Trends Biotechnol 23:420–428
5. Hirst ER, Yuan YJ, Xu WL, Bronlund JE (2008) Bond-rupture immunosensors—a review. Biosens Bioelectron 23:1759–1768
6. Wanekaya AK, Chen W, Mulchandani A (2008) Recent biosensing developments in environmental security. J Environ Monit 10:703–712
7. Sadik OA, Aluoch AO, Zhou AL (2009) Status of biomolecular recognition using electrochemical techniques. Biosens Bioelectron 24:2749–2765
8. Sozer N, Kokini JL (2009) Nanotechnology and its applications in the food sector. Trends Biotechnol 27:82–89
9. Castillo J, Gaspar S, Leth S, Niculescu M, Mortari A, Bontidean I, Soukharev V, Dorneanu SA, Ryabov AD, Csoregi E (2004) Biosensors for life quality—design, development and applications. Sens Actuators B Chem 102:179–194
10. Wei D, Bailey MJA, Andrew P, Ryhanen T (2009) Electrochemical biosensors at the nanoscale. Lab Chip 9:2123–2131
11. Lu QZ, Lin HL, Ge ST, Luo SL, Cai QY, Grimes CA (2009) Wireless, remote-query, and high sensitivity *Escherichia coli* O157:H7 biosensor based on the recognition action of Concanavalin A. Anal Chem 81:5846–5850
12. McMahon CP, Rocchitta G, Kirwan SM, Killoran SJ, Serra PA, Lowry JP, O'Neill RD (2007) Oxygen tolerance of an implantable polymer/enzyme composite glutamate biosensor displaying polycation-enhanced substrate sensitivity. Biosens Bioelectron 22:1466–1473
13. Schuvailo OM, Gaspar S, Soldatkin AP, Csoregi E (2007) Ultramicrobiosensor for the selective detection of glutamate. Electroanalysis 19:71–78
14. Lowry JP, O'Neill RD (2006) Neuroanalytical chemistry *in vivo* using electrochemical sensors. In: Grimes CA, Dickey EC, Pishko MV (eds) Encyclopedia of sensors. American Scientific, Los Angeles, CA, pp 501–524
15. Wilson GS, Gifford R (2005) Biosensors for real-time in vivo measurements. Biosens Bioelectron 20:2388–2403
16. O'Neill RD, Lowry JP, Mas M (1998) Monitoring brain chemistry *in vivo*: voltammetric techniques, sensors and behavioral applications. Crit Rev Neurobiol 12:69–127
17. Pantano P, Kuhr WG (1995) Enzyme-modified microelectrodes for in vivo neurochemical measurements. Electroanalysis 7:405–416
18. Watson CJ, Venton BJ, Kennedy RT (2006) In vivo measurements of neurotransmitters by microdialysis sampling. Anal Chem 78:1391–1399
19. Lee GJ, Park JH, Park HK (2008) Microdialysis applications in neuroscience. Neurol Res 30:661–668
20. Baldini F (2010) Microdialysis-based sensing in clinical applications. Anal Bioanal Chem 397:909–916
21. Duff A, O'Neill RD (1994) Effect of probe size on the concentration of brain extracellular uric acid monitored with carbon paste electrodes. J Neurochem 62:1496–1502
22. O'Neill RD, Gonzalez-Mora JL, Boutelle MG, Ormonde DE, Lowry JP, Duff A, Fumero B, Fillenz M, Mas M (1991) Anomalously high concentrations of brain extracellular uric acid detected with chronically implanted probes: implications for *in vivo* sampling techniques. J Neurochem 57:22–29
23. Peters JL, Miner LH, Michael AC, Sesack SR (2004) Ultrastructure at carbon fiber microelectrode implantation sites after acute

voltammetric measurements in the striatum of anesthetized rats. J Neurosci Methods 137:9–23

24. Rutherford EC, Pomerleau F, Huettl P, Stromberg I, Gerhardt GA (2007) Chronic second-by-second measures of L-glutamate in the central nervous system of freely moving rats. J Neurochem 102:712–722
25. John J, Ramanathan L, Siegel JM (2008) Rapid changes in glutamate levels in the posterior hypothalamus across sleep-wake states in freely behaving rats. Am J Physiol Regul Integr Comp Physiol 295:R2041–R2049
26. Calia G, Rocchitta G, Migheli R, Puggioni GM, Spissu Y, Bazzu G, Mazzarello V, Lowry JP, O'Neill RD, Desole MS, Serra PA (2009) Biotelemetric monitoring of brain neurochemistry in conscious rats, using microsensors and biosensors. Sensors (Basel) 9:2511–2523
27. Dash MB, Douglas CL, Vyazovskiy VV, Cirelli C, Tononi G (2009) Long-term homeostasis of extracellular glutamate in the rat cerebral cortex across sleep and waking states. J Neurosci 29:580–589
28. Tian FM, Gourine AV, Huckstepp RTR, Dale N (2009) A microelectrode biosensor for real time monitoring of L-glutamate release. Anal Chim Acta 645:86–91
29. Pernot P, Mothet JP, Schuvailo O, Soldatkin A, Pollegioni L, Pilone M, Adeline MT, Cespuglio R, Marinesco S (2008) Characterization of a yeast D-amino acid oxidase microbiosensor for D-serine detection in the central nervous system. Anal Chem 80:1589–1597
30. Burmeister JJ, Gerhardt GA (2003) Ceramic-based multisite microelectrode arrays for in vivo electrochemical recordings of glutamate and other neurochemicals. Trends Anal Chem 22:498–502
31. Lowry JP, Ryan MR, O'Neill RD (1998) Behaviourally induced changes in extracellular levels of brain glutamate monitored at 1 s resolution with an implanted biosensor. Anal Commun 35:87–89
32. O'Neill RD (1994) Microvoltammetric techniques and sensors for monitoring neurochemical dynamics in vivo—a review. Analyst 119:767–779
33. Hirano A, Sugawara M (2006) Receptors and enzymes for medical sensing of L-glutamate. Mini Rev Med Chem 6:1091–1100
34. Hamdi N, Wang JJ, Walker E, Maidment NT, Monbouquette HG (2006) An electroenzymatic L-glutamate microbiosensor selective against dopamine. J Electroanal Chem 591:33–40
35. Chakraborty S, Raj CR (2007) Amperometric biosensing of glutamate using carbon nanotube based electrode. Electrochem Commun 9:1323–1330
36. Ahuja T, Mir IA, Kumar D, Rajesh (2007) Biomolecular immobilization on conducting polymers for biosensing applications. Biomaterials 28:791–805
37. Lojou E, Bianco P (2006) Application of the electrochemical concepts and techniques to amperometric biosensor devices. J Electroceram 16:79–91
38. Wilson GS, Hu YB (2000) Enzyme based biosensors for in vivo measurements. Chem Rev 100:2693–2704
39. Rodriguez-Mozaz S, de Alda MJL, Barcelo D (2006) Biosensors as useful tools for environmental analysis and monitoring. Anal Bioanal Chem 386:1025–1041
40. O'Neill RD (1993) Sensor–tissue interactions in neurochemical analysis with carbon paste electrodes *in vivo*. Analyst 118:433–438
41. Ahmad F, Christenson A, Bainbridge M, Yusof APM, Ab Ghani S (2007) Minimizing tissue–material interaction in microsensor for subcutaneous glucose monitoring. Biosens Bioelectron 22:1625–1632
42. McMahon CP, O'Neill RD (2005) Polymer–enzyme composite biosensor with high glutamate sensitivity and low oxygen dependence. Anal Chem 77:1196–1199
43. Koschwanez HE, Reichert WM (2007) In vitro, in vivo and post explantation testing of glucose-detecting biosensors: current methods and recommendations. Biomaterials 28:3687–3703
44. Mitala JJ, Michael AC (2006) Improving the performance of electrochemical microsensors based on enzymes entrapped in a redox hydrogel. Anal Chim Acta 556:326–332
45. McAteer K, O'Neill RD (1996) Strategies for decreasing ascorbate interference at glucose oxidase-modified poly(o-phenylenediamine)-coated electrodes. Analyst 121:773–777
46. Lin L, Guthrie JT (2006) Novel oxygen-enhanced membrane assemblies for biosensors. J Memb Sci 278:173–180
47. McMahon CP, Killoran SJ, O'Neill RD (2005) Design variations of a polymer–enzyme composite biosensor for glucose: enhanced analyte sensitivity without increased oxygen dependence. J Electroanal Chem 580:193–202
48. Lowry JP, O'Neill RD, Boutelle MG, Fillenz M (1998) Continuous monitoring of extracellular glucose concentrations in the striatum of freely moving rats with an implanted glucose biosensor. J Neurochem 70:391–396

49. Boutelle MG, Stanford C, Fillenz M, Albery WJ, Bartlett PN (1986) An amperometric enzyme electrode for monitoring brain glucose in the freely moving rat. Neurosci Lett 72:283–288
50. Xin Q, Wightman RM (1997) Enzyme modified amperometric sensors for choline and acetylcholine with tetrathiafulvalene tetracyanoquinodimethane as the electron-transfer mediator. Anal Chim Acta 341:43–51
51. Jung CC, Hall EAH (1995) Impedance analysis of NMP–TCNQ at glucose oxidase interfaces. Anal Chem 67:2393–2400
52. Centonze D, Losito I, Malitesta C, Palmisano F, Zambonin PG (1997) Electrochemical immobilisation of enzymes on conducting organic salt electrodes: characterisation of an oxygen independent and interference-free glucose biosensor. J Electroanal Chem 435:103–111
53. Lowry JP, O'Neill RD (1992) Strategies for reducing ascorbate interference at glucose oxidase modified conducting organic salt electrodes. J Electroanal Chem 334:183–194
54. Rocchitta G, McMahon CP, Serra PA, Kirwan SM, Bolger FB, Lowry JP, O'Neill RD (2006) Significant enhancement of glutamate biosensor sensitivity, using a polycation-modified polymer/enzyme composite sensing layer. In: Di Chiara G, Carboni E, Valentini V, Acquas E, Bassareo V, Cadoni C (eds) Monitoring molecules in neuroscience. University of Cagliari, Cagliari, pp 331–333
55. Oldenziel WH, Dijkstra G, Cremers TIFH, Westerink BHC (2006) In vivo monitoring of extracellular glutamate in the brain with a microsensor. Brain Res 1118:34–42
56. Garguilo MG, Michael AC (1996) Amperometric microsensors for monitoring choline in the extracellular fluid of brain. J Neurosci Methods 70:73–82
57. Hu Y, Mitchell KM, Albahadily FN, Michaelis EK, Wilson GS (1994) Direct measurement of glutamate release in the brain using a dual enzyme-based electrochemical sensor. Brain Res 659:117–125
58. Kulagina NV, Shankar L, Michael AC (1999) Monitoring glutamate and ascorbate in the extracellular space of brain tissue with electrochemical microsensors. Anal Chem 71:5093–5100
59. Oka T, Tasaki C, Sezaki H, Sugawara M (2007) Implantation of a glass capillary-based enzyme electrode in mouse hippocampal slices for monitoring of L-glutamate release. Anal Bioanal Chem 388:1673–1679
60. Oldenziel WH, Dijkstra G, Cremers TIFH, Westerink BHC (2006) Evaluation of hydrogel-coated glutamate microsensors. Anal Chem 78:3366–3378
61. Kulagina NV, Michael AC (2003) Monitoring hydrogen peroxide in the extracellular space of the brain with amperometric microsensors. Anal Chem 75:4875–4881
62. Schmidtke DW, Heller A (1998) Accuracy of the one-point in vivo calibration of "wired" glucose oxidase electrodes implanted in jugular veins of rats in periods of rapid rise and decline of the glucose concentration. Anal Chem 70:2149–2155
63. O'Neill RD, Fillenz M, Sundstrom L, Rawlins JNP (1984) Voltammetrically monitored brain ascorbate as an index of excitatory amino acid release in the unrestrained rat. Neurosci Lett 52:227–233
64. Burmeister JJ, Palmer M, Gerhardt GA (2003) Ceramic-based multisite microelectrode array for rapid choline measures in brain tissue. Anal Chim Acta 481:65–74
65. Konradsson-Geuken A, Gash CR, Alexander K, Pomerleau F, Huettl P, Gerhardt GA, Bruno JP (2009) Second-by-second analysis of alpha 7 nicotine receptor regulation of glutamate release in the prefrontal cortex of awake rats. Synapse 63:1069–1082
66. Hascup KN, Hascup ER, Pomerleau F, Huettl P, Gerhardt GA (2008) Second-by-second measures of L-glutamate in the prefrontal cortex and striatum of freely moving mice. J Pharmacol Exp Ther 324:725–731
67. Parikh V, Kozak R, Martinez V, Sarter M (2007) Prefrontal acetylcholine release controls cue detection on multiple timescales. Neuron 56:141–154
68. Hu YB, Wilson GS (1997) Rapid changes in local extracellular rat brain glucose observed with an in vivo glucose sensor. J Neurochem 68:1745–1752
69. Hu YB, Wilson GS (1997) A temporary local energy pool coupled to neuronal activity: fluctuations of extracellular lactate levels in rat brain monitored with rapid-response enzyme-based sensor. J Neurochem 69:1484–1490
70. Walker E, Wang J, Hamdi N, Monbouquette HG, Maidment NT (2007) Selective detection of extracellular glutamate in brain tissue using microelectrode arrays coated with over-oxidized polypyrrole. Analyst 132:1107–1111
71. Salazar P, Martin M, Roche R, Gonzalez-Mora JL, O'Neill RD (2010) Microbiosensors for glucose based on Prussian blue modified carbon fiber electrodes for *in vivo* monitoring in the central nervous system. Biosens Bioelectron 26:748–753
72. Hocevar SB, Zivin M, Milutinovic A, Hawlina M, Hutton EA, Ogorevc B (2006) Simultaneous

in vivo measurement of dopamine, serotonin and ascorbate in the striatum of experimental rats using voltammetric microprobe. Front Biosci 11:2782–2789
73. Patel BA, Arundell M, Parker KH, Yeoman MS, O'Hare D (2006) Detection of nitric oxide release from single neurons in the pond snail, Lymnaea stagnalis. Anal Chem 78:7643–7648
74. Ferreira NR, Ledo A, Frade JG, Gerhardt GA, Laranjinha J, Barbosa RM (2005) Electrochemical measurement of endogenously produced nitric oxide in brain slices using Nafion/o-phenylenediamine modified carbon fiber microelectrodes. Anal Chim Acta 535:1–7
75. Fillenz M, Lowry JP (1998) The relation between local cerebral blood flow and extracellular glucose concentration in rat striatum. Exp Physiol 83:233–238
76. Malitesta C, Palmisano F, Torsi L, Zambonin PG (1990) Glucose fast-response amperometric sensor based on glucose oxidase immobilized in an electropolymerized poly(o-phenylenediamine) film. Anal Chem 62:2735–2740
77. Sasso SV, Pierce RJ, Walla R, Yacynych AM (1990) Electropolymerized 1,2-diaminobenzene as a means to prevent interferences and fouling and to stabilize immobilized enzyme in electrochemical biosensors. Anal Chem 62:1111–1117
78. Lowry JP, McAteer K, El Atrash SS, Duff A, O'Neill RD (1994) Characterization of glucoseoxidase-modifiedpoly(phenylenediamine)-coated electrodes *in vitro* and *in vivo*: homogeneous interference by ascorbic acid in hydrogen peroxide detection. Anal Chem 66:1754–1761
79. Palmisano F, Centonze D, Malitesta C, Zambonin PG (1995) Electrochemical immobilization of enzymes on conducting organic salt electrodes: preparation of an oxygen independent and interference-free glucose biosensor. J Electroanal Chem 381:235–237
80. Wang J, Chen L, Liu J, Lu F (1996) Enhanced selectivity and sensitivity of first-generation enzyme electrodes based on the coupling of rhodinized carbon paste transducers and permselective poly(o-phenylenediamine) coatings. Electroanalysis 8:1127–1130
81. Bartlett PN, Wang JH, James W (1998) Measurement of low glucose concentrations using a microelectrochemical enzyme transistor. Analyst 123:387–392
82. Reyes De Corcuera JI, Cavalieri RP, Powers JR (2005) Improved platinization conditions produce a 60-fold increase in sensitivity of amperometric biosensors using glucose oxidase immobilized in poly-o-phenylenediamine. J Electroanal Chem 575:229–241
83. Serra PA, Rocchitta G, Bazzu G, Manca A, Puggioni GM, Lowry JP, O'Neill RD (2007) Design and construction of a low cost single-supply embedded telemetry system for amperometric biosensor applications. Sens Actuators B 122:118–126
84. Rothwell SA, Killoran SJ, O'Neill RD (2010) Enzyme immobilization strategies and electropolymerization conditions to control sensitivity and selectivity parameters of a polymer–enzyme composite glucose biosensor. Sensors 10:6439–6462
85. Salazar P, Martin M, Roche R, O'Neill RD, Gonzalez-Mora JL (2010) Prussian Blue-modified microelectrodes for selective transduction in enzyme-based amperometric microbiosensors for *in vivo* neurochemical monitoring. Electrochim Acta 55:6476–6484
86. Dempsey E, Wang J, Smyth MR (1993) Electropolymerised o-phenylenediamine film as means of immobilising lactate oxidase for a L-lactate biosensor. Talanta 40:445–451
87. Palmisano F, Centonze D, Zambonin PG (1994) An in situ electrosynthesized amperometric biosensor based on lactate oxidase immobilized in a poly-o-phenylenediamine film: determination of lactate in serum by flow injection analysis. Biosens Bioelectron 9:471–479
88. Schuvailo OM, Soldatkin OO, Lefebvre A, Cespuglio R, Soldatkin AP (2006) Highly selective microbiosensors for in vivo measurement of glucose, lactate and glutamate. Anal Chim Acta 573:110–116
89. Cooper JM, Foreman PL, Glidle A, Ling TW, Pritchard DJ (1995) Glutamate oxidase enzyme electrodes: microsensors for neurotransmitter determination using electrochemically polymerized permselective films. J Electroanal Chem 388:143–149
90. Ryan MR, Lowry JP, O'Neill RD (1997) Biosensor for neurotransmitter L-glutamic acid designed for efficient use of L-glutamate oxidase and effective rejection of interference. Analyst 122:1419–1424
91. Yao T, Suzuki S, Nakahara T, Nishino H (1998) Highly sensitive detection of L-glutamate by on-line amperometric micro-flow analysis based on enzymatic substrate recycling. Talanta 45:917–923
92. O'Neill RD, Chang SC, Lowry JP, McNeil CJ (2004) Comparisons of platinum, gold, palladium and glassy carbon as electrode materials in the design of biosensors for glutamate. Biosens Bioelectron 19:1521–1528

93. McMahon CP, Rocchitta G, Serra PA, Kirwan SM, Lowry JP, O'Neill RD (2006) Control of the oxygen dependence of an implantable polymer/enzyme composite biosensor for glutamate. Anal Chem 78:2352–2359
94. Govindarajan S, McNeil CJ (2009) Decreasing the polymerization potential improves the selectivity of PPD-coated disc biosensors for glutamate. Anal Lett 42:739–745
95. Rothwell SA, Kinsella ME, Zain ZM, Serra PA, Rocchitta G, Lowry JP, O'Neill RD (2009) Contributions by a novel edge effect to the permselectivity of an electrosynthesized polymer for microbiosensor applications. Anal Chem 81:3911–3918
96. Salazar P, Martin M, Roche R, Gonzalez-Mora JL, O'Neill RD (2010) Glutamate biosensor based on Prussian Blue modified carbon fiber microelectrodes for in-vivo neurochemical monitoring. In: Westerink B, Clinckers R, Smolders I, Sarre S, Michotte Y (eds) Monitoring molecules in neuroscience. Vrije Universiteit Brussel, Brussels, pp 292–294
97. Yao T, Suzuki S, Nishino H, Nakahara T (1996) On-line amperometric assay of glucose, L-glutamate, and acetylcholine using microdialysis probes and immobilized enzyme reactors. Electroanalysis 7:1114–1117
98. Zain ZM, O'Neill RD, Lowry JP, Pierce KW, Tricklebank M, Dewa A, Ab Ghani S (2010) Development of an implantable D-serine biosensor for in vivo monitoring using mammalian D-amino acid oxidase on a poly(o-phenylenediamine) and Nafion-modified platinum–iridium disk electrode. Biosens Bioelectron 25:1454–1459
99. O'Brien KB, Killoran SJ, O'Neill RD, Lowry JP (2006) Development of a catalase based biosensor for the detection of endogenous hydrogen peroxide in brain extracellular fluid. In: Di Chiara G, Carboni E, Valentini V, Acquas E, Bassareo V, Cadoni C (eds) Monitoring molecules in neuroscience. University of Cagliari, Cagliari, pp 325–327
100. McMahon CP, Rocchitta G, Serra PA, Kirwan SM, Lowry JP, O'Neill RD (2006) The efficiency of immobilised glutamate oxidase decreases with surface enzyme loading: an electrostatic effect, and reversal by a polycation significantly enhances biosensor sensitivity. Analyst 131:68–72
101. Myler S, Eaton S, Higson SPJ (1997) Poly(o-phenylenediamine) ultra-thin polymer-film composite membranes for enzyme electrodes. Anal Chim Acta 357:55–61
102. Ohnuki Y, Matsuda H, Ohsaka T, Oyama N (1983) Permselectivity of films prepared by electrochemical oxidation of phenol and amino-aromatic compounds. J Electroanal Chem 158:55–67
103. Craig JD, O'Neill RD (2003) Comparison of simple aromatic amines for electrosynthesis of permselective polymers in biosensor fabrication. Analyst 128:905–911
104. Dai YQ, Zhou DM, Shiu KK (2006) Permeability and permselectivity of polyphenylenediamine films synthesized at a palladium disk electrode. Electrochim Acta 52:297–303
105. Hamdi N, Wang JJ, Monbouquette HG (2005) Polymer films as permselective coatings for H2O2-sensing electrodes. J Electroanal Chem 581:258–264
106. Mitchell KM (2004) Acetylcholine and choline amperometric enzyme sensors characterized in vitro and in vivo. Anal Chem 76:1098–1106
107. Wang J, Chen L, Luo DB (1997) Electrocatalytic detection of hydrogen peroxide at a poly(m-phenylenediamine)-modified carbon paste electrode and its use for biosensing of glucose. Anal Commun 34:217–219
108. Rhemrev-Boom MM, Jonker MA, Venema K, Jobst G, Tiessen R, Korf J (2001) On-line continuous monitoring of glucose or lactate by ultraslow microdialysis combined with a flow-through nanoliter biosensor based on poly(m-phenylenediamine) ultra-thin polymer membrane as enzyme electrode. Analyst 126:1073–1079
109. Yang QL, Atanasov P, Wilkins E (1999) Needle-type lactate biosensor. Biosens Bioelectron 14:203–210
110. Yang QL, Atanasov P, Wilkins E (1997) A novel amperometric transducer design for needle-type implantable biosensor applications. Electroanalysis 9:1252–1256
111. Netchiporouk LI, Shram NF, Jaffrezic-Renault N, Martelet C, Cespuglio R (1996) In vivo brain glucose measurements: differential normal pulse voltammetry with enzyme-modified carbon fiber microelectrodes. Anal Chem 68:4358–4364
112. Lowry JP, O'Neill RD (1994) Partial characterization *in vitro* of glucose oxidase-modified poly(phenylenediamine)-coated electrodes for neurochemical analysis *in vivo*. Electroanalysis 6:369–379
113. Ekinci E, Ogunc ST, Karagozler AE (1998) Electrochemical poly(1,3-phenylenediamine) synthesis as enzyme immobilization media. J Appl Polym Sci 68:145–152
114. Soldatkin OO, Schuvailo OM, Marinesco S, Cespuglio R, Soldatkin AR (2009)

Microbiosensor based on glucose oxidase and hexokinase co-immobilised on platinum microelectrode for selective ATP detection. Talanta 78:1023–1028
115. Li XG, Huang MR, Duan W, Yang YL (2002) Novel multifunctional polymers from aromatic diamines by oxidative polymerisations. Chem Rev 102:2925–3030
116. Zhou DM, Dai YQ, Shiu KK (2010) Poly(phenylenediamine) film for the construction of glucose biosensors based on platinized glassy carbon electrode. J Appl Electrochem 40:1997–2003
117. McMahon CP, Killoran SJ, Kirwan SM, O'Neill RD (2004) The selectivity of electrosynthesised polymer membranes depends on the electrode dimensions: implications for biosensor applications. J Chem Soc Chem Commun: 2128–30
118. Killoran SJ, O'Neill RD (2008) Characterization of permselective coatings electrosynthesized on Pt–Ir from the three phenylenediamine isomers for biosensor applications. Electrochim Acta 53:7303–7312
119. Dixon BM, Lowry JP, O'Neill RD (2002) Characterization *in vitro* and *in vivo* of the oxygen dependence of an enzyme/polymer biosensor for monitoring brain glucose. J Neurosci Methods 119:135–142
120. Lowry JP, Miele M, O'Neill RD, Boutelle MG, Fillenz M (1998) An amperometric glucose-oxidase/poly(o-phenylenediamine) biosensor for monitoring brain extracellular glucose: *in vivo* characterisation in the striatum of freely-moving rats. J Neurosci Methods 79:65–74
121. Oldenziel WH, van der Zeyden M, Dijkstra G, Ghijsen WEJM, Karst H, Cremers TIFH, Westerink BHC (2007) Monitoring extracellular glutamate in hippocampal slices with a microsensor. J Neurosci Methods 160:37–44
122. Baronas R, Ivanauskas F, Ivanauskas F, Kulys J (2004) The effect of diffusion limitations on the response of amperometric biosensors with substrate cyclic conversion. J Math Chem 35:199–213
123. Bernhardt PV (2006) Enzyme electrochemistry—biocatalysis on an electrode. Aust J Chem 59:233–256
124. Gooding JJ, Hall EAH, Hibbert DB (1998) From thick films to monolayer recognition layers in amperometric enzyme electrodes. Electroanalysis 10:1130–1136
125. Cooper JM, Pritchard DJ (1994) Biomolecular sensors for neurotransmitter determination—electrochemical immobilization of glutamate oxidase at microelectrodes in a poly(o-phenylenediamine) film. J Mater Sci Mater Electron 5:111–116
126. Gooding JJ, Hall EAH (1996) Parameters in the design of oxygen detecting oxidase enzyme electrodes. Electroanalysis 8:407–413
127. Leypoldt JK, Gough DA (1984) Model of a two-substrate enzyme electrode for glucose. Anal Chem 56:2896–2904
128. Goodsell DS (2006) Glucose oxidase: molecule of the month. RCSB Protein Data Bank. http://www.pdb.org
129. Arima J, Sasaki C, Sakaguchi C, Mizuno H, Tamura T, Kashima A, Kusakabe H, Sugio S, Inagaki K (2009) Structural characterization of L-glutamate oxidase from Streptomyces sp X-119-6. FEBS J 276:3894–3903
130. Compagnone D, Federici G, Bannister JV (1996) A new conducting polymer glucose sensor based on polythianaphthene. Electroanalysis 7:1151–1155
131. Zhang YN, Wilson GS (1993) In vitro and in vivo evaluation of oxygen effects on a glucose oxidase based implantable glucose sensor. Anal Chim Acta 281:513–520
132. Siesjo BK (1978) Brain energy metabolism. Wiley, Chichester
133. Nair PK, Buerk DG, Halsey JH Jr (1987) Comparison of oxygen metabolism and tissue pO_2 in cortex and hippocampus. Stroke 18:616–622
134. Murr R, Berger S, Schuerer L, Peter K, Baethmann A (1994) A novel, remote-controlled suspension device for brain tissue PO_2 measurements with multiwire surface electrodes. Pflugers Arch 426:348–350
135. Clark LC Jr, Misrahy G, Fox RP (1958) Chronically implanted polarographic electrodes. J Appl Physiol 13:85–91
136. Bazzu G, Puggioni GM, Dedola S, Calia G, Rocchitta G, Migheli R, Desole MS, Lowry JP, O'Neill RD, Serra PA (2009) Real-time monitoring of brain tissue oxygen using a miniaturized biotelemetric device implanted in freely-moving rats. Anal Chem 81:2235–2241
137. Wang J, Lu F (1998) Oxygen-rich oxidase enzyme electrodes for operation in oxygen-free solutions. J Am Chem Soc 120:1048–1050
138. Moussy F, Jakeway S, Harrison DJ, Rajotte RV (1994) In vitro and in vivo performance and lifetime of perfluorinated ionomer-coated glucose sensors after high-temperature curing. Anal Chem 66:3882–3888
139. Eyre JA, Stuart AG, Forsyth RJ, Heaviside D, Bartlett K (1994) Glucose export from the

brain in man: evidence for a role for astrocytic glycogen as a reservoir of glucose for neural metabolism. Brain Res 635:349–352
140. Hofer RE, Lanier WL (1991) The effects of insulin infusion on plasma and brain glucose in hyperglycemic diabetic rats: a comparison with placebo-treated diabetic and nondiabetic rats. Anesthesiology 75:673–678
141. Lowry JP, Fillenz M (2001) Real-time monitoring of brain energy metabolism in vivo using microelectrochemical sensors: the effects of anesthesia. Bioelectrochemistry 54:39–47
142. Berners MOM, Boutelle MG, Fillenz M (1994) On-line measurement of brain glutamate with an enzyme/polymer-coated tubular electrode. Anal Chem 66:2017–2021
143. Kirwan SM, Rocchitta G, McMahon CP, Craig JD, Killoran SJ, O'Brien KB, Serra PA, Lowry JP, O'Neill RD (2007) Modifications of poly(*o*-phenylenediamine) permselective layer on Pt–Ir for biosensor application in neurochemical monitoring. Sensors 7:420–437
144. Zhang Y, Wilson GS (1993) Electrochemical oxidation of H_2O_2 on Pt and Pt + Ir electrodes in physiological buffer and its applicability to H_2O_2-based biosensors. J Electroanal Chem 345:253–271
145. Lingane JJ, Lingane PJ (1963) Chronopotentiometry of hydrogen peroxide with a platinum wire electrode. J Electroanal Chem 5:411–419
146. Ikeda T, Schmehl R, Denisevich P, Willman K, Murray RW (1982) Permeation of electroactive solutes through ultrathin polymeric films on electrode surfaces. J Am Chem Soc 104:2683–2691
147. Saveant JM (1991) Permeation through polymer coatings on electrodes: membrane versus pinhole permeation. J Electroanal Chem 302:91–101
148. Pyati R, Murray RW (1994) Permeability and partitioning of ferrocene ethylene oxide and propylene oxide oligomers into electropolymerized films from acetonitrile and polyether solutions. J Phys Chem 98:11129–11135
149. Centonze D, Malitesta C, Palmisano F, Zambonin PG (1994) Permeation of solutes through an electropolymerized ultrathin poly-o-phenylenediamine film used as an enzyme-entrapping membrane. Electroanalysis 6:423–429
150. Murphy LJ (1998) Reduction of interference response at a hydrogen peroxide detecting electrode using electropolymerized films of substituted naphthalenes. Anal Chem 70:2928–2935
151. Craig JD, O'Neill RD (2003) Electrosynthesis and permselective characterisation of phenol-based polymers for biosensor applications. Anal Chim Acta 495:33–43
152. Rothwell SA, McMahon CP, O'Neill RD (2010) Effects of polymerization potential on the permselectivity of poly(o-phenylenediamine) coatings deposited on Pt–Ir electrodes for biosensor applications. Electrochim Acta 55:1051–1060
153. Dai YQ, Shiu KK (2004) Highly sensitive amperometric glucose biosensor based on glassy carbon electrode with copper/palladium coating. Electroanalysis 16:1806–1813
154. Carelli I, Chiarotto I, Curulli A, Palleschi G (1996) Electropolymerization of hydroxybenzene and aminobenzene isomers on platinum electrodes to assemble interference-free electrochemical biosensors. Electrochim Acta 41:1793–1800
155. Fu YC, Chen C, Xie QJ, Xu XH, Zou C, Zhou QM, Tan L, Tang H, Zhang YY, Yao SZ (2008) Immobilization of enzymes through one-pot chemical preoxidation and electropolymerization of dithiols in enzyme-containing aqueous suspensions to develop biosensors with improved performance. Anal Chem 80:5829–5838
156. Miele M, Fillenz M (1996) In vivo determination of extracellular brain ascorbate. J Neurosci Methods 70:15–19
157. De Simoni MG, De Luigi A, Imeri L, Algeri S (1989) An optoelectronic transmission system for telemetry of in vivo voltammetric signals. J Neurosci Methods 29:304
158. Kagohashi M, Nakazato T, Yoshimi K, Moizumi S, Hattori N, Kitazawa S (2008) Wireless voltammetry recording in unanesthetised behaving rats. Neurosci Res 60:120–127
159. Rocchitta G, Migheli R, Dedola S, Calia G, Desole MS, Miele E, Lowry JP, O'Neill RD, Serra PA (2007) Development of a distributed, fully automated, bidirectional telemetry system for amperometric microsensor and biosensor applications. Sens Actuators B 126:700–709
160. Parkin MC, Hopwood SE, Strong AJ, Boutelle MG (2003) Resolving dynamic changes in brain metabolism using biosensors and on-line microdialysis. Trends Anal Chem 22:487–497
161. Belluzo MS, Ribone ME, Lagier CM (2008) Assembling amperometric biosensors for clinical diagnostics. Sensors 8:1366–1399
162. Kudo H, Yagi T, Chu MX, Saito H, Morimoto N, Iwasaki Y, Akiyoshi K, Mitsubayashi K (2008) Glucose sensor using a phospholipid

polymer-based enzyme immobilization method. Anal Bioanal Chem 391:1269–1274
163. Nikolaus N, Strehlitz B (2008) Amperometric lactate biosensors and their application in (sports) medicine, for life quality and wellbeing. Microchim Acta 160:15–55
164. Georganopoulou DG, Carley R, Jones DA, Boutelle MG (2000) Development and comparison of biosensors for in-vivo applications. Faraday Discuss 116:291–303
165. Albery WJ, Boutelle MG, Galley PT (1992) The dialysis electrode—a new method for invivo monitoring. J Chem Soc Chem Commun 12:900–901
166. Shimura T, Karadi Z, Yamamoto T (1997) Facilitation of glutamate release in the ventromedial division of the globus pallidus during palatable taste stimulation in freely moving rats: Real-time measurement. Neurosci Res 28:281–284
167. Zhang MN, Mao LQ (2005) Enzyme-based amperometric biosensors for continuous and on-line monitoring of cerebral extracellular microdialysate. Front Biosci 10:345–352
168. Lin YQ, Zhu NN, Yu P, Su L, Mao LQ (2009) Physiologically relevant online electrochemical method for continuous and simultaneous monitoring of striatum glucose and lactate following global cerebral ischemia/reperfusion. Anal Chem 81:2067–2074
169. Mannino S, Brenna O, Buratti S, Cosio MS (1997) Microdialysis-bioreactor for on-line monitoring of glucose in food samples. Electroanalysis 9:1337–1340
170. John R, John MJ (1998) Evaluation of the microdialysis electrode concept as a generic biosensing technology. Electroanalysis 10:1125–1129
171. Lee GJ, Choi SK, Eo YH, Lim JE, Han JH, Choi S, Park JH, Son DH, Hong S, Oh BS, Park HK (2009) Comparison study between dialysis electrode and CNT probe for neurotransmitter monitoring. Biochip J 3:82–86
172. Murphy LJ, Galley PT (1994) Measurement in vitro of human plasma glycerol with a hydrogen peroxide detecting microdialysis enzyme electrode. Anal Chem 66:4345–4353
173. Zilkha E, Obrenovitch TP, Koshy A, Kusakabe H, Bennetto HP (1995) Extracellular glutamate: on-line monitoring using microdialysis coupled to enzyme-amperometric analysis. J Neurosci Methods 60:1–9
174. Asai S, Iribe Y, Kohno T, Ishikawa K (1996) Real time monitoring of biphasic glutamate release using dialysis electrode in rat acute brain ischemia. Neuroreport 7:1092–1096
175. Yao T, Yano T, Nishino H (2004) Simultaneous in vivo monitoring of glucose, L-lactate, and pyruvate concentrations in rat brain by a flow-injection biosensor system with an on-line microdialysis sampling. Anal Chim Acta 510:53–59
176. Fumero B, Guadalupe T, Valladares F, Mora F, O'Neill RD, Mas M, Gonzalez-Mora JL (1994) Fixed versus removable microdialysis probes for *in vivo* neurochemical analysis: implications for behavioral studies. J Neurochem 63:1407–1415
177. Dungel P, Long N, Yu B, Moussy Y, Moussy F (2008) Study of the effects of tissue reactions on the function of implanted glucose sensors. J Biomed Mater Res A 85A:699–706
178. Mitala CM, Wang YX, Borland LM, Jung M, Shand S, Watkins S, Weber SG, Michael AC (2008) Impact of microdialysis probes on vasculature and dopamine in the rat striatum: a combined fluorescence and voltammetric study. J Neurosci Methods 174:177–185
179. Williams DF (2008) On the mechanisms of biocompatibility. Biomaterials 29:2941–2953
180. Hascup ER, AfBjerken S, Hascup KN, Pomerleau F, Huettl P, Stromberg I, Gerhardt GA (2009) Histological studies of the effects of chronic implantation of ceramic-based microelectrode arrays and microdialysis probes in rat prefrontal cortex. Brain Res 1291:12–20
181. Polikov VS, Su EC, Ball MA, Hong JS, Reichert WM (2009) Control protocol for robust in vitro glial scar formation around microwires: essential roles of bFGF and serum in gliosis. J Neurosci Methods 181:170–177

Part II

Case Studies

Chapter 7

Monitoring Extracellular Glutamate in the Brain by Microdialysis and Microsensors

Si Qin, Miranda (van der Zeyden) Evering, Ngabi Wahono, Thomas I.F.H. Cremers, and Ben H.C. Westerink

Abstract

Recording basal levels of extracellular glutamate by microdialysis in the brain of freely moving animals is of limited value as the result does not fulfill the criteria of synaptic origin (TTX or calcium dependency). In this chapter we discuss whether the recently developed microsensor technique provides an alternative for sampling glutamate by microdialysis. Although glutamate microsensors show advantages in dimensional and temporal resolution, the properties of the detected glutamate do not differ basically from glutamate in microdialysate. The results suggest that glutamate detected by both microdialysis and microsensoring methods is closely regulated by non-neuronal cells. However, there is some evidence that evoked glutamate release recorded by microsensors (e.g., by elevated potassium or behavioral simulation), in the time frame of seconds, might reflect neuronal release.

Key words Glutamate, In vivo detection, Microdialysis, Microsensoring

1 Introduction

Glutamate, which is the ionized form of glutamic acid, is the main excitatory neurotransmitter in the central nervous system (CNS). Glutamate is present in ~60 % of neurons and 40 % of synapses (1). Glutamate acting at the synaptic level is intimately linked to synaptic plasticity, neuronal development, and degeneration. The transmitter therefore plays an important role in a wide range of brain functions, including memory, cognition, and emotion. Abnormal changes or dysregulation in glutamate transmission may induce excitotoxicity, which is closely implicated in neuropathologies ranging from psychiatric to neurodegenerative disorders (2) such as stress, depression, schizophrenia, epilepsy, stroke, Parkinson's disease, and Alzheimer's disease. In addition, glutamate is implicated in the working mechanisms of a variety of centrally acting

Stéphane Marinesco and Nicholas Dale (eds.), *Microelectrode Biosensors*, Neuromethods, vol. 80,
DOI 10.1007/978-1-62703-370-1_7,

drugs (3–6). It is therefore important to understand the properties of glutamatergic transmission in more detail.

Our knowledge about the function and significance of extracellular glutamate in the conscious brain is still poor. One of the reasons for our lack of understanding is that glutamate has multiple roles in brain physiology. Apart from its transmitter function glutamate is the precursor of the main inhibitory neurotransmitter GABA, it detoxifies ammonia in brain tissue, and it delivers the amino acid building block for the synthesis of proteins, peptides, and crucial metabolites such as glutathione.

A variety of techniques are now available for sampling extracellular levels of glutamate in the brain of conscious animals. These methods include push–pull perfusion (7), microdialysis, microsensors (8–11), or direct sampling of extracellular fluid (ECF) (12). In this chapter we will focus on recently developed glutamate microsensors and compare this method with the currently much used microdialysis technique.

2 Extracellular Glutamate Concentrations

Brain tissue contains high concentrations of glutamate: ~5–15 mmol/kg (13, 14). Most of this glutamate is found in neurons. The glutamate concentrations in the cytoplasm of glutamatergic neurons are estimated to be 5–10 mM (13, 15, 16). Neuronal cytoplasmic glutamate concentrations are highest in axon terminals, where glutamate levels are 2–3 times of those in cell bodies or dendrites (16). Glutamate concentrations in vesicles are close to 0.1 M. A summary of glutamate concentrations in body and brain is shown in Table 1 (13, 14, 17, 20, 21). The affinities for the various subtypes of glutamate receptors are shown in Table 2.

Table 1
Glutamate concentrations in the brain

Brain tissue (homogenate)	10 mmol/L (17) or 5–15 mmol/kg (13)
CSF	<1 μmol/L (17); 10 μmol/L (13) or 11.4 μmol/L (18)
Brain extracellular fluid (ECF)	0.5–2 μmol/L (17) or 3–4 μmol/L (13)
Plasma	30–100 μmol/L
Red blood cell cytoplasmic glutamate	~500 μmol/L
Synaptic cleft	25 nmol/L (19) or 2–1,000 μmol/L (17)
Synaptic vesicle	100 mmol/L
Neurons cytoplasm	5–10 mM
Astrocytes	2–3 mM

Table 2
Affinities for glutamate receptors in the brain (EC_{50}, ED_{50}, IC_{50})

Glutamate receptors	EC_{50}, ED_{50}, IC_{50} (mmol/L)	References
GLT-1	1–20	
NMDA receptors	0.4–1.8	(22)
	2.3	(23)
	2.5–3	(17)
	2–4	(14)
AMPA receptors	500–560	(22)
	~16	(23)
	200–500	(17)
	700–1,000	(14)
KA receptors	500–5,900	(22)
	300–800	(14)
mGluR1	11–56	(24)
	9–13	(25)
mGluR2	4–20	(24, 25)
mGluR3	4–5	(24, 25)
mGluR4	3–20	(25)
	~27	(24)
mGluR5	3–10	(24, 25)
mGluR6	16	(24, 25)
mGluR 7	1,000	(24, 25)
mGluR 8	0.02	(25)

To assure a high signal-to-noise ratio for neurotransmission and to avoid excitotoxic actions of glutamate on neurons, the extrasynaptic ECF concentrations of glutamate are believed to be maintained between 1 and 3 μM (26–28). To achieve this, high-affinity glutamate transporters (EAAT1-5) are localized in the vicinity of the synapse (28, 29). The efficacy of the transporters is impressive; it has been calculated that glutamate transporters have the potency to bring concentrations in the ECF locally down to approximately 20 nM (30, 31). The glutamate transporters are present on glutamate neurons as well as astroglial cells, but it seems that the astroglial cells play a more dominant role in this respect. As the transporters are not evenly distributed over the glial membranes, it is to be expected that the glutamate concentration will also display a heterogeneous concentration in the ECF. Moreover, there is accumulating evidence that astrocytes actively release glutamate, which means that at certain sites, in the ECF glutamate release might dominate over uptake. In general our understanding of the distribution of glutamate in the extracellular space is yet far from clear.

Glutamate concentrations in the extracellular space measured by means of quantitative microdialysis are estimated to be 0.2–4 μM (18, 32–34). Microsensor studies suggest even higher glutamate concentrations: between 1 and 50 μM (9, 11, 35). However by using in vitro techniques much lower values such as 25 nM (19), 27–33 nM (36), and 83–87 nM (37) were reported. The extracellular glutamate concentrations in the brain as measured with different methods are summarized in Table 3.

A possible explanation for the reported high extracellular basal levels by studies using microdialysis or microsensors may be found in the fact (36) that the transporter efficacy around the probe or sensor is affected by tissue damage caused during the implantation. Another possible explanation for the relatively high glutamate values detected with both in vivo methods might relate to the properties of glial cells. As the number of glial cells dominates the number of neurons it is not unlikely that microdialysis and microsensors measure extracellular glutamate levels that are closely regulated by glial cells. The latter assumption is supported by the fact that glutamate detected by either microdialysis and microsensoring methods does not or hardly fulfill the classical neuronal release criteria (this will be further discussed in Sect. 4). In this regard it is of interest that various authors provided evidence that glutamate is functionally released from glial cells. During the last decade a large number of cellular mechanisms have been described showing that astrocytes are capable of releasing glutamate: vesicular release (56), non-vesicular release from cystine–glutamate exchangers (32, 57), reversal of glutamate transporters (58, 59), astrocytic gap junction hemichannels (60, 61), and $P2X_7$ receptors (62). In addition release involving the activation of anion channels (63, 64), volume sensitive organic anion channels (VSOAC) (65), and hydrolysis of *N*-aceltylaspartyl-glutamate (66) have been described. However, it is emphasized that the contribution of these mechanisms to the extracellular compartment sampled by microdialysis probes or microsensors has not yet been established.

3 Microdialysis of Glutamate

The most common method to measure extracellular glutamate in conscious animals has been the microdialysis technique (67–69). This approach implies implantation of a concentric probe into the brain. The probe consists of a semipermeable membrane that surrounds two tiny cannulae, through which fluid flows in and out of the probe (see Fig. 1a). Microdialysis is a sampling technique based on a size-selective diffusion of the analyte through a semipermeable membrane. Molecules smaller than the pore size (5–30 kDa) may permeate into the perfusion medium following the concentration gradient, while molecules larger than the membrane cutoff

Table 3
Extracellular glutamate concentration according to various authors

Glutamate concentration	Area	Method	References
Theoretically			
~2 nM ~2 μM		Theoretically according to thermodynamic coupling of Na^+ and K^+ gradients to glutamate transport	(30, 31)
		EC_{50} of the NMDA receptor	(23)
In vitro measured with glutamate sensors			
1.7 ± 1.0 μM	Acute slices rat Hippocampal (CA1)	Implanted carbon fibre hydrogel-coated E	(38, 39)
1.0 ± 0.1 to 10 ± 2 μM	Acute slices mouse Hippocampal (CA1)	Implanted capillary E	(40)
10 ± 5 to 20 ± 12 μM	Acute slices mouse Hippocampal (DG)	Implanted capillary E	(40)
30 ± 2 μM	Acute slices mouse Hippocampal(DG)	Implanted capillary E	(40)
In vivo under anesthesia measured with microdialysis			
1.97 ± 0.7 μM	Anesthetized rat Striatum	Push–pull probe Microdialysis	(41)
1.76 ± 0.15 μM	Anesthetized rat Striatum	Fuse silica tubing combined with capillary electrophoresis	(42)
~1.3–2.7 μM	Anesthetized rat Striatum	Microfluid chips combined with low-flow push–pull Microdialysis	(43)
~2 μM	Anesthetized rat Hippocampus	Microdialysis	(33)
2.9 ± 0.38 μM	Anesthetized rat Hippocampus	Microdialysis	(18)
In vivo in freely moving measured with microdialysis			
1.2 ± 0.1 to 1.8 ± 0.1 μM	Freely moving mice Striatum	Microdialysis	(44)
2.75 ± 0.1 to 4.2 ± 0.1 μM	Freely moving mice Striatum	Microdialysis	(45)
1.33 ± 0.29 to 3.47 ± 0.67 μM (~2 μM)	Freely moving rat Striatum	Microdialysis	(32)
0.24 ± 0.04 μM	Freely moving rat Striatum	Microdialysis	(46)

(continued)

Table 3
(continued)

Glutamate concentration	Area	Method	References
0.48 ± 0.09 to 0.64 ± 0.18 μM	Freely moving rat Striatum	Microdialysis	(47, 48)
1.3 ± 0.3 to 1.6 ± 0.2 μM	Freely moving rat Striatum	Microdialysis	(49)
8.64 ± 0.5 μM	Freely moving rat PFC	Microdialysis	(50)
~2 μM	Freely moving rat PFC	Microdialysis	(51)
In vivo under anesthesia measured with glutamate sensors			
2.0 ± 0.5 μM	Anesthetized rat Chloral hydrate Striatum	Pt E	(52)
18.2 ± 9.3 μM	Anesthetized rat Equitensine Striatum	Carbon fibre hydrogel-coated E	(11, 38)
29.0 ± 9.0 μM	Anesthetized rat Chloral hydrate Striatum	Carbon fibre hydrogel-coated E	(10)
1–5 μM	Anesthetized rat Urethane Dentate gyrus	Pt–Ir E	(9)
In vivo under anesthesia measured with microelectrode array (MEA)			
1.4 ± 0.2 μM	Anesthetized rat Urethane Striatum	MEA	(53)
1.6 ± 0.3 μM	Anesthetized rat Urethane Frontal cortex	MEA	(53)
18.9 ± 0.8 to 41.7 ± 1.2 μM	Anesthetized rat Urethane Striatum	MEA	(54)
In vivo freely moving measured with MEA			
5.0 ± 1.2 μM	Freely moving mice Striatum	MEA	(55)
3.3 ± 1.0 μM	Freely moving mice PFC	MEA	(55)
7.3 ± 0.9 μM	Freely moving rats Striatum	MEA	(35)
44.9 ± 4.7 μM	Freely moving rats PFC	MEA	(35)

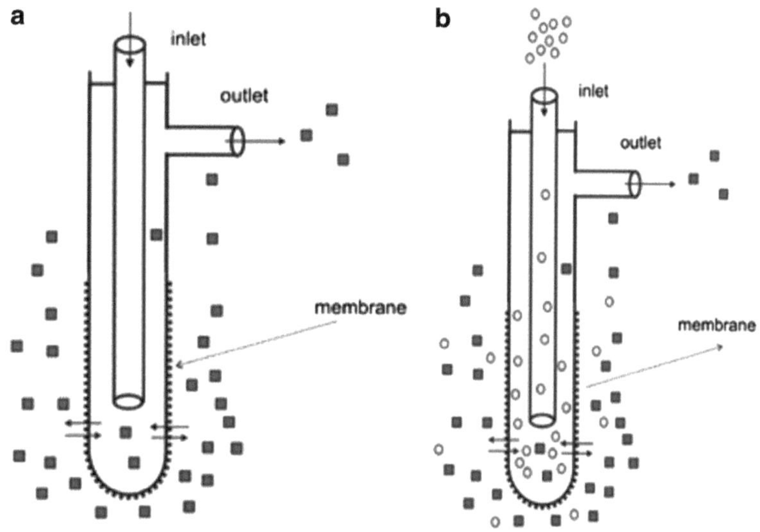

Fig. 1 *The principle of a microdialysis probe*. (**a**) Neurotransmitters (*filled squares*) are sampled by the principle of dialysis and leave the brain via the outlet; (**b**) chemical compounds (*open circles*) are infused during recording of neurotransmitters

(e.g., proteins) are excluded. Mass transport through the dialysis membrane is driven by the concentration gradient of a particular compound between the perfused fluid inside the microdialysis probe and the extracellular fluid and depends on the flow velocity of the fluid.

The microdialysis method has several advantages. It represents a straightforward technique that is not too difficult to establish on a routine basis. The method provides cell and protein free samples that can be directly analyzed without further purification. Microdialysis sampling hardly changes the net fluid balance of the surrounding matrix. Because of the limited fluid loss, samples can be collected continuously for hours or days from a single freely moving animal. Most importantly, each animal can serve as its own control; therefore by using microdialysis it requires fewer subjects than with some other methods, such as direct tissue assay (70). Another advantage of the microdialysis method is that pharmacological compounds can be easily applied to the sampled area (see Fig. 1b), which prompt that the effect of an experimental compound on local neurochemistry can be directly recorded. The latter method is often used to identify the origin and properties of the recorded transmitters.

The criteria for determining neuronal release in the extracellular fluid is based on the typical properties of neurons, which are expected to be sensitive to potassium (K^+) depolarization, sodium (Na^+) channel blockage, e.g., induced by tetrodotoxin (TTX), removal of calcium (Ca^{2+}), and depletion of presynaptic vesicles by local administration of the selective neurotoxin such as α-latrotoxin. For dopamine, noradrenaline, serotonin, and acetylcholine it was established that under a wide variety of experimental conditions, dialysate contents of these transmitters are derived from exocytotic processes triggered by action potentials (69, 71–74). The dialysate levels of these transmitters decrease immediately when neuronal release mechanisms are inhibited (e.g., by omission of calcium from the perfusion fluid or blockage of voltage-gated sodium channels by infusion of TTX).

The microdialysis method certainly has also limitations. The technique is rather invasive (as the diameter of a microdialysis probe is about 250 μm) and will therefore induce extensive tissue damage. Initial studies of the damage caused by inserting microdialysis probes showed that blood flow and glucose metabolism decreased in the vicinity of the probe immediately after implantation (within 2 h) but recovered to nearly normal levels within 24 h (75). Several histological studies of the tissue damage associated with implantation of a dialysis probe for 1–3 days have showed that the probe creates a disrupted zone of tissue as inflammation, gliosis, and swollen axons were present within 1.4 mm away from the probe (75–77). Scar tissue may represent a substantial barrier for the diffusion of neuron-released glutamate. Another disadvantage of the microdialysis method is its poor temporal resolution. Although occasionally rapid sampling times down to a few seconds have been achieved (78, 79) collection times typically vary between 1 and 30 min, which means that microdialysis provides information on a timescale considerably slower than many dynamic processes in the CNS, Most importantly, the dialysis process often does not reach equilibrium and will therefore cause dilution of analyte levels leading to a low recovery of the compound (often <10 %). The latter disadvantage was recently overcome with the introduction of a quantitative microdialysis probe (80).

4 The Origin of Glutamate in Microdialysates

In contrast with results obtained on monoamines it has repeatedly been shown that glutamate in microdialysates does not fulfill the classical release criteria for exocytotic release. Glutamate levels do not respond to infusion of TTX or calcium-free perfusion fluid (69, 81). TTX can efficiently suppress the action potential, which is the main trigger of neuronal release of neurotransmitters; however, the compound does not necessarily affect glial release of glutamate since it is not action potential dependent. Basal levels of glutamate detected by microdialysis might originate from the various non-neuronal pools that have been described (see above). It is assumed that astrocytes have a major contribution to the sampled transmitter. Infusion of the glutamate high-affinity transport blocker TBOA, blocking the uptake of glutamate by glial cells, resulted in a pronounced increase of glutamate. However, even this increase in glutamate was not TTX dependent (82), which indicates that the accumulated extracellular glutamate also under these conditions was derived from non-exocytotic sources.

One more explanation for the difficulty to detect glutamate of synaptic origin in microdialysis samples is the fact that the "spill-over" of glutamate from synaptic cleft to the extra-synaptic space is limited even though they are in spatial continuity. During glutamate neurotransmission, millimolar glutamate is released within a time frame of millisecond into the synaptic cleft (5–50 nm). However under certain conditions, the transmitter seems to be able to escape the synaptic cleft to react with extra- and heterosynaptic receptors in the peri-synaptic environment (83, 84). Various authors therefore attempted to detect synaptic glutamate during excessive activity of glutamatergic neurons. This was achieved by means of chemical, electrical, or behavioral stimulation of glutamate transmission. In addition some researchers succeeded to improve the temporal resolution of the microdialysis sampling to 1 min and even seconds by using capillary electrophoresis with laser-induced fluorescence detection (79, 85). Chemical stimulation includes the infusion of excitatory agents via the microdialysis probe to increase the release of glutamate. For example, the potassium channel blocker 4-aminopyridine prolongs the opening of the potassium channels. Infusion of high doses of this compound resulted in an increase in glutamate levels in microdialysates, which was largely TTX dependent (86).

A different but rather successful approach to detect synaptic glutamate was to use a dual-probe microdialysis stimulation method. During this approach the cell body area of glutamatergic neurons is chemically stimulated by infusion of excitatory agents via the microdialysis probe, and the transmitter released in the terminal area is then monitored simultaneously. In this regard, application of substance P in the substantia nigra increased extracellular levels of glutamate in the striatum (87). In comparable experiments, local administration of NMDA in the frontal cortex increased glutamate levels in the striatum (88, 89). In other cases, electrical stimulation of the cell body areas of glutamate pathways has been successfully applied to increase extracellular glutamate levels in microdialysates of the corresponding nerve terminal areas. For example, stimulation of the medulla, sciatic nerve, or raphe magnus increased glutamate in areas of the spinal cord (90–92). Electrical stimulation of the nervus vagus increased glutamate levels in the nucleus tractus solitarius (93), whereas stimulation of the nucleus paragigantocellularis increased glutamate in the ipsilateral locus coeruleus (94). Tucci et al. (95) demonstrated that electrical stimulation of the prefrontal cortex increased extracellular glutamate in the nucleus accumbens.

5 Glutamate Microsensors

Electrochemical microsensors based on cyclic voltammetry or constant amperometry have been developed to measure neurotransmitters and modulators that are readily oxidized at electrodes. Microsensors have been particularly useful in the measurement of dopamine, serotonin, purines, and nitric oxide levels in both anesthetized and conscious animals (96–98). For analytes that are not intrinsically electrochemically active (such as glutamate) specific enzymes need to be immobilized onto the surface of electrodes. In this approach a neurotransmitter is detected by monitoring the consumption of an electroactive enzyme co-substrate, such as oxygen, or the formation of an electroactive product, such as hydrogen peroxide. A considerable number of enzymes, mainly oxidases and dehydrogenases, have been successfully incorporated into the design of electrochemical sensors.

Glutamate microsensors are based on immobilized glutamate oxidase that converts glutamate into H_2O_2, which is then detected amperometrically at a constant electric potential. A range of immobilized enzyme-based electrochemical biosensors for glutamate detection, based on this principle, are now available for directly monitoring glutamate in the extracellular fluid (9, 10, 52, 99, 100). Some research groups have already reported the application of their glutamate sensors in freely moving animals (35, 55, 101). The various types of enzyme-based electrochemical glutamate sensors that have been applied in vivo are described in Table 4.

According to their working mechanisms and electrochemical mediators, the amperometric sensors are classified as *first-generation* sensors (using natural enzyme products, such as O_2 or H_2O_2) or *second-generation* sensors (using additional redox mediators) (102–104). As shown in Fig. 2, glutamate first-generation sensors like the platinum (Pt)-needle microsensors detect the produced H_2O_2 directly by applying a high potential (e.g., +700 mV) on the electrode (105). Second-generation sensors, such as the hydrogel-coated carbon fibre electrode (CFE) microsensors, contain an additional redox mediator (osmium redox polymer) for detection of H_2O_2 (10, 38). In this way, the H_2O_2 is reduced by the redox mediator and then the oxidized form of the redox mediator is electrochemically detected on the electrode surface at a relatively low potential (e.g., −150 mV). The latter approach results in sensors with less sensitivity for interfering compounds that are electrochemically active at higher potentials. However, the oxygen dependency could be an issue for this type of sensors, since the oxygen content of the brain may not be sufficiently applied for the continuous redox reactions.

The difference in working principle and associated components as well as geometry and size of biosensor influence the performance

Table 4
Different types of enzyme-based amperometric glutamate microbiosensor

Research group	Electrode	Characteristics of sensor in vitro					Generation	References
		Surface (mm^2)	Sensitivity (nA/μM)	Glutamate detection limit (μM)	Potential (mV) vs. Ag/AgCl	Response time (s)		
Gerhardt	Pt electrode array	0.005	0.016 ± 0.001	1.82 ± 0.17	+700	~1	First	(53)
Michael	CFE	0.0095–0.0126	0.0034 ± 0.001	1 ~ 3	−100	20 ~ 40	Second	(10)
Shim	CP-coated Pt cylinder	0.024	14.0 ± 0.2	0.1 ± 0.03	+450	~10	First	(52)
Soldatkin	Ru-coated CFE	0.048	0.029	2.5	+400	–	First	(99)
Westerink	CFE	0.0095–0.016	0.0055 ± 0.00007	5	−150	~8	Second	(38)
Wilson	Pt/Ir	0.183	0.1	2	+600	~1	First	(9)

CFE carbon fibre electrode, *CP* contact polymer, *Ru* ruthenium, *SCE* saturated calomel electrode, *Ir* iridium

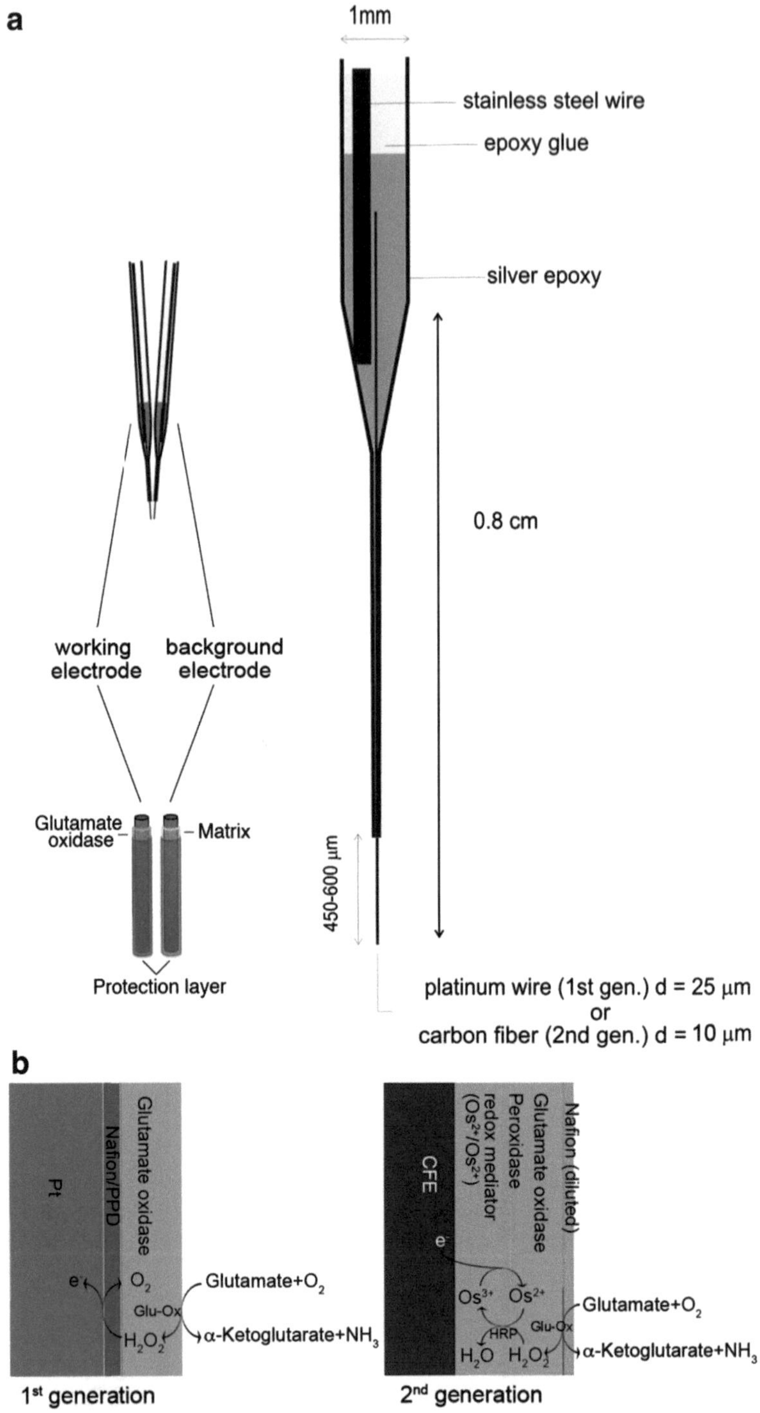

Fig. 2 *First- and second-generation sensor*. (**a**) Depicts the design of the sensors and (**b**) represents the various layers

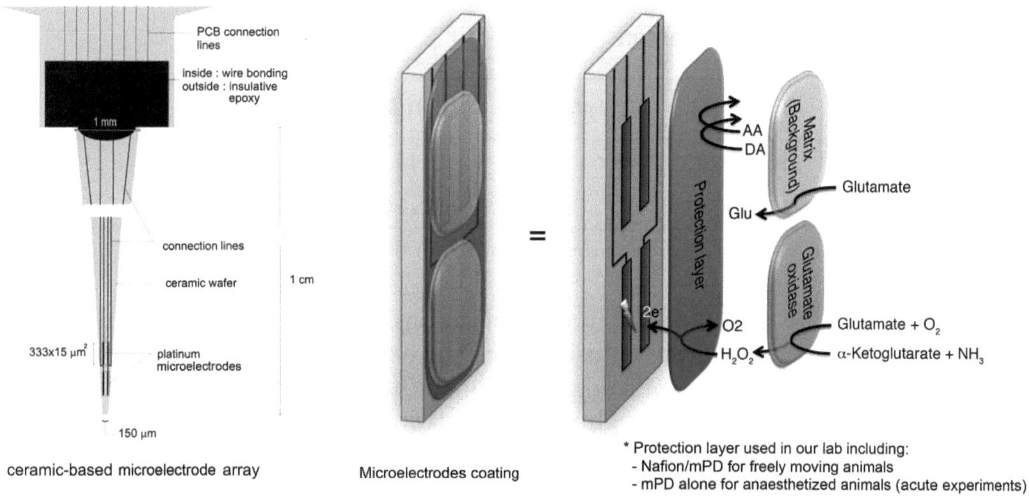

Fig. 3 Illustration of ceramic-based microelectrode array and coating layers

of the microsensor and consequently the detection of glutamate in the brain (106, 107).

In our laboratory we have recently compared (104) different types of glutamate sensors. We studied a first-generation Pt-wire-needle microsensor (25 μm in diameter) and a second-generation hydrogel-coated carbon fiber electrode (CFE) microsensor (10 μm in diameter) (10, 38). The design of these two types of sensors is depicted in Fig. 2. Note that after coating the hydrogel, the second-generation sensor became much thicker; the diameter of the whole sensor was about 25 μm, which is comparable with the first-generation Pt-wire sensors. The sensitivity and selectivity and response time for glutamate of these sensors were evaluated ex vivo in the same calibrations system. The third sensor type that we studied was the commercially available first-generation microelectrode array (MEA) (8, 53), shown in Fig. 3. For In vivo evaluation all the types of sensors were implanted in the brain of anesthetized rats. Recordings were made within hours after implantation.

5.1 Sensitivity

The sensitivity of Pt wire first-generation sensor (with surface of 0.04 mm^2) was 12.40 ± 2.08 pA/μM ($n = 18$) and that of hydrogel second-generation sensor (with surface of 0.04 mm^2) was 4.31 ± 0.50 pA/μM ($n = 10$). The detection limit for the Pt wire first-generation sensor was <0.5 μM, while the hydrogel-coated second-generation sensor could not generate linear amperometric currents when the glutamate concentration was below 5 μM (108). Overall the hydrogel-coated second-generation glutamate sensor was significantly less sensitive than the Pt wire first generation, which could be due to the difference in amount of glutamate oxidase used in the enzyme coating mixtures, the difference in electrode surface

Table 5
Selectivity of both generations of sensors

	AA 200 μM	DA 5 μM	Cys 50 μM	UA 50 μM	DOPAC 20 μM
First generation	115.4 ± 6.7	108.6 ± 12.0	92.2 ± 15.3	105.4 ± 17.8	99.7 ± 9.2
Second generation	75.2 ± 3.5	82.1 ± 7.8	128.7 ± 10.6	53.4 ± 19.6	84.3 ± 35.3

The values were expressed as percentage of the current generated with 100 μM glutamate

area, or different applications of protective layer which could affect the diffusion rate. The oxygen reproduced on the electrode surface of first-generation sensor could compensate for the initial reaction of glutamate oxidation; however, for hydrogel second-generation sensor oxygen is totally provided from the working environment.

5.2 Selectivity

The brain contains various electrochemically active compounds, such as ascorbic acid (AA), uric acid (UA), cysteine (Cys), dopamine (DA), and its main metabolite 3,4-dihydroxyphenylacetic acid (DOPAC). Glutamate microsensors can pick up nonspecific signals from these interfering compounds. One way to improve the specificity of glutamate microsensors from these interfering signals is to subtract the recorded current produced by a background sensor that contains no glutamate oxidase. However, additional means to reduce interference are still necessary. First-generation sensors often contain a semi-selective membrane to exclude interfering compounds. In our laboratory the selectivity of the first-generation glutamate sensor improved significantly when the Pt wire was coated with Nafion and poly-phenylenediamine (PPD) (109). The second-generation sensor that we used included ascorbate oxidase in the hydrogel enzyme layer to prevent AA reacting directly with the electrode (108). This hydrogel-based sensor also benefits from the lower detection potential (−150 mV) due to the presence of redox mediator, which reduces the likelihood of other easily oxidizable agents becoming oxidized on the electrodes Our experiments, shown in Table 5, indicated that the first-generation microsensors were hardly influenced by any of the interfering substrates; however, the hydrogel second-generation sensor was more affected by interfering substrates.

5.3 Response Time

The response time for glutamate in vitro, defined as the time lasting from the injection until the peak value of glutamate, was 3–4 s for the Pt-wire first-generation sensor. The response time of hydrogel-coated microsensors was about 9 s (38).

Taken together it is concluded that the first-generation sensors displayed a faster, more sensitive, and more selective response to glutamate than the second-generation sensor.

5.4 Biofouling

The microsensors were calibrated before and after in vivo experiments, which lasted 2–3 h. The sensitivity of first-generation Pt sensor to glutamate decreased 57.9 % after in vivo application, while that of hydrogel sensor dropped even more significantly: 72.6 %. As a consequence the lowest detection level increased significantly for both microsensors. Biofouling is a likely explanation for the decrease in sensitivity. Damaged brain tissue, phagocytes, and active biological substance such as proteinase might attach to the coating layers of the sensor electrodes, causing degradation of glutamate oxidase and limiting diffusion of redox products such as H_2O_2 to the electrodes. Consequently the absolute glutamate concentrations should be calculated by using the sensitivity measured post-in vivo experiments to correct for the biofouling artifact. The most pronounced decrease of signal strength typically occurs within the first 0.5–1 h after applying potential, which is named "stabilization period". The decline of sensitivity of sensors probably is the main reason for the decrease of the current signals. The described experiment always started after the stabilization period.

Importantly there was no significant change of the interference for the first-generation microsensors before and after in vivo application by any of the mentioned interfering substrates. On the other hand, the hydrogel-coated second-generation sensor was more sensitive for interfering substrates after implantation. UA and DOPAC showed the most pronounced interference during glutamate detection (38). These results suggest that, also during in vivo conditions, the first-generation sensors are more resistant to possible interfering substance.

6 Comparison of Microdialysis and Microsensors

An interesting question is whether microdialysis and microsensors detect different glutamate pools. Here we describe a series of in vivo experiments in order to compare the two methods.

6.1 Basal Extracellular Glutamate Concentrations

Basal extracellular glutamate levels for needle shape Pt-wire first-generation sensor and the hydrogel-based second-generation sensor in the prefrontal cortex (PFC) of anesthetized rat were rather similar: 40–70 μM after correction of the biofouling using post-in vivo calibration. The commercial MEA sensor detected basal glutamate concentrations around 10 μM. The variation in basal glutamate levels can be partly explained by the geometric differences of these sensors. The needle shape microsensors have a smaller size of electrode (10–25 μm in diameter and 500 μm in length), whereas the MEAs have larger dimensions (with a tip of 50–150 μm by 100 μm) (53). Extracellular glutamate measured by means of quantitative microdialysis is estimated to be 0.2–4 μM (32–34) (Table 3).

Basal glutamate levels detected with microsensors are in general higher than those detected with microdialysis; in our laboratory basal glutamate concentrations could reach about 50 μM. This might be explained by the smaller size of platinum wire electrodes, which allows the detection of glutamate pools closer to synaptic spaces or glial glutamate transporters where the highest concentrations of glutamate are found in the extracellular space. In addition, the readout value of the glutamate concentration by microsensor or microdialysis is an averaged concentration of various glutamate pools around the detection device; therefore the closer the detecting device located to glutamate release sites, the higher readout value would be obtained.

A recent published immunocytochemistry paper that studied implantation of microsensors (MEAs) and microdialysis probes (110) showed that 1 day after implantation the glial cells around microsensors or microdialysis probes are not significantly activated compared to control. In addition glial activations induced by the implantation of microsensors and microdialysis appeared to be comparable. This study also showed that the implantation of microdialysis probes might cause more damage to glutamatergic synapses than biosensors do. This could also be the reason that microdialysis probes detect less amount of basal glutamate than microsensors.

6.2 Glutamate Release Stimulated by Elevated Potassium In Vivo

Local administration of an elevated concentration of potassium ions in the brain can induce depolarization of neurons, which causes neurotransmitter to be released from the synapses via exocytosis. Both generations of microsensors were able to detect this increase of extracellular glutamate. Potassium-induced increases in glutamate were comparable for the first-generation Pt wire and MEAs. The first-generation sensors responded faster and detected a more pronounced increase of glutamate levels than the hydrogel-coated second-generation sensors. These results are in line with results from microdialysis experiments (69) (Fig. 4a).

One of the major advantages of microsensoring is its high temporary resolution. As shown in the elevated potassium experiments, microdialysis needs 20–30 min infusion periods to collect enough sample for analysis, while microsensors are able to detect glutamate levels every second (Fig. 4b).

6.3 Effects of TTX on Basal Extracellular Glutamate

To investigate whether basal extracellular glutamate is of synaptic origin, the sodium channel blocker TTX was locally applied in order to block neuronal depolarization. In contrast to the virtually complete and fast decrease in levels of various neurotransmitters (such as the monoamines and acetylcholine) following infusion of TTX, there is general agreement that basal glutamate levels in microdialysis experiments do not respond to TTX (69, 111, 112) (Fig. 5a) nor by blocking exocytosis with botulium A or tetanus neurotoxin (113).

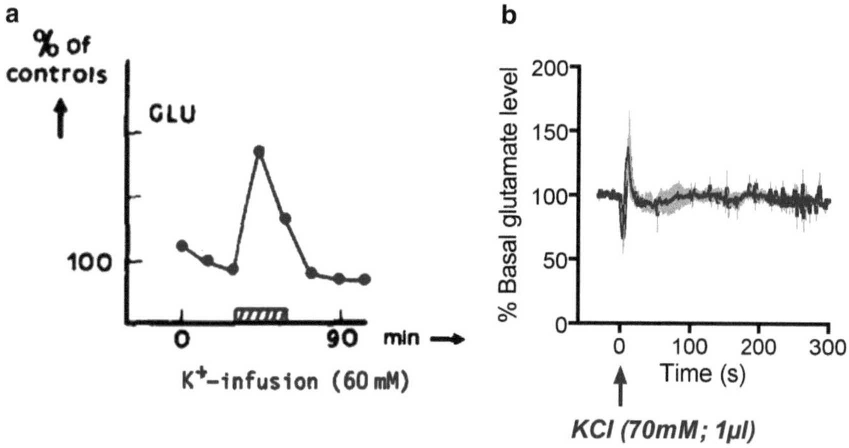

Fig. 4 *Effect of high K^+ on extracellular glutamate levels in freely moving rat.* The glutamate was detected with microdialysis (**a**); or with MEA microsensors (**b**) at basal levels of 7.20 ± 0.54 μM. *Figure (a) was adapted from previous work of Westerink et al. published in Life Science* (69)

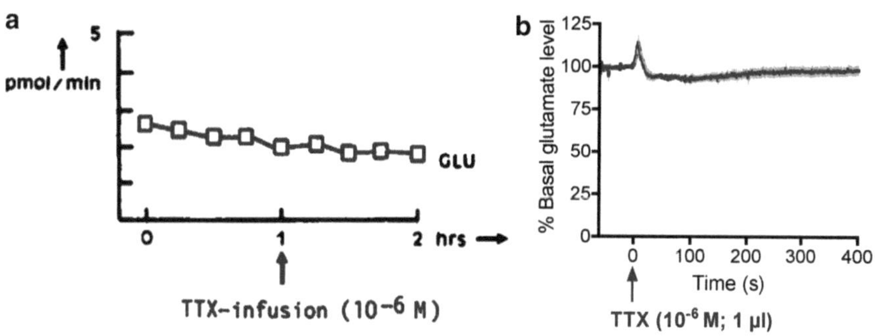

Fig. 5 *Effect of TTX on extracellular glutamate levels* through infusion in a microdialysis experiment (**a**) or local injection detected with MEA (**b**) at basal levels of 15.20 ± 6.71 μM. *Figure (a) was adapted from previous work of Westerink et al. published in Life Science* (69)

In recent microsensor experiments carried out in our laboratory, we noticed a small but not statistically significant decrease in glutamate concentration after application of TTX in prefrontal cortex (PFC), about 15–25 % of basal levels (Fig. 5b). Other microsensor studies have reported different results and much variation: 10–90 % decrease of glutamate during TTX infusion (10, 11, 114). From these data we may conclude that a minor part of basal glutamate detected by microsensors might be derived from synaptic processes.

6.4 Extracellular Glutamate During Transporter Inhibition

Local application of DL-TBOA, which is a blocker of glutamate transporters EAAT1-3 (i.e., GLAST, GLT-1, and EAAC1), is expected to induce a significant increase in extracellular glutamate. A pronounced increase of glutamate was shown during TBOA

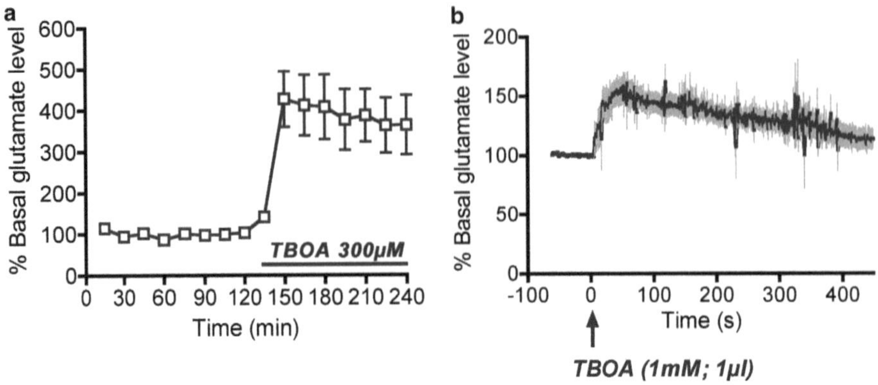

Fig. 6 *Effect of TBOA on extracellular glutamate levels.* (**a**) Effect of TBOA-infusion recorded with microdialysis; (**b**) effect of TBOA on glutamate levels detected with MEA at basal values of 15.83 ± 11.18 μM

infusion in microdialysis experiments as well as in microsensor experiments (about 150 %). A typical result is shown in Fig. 6. Note that the detected increase in glutamate levels was different for the two techniques, which is due to the different timescale of the application of TBOA. In the microdialysis experiments, the TBOA was locally infused during 2 h; while in the microsensor experiments, 1 μl of TBOA was injected into the detection area within 1 s.

It is emphasized that the TBOA-induced increase of glutamate recorded with microdialysis as well as with microsensors (preliminary results from our laboratory) was not TTX dependent (82).

In summary, microdialysis and microsensors are able to detect high potassium-induced neuronal glutamate release, but both methods have difficulties to detect reproducibly nerve-impulse-derived glutamate release from neuronal synapses. As extracellular glutamate basal levels are carefully controlled by high-affinity glutamate transporters, glial cells might dominate as source for basal glutamate levels. Compared to microdialysis, the major advantage of microsensors is the high temporal resolution.

7 Extracellular Glutamate Recorded During Behavioral Experiments

The question whether neuron-released glutamate can reach extra-synaptic sources is still a matter of debate, although various authors have provided evidence that under certain conditions (for example, during behavioral activation) impulse-flow-dependent glutamate can reach extra-synaptic sites (104). In this regard it is of interest that certain types of behavioral stimulations are accompanied by characteristic changes (often an increase) in extracellular levels of glutamate in microdialysis experiments. In several studies, extracellular glutamate was recorded during restraint stress.

Substantial increases in glutamate levels, 200–450 % of basal values lasting for 10–20 min, were reported in dialysates derived from the prefrontal cortex, hippocampus, nucleus accumbens, and striatum (112). Photic stimulation of the conscious rat increased extracellular glutamate in the visual cortex to 200 % of controls for at least 3 min (115) and formalin injection into the hind paw resulted in an increase of extracellular glutamate that lasted for 2 min in certain subareas of the hypothalamus (preoptic area but not in lateral and ventromedial hypothalamus) (116). Results with a different timescale were presented by Rossell et al. (79), who stimulated a whisker of the rat and sampled glutamate in the motor cortex. By using capillary electrophoresis coupled to laser-induced fluorescence detection, they were able to determine glutamate in 1-s microdialysis samples. Upon whisker stimulation, glutamate was increased only in the first 1-s sample; in the 2-s sample glutamate again fell back to control values, although the whisker stimulation was continued. The results of the 1-s lasting increase in glutamate content contrast with the 2–20-min lasting increases in glutamate that were observed in the behavioral experiments discussed above. These data demonstrated the importance of the timescale in detecting physiological changes in extracellular glutamate. It is likely that glutamate displaying sub-second increases reflects a different origin than glutamate that was enhanced in behavioral experiments lasting several minutes.

An intriguing question is whether the above-described behavioral-evoked glutamate content is calcium and TTX dependent. It was found that handling stress induced increases in glutamate that persisted in the presence of TTX (81). Grooming—induced by dropping water onto the rat's snout—increased extracellular glutamate in the striatum to about 200 % of controls (117). However, a similar increase was seen when the experiments were performed in the presence of TTX. In contrast, there is a report about feeding-induced increase in extracellular glutamate in the nucleus accumbens that was found to be TTX dependent (118). In a study that used formalin injection into the hind paw described increases in glutamate that were both TTX and calcium dependent (116). However, the specificity of this observation is questioned by the fact that other amino acids, such as arginine and aspartate, also responded in a calcium- and TTX-dependent manner.

The two commercial available first-generation sensors (MEA sensor and the Pt-needle Pinnacle sensor) are designed for recordings in conscious animals. The Pt-needle-based first-generation sensor is at present applied in our laboratory in conscious animals. Given the fragility of hydrogel second-generation CFE sensors, they are not suitable to be applied to freely moving animals. In the near future behavioral studies using glutamate microsensors are to be expected. An intriguing question is whether the behavior-induced fast changes in extracellular glutamate are related to synaptic events.

8 Conclusions

In this review we have compared the two generations of glutamate microsensors. It was found that the hydrogel second-generation sensor responded much slower when compared to the first-generation Pt-based sensors (based on Pt-wire as well as the MEA). Moreover the hydrogel-based sensor appeared fragile, making it difficult to handle and implant during surgery. Combined with their lower sensitivity and less resistance toward interfering compounds, the hydrogel-based sensor is apparently not the first choice to detect extracellular glutamate in brain tissue.

Next we have compared and discussed results from glutamate microsensors experiments with earlier published comparable microdialysis studies. In general microsensors detect substantially higher basal values of extracellular glutamate than microdialysis probes. The reason for this is presently unknown.

There is an ongoing discussion in the literature whether the recorded extracellular glutamate relates to neuronal activity. Where microdialysis literature firmly concluded that the sampled glutamate was TTX and calcium independent, some published glutamate sensoring experiments contradicted these notions (10, 11, 114). However recent experiments performed with various types of glutamate microsensors in our laboratory could not reproduce these findings. It is concluded that both detection methods sample mainly glutamate from sources not directly related to synaptic events. However this conclusion is only valid for "basal levels" of glutamate. Evidence has been provided that during excessive stimulation of glutamate neurons (e.g., during certain behavioral conditions) part of the detected glutamate might be related to neuronal exocytotic release. To detect this type of release, the used method should be able to record glutamate levels in (sub)seconds. The presently available microsensors are the means to detect these fast changes, as the currently used microdialysis protocols do not provide this high temporal resolution. In this regard it is of interest to note that the tail-pinch-induced increase in extracellular glutamate—detected by the MEA's—was TTX dependent (114).

It is tempting to speculate that changes of glutamate within seconds or subseconds time frame might be related to synaptic activity, whereas changes in minutes might represent non-exocytotic release from yet unknown probably astroglial sources. It is emphasized that in the past decade several non-exocytotic release mechanisms from astrocytes are described. To what extent these release mechanisms are represented in the glutamate detected by microsensors or microdialysis is presently unknown.

References

1. Javitt DC (2007) Glutamate and schizophrenia: phencyclidine. N-methyl-D-aspartate receptors, and dopamine-glutamate interactions. Int Rev Neurobiol 78:69–108
2. Gardoni F, Di Luca M (2006) New targets for pharmacological intervention in the glutamatergic synapse. Eur J Pharmacol 545:2–10
3. Goff DC, Coyle JT (2001) The emerging role of glutamate in the pathophysiology and treatment of schizophrenia. Am J Psychiatry 158:1367–1377
4. Meldrum BS (1994) The role of glutamate in epilepsy and other CNS disorders. Neurology 44:S14–S23
5. Amiel JM, Mathew SJ (2007) Glutamate and anxiety disorders. Curr Psychiatry Rep 9:278–283
6. Javitt DC (2004) Glutamate as a therapeutic target in psychiatric disorders. Mol Psychiatry 9(984–997):979
7. Gaddum J (1961) Push-pull cannulae. J Physiol 155:1P–2P
8. Burmeister JJ, Pomerleau F, Palmer M, Day BK, Huettl P, Gerhardt GA (2002) Improved ceramic-based multisite microelectrode for rapid measurements of L-glutamate in the CNS. J Neurosci Methods 119:163–171
9. Hu Y, Mitchell KM, Albahadily FN, Michaelis EK, Wilson GS (1994) Direct measurement of glutamate release in the brain using a dual enzyme-based electrochemical sensor. Brain Res 659:117–125
10. Kulagina NV, Shankar L, Michael AC (1999) Monitoring glutamate and ascorbate in the extracellular space of brain tissue with electrochemical microsensors. Anal Chem 71:5093–5100
11. Oldenziel WH, Dijkstra G, Cremers TI, Westerink BH (2006) In vivo monitoring of extracellular glutamate in the brain with a microsensor. Brain Res 1118:34–42
12. Bowser MT, Kennedy RT (2001) In vivo monitoring of amine neurotransmitters using microdialysis with on-line capillary electrophoresis. Electrophoresis 22:3668–3676
13. Danbolt NC (2001) Glutamate uptake. Prog Neurobiol 65:1–105
14. Featherstone DE, Shippy SA (2008) Regulation of synaptic transmission by ambient extracellular glutamate. Neuroscientist 14:171–181
15. Osen KK, Storm-Mathisen J, Ottersen OP, Dihle B (1995) Glutamate is concentrated in and released from parallel fiber terminals in the dorsal cochlear nucleus: a quantitative immunocytochemical analysis in guinea pig. J Comp Neurol 357:482–500
16. Ottersen OP (1989) Postembedding immunogold labelling of fixed glutamate: an electron microscopic analysis of the relationship between gold particle density and antigen concentration. J Chem Neuroanat 2:57–66
17. Meldrum BS (2000) Glutamate as a neurotransmitter in the brain: review of physiology and pathology. J Nutr 130:1007S–1015S
18. Lerma J, Herranz AS, Herreras O, Abraira V, Martin del Rio R (1986) In vivo determination of extracellular concentration of amino acids in the rat hippocampus. A method based on brain dialysis and computerized analysis. Brain Res 384:145–155
19. Herman MA, Jahr CE (2007) Extracellular glutamate concentration in hippocampal slice. J Neurosci 27:9736–9741
20. Clements JD, Lester RA, Tong G, Jahr CE, Westbrook GL (1992) The time course of glutamate in the synaptic cleft. Science 258:1498–1501
21. Nedergaard M, Takano T, Hansen AJ (2002) Beyond the role of glutamate as a neurotransmitter. Nat Rev Neurosci 3:748–755
22. Dingledine R, Borges K, Bowie D, Traynelis SF (1999) The glutamate receptor ion channels. Pharmacol Rev 51:7–62
23. Patneau DK, Mayer ML (1990) Structure-activity relationships for amino acid transmitter candidates acting at N-methyl-D-aspartate and quisqualate receptors. J Neurosci 10:2385–2399
24. Pin JP, Duvoisin R (1995) The metabotropic glutamate receptors: structure and functions. Neuropharmacology 34:1–26
25. Conn PJ, Pin JP (1997) Pharmacology and functions of metabotropic glutamate receptors. Annu Rev Pharmacol Toxicol 37: 205–237
26. Barbour B, Keller BU, Llano I, Marty A (1994) Prolonged presence of glutamate during excitatory synaptic transmission to cerebellar Purkinje cells. Neuron 12:1331–1343
27. Turecek R, Trussell LO (2000) Control of synaptic depression by glutamate transporters. J Neurosci 20:2054–2063
28. Diamond JS, Jahr CE (1997) Transporters buffer synaptically released glutamate on a submillisecond time scale. J Neurosci 17:4672–4687
29. Lehre KP, Danbolt NC (1998) The number of glutamate transporter subtype molecules at glutamatergic synapses: chemical and stereo-

logical quantification in young adult rat brain. J Neurosci 18:8751–8757
30. Levy LM, Warr O, Attwell D (1998) Stoichiometry of the glial glutamate transporter GLT-1 expressed inducibly in a Chinese hamster ovary cell line selected for low endogenous Na+-dependent glutamate uptake. J Neurosci 18:9620–9628
31. Zerangue N, Kavanaugh MP (1996) Flux coupling in a neuronal glutamate transporter. Nature 383:634–637
32. Baker DA, Xi ZX, Shen H, Swanson CJ, Kalivas PW (2002) The origin and neuronal function of in vivo nonsynaptic glutamate. J Neurosci 22:9134–9141
33. Montiel T, Camacho A, Estrada-Sanchez AM, Massieu L (2005) Differential effects of the substrate inhibitor l-trans-pyrrolidine-2,4-dicarboxylate (PDC) and the non-substrate inhibitor DL-threo-beta-benzyloxyaspartate (DL-TBOA) of glutamate transporters on neuronal damage and extracellular amino acid levels in rat brain in vivo. Neuroscience 133:667–678
34. Nyitrai G, Kekesi KA, Juhasz G (2006) Extracellular level of GABA and Glu: in vivo microdialysis-HPLC measurements. Curr Top Med Chem 6:935–940
35. Rutherford EC, Pomerleau F, Huettl P, Stromberg I, Gerhardt GA (2007) Chronic second-by-second measures of L-glutamate in the central nervous system of freely moving rats. J Neurochem 102:712–722
36. Cavelier P, Attwell D (2005) Tonic release of glutamate by a DIDS-sensitive mechanism in rat hippocampal slices. J Physiol 564:397–410
37. Le Meur K, Galante M, Angulo MC, Audinat E (2007) Tonic activation of NMDA receptors by ambient glutamate of non-synaptic origin in the rat hippocampus. J Physiol 580:373–383
38. Oldenziel WH, Dijkstra G, Cremers TI, Westerink BH (2006) Evaluation of hydrogel-coated glutamate microsensors. Anal Chem 78:3366–3378
39. Oldenziel WH, van der Zeyden M, Dijkstra G, Ghijsen WE, Karst H, Cremers TI, Westerink BH (2007) Monitoring extracellular glutamate in hippocampal slices with a microsensor. J Neurosci Methods 160:37–44
40. Oka T, Tasaki C, Sezaki H, Sugawara M (2007) Implantation of a glass capillary-based enzyme electrode in mouse hippocampal slices for monitoring of L-glutamate release. Anal Bioanal Chem 388:1673–1679
41. Kottegoda S, Shaik I, Shippy SA (2002) Demonstration of low flow push-pull perfusion. J Neurosci Methods 121:93–101
42. Kennedy RT, Thompson JE, Vickroy TW (2002) In vivo monitoring of amino acids by direct sampling of brain extracellular fluid at ultralow flow rates and capillary electrophoresis. J Neurosci Methods 114:39–49
43. Cellar NA, Burns ST, Meiners JC, Chen H, Kennedy RT (2005) Microfluidic chip for low-flow push-pull perfusion sampling in vivo with on-line analysis of amino acids. Anal Chem 77:7067–7073
44. Shakil SS, Holmer HK, Moore C, Abernathy AT, Jakowec MW, Petzinger GM, Meshul CK (2005) High and low responders to novelty show differential effects in striatal glutamate. Synapse 58:200–207
45. Holmer HK, Keyghobadi M, Moore C, Meshul CK (2005) L-dopa-induced reversal in striatal glutamate following partial depletion of nigrostriatal dopamine with 1-methyl-4-phenyl-1,2,3,6-tetrahydropyridine. Neuroscience 136:333–341
46. Di Cara B, Dusticier N, Forni C, Lievens JC, Daszuta A (2001) Serotonin depletion produces long lasting increase in striatal glutamatergic transmission. J Neurochem 78:240–248
47. Segovia G, Porras A, Del Arco A, Mora F (2001) Glutamatergic neurotransmission in aging: a critical perspective. Mech Ageing Dev 122:1–29
48. Segovia G, Del Arco A, Mora F (1997) Endogenous glutamate increases extracellular concentrations of dopamine. GABA, and taurine through NMDA and AMPA/kainate receptors in striatum of the freely moving rat: a microdialysis study. J Neurochem 69: 1476–1483
49. Miele M, Berners M, Boutelle MG, Kusakabe H, Fillenz M (1996) The determination of the extracellular concentration of brain glutamate using quantitative microdialysis. Brain Res 707:131–133
50. Rocha L, Briones M, Ackermann RF, Anton B, Maidment NT, Evans CJ, Engel J Jr (1996) Pentylenetetrazol-induced kindling: early involvement of excitatory and inhibitory systems. Epilepsy Res 26:105–113
51. Boatell ML, Bendahan G, Mahy N (1995) Time-related cortical amino acid changes after basal forebrain lesion: a microdialysis study. J Neurochem 64:285–291
52. Rahman MA, Kwon NH, Won MS, Choe ES, Shim YB (2005) Functionalized conducting polymer as an enzyme-immobilizing substrate:

an amperometric glutamate microbiosensor for in vivo measurements. Anal Chem 77: 4854–4860

53. Day BK, Pomerleau F, Burmeister JJ, Huettl P, Gerhardt GA (2006) Microelectrode array studies of basal and potassium-evoked release of L-glutamate in the anesthetized rat brain. J Neurochem 96:1626–1635
54. Nickell J, Pomerleau F, Allen J, Gerhardt GA (2005) Age-related changes in the dynamics of potassium-evoked L-glutamate release in the striatum of Fischer 344 rats. J Neural Transm 112:87–96
55. Hascup KN, Hascup ER, Pomerleau F, Huettl P, Gerhardt GA (2008) Second-by-second measures of L-glutamate in the prefrontal cortex and striatum of freely moving mice. J Pharmacol Exp Ther 324:725–731
56. Bezzi P, Gundersen V, Galbete JL, Seifert G, Steinhauser C, Pilati E, Volterra A (2004) Astrocytes contain a vesicular compartment that is competent for regulated exocytosis of glutamate. Nat Neurosci 7:613–620
57. Moran MM, McFarland K, Melendez RI, Kalivas PW, Seamans JK (2005) Cystine/glutamate exchange regulates metabotropic glutamate receptor presynaptic inhibition of excitatory transmission and vulnerability to cocaine seeking. J Neurosci 25:6389–6393
58. Fillenz M (1995) Physiological release of excitatory amino acids. Behav Brain Res 71: 51–67
59. Levi G, Raiteri M (1993) Carrier-mediated release of neurotransmitters. Trends Neurosci 16:415–419
60. Bennett MV, Contreras JE, Bukauskas FF, Saez JC (2003) New roles for astrocytes: gap junction hemichannels have something to communicate. Trends Neurosci 26:610–617
61. Ye ZC, Wyeth MS, Baltan-Tekkok S, Ransom BR (2003) Functional hemichannels in astrocytes: a novel mechanism of glutamate release. J Neurosci 23:3588–3596
62. Duan S, Anderson CM, Keung EC, Chen Y, Swanson RA (2003) P2X7 receptor-mediated release of excitatory amino acids from astrocytes. J Neurosci 23:1320–1328
63. Kimelberg HK, Goderie SK, Higman S, Pang S, Waniewski RA (1990) Swelling-induced release of glutamate, aspartate, and taurine from astrocyte cultures. J Neurosci 10: 1583–1591
64. Rutledge EM, Kimelberg HK (1996) Release of (3H)-D-aspartate from primary astrocyte cultures in response to raised external potassium. J Neurosci 16:7803–7811
65. Basarsky TA, Feighan D, MacVicar BA (1999) Glutamate release through volume-activated channels during spreading depression. J Neurosci 19:6439–6445
66. Blakely RD, Robinson MB, Thompson RC, Coyle JT (1988) Hydrolysis of the brain dipeptide N-acetyl-L-aspartyl-L-glutamate: subcellular and regional distribution, ontogeny, and the effect of lesions on N-acetylated-alpha-linked acidic dipeptidase activity. J Neurochem 50:1200–1209
67. Ungerstedt U (1991) Microdialysis–principles and applications for studies in animals and man. J Intern Med 230:365–373
68. Westerink B, Rea K, Oldenziel W (2007) Microdialysis of glutamate and GABA in the brain: analysis and interpretation. In: Westerink BHC, Cremers TIFH (eds) Handbook of microdialysis. Elsevier, The Netherlands
69. Westerink BH, Damsma G, Rollema H, De Vries JB, Horn AS (1987) Scope and limitations of in vivo brain dialysis: a comparison of its application to various neurotransmitter systems. Life Sci 41:1763–1776
70. Li Z, Boubriak OA, Urban JP, Cui ZF (2006) Microdialysis for monitoring the process of functional tissue culture. Int J Artif Organs 29:858–865
71. Consolo S, Wu CF, Fiorentini F, Ladinsky H, Vezzani A (1987) Determination of endogenous acetylcholine release in freely moving rats by transstriatal dialysis coupled to a radioenzymatic assay: effect of drugs. J Neurochem 48:1459–1465
72. Imperato A, Di Chiara G (1984) Trans-striatal dialysis coupled to reverse phase high performance liquid chromatography with electrochemical detection: a new method for the study of the in vivo release of endogenous dopamine and metabolites. J Neurosci 4:966–977
73. Kalen P, Strecker RE, Rosengren E, Bjorklund A (1988) Endogenous release of neuronal serotonin and 5-hydroxyindoleacetic acid in the caudate-putamen of the rat as revealed by intracerebral dialysis coupled to high-performance liquid chromatography with fluorimetric detection. J Neurochem 51: 1422–1435
74. L'Heureux R, Dennis T, Curet O, Scatton B (1986) Measurement of endogenous noradrenaline release in the rat cerebral cortex in vivo by transcortical dialysis: effects of drugs affecting noradrenergic transmission. J Neurochem 46:1794–1801
75. Benveniste H, Diemer NH (1987) Cellular reactions to implantation of a microdialysis tube in the rat hippocampus. Acta Neuropathol 74:234–238

76. Clapp-Lilly KL, Roberts RC, Duffy LK, Irons KP, Hu Y, Drew KL (1999) An ultrastructural analysis of tissue surrounding a microdialysis probe. J Neurosci Methods 90:129–142
77. Zhou F, Braddock JF, Hu Y, Zhu X, Castellani RJ, Smith MA, Drew KL (2002) Microbial origin of glutamate, hibernation and tissue trauma: an in vivo microdialysis study. J Neurosci Methods 119:121–128
78. Rada P, Mendialdua A, Hernandez L, Hoebel BG (2003) Extracellular glutamate increases in the lateral hypothalamus during meal initiation and GABA peaks during satiation: microdialysis measurement every 30 s. Behav Neurosci 117:222
79. Rossell S, Gonzalez LE, Hernandez L (2003) One-second time resolution brain microdialysis in fully awake rats. Protocol for the collection, separation and sorting of nanoliter dialysate volumes. J Chromatogr B Analyt Technol Biomed Life Sci 784:385–393
80. Cremers TI, de Vries MG, Huinink KD, van Loon JP, v d Hart M, Ebert B, Westerink BH, De Lange EC (2009) Quantitative microdialysis using modified ultraslow microdialysis: direct rapid and reliable determination of free brain concentrations with the MetaQuant technique. J Neurosci Methods 178: 249–254
81. Timmerman W, Westerink BH (1997) Brain microdialysis of GABA and glutamate: what does it signify? Synapse 27:242–261
82. Xi ZX, Shen H, Baker DA, Kalivas PW (2003) Inhibition of non-vesicular glutamate release by group III metabotropic glutamate receptors in the nucleus accumbens. J Neurochem 87:1204–1212
83. Arnth-Jensen N, Jabaudon D, Scanziani M (2002) Cooperation between independent hippocampal synapses is controlled by glutamate uptake. Nat Neurosci 5:325–331
84. Rusakov DA, Lehre KP (2002) Perisynaptic asymmetry of glia: new insights into glutamate signalling. Trends Neurosci 25: 492–494
85. Lada MW, Vickroy TW, Kennedy RT (1997) High temporal resolution monitoring of glutamate and aspartate in vivo using microdialysis on-line with capillary electrophoresis with laser-induced fluorescence detection. Anal Chem 69:4560–4565
86. Pena F, Tapia R (2000) Seizures and neurodegeneration induced by 4-aminopyridine in rat hippocampus in vivo: role of glutamate- and GABA-mediated neurotransmission and of ion channels. Neuroscience 101:547–561
87. Reid MS, Herrera-Marschitz M, Kehr J, Ungerstedt U (1990) Striatal dopamine and glutamate release: effects of intranigral injections of substance P. Acta Physiol Scand 140:527–537
88. Dijk SN, Francis PT, Stratmann GC, Bowen DM (1995) NMDA-induced glutamate and aspartate release from rat cortical pyramidal neurones: evidence for modulation by a 5-HT1A antagonist. Br J Pharmacol 115: 1169–1174
89. Palmer AM, Hutson PH, Lowe SL, Bowen DM (1989) Extracellular concentrations of aspartate and glutamate in rat neostriatum following chemical stimulation of frontal cortex. Exp Brain Res 75:659–663
90. Kapoor V, Nakahara D, Blood RJ, Chalmers JP (1990) Preferential release of neuroactive amino acids from the ventrolateral medulla of the rat in vivo as measured by microdialysis. Neuroscience 37:187–191
91. Paleckova V, Palecek J, McAdoo DJ, Willis WD (1992) The non-NMDA antagonist CNQX prevents release of amino acids into the rat spinal cord dorsal horn evoked by sciatic nerve stimulation. Neurosci Lett 148:19–22
92. Sorkin LS, McAdoo DJ (1993) Amino acids and serotonin are released into the lumbar spinal cord of the anesthetized cat following intradermal capsaicin injections. Brain Res 607:89–98
93. Allchin RE, Batten TF, McWilliam PN, Vaughan PF (1994) Electrical stimulation of the vagus increases extracellular glutamate recovered from the nucleus tractus solitarii of the cat by in vivo microdialysis. Exp Physiol 79:265–268
94. Liu N, Ho IK, Rockhold RW (1999) Contribution of glutamatergic systems in locus coeruleus to nucleus paragigantocellularis stimulation-evoked behavior. Pharmacol Biochem Behav 63:555–567
95. Tucci S, Contreras Q, Paez X, Gonzalez L, Rada P, Hernandez L (2000) Medial prefrontal transection enhances social interaction. II: Neurochemical studies. Brain Res 887: 259–265
96. Clement P, Sarda N, Cespuglio R, Gharib A (2004) Changes occurring in cortical NO release and brain NO-synthases during a paradoxical sleep deprivation and subsequent recovery in the rat. J Neurochem 90: 848–856
97. Giros B, Jaber M, Jones SR, Wightman RM, Caron MG (1996) Hyperlocomotion and indifference to cocaine and amphetamine in

mice lacking the dopamine transporter. Nature 379:606–612

98. Xin Q, Wightman RM (1998) Simultaneous detection of catecholamine exocytosis and Ca2+ release from single bovine chromaffin cells using a dual microsensor. Anal Chem 70:1677–1681
99. Schuvailo OM, Soldatkin OO, Lefebvre A, Cespuglio R, Soldatkin AP (2006) Highly selective microbiosensors for in vivo measurement of glucose, lactate and glutamate. Anal Chim Acta 573–574:110–116
100. Pomerleau F, Day BK, Huettl P, Burmeister JJ, Gerhardt GA (2003) Real time in vivo measures of L-glutamate in the rat central nervous system using ceramic-based multisite microelectrode arrays. Ann NY Acad Sci 1003:454–457
101. Calia G, Rocchitta G, Migheli R, Puggioni G, Spissu Y, Bazzu G, Mazzarello V, Lowry J, O'Neill R, Desole M (2009) Biotelemetric monitoring of brain neurochemistry in conscious rats using microsensors and biosensors. Sensors 9:2511–2523
102. Schuhmann W (2002) Amperometric enzyme biosensors based on optimised electron-transfer pathways and non-manual immobilisation procedures. J Biotechnol 82:425–441
103. Wilson GS, Hu Y (2000) Enzyme-based biosensors for in vivo measurements. Chem Rev 100:2693–2704
104. Qin S, van der Zeyden M, Oldenziel WH, Cremers TI, Westerink BH (2008) Microsensors for in vivo measurement of glutamate in brain tissue. Sensors 8: 6860–6884
105. Burmeister JJ, Moxon K, Gerhardt GA (2000) Ceramic-based multisite microelectrodes for electrochemical recordings. Anal Chem 72:187–192
106. McMahon CP, Rocchitta G, Serra PA, Kirwan SM, Lowry JP, O'Neill RD (2006) Control of the oxygen dependence of an implantable polymer/enzyme composite biosensor for glutamate. Anal Chem 78:2352–2359
107. Rothwell SA, Kinsella ME, Zain ZM, Serra PA, Rocchitta G, Lowry JP, O'Neill RD (2009) Contributions by a novel edge effect to the permselectivity of an electrosynthesized polymer for microbiosensor applications. Anal Chem 81:3911–3918
108. Oldenziel WH, de Jong LA, Dijkstra G, Cremers TI, Westerink BH (2006) Improving the performance of glutamate microsensors by purification of ascorbate oxidase. Anal Chem 78:2456–2460
109. Wahono N, Qin S, Oomen P, Cremers TI, de Vries MG, Westerink BH (2012) Evaluation of permselective membranes for optimization of intracerebral amperometric glutamate biosensors. Biosens Bioelectron 33:260–266
110. Hascup ER, af Bjerken S, Hascup KN, Pomerleau F, Huettl P, Stromberg I, Gerhardt GA (2009) Histological studies of the effects of chronic implantation of ceramic-based microelectrode arrays and microdialysis probes in rat prefrontal cortex. Brain Res 1291:12–20
111. Hashimoto A, Oka T, Nishikawa T (1995) Extracellular concentration of endogenous free D-serine in the rat brain as revealed by in vivo microdialysis. Neuroscience 66:635–643
112. Moghaddam B (1993) Stress preferentially increases extraneuronal levels of excitatory amino acids in the prefrontal cortex: comparison to hippocampus and basal ganglia. J Neurochem 60:1650–1657
113. Jabaudon D, Shimamoto K, Yasuda-Kamatani Y, Scanziani M, Gahwiler BH, Gerber U (1999) Inhibition of uptake unmasks rapid extracellular turnover of glutamate of nonvesicular origin. Proc Natl Acad Sci USA 96:8733–8738
114. Hascup ER, Hascup KN, Stephens M, Pomerleau F, Huettl P, Gratton A, Gerhardt GA (2010) Rapid microelectrode measurements and the origin and regulation of extracellular glutamate in rat prefrontal cortex. J Neurochem 115:1608–1620
115. Reyes E, Rossell S, Paredes D, Rada P, Tucci S, Gonzalez LE, Hernandez L (2002) Haloperidol abolished glutamate release evoked by photic stimulation of the visual cortex in rats. Neurosci Lett 327:149–152
116. Silva E, Hernandez L, Quinonez B, Gonzalez LE, Colasante C (2004) Selective amino acids changes in the medial and lateral preoptic area in the formalin test in rats. Neuroscience 124:395–404
117. Miele M, Boutelle MG, Fillenz M (1996) The source of physiologically stimulated glutamate efflux from the striatum of conscious rats. J Physiol 497(Pt 3):745–751
118. Saulskaya NB, Mikhailova MO (2002) Feeding-induced decrease in extracellular glutamate level in the rat nucleus accumbens: dependence on glutamate uptake. Neuroscience 112:791–801

Chapter 8

Sub-Second Measurements of Glutamate and Other Neurotransmitter Signaling Using Enzyme-Based Ceramic Microelectrode Arrays

Erin R. Hascup, Kevin N. Hascup, Pooja M. Talauliker, David A. Price, Francois Pomerleau, Jorge E. Quintero, Peter Huettl, Alain Gratton, Ingrid Strömberg, and Greg A. Gerhardt

Abstract

We have set out to develop a novel, implantable microelectrode array that has the capabilities to detect neurotransmitters with enhanced sensitivity, selectivity, and temporal sampling capabilities compared to other current technologies. We have shown that this device maintains recording performance during chronic measurements of extracellular neurotransmitter levels for at least 7 days postimplantation, single-unit neuronal activity for as long as 6 months, and provides enhanced biocompatibility compared to current technologies. As we continue to refine and improve our recording capability, we are able to incorporate the chronic microelectrode array technology into multimodal experimental paradigms, such as behavioral testing, pharmacological intervention (local and systemic), or combined measurements of neurotransmitter levels and neuronal activity (local field potential). Furthermore, the improvements made with the microelectrode technology discussed in this chapter have the potential to conduct longitudinal analyses that can benefit a wide range of translational efforts, including studies on learning and memory, aging, neurodegenerative disease progression, and traumatic brain injury neuropathology.

Key words Biosensor, Chronic, Glutamate, Histology, Behavior, Neurochemistry

1 Introduction

The ability to combine direct neurotransmitter measures with awake behaving paradigms provides a valuable means to correlate neurochemical data with real-time changes in central nervous system (CNS) function. Because behavioral assessments can require extended animal training, an investigative or diagnostic technique that allows chronic neurotransmitter measurements over multiple days or weeks is needed. Historically, microdialysis has been the gold standard for in vivo neurochemical measurements during

Stéphane Marinesco and Nicholas Dale (eds.), *Microelectrode Biosensors*, Neuromethods, vol. 80,
DOI 10.1007/978-1-62703-370-1_8, © Springer Science+Business Media, LLC 2013

behavior. However, the routine sampling rate of this technique (5–20 min) is a significant limitation for measuring the rapid dynamics (milliseconds–seconds) of fast signaling neurotransmitters such as L-glutamate and acetylcholine in the brain (1–3). Although recent advancements have allowed for subminute microdialysis measures, this methodology is not practical for routine use in most laboratories due to the small sample sizes and/or requirement for on-line HPLC techniques (4–6). Furthermore, chronic microdialysis experiments are generally limited to 6–8 h in duration, although in some instances may be longer (1, 7). Additionally, the large size of the microdialysis probes (1–4 mm length, 0.2–0.3 diameter) produces considerable tissue damage (7). Alternate technologies with improved spatial-temporal resolution have been developed using electrochemical methods with microelectrodes. Both carbon fiber-based microelectrodes and hydrogel-based microelectrodes have been successfully used for measuring rapid changes in CNS levels of dopamine and other electroactive monoamines (8, 9) but have had limited success for monitoring other non-electroactive neurotransmitters such as glutamate and acetylcholine largely due to the inherent lack of self-referencing capabilities, limits of detection, or temporal resolution. Our laboratory and others have made great strides towards developing biosensors capable of measuring these non-electroactive neurotransmitters (10–12).

We have taken the features of both techniques and incorporated refinements to develop an advanced technology using ceramic-based microelectrode arrays (MEAs) for chronic neurotransmitter measurements in rodents and nonhuman primates (13, 14). As discussed in Chap. 2, there are numerous neurochemicals that can be measured using the MEA technology in conjunction with constant voltage amperometry and the Fast Analytical Sensing Technology (FAST16mkII; Quanteon, LLC, Nicholasville, KY) recording system. While the experiments and examples discussed in this chapter can be applied to other neurochemicals, we have the most extensive experience measuring the major excitatory neurotransmitter in the CNS, L-glutamate (13–15). Therefore, in this chapter we will focus on glutamatergic neurotransmission. The current state of the technology allows for reliable glutamate measurements for over 7–10 days in the striatum, prefrontal cortex, or hippocampus (13, 14) in awake mice and rats. Our MEAs also have the capability to record single-unit neuronal activity (16) as well as local field potentials (17, 18). In fact, chronic MEAs implanted into the rat hippocampus can measure multiple single-unit activity for as long as 6 months (16). In this chapter, we will identify and discuss technical issues associated with these recordings in awake rats and mice using ceramic enzyme-based MEAs. The primary focus will be on the self-referencing recording technique (see below for a description), but we will also discuss the biocompatibility of the MEAs, which is a key issue surrounding the use of intracerebral devices. In addition, examples of the current capabilities of the MEAs are incorporated throughout the chapter.

2 Materials and Methods

2.1 MEA Preparation and Chronic Implantation

2.1.1 Ceramic MEA Preparation

While the MEAs used for freely moving animals are a fabricated product and ready for use in anesthetized recordings, they are also designed to be readily modified into the pedestal MEA assembly (a modified MEA consisting of the MEA, varnished copper wires, gold plated amphenols, a reference electrode, and a miniature connector) discussed here. MEAs are manufactured, assembled, and selected for in vivo recordings using previously described performance criteria (19, 20). Preparation of the MEA for in vivo recording has been thoroughly documented (20, 21). For glutamate recordings, the platinum (Pt) recording sites are dip coated with Nafion® (Sigma-Aldrich Corp., St. Louis, MO; catalogue #27470-4) to repel anions or electroplated with meta-phenylenediamine (mPD; Acros #33360-50 g-f mPD) as previously described (14, 22). One of the MEA recording sites or recording pairs is coated with an L-glutamate oxidase (GluOx; EC 1.4.3.11; US Biological, Swampscott, MA; catalogue #G4001-01) coating solution as previously described (23). In order to cross-link GluOx and increase its adhesion to the MEA, glutaraldehyde (Sigma-Aldrich Corp., St. Louis, MO; catalogue #G5882) is added with bovine serum albumin (BSA; Sigma-Aldrich Corp., St. Louis, MO; catalogue #A3059) to stabilize the GluOx solution on the surface of the MEA. The GluOx layer is required to measure glutamate, as it enzymatically breaks down glutamate into α-ketoglutarate and produces the electroactive reporter molecule, hydrogen peroxide (H_2O_2). When a potential of +0.7 V vs. a Ag/AgCl reference electrode is applied to the MEA surfaces, the reporter molecule, H_2O_2, is oxidized upon contact with the Pt recording surface yielding two electrons. The resulting current is then amplified and digitally recorded by the FAST16mkII recording system. The remaining Pt recording sites we term "sentinel sites," because they are coated in a similar manner as the glutamate recording sites, with the exception that the coating solution does not contain GluOx (Fig. 1). As previously discussed, the sentinel sites are not capable of converting glutamate to the H_2O_2 reporter molecules but are capable of detecting electroactive interferents not repelled by Nafion® or mPD. When the current recorded on the sentinel sites is subtracted from the current on the glutamate recording sites, the resulting signal represents the tonic or resting levels of glutamate in the extracellular space of the brain (13, 15, 22).

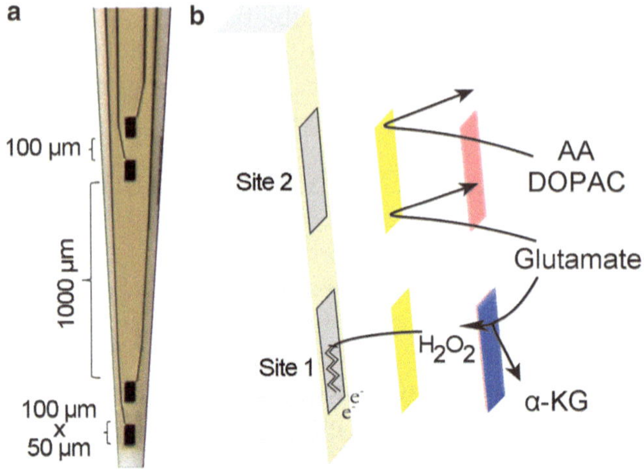

Fig. 1 (a) Photomicrograph of ceramic MEA tip (R2 layout) with latest design features and (**b**) schematic diagram of enzyme layers for (Site 1) active recording site with the GluOx/BSA/glutaraldehyde coat in *blue* and (Site 2) sentinel site with the BSA/glutaraldehyde coat in *pink* for the R2 MEA. *Yellow* indicates exclusion layer, either Nafion® or mPD

2.1.2 Ceramic MEA Pedestal Preparation

The MEA is modified for use in freely moving animals as previously described (13–15). Briefly, the pedestal is the system component that is chronically implanted intracerebrally and contains the connector, the MEA, and reference electrode. During recording sessions, the pedestal is connected to a miniaturized 4-channel amplifier (Rat Hat) that, in turn, is connected to a low torque commutator (Quanteon, LLC). Ceramic-based MEAs are prepared for freely moving recordings by first cleaning the MEA with isopropyl alcohol and ddH_2O. MEAs are dried for 5 min at 180 °C to make sure there is no residual water on the sites. Next, the MEAs are coated with Nafion® and cured at 180 °C for 4 min before applying coating solutions (with or without GluOx) as previously described (13–15). Alternatively, the MEA can be electroplated with mPD to block interferents (14). To date, we have had the best experience with Nafion® for recording periods as long as 17 days or more (13, 14).

In order to attach the miniature connector to the modified MEA for freely moving recordings (together termed the pedestal), connecting wires (30 AWG varnished copper wire, RadioShack, Fort Worth, TX) are used. The copper wires are tucked around the miniature connector (Ginder Scientific, Ottawa, ON #GS09PLG-220), and the MEA tip is carefully positioned parallel to the connector. Correct alignment of the MEA shank is essential for accurate MEA placement into the brain. If the MEA is slightly askew or angled, the actual recording site location may be off from the desired stereotactic coordinates. Waterproof 5 min epoxy (The Original Super Glue Corporation, Rancho Cucamonga, CA) is used to secure the paddle and wires to the round connector, making sure that the exposed part of the amphenols on the end of the connector closest to the MEA tip and wires are covered with epoxy. This prevents moisture from penetrating the pedestal. The completed assembly is allowed to cure at room temperature for a minimum of 24 h.

2.1.3 Pedestal/Cannula Assembly

A major capability of these techniques involves locally applying drugs or chemicals into the extracellular space near the MEA for studies addressing neurotransmitter release and uptake or pharmacological regulation of neurotransmission in a specific brain region (15). For these studies, a 26-gauge stainless steel guide cannula (Plastics One, Roanoke, VA; catalogue #C315GRL) can be attached to the MEA assembly using sticky wax (Kerr Corp., Orange, CA; #00625). A stainless steel dummy cannula (Plastics One, Roanoke, VA; #C315IDC), 1 mm longer than the guide cannula, is inserted into the guide cannula and screwed into place, such that the dummy cannula tip is aimed among the Pt recording sites at a distance of ~100 μm above the MEA tip. Ejections of substances can be performed through a 33-gauge internal cannula (Plastics One, Roanoke, VA; #C315I) that protrudes 1 mm past the guide cannula. The stainless steel dummy cannula should remain in the guide whenever the internal cannula is not in use to prevent blockage. While this method has unique advantages for in vivo recordings, it should be noted that the cannula assembly can add additional variables to the study design that should be considered, such as movement of fluids up the cannula shank during ejections, possible contamination of the brain, and possible additional tissue damage. These aspects have been recently discussed (15) and must be considered when using this method. Routine histological evaluation of all targeted coordinates is required to confirm the stereotactic recording site and to ensure tissue integrity around the recording site.

2.1.4 MEA Calibration

MEAs are calibrated to determine their sensitivity and selectivity for glutamate against ascorbic acid and other potential interferents before implantation (14, 15). Briefly, constant potential amperometry is performed using the FAST16mkII system in combination with a headstage current-to-voltage converter of 2 pA/mV (operated at a constant applied potential of +0.7 V vs. a Ag/AgCl reference electrode). The tip of the pedestal MEA is placed in a continuously stirred solution of 0.05 M PBS (pH 7.4). A recirculating water bath (Gaymar Industries, Inc., Orchard Park, NY; catalogue #TP500) is used to maintain a constant buffer temperature of 37 °C allowing the enzyme layer to function optimally relative to in vivo conditions. Calibrations are performed using final buffer concentrations of 250 μM ascorbic acid and 20, 40, and 60 μM glutamate through additions of aliquots of 20 mM stock solutions. Selectivity ratios for glutamate over ascorbic acid are calculated in addition to the slope (sensitivity), limit of detection (LOD), and linearity (R^2) for glutamate for all MEAs. Prior to building the pedestal, the MEAs are evaluated to determine the consistency between the four Pt recording sites using H_2O_2 (8.8 μM final concentration) as the test substance. Once the pedestal is ready and the MEA is enzyme coated, H_2O_2 is used at the end of the glutamate calibration to verify again the integrity of the Pt recording sites. MEAs are used only if the glutamate responding sites and the corresponding sentinel site response to H_2O_2 is within 20 % of each other. Finally, all substances used for the experiment, locally or systemically, should be added as test substances to ensure that they are not electrochemically active at +0.7 V vs. Ag/AgCl to confirm they do not interfere with the glutamate recordings.

2.1.5 Reference Electrodes for In Vivo Recordings

Perhaps one of the most overlooked requirements for an in vivo electrochemical reference electrode is that it must hold its half-cell potential for the duration of the experiment. This is an absolute requirement for accurate and stable in vivo electrochemical measurements (24). The miniature Ag/AgCl reference electrodes we use for in vivo studies have proven to be very reliable. They are prepared by stripping 2 mm of Teflon from each end of a Teflon-coated silver wire (200 μm bare, 275 μm coated; A-M Systems, Inc., Carlsberg, WA; #786500). One end is soldered to a gold-plated amphenol (Ginder Scientific, Ottawa, ON; #220-S02), and the other end is coated with AgCl in a plating bath. Plating is accomplished by placing the tip of the stripped silver wire into a 1.0 M HCl plating bath saturated with NaCl containing a Pt wire as a counter electrode and applying +9 V DC using a power supply to the silver wire vs. the Pt wire for ~10 min. When successfully plated, the tip of the plated silver wire should be a light gray in color. The resulting Ag/AgCl electrode is soaked in 3 M NaCl for at least 1–2 h and kept out of direct light until use.

2.1.6 Electrode Implantations

After a successful calibration of the MEA, rats or mice are anesthetized with ~2 % isoflurane and placed in a stereotaxic apparatus (Kopf Instruments, Tujunga, CA). Animal body temperature is maintained at 37 °C with a heating pad (Gaymar Industries, Inc., Orchard Park, NY; catalogue #TP500), and the animals' eyes are lubricated with artificial tears (The Butler Company, Columbus, OH; #007312) to help maintain moisture and prevent infection. Prior to an incision, the skin directly on top of the animal's head is wiped with Betadine® solution to keep the incision area clean and to prevent infection, and then reflected, making as small an incision as possible. First, three small holes are drilled in the skull in the quadrants adjacent to where the MEA is implanted for placement of stainless steel skull screws. A fourth hole is drilled contralateral from the recording site for insertion of the Ag/AgCl reference electrode wire. Next, three small stainless steel screws

(Small Parts, Inc., Miami Lakes, FL; #B000FN0K02) are threaded into the skull to serve as anchors and care is taken so that the screw tips do not touch brain tissue or dura. A small (up to 2 mm × 2 mm) craniotomy is performed over the recording area and a glutamate selective MEA pedestal assembly is implanted based on stereotactic coordinates (25). The assembly is secured with approximately four layers of dental acrylic (Lang Dental MFG, Wheeling, IL; catalogue #1406), making sure to cover as much of the pedestal assembly as possible. The dental acrylic should have a smooth texture, and excess acrylic should be removed from the skin surface to avoid postsurgical irritation, which may cause the rat to scratch its head and potentially damage the implant.

2.2 Amperometric Recordings Utilizing Self-Referencing MEAs

2.2.1 Importance of Self-Referencing MEAs

The fabrication process for our MEAs (see Chap. 2 for more detail) allows for precise control of the geometric layout of multiple Pt recording sites within a small area. Having multiple recording sites in close proximity is particularly important because it allows us to use a technique we call self-referencing to (1) remove chemical interferents that may contribute to a portion of the analyte signal, (2) subtract noise from the signal, and (3) measure both tonic levels and transient changes (22). This is a unique recording feature of the multisite MEAs that cannot be carried out using single microelectrode recordings and is a major strength of the MEA technology. One drawback to the geometrical layout of the current MEAs is the inability to adequately coat side-by-side pairs with different solutions (i.e., one recording and one self-referencing). We are addressing this deficiency twofold. First, we are developing a micro-coater that will be capable of differentially coating side-by-side pairs. Second, we are developing a MEA that will have geometrically identical recording sites on both the front and back of the electrode which would allow for one side to be the recording side and the reverse side to be the self-referencing side.

The principle of the self-referencing MEA recording technology is that one recording site is capable of recording the compound of interest, such as L-glutamate, while an adjacent site cannot (see Fig. 1) (13, 22, 26). This is achieved by an enzyme that is very selective for glutamate. GluOx is cross-linked to the surface of the microelectrode (site 1 shown in Fig. 1b) such that when glutamate contacts the enzyme layer, GluOx converts glutamate into α-ketoglutarate plus the reporter molecule H_2O_2. The H_2O_2 is oxidized at the Pt microelectrode with the resulting measured current being directly proportional to the concentration of glutamate. Recently, we have further improved the selectivity of the microelectrodes by coating all MEA Pt surfaces with mPD. This produces a size exclusion layer, which repels all molecules larger than peroxide. This has greatly increased the selectivity of the MEAs such that they do not measure larger aromatic molecules such as dopamine, serotonin, norepinephrine, and other molecules including uric acid, ascorbic acid, 3,4-dihydroxyphenylacetic acid (DOPAC), or other potential interferents. However, some compounds like nitric oxide or other unknown electrochemically active substances can potentially penetrate the mPD layer of the MEA and be detected. This has been a major limitation of prior work by many laboratories using a variety of electrode types. The concept of our ceramic-based arrays involves using an adjacent sentinel recording site for self-referencing (site 2 in Fig. 1b), which not only removes unwanted interferents but also more importantly the non-Faradaic background current. This sentinel site contains all of the materials identical to the recording MEA but does not have the oxidase enzyme cross-linked to the surface. Our previous studies support that the mPD coatings and enzyme layers readily adhere to these surfaces (13, 14). Thus, when we subtract the sentinel response (Fig. 1b, site 2) from the active recording response (Fig. 1b, site 1), we get a nearly specific measure of glutamate. This methodology is capable of measuring glutamate concentrations as low as 0.1 μM ($S/N=3$, $n=20$) with a spatial resolution of 50 μm × 100 μm and a tip width of 65 μm (for the R2 MEA design shown in Fig. 1). MEAs have been used for numerous studies for chronic recordings in rats and mice for up to 17 days postimplantation (13–15, 21). In addition, with improved MATLAB®-based software, we can now readily analyze 24-h+ recording periods in awake rats and can measure "spontaneous transients" of glutamate and potassium-induced phasic changes of glutamate in awake animals.

2.3 Applications to Biological Systems

2.3.1 Studies in Freely Moving Rats and Mice

Recordings of glutamate and other neurotransmitters in freely moving animals have several distinct advantages. This approach allows for monitoring glutamate neurotransmission without the effects of anesthesia, which has been shown to alter mostly tonic glutamate levels and in some circumstances phasic glutamate release dynamics (13, 27, 28). More importantly neurochemical measurements can be coupled to behavioral studies to investigate changes in glutamate that are related to inherent or learned behaviors. Additionally, the MEAs can reliably record glutamate for at least 1 week in freely moving animals (13, 14). This means that multiple independent or continuous measures can be more readily made over repeated days as compared to most microdialysis studies. Furthermore, this has direct application to behavioral paradigms or disease state models where there is the likelihood that tonic and/or phasic changes in neurotransmitters or neurochemicals may be altered over the time course of the experiment, such as studies in addiction, sensitization, circadian rhythms, epilepsy, and traumatic brain injury.

2.3.2 Tonic Release of Neurotransmitters

As mentioned earlier, another advantage of the self-referencing technique is the ability to reliably measure the low levels of tonic neurotransmitter release on a sub-second basis due to the close proximity of the recording sites to one another. Tonic, or resting, measures refer to the level of neurotransmitters constantly found in the extracellular space that contribute to CNS homeostasis. Even though the MEAs have exclusion layers to improve the selectivity, interferents such as ascorbic acid, can exist in the extracellular space of the brain at concentrations 100–1,000 times higher than the analyte of interest in vivo and may contribute to the signal associated with tonic measures (29, 30). However, use of the self-referencing technique allows for essentially all interferents to be subtracted from recorded signals, resulting in measures of true tonic neurotransmitter levels.

Resting levels of glutamate in the rat can be affected by drugs that alter neuronal release and glial/neuronal uptake of glutamate. Regulation of tonic glutamate levels in the CNS involves a variety of both neuronal and glial processes. Using the MEA technology, our recent pharmacological studies in awake rats have shed light on the origins of tonic and phasic glutamate signals (13, 15). Results obtained from the rat prefrontal cortex are seen in Fig. 2. Compared to vehicle, local application of the Na^+ channel blocker TTX (3 μl, 100 μM) or the Ca^{2+} channel blocker ω-conotoxin (1 μl, 3 μM) produced a significant decrease (~40–50 %) in resting glutamate. Additionally, we examined the contribution of metabotropic glutamate receptors group 2/3 ($mGluR_{2/3}$), which resides primarily on presynaptic neurons in the prefrontal cortex and act as inhibitory autoreceptors, to resting glutamate levels. The $mGluR_{2/3}$ antagonist LY341495 (3 μl, 100 μM) produced a ~35 % increase in resting levels, while the $mGluR_{2/3}$ agonist LY379268 (3 μl, 100 μM) caused a ~20 % decrease, supporting a significant neuronal contribution to the glutamate signal. Local administration of the nonselective glutamate transporter inhibitor, D,L-*threo*-β-benzyloxy aspartate (TBOA; 1 μl, 100 μM), produced a ~120 % increase in resting levels, supporting the idea of regulation of extracellular glutamate levels by glutamate transporters, located mostly on glia. Using microdialysis, it has been suggested that the cystine/glutamate exchanger is involved in glutamate homeostasis (31), but in our paradigm, local application of the cystine/glutamate antiporter inhibitor (*S*)-4-carboxyphenylglycine (CPG; 3 μl, 50 μM) did not significantly alter resting levels. Thus, the MEA technology is capable of recording extracellular resting glutamate levels that are regulated by neuronal and non-neuronal processes. However, it is important to note that the neuronal dependence of these measures far exceeds that previously observed with microdialysis (3).

A key issue surrounding the use of the MEA technology is the concentrations that are measured by the technique as compared to microdialysis. Our recent publications support that we can reliably measure resting glutamate levels over at least 7–10 days in vivo (13, 14). However, the MEAs measure concentration values of resting glutamate in the range of 7–13 μM, which contrasts with microdi-

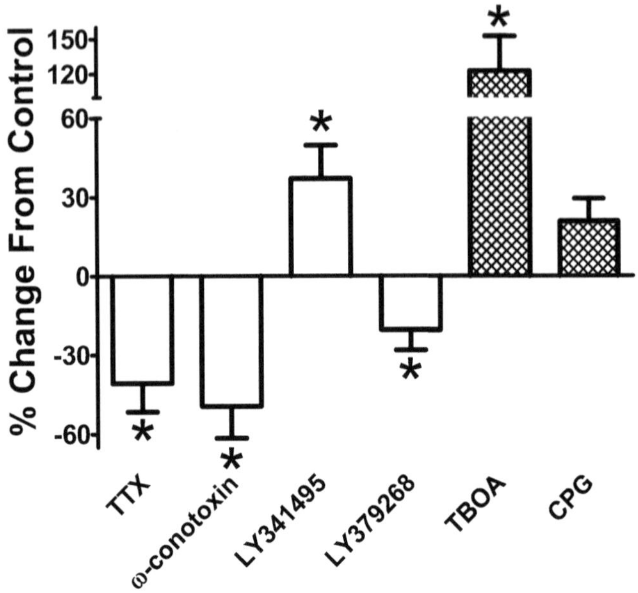

Fig. 2 Effects of local application of drugs that affect neuronal (*open bars*) and glial (*hatched bars*) regulation of glutamate in the rat prefrontal cortex. Note that TTX, ω-conotoxin, and LY379268 all decreased resting glutamate levels, supporting a significant neuronal origin of the resting glutamate signal. The increase produced by LY341495, the $mGluR_{2/3}$ antagonist, also supports a neuronal origin of the resting glutamate signal. TBOA affects glutamate uptake primarily by affecting glutamate uptake into glia. *$p < 0.05$

alysis studies reporting levels in the range of 0.3–2 μM (32–36). In contrast to microdialysis, the MEA technology involves second-by-second measures with a relatively small microelectrode that resides directly within the extracellular space of the brain—residing closer to glutamatergic synapses and causing less tissue damage (16, 37). MEA signals should be more accurate, in general, due to less dilution and less loss of signal due to diffusion (38). More importantly, microdialysis methods typically exhibit in vitro recoveries of 10–20 %. Thus, the MEA signals are 5–10 times greater than those seen with microdialysis, which is expected from the technology. Furthermore, the larger resting levels measured with the MEAs are not considered excitotoxic based on the known affinity of the high affinity glial and neuronal glutamate transporters, which are in the 1–100 μM range, depending on the assay and subtype (39).

2.3.3 Phasic Release of Neurotransmitters

Synchronized or burst firing of neuronal activity due to naturally occurring behaviors can lead to phasic release of neurotransmitters. As with an artificial stimulus, such as pharmacological depolarization with KCl or electrical stimulation, behavioral events can lead to the release of a multitude of neurotransmitters. We have recently shown that tonic (resting) and phasic (transient release) MEA recordings are dissociable events. For example, initial MEA studies conducted in rats under anesthesia at various ages revealed that tonic levels of glutamate are often minimally affected by normal aging, whereas phasic glutamate release stimulated by local application of potassium results in dynamic age-related decreases in presynaptic release and glial-mediated clearance of glutamate (23, 40, 41). The rapid temporal resolution of the MEA technology may explain why previous results using in vivo microdialysis have not consistently shown age-related changes in glutamate signaling. Our more recent studies have demonstrated that we can measure potassium-evoked release, resting levels, and spontaneous release of glutamate in awake rats. As seen in Fig. 3a, potassium-evoked signals in awake animals are highly reproducible (3.3 ± 0.7 μM, $n=4$). In addition, we have recently discovered that spontaneous transients of glutamate in nucleus accumbens, striatum, and prefrontal cortex occur in awake rats. A representative glutamate transient, determined with the self-referencing technique, is shown in Fig. 3b.

We have recently collected data over a 12/12 h light/dark cycle in rats ($n=4$) with MEAs placed in the prefrontal cortex. As seen in Fig. 4a, resting levels of glutamate maintain a remarkably homeostatic level during light and dark phases somewhat analogous to what was previously observed (42). Using a new analysis package developed in MATLAB® software, we are now able to quantify the phasic release of glutamate. The data are presented in a histogram-type plot showing the amplitudes of spontaneous glutamate events per 30 min bin during the light or dark phases and shown in Fig. 4b. The number of spontaneous transients of glutamate release average ~30 per half hour or 1/min, with amplitudes ranging from 0.5 to 10.0 μM across both the light and dark cycles.

2.3.4 Chronic Implantation Histopathology and Biocompatibility

While our current MEAs can reliably measure glutamate in awake animals for 7–10 days, we and others would like to investigate neurochemical signaling for months and possibly years in a single animal. To date, we are able to perform single-unit neuronal recordings with the MEAs for at least 6 months (unpublished observations, (16)), and we can reliably detect the reporter H_2O_2 for at least 90 days (unpublished observation). The fact that enzyme-coated MEAs are routinely reliable for only 7–10 days likely indicates a change in the enzyme layer, as our recent histology studies and single-unit recordings support a minimal change in the microenvironment surrounding the MEA in the CNS (16). The MEA is predominantly composed of ceramic (Al_2O_3), which has been shown to enhance integration with host tissue (e.g., orthopedic prostheses) when the roughness value of the implant surface is equivalent to the dimensions of bone-forming cells (50–400 μm) (43). However, studies with ceramic substrates on CNS biocompatibility

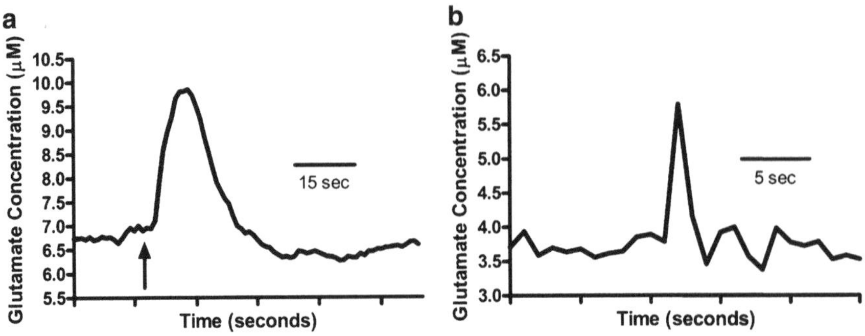

Fig. 3 (**a**) Potassium-evoked phasic release of glutamate in the CA1 of the hippocampus of an awake rat. (**b**) Spontaneous transient release of glutamate recorded from the prefrontal cortex of an awake rat. Note the faster time course but similar amplitude to the potassium-evoked signal

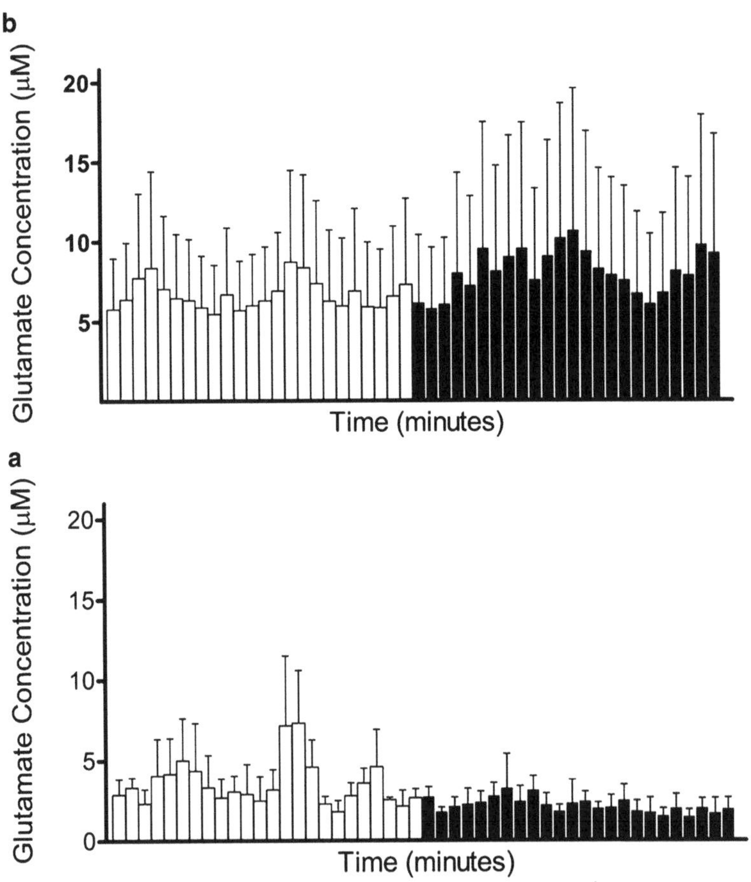

Fig. 4 (**a**) Histograms showing average amplitude of tonic (resting) glutamate per 30 min over a 12 h light/dark cycle from four rats. (**b**) Average amplitude of the phasic glutamate release (spontaneous transients) per 30 min over 24 h. *Open bars* indicate recordings during the light cycle, while *black bars* indicate recordings during the dark cycle. Measures were carried out in the prefrontal cortex with our new recording and analysis system using MATLAB®-based software

are lacking, because the ceramic-based neural devices are a very recent development (13, 14). We have begun to examine the tissue effects of our chronically implanted (1–6 months) MEAs in rats and have seen minimal damage to the surrounding brain tissue in the prefrontal cortex. Observed changes in astrocytes (GFAP) and microglia (Iba1) (13, 16)—the two major cell types participating in the inflammatory reaction to tissue injury following electrode insertion (44, 45) show minimal increases due to MEA implantation. For glutamate investigations, glial changes in the recording environment of the MEA could present a challenge due to the presence of glutamate transporters located on the cell membrane of astrocytes, as they are the primary mechanism by which glutamate is cleared from the extracellular environment (39). In turn, this could affect both tonic and phasic levels of glutamate. While microglia accumulate around the electrode in the days and weeks following implantation (46, 47), we have seen evidence for stabilization of microglia after the MEAs are in vivo for 1 week. Additionally, it has been shown that astrocytes and microglia may encapsulate other types of microelectrodes (silicon, Pt wire references) (44, 45, 48–53), thereby, forming a physical barrier that limits electrode access to healthy neurons (52). However, if this were the case with MEAs, this would preclude recordings of single-unit activity, which we are able to measure for up to 6 months in vivo. The most notable damage we observed with chronic implantation of the ceramic MEAs in the prefrontal cortex is the formation of a relatively small glial scar that is stable after 1 month in vivo and maintains a similar level at 6 months postimplantation (Fig. 6). This is in marked contrast to the effects that occur from other chronically implanted devices (i.e., silicon microelectrodes or microdialysis probes) (16, 50). Recent studies of chronic hippocampal implants have further confirmed this finding (Fig. 5). Following MEA implantation, we only observed a significant increase in GFAP staining in the hippocampus. GFAP staining levels remained unchanged in the prefrontal cortex over the 6-month period in the prior study. Iba1 levels remained unchanged across all time points and structures and there are no differences in staining levels of each marker between structures (Fig. 6). We are encouraged by the lack of damage from these devices and continue to examine what factors (ceramic material, shape, etc.) contribute to the limited tissue changes observed with ceramic MEAs.

In order to improve the long-term performance of the MEAs, we must determine how the MEA surfaces are altered with chronic implantation. Under acute experimental conditions, high quality signal generation is dependent upon electrode sensitivity and selectivity for the analyte of interest as well as total electroactive area and reaction kinetics. To improve these recording parameters, knowledge about intrinsic physical and electrochemical properties of the Pt recording surface such as topography, electrical behavior,

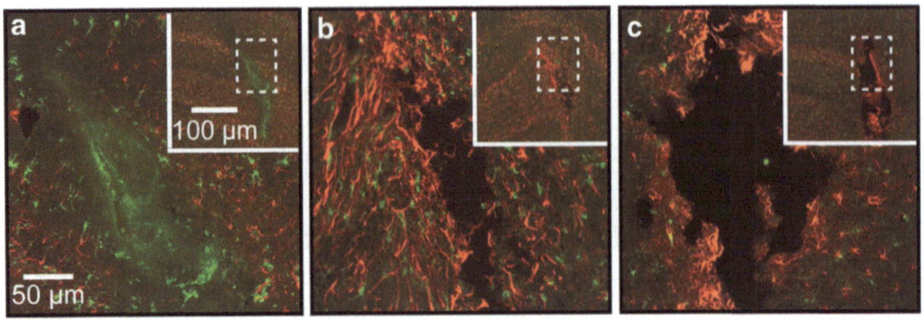

Fig. 5 Postimplantation fluorescent labeling of astrocytes and microglia in the hippocampus. Tissue sections were labeled to identify astrocytes (GFAP, *red*) and microglia (Iba1, *green*) at (**a**) 1 day, (**b**) 8 weeks, and (**c**) 16 weeks after MEA implantation into the hippocampus. Scale bar = 50 μm. Images shown *inset* are low magnification photomicrographs showing the MEA tract within the hippocampus. Scale bar = 100 μm

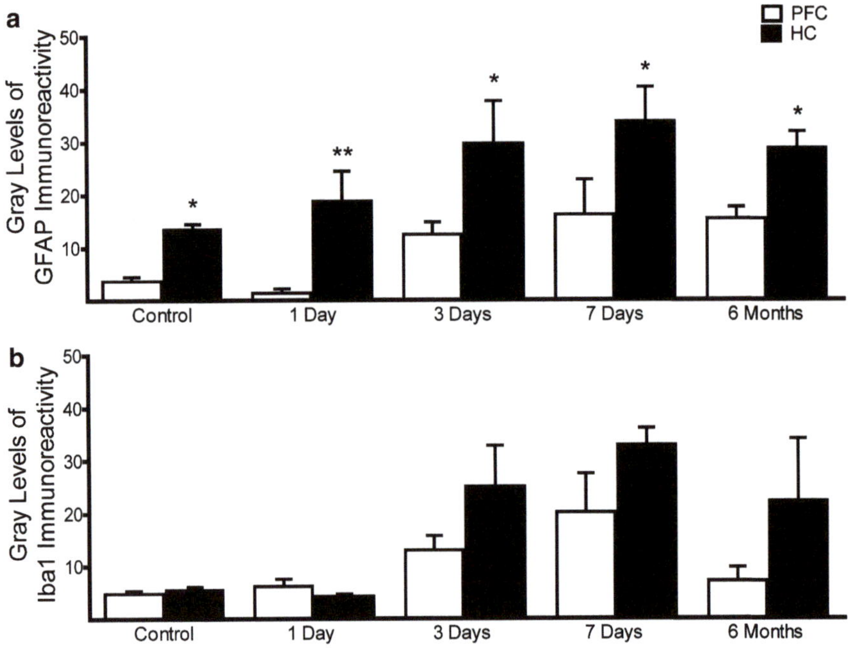

Fig. 6 Interregional comparisons of tissue reactivity to the MEA. Gray levels of GFAP (**a**) and Iba1 (**b**) immunoreactivity were measured adjacent to the MEA tract in the prefrontal cortex (PFC, shown in *white*) and hippocampus (HC, shown in *black*) in two separate groups of rats. Comparisons vs. control tissue involved a one-way ANOVA with a Bonferroni's posttest. At all time points including control tissue, GFAP immunoreactivity was significantly elevated in the hippocampus due to a greater density of astrocytes in this region. $^{*}p<0.05$

chemical reactivity, and molecular composition is necessary. Pt is widely used in amperometric electrodes, both planar and wire based, because of its biocompatibility and high catalytic activity for amperometric measurements of H_2O_2 (24). As previously described, immobilized enzymes convert the analyte of interest into H_2O_2 reporter molecules that are subsequently oxidized on contact with the recording surface where a potential is applied.

It is generally held that the reaction mechanism of H_2O_2 oxidation on Pt surfaces is independent of electrode geometry. However, this similarity may not translate to equivalent recording performance per unit area (54). We recently published a review of the current scientific literature, which revealed that the recording performance per unit area varies for the detection of H_2O_2, glutamate and glucose at different Pt-based electrodes (54). MEA performance rivaled or exceeded that of other electrode subtypes for each analyte (54). For example, the sensitivity of Pt disk electrodes (geometric area ranging from 2 to 490 mm^2) for H_2O_2 detection has been reported by other groups and ranged from 490 to 554 pA μM^{-1} mm^{-2} (54). Thin film MEAs with a geometric area of 0.0025 mm^2 had a sensitivity of 4,880 pA μM^{-1} mm^{-2} for the detection of H_2O_2 (54). Knowledge of the MEA surfaces before and after implantation will be important for improving the long-term performance of the device in vivo. We recently carried out a series of studies that for the first time characterized the intrinsic physical properties and surface topography of an implantable ceramic-based neural device for intracerebral monitoring of neurotransmitter dynamics (54).

Because of the unique processes used in the fabrication of our ceramic-based MEAs (i.e., photolithography and sputter-coated thin film Pt; see K.N. Hascup et al., Chap. 2), we hypothesized that the intrinsic physical properties of the device would provide insight into the previously mentioned unique performance capabilities. We used complimentary investigative techniques, including scanning electron microscopy (SEM) and atomic force microscopy (AFM), to evaluate the surface topography of the MEA device prior to implantation (54). The ceramic substrate, protective layer of polyimide, and Pt recording sites are clearly identified on scanning electron micrographs (Fig. 7a). The uniform scan intensity of the polyimide layer supports that it has a homogenous surface composition for eliminating cross-talk by providing protection of the connecting lines from the aqueous environment. Importantly, Pt recording sites are present in a microwell structure (Fig. 7b). The microwell is a unique feature of the MEA and is an important aspect with regard to the immobilized enzyme layer. Because well-type structures are advantageous with regard to protein immobilization (55), the microwell is important to our continued investigations aimed at refining MEA coating strategies by targeting and restricting enzyme solutions to the microwell structure. Following photolithographic fabrication, individual MEAs are cut out using a computer-controlled diamond saw. In turn, the ceramic sides of the MEA retain a rough appearance (Fig. 7c, d). Perhaps one of the most important findings from our study was the demonstration of nonuniform elevations and depressions on the thin film Pt recording sites, which had a nanoscale roughness (54). A representative 225 μm^2 area of the Pt recording surface used in AFM studies is

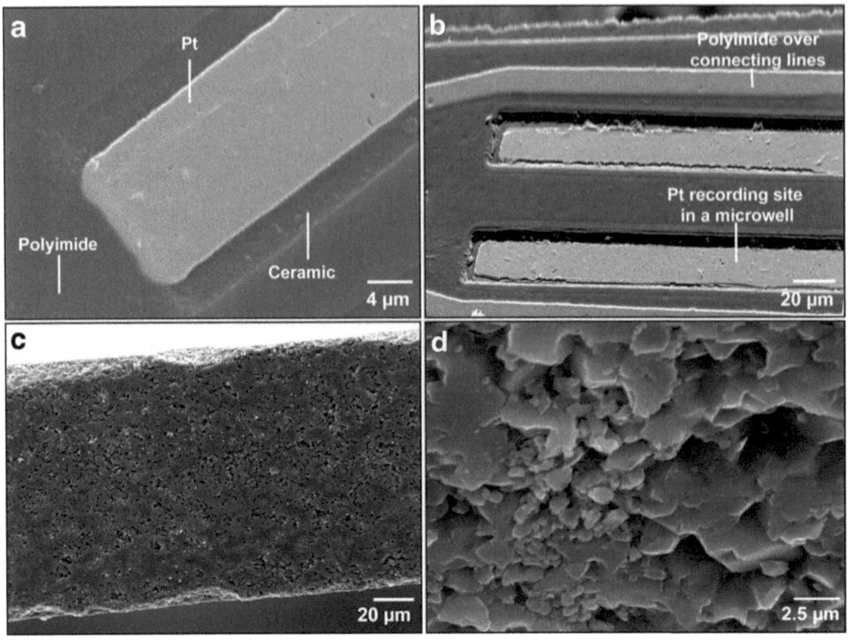

Fig. 7 Scanning electron micrographs reveal general features of the MEA surfaces. (**a**) Micrograph of the front surface of a MEA indicating the ceramic substrate, insulating layer of polyimide, and Pt recording site. (**b**) Micrograph of the front surface of the MEA demonstrating the presence of a Pt recording site in a microwell structure. (**c**) Low and (**d**) high magnification micrograph of the ceramic side of the MEA showing the porosity of the surface

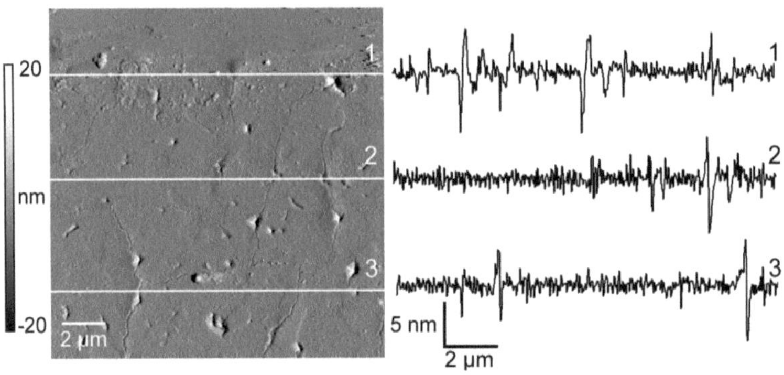

Fig. 8 **(a)** Atomic force microscopy micrograph of a 225 μm^2 sampled region of a Pt recording site indicating the area used for subsequent line scan analysis. (**b**) Line scan analysis demonstrates the presence of nanosized changes in Pt surface height on the MEA surface

shown in Fig. 8a. Line scan analysis demonstrates the presence of nano-sized peaks and indentations on the Pt recording site (Fig. 8b). The presence of nanoscale surface features on the Pt recording sites is important as this produces an electroactive area that is greater than the geometric area of the recording surface. The fabrication of MEAs with photolithography and sputter-coated Pt produces a precise and reproducible neural device. However, there are subtle differences in the nanoscale features between individual Pt recording sites on the same MEA device. Thus, the empirical finding of nonuniform nano-sized elevations and depressions in the Pt recording surface is a step towards explaining (1) the performance abilities of the MEA relative to other Pt electrode types and (2) variability in the sensitivity of individual recording sites on the same MEA device. To this end, our laboratory is currently investigating the intrinsic physical properties and chemistry of the MEA device following electrodeposition of platinum–iridium (Pt–Ir) on the native thin film Pt surface. Electrodeposition of Pt–Ir enhances the nanofeatures of the recording surface and will be instrumental to elucidating the contribution of nano-sized features to device performance.

Recently, tissue adhesions to MEA surfaces were examined upon MEA removal from perfusion-fixed tissue following 3 weeks of chronic implantation in the rat hippocampus. SEM imaging was used in a high-vacuum environment to assess the adhesion of biological material to the MEAs. Although attached biological material was observed for all MEAs following their removal, the distribution of tissue was variable between individual MEAs. In general, tissue adhesion to the front surface of the MEA (Fig. 9a) was much less when compared to the rough edges of the ceramic sides (Fig. 9b). The high-magnification image shown in Fig. 9b indicates a multilayer cluster of cellular debris with a smooth appearance. The majority of the observed biological material was found along the rough ceramic sides of the MEA (Fig. 9a, b, which contains an extended network of ridges and cavities. Tissue adhesion to the MEAs was minimal compared to prior reports of extensive encapsulation of recordings arrays (44, 45, 48–52).

We also processed freshly removed MEAs to determine the presence of astrocytes on the MEA surfaces with confocal, fluorescent microscopy 4–6 months after implantation. Except for an isolated region on one MEA (Fig. 10a, b), high magnification images of the GFAP-immunoreactive region showed fragmentation that resembled astrocytic processes and not whole cells on the MEA surfaces (Fig. 10b). The smooth multilayer material observed with SEM (Fig. 10b) was also observed with confocal microscopy, which was made up of a thin layer that covered most of the MEA (Fig. 10c, d). Although there were numerous GFAP-immunoreactive spots—the morphology and identity of this material requires further investigation. In keeping with the SEM images, the majority of the observed biological material consisted of isolated clusters along the rough ceramic edges and did not show reactivity to GFAP (Fig. 10c, d). This non-astrocytic tissue adhesion may be indicative of acceptance of the MEA implantation by the host tissue.

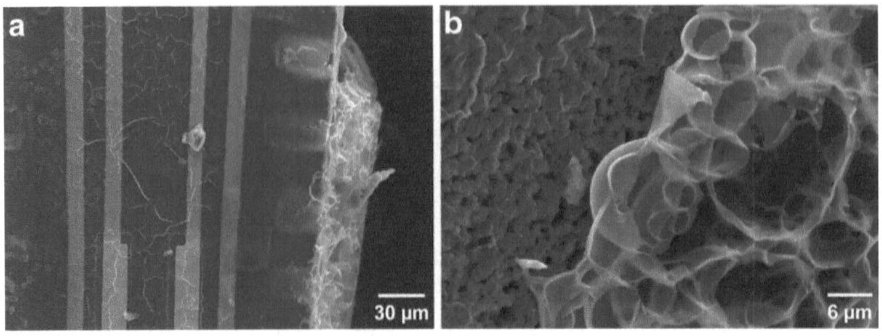

Fig. 9 Scanning electron micrographs of MEA surfaces upon removal 3 weeks postimplantation. (**a**) Micrograph showing the adhesion of biological material to the front surface of the MEA. Tissue debris can also be seen on the side of the MEA. (**b**) High magnification micrograph showing the adhesion of a smooth, multilayer of tissue on the ceramic side of an MEA

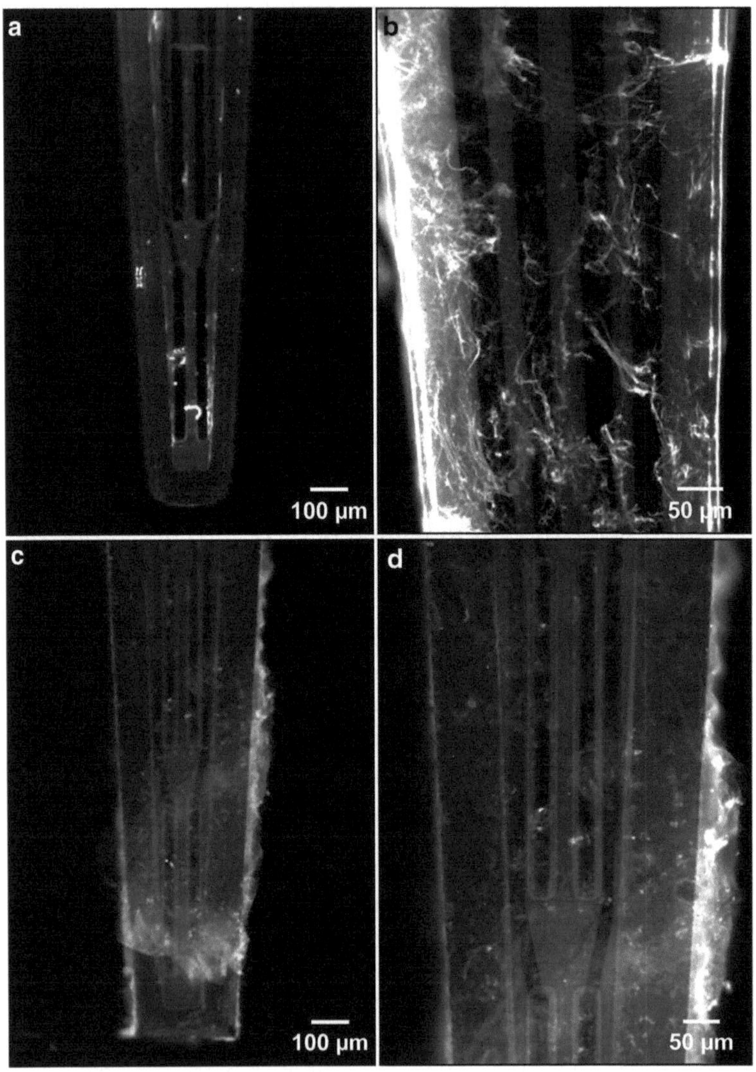

Fig. 10 GFAP-immunoreactivity on MEA surfaces was investigated using confocal microscopy 3 weeks postimplantation. (**a**) Low and (**b**) high magnification images for a MEA that had isolated areas of GFAP-immunoreactive staining. (**c**) Low and (**d**) high magnification images for a typical MEA, which had a lack of GFAP-immunoreactivity after removal from the brain

3 Summary

Our efforts to adapt the MEA technology for chronic, in vivo investigations in rodents and non-human primates have resulted in a novel, indwelling device that (1) demonstrates enhanced neurotransmitter sensitivity, selectivity and sub-second sampling capabilities, (2) maintains recording performance during chronic measurements of extracellular neurotransmitter levels (days) and single-unit neuronal activity (months), and (3) produces a stable, low level tissue response as seen from postimplantation evaluation of neuroinflammation and reactive astrocytosis. On the basis of nanoscale examinations of MEA topography and longitudinal studies of the brain–MEA interface, we hypothesize that the intrinsic physical properties of the MEA likely contribute to its superior performance. We continue to refine and improve our recording capability; however, the accomplishments discussed here are important for several reasons. First, the wealth of information and recording reliability of our technology is unmatched by parallel investigative techniques. Second, the chronic MEA technology can be incorporated into multimodal experimental paradigms, such as behavioral testing, pharmacological intervention (both local and systemic), or combined measurements of neurotransmitter levels and neuronal activity. Third, and perhaps most important, the potential to conduct longitudinal analyses can benefit a wide range of translational efforts, including studies on brain development, aging, neurodegenerative disease progression, and injury-related neuropathology.

References

1. Kennedy RT, Thompson JE, Vickroy TW (2002) In vivo monitoring of amino acids by direct sampling of brain extracellular fluid at ultralow flow rates and capillary electrophoresis. J Neurosci Methods 114:39–49
2. Watson CJ, Venton BJ, Kennedy RT (2006) In vivo measurements of neurotransmitters by microdialysis sampling. Anal Chem 78: 1391–1399
3. Timmerman W, Westerink BH (1997) Brain microdialysis of GABA and glutamate: what does it signify? Synapse 27:242–261
4. Lada MW, Vickroy TW, Kennedy RT (1997) High temporal resolution monitoring of glutamate and aspartate in vivo using microdialysis on-line with capillary electrophoresis with laser-induced fluorescence detection. Anal Chem 69:4560–4565
5. Tucci S, Rada P, Sepulveda MJ, Hernandez L (1997) Glutamate measured by 6-s resolution brain microdialysis: capillary electrophoretic and laser-induced fluorescence detection application. J Chromatogr B Biomed Sci Appl 694: 343–349
6. Rossell S, Gonzalez LE, Hernandez L (2003) One-second time resolution brain microdialysis in fully awake rats. Protocol for the collection, separation and sorting of nanoliter dialysate volumes. J Chromatogr B Analyt Technol Biomed Life Sci 784:385–393
7. Clapp-Lilly KL, Roberts RC, Duffy LK, Irons KP, Hu Y, Drew KL (1999) An ultrastructural analysis of tissue surrounding a microdialysis probe. J Neurosci Methods 90:129–142
8. Westerink RH (2004) Exocytosis: using amperometry to study presynaptic mechanisms of neurotoxicity. Neurotoxicology 25: 461–470
9. Suaud-Chagny MF, Cespuglio R, Rivot JP, Buda M, Gonon F (1993) High sensitivity measurement of brain catechols and indoles in vivo using electrochemically treated carbon-fiber electrodes. J Neurosci Methods 48: 241–250

10. Hu Y, Mitchell KM, Albahadily FN, Michaelis EK, Wilson GS (1994) Direct measurement of glutamate release in the brain using a dual enzyme-based electrochemical sensor. Brain Res 659:117–125
11. Oldenziel WH, Dijkstra G, Cremers TI, Westerink BH (2006) In vivo monitoring of extracellular glutamate in the brain with a microsensor. Brain Res 1118:34–42
12. Tian F, Gourine AV, Huckstepp RT, Dale N (2009) A microelectrode biosensor for real time monitoring of L-glutamate release. Anal Chim Acta 645:86–91
13. Rutherford EC, Pomerleau F, Huettl P, Stromberg I, Gerhardt GA (2007) Chronic second-by-second measures of L-glutamate in the central nervous system of freely moving rats. J Neurochem 102:712–722
14. Hascup KN, Hascup ER, Pomerleau F, Huettl P, Gerhardt GA (2008) Second-by-second measures of L-glutamate in the prefrontal cortex and striatum of freely moving mice. J Pharmacol Exp Ther 324:725–731
15. Hascup ER, Hascup KN, Stephens M, Pomerleau F, Huettl P, Gratton A, Gerhardt GA (2010) Rapid microelectrode measurements and the origin and regulation of extracellular glutamate in rat prefrontal cortex. J Neurochem 115:1608–1620
16. Hascup ER, af Bjerken S, Hascup KN, Pomerleau F, Huettl P, Stromberg I, Gerhardt GA (2009) Histological studies of the effects of chronic implantation of ceramic-based microelectrode arrays and microdialysis probes in rat prefrontal cortex. Brain Res 1291:12–20
17. Zhang H, Lin SC, Nicolelis MA (2010) Spatiotemporal coupling between hippocampal acetylcholine release and theta oscillations in vivo. J Neurosci 30:13431–13440
18. Zhang H, Lin SC, Nicolelis MA (2009) Acquiring local field potential information from amperometric neurochemical recordings. J Neurosci Methods 179:191–200
19. Burmeister JJ, Moxon K, Gerhardt GA (2000) Ceramic-based multisite microelectrodes for electrochemical recordings. Anal Chem 72:187–192
20. Burmeister JJ, Pomerleau F, Palmer M, Day BK, Huettl P, Gerhardt GA (2002) Improved ceramic-based multisite microelectrode for rapid measurements of L-glutamate in the CNS. J Neurosci Methods 119:163–171
21. Hascup KN, Rutherford EC, Quintero JE, Day BK, Nickell JR, Pomerleau F, Huettl P, Burmeister JJ, Gerhardt GA (2006) Second-by-second measures of L-glutamate and other neurotransmitters using enzyme-based microelectrode arrays. In: Michael AC, Borland LM (eds) Electrochemical methods for neuroscience. CRC, Boca Raton, FL, pp 407–450
22. Burmeister JJ, Gerhardt GA (2001) Self-referencing ceramic-based multisite microelectrodes for the detection and elimination of interferences from the measurement of L-glutamate and other analytes. Anal Chem 73:1037–1042
23. Nickell J, Pomerleau F, Allen J, Gerhardt GA (2005) Age-related changes in the dynamics of potassium-evoked L-glutamate release in the striatum of Fischer 344 rats. J Neural Transm 112:87–96
24. Gerhardt GA, Burmeister JJ (2000) In vivo voltammetry for chemical analysis of the nervous system. In: Meyers RA (ed) Encyclopedia of analytical chemistry: instrumentation and applications. Wiley, Chichester
25. Paxinos G, Watson C (2008) The rat brain in stereotaxic coordinates: compact, 6th edn. Academic, New York, NY
26. Day BK, Pomerleau F, Burmeister JJ, Huettl P, Gerhardt GA (2006) Microelectrode array studies of basal and potassium-evoked release of L-glutamate in the anesthetized rat brain. J Neurochem 96:1626–1635
27. Liachenko S, Tang P, Somogyi GT, Xu Y (1998) Comparison of anaesthetic and non-anaesthetic effects on depolarization-evoked glutamate and GABA release from mouse cerebrocortical slices. Br J Pharmacol 123:1274–1280
28. Liachenko S, Tang P, Somogyi GT, Xu Y (1999) Concentration-dependent isoflurane effects on depolarization-evoked glutamate and GABA outflows from mouse brain slices. Br J Pharmacol 127:131–138
29. Adams RN (1990) In vivo electrochemical measurements in the CNS. Prog Neurobiol 35:297–311
30. Martin KF, Marsden CA (1987) In vivo electrochemistry–principles and applications. Life Sci 41:865–868
31. Melendez RI, Vuthiganon J, Kalivas PW (2005) Regulation of extracellular glutamate in the prefrontal cortex: focus on the cystine glutamate exchanger and group I metabotropic glutamate receptors. J Pharmacol Exp Ther 314:139–147
32. Ballini C, Corte LD, Pazzagli M, Colivicchi MA, Pepeu G, Tipton KF, Giovannini MG (2008) Extracellular levels of brain aspartate, glutamate and GABA during an inhibitory avoidance response in the rat. J Neurochem 106:1035–1043
33. Clinckers R, Gheuens S, Smolders I, Meurs A, Ebinger G, Michotte Y (2005) In vivo modulatory action of extracellular glutamate on the anticonvulsant effects of hippocampal dopamine and serotonin. Epilepsia 46:828–836
34. Giovannini MG, Rakovska A, Benton RS, Pazzagli M, Bianchi L, Pepeu G (2001) Effects

of novelty and habituation on acetylcholine. GABA, and glutamate release from the frontal cortex and hippocampus of freely moving rats. Neuroscience 106:43–53
35. Segovia G, Yague AG, Garcia-Verdugo JM, Mora F (2006) Environmental enrichment promotes neurogenesis and changes the extracellular concentrations of glutamate and GABA in the hippocampus of aged rats. Brain Res Bull 70:8–14
36. Ueda Y, Tsuru N (1995) Simultaneous monitoring of the seizure-related changes in extracellular glutamate and gamma-aminobutyric acid concentration in bilateral hippocampi following development of amygdaloid kindling. Epilepsy Res 20:213–219
37. Bungay PM, Newton-Vinson P, Isele W, Garris PA, Justice JB (2003) Microdialysis of dopamine interpreted with quantitative model incorporating probe implantation trauma. J Neurochem 86:932–946
38. Rice ME, Cragg SJ (2008) Dopamine spillover after quantal release: rethinking dopamine transmission in the nigrostriatal pathway. Brain Res Rev 58:303–313
39. Danbolt NC (2001) Glutamate uptake. Prog Neurobiol 65:1–105
40. Nickell J, Salvatore MF, Pomerleau F, Apparsundaram S, Gerhardt GA (2007) Reduced plasma membrane surface expression of GLAST mediates decreased glutamate regulation in the aged striatum. Neurobiol Aging 28:1737–1748
41. Stephens ML, Quintero JE, Pomerleau F, Huettl P, Gerhardt GA (2009) Age-related changes in glutamate release in the CA3 and dentate gyrus of the rat hippocampus. Neurobiol Aging 32:811–820
42. Dash MB, Douglas CL, Vyazovskiy VV, Cirelli C, Tononi G (2009) Long-term homeostasis of extracellular glutamate in the rat cerebral cortex across sleep and waking states. J Neurosci 29:620–629
43. Hulbert SF, Morrison SJ, Klawitter JJ (1972) Tissue reaction to three ceramics of porous and non-porous structures. J Biomed Mater Res 6:347–374
44. Biran R, Martin DC, Tresco PA (2007) The brain tissue response to implanted silicon microelectrode arrays is increased when the device is tethered to the skull. J Biomed Mater Res A 82:169–178
45. Biran R, Martin DC, Tresco PA (2005) Neuronal cell loss accompanies the brain tissue response to chronically implanted silicon microelectrode arrays. Exp Neurol 195:115–126
46. Turner KL, Doherty MK, Heering HA, Armstrong FA, Reid GA, Chapman SK (1999) Redox properties of flavocytochrome c3 from Shewanella frigidimarina NCIMB400. Biochemistry 38:3302–3309
47. Turner JA, Lee JS, Martinez O, Medlin AL, Schandler SL, Cohen MJ (2001) Somatotopy of the motor cortex after long-term spinal cord injury or amputation. IEEE Trans Neural Syst Rehabil Eng 9:154–160
48. Turner JN, Shain W, Szarowski DH, Andersen M, Martins S, Isaacson M, Craighead H (1999) Cerebral astrocyte response to micromachined silicon implants. Exp Neurol 156:33–49
49. Leung BK, Biran R, Underwood CJ, Tresco PA (2008) Characterization of microglial attachment and cytokine release on biomaterials of differing surface chemistry. Biomaterials 29:3289–3297
50. Szarowski DH, Andersen MD, Retterer S, Spence AJ, Isaacson M, Craighead HG, Turner JN, Shain W (2003) Brain responses to micromachined silicon devices. Brain Res 983:23–35
51. Spataro L, Dilgen J, Retterer S, Spence AJ, Isaacson M, Turner JN, Shain W (2005) Dexamethasone treatment reduces astroglia responses to inserted neuroprosthetic devices in rat neocortex. Exp Neurol 194:289–300
52. Polikov VS, Tresco PA, Reichert WM (2005) Response of brain tissue to chronically implanted neural electrodes. J Neurosci Methods 148:1–18
53. Rousche PJ, Normann RA (1998) Chronic recording capability of the Utah Intracortical Electrode Array in cat sensory cortex. J Neurosci Methods 82:1–15
54. Talauliker PM, Price DA, Burmeister JJ, Nagari S, Quintero JE, Gerhardt GA (2011) Ceramic based microelectrode arrays: recording surface characteristics and topographical analysis. J Neurosci Methods 198:222–229
55. Jung E, Manessis D, Neumann A, Böttcher L, Braun T, Bauer J, Reichl H, Iafelice B, Destro F, Gambari R (2008) Lamination and laser structuring for a microwell array. Microsyst Technol 14:5

Chapter 9

Regulation of Extracellular Concentrations of D-Serine in the Central Nervous System Revealed by D-Amino Acid Oxidase Microelectrode Biosensors

Pierre Pernot, Caroline Maucler, and Stéphane Marinesco

Abstract

D-serine is an endogenous co-agonist of *N*-methyl D-aspartate receptors, and has been implicated in several neurological and psychiatric disorders. Correspondingly, D-serine signaling may represent an important pharmacological target for treating these diseases. However, the biochemical mechanisms that control extracellular concentrations of D-serine in vivo remain unclear. Therefore, we have developed biosensors to monitor D-serine levels in vivo. These biosensors are cylindrical platinum microelectrodes that have been coated with a layer of poly-*m*-phenylenediamine and then a membrane of immobilized D-amino acid oxidase (DAAO) obtained from the yeast, *Rhodotorula gracilis*. Using these microelectrode biosensors, extracellular levels of D-serine in the rat frontal cortex were investigated. D-serine was detected in micromolar concentrations in the interstitial fluid, and was observed to diffuse slowly through the blood–brain barrier following peripheral administration of D-serine. By coupling these biosensor measurements with HPLC analysis of whole tissue D-serine contents, we have showed that intracellular stores represent more than 99 % of the total brain D-serine contents. Therefore, the development of DAAO microelectrode biosensors has already improved our ability to understand D-serine regulation in vivo, and holds great potential for future studies.

Key words Serine racemase, NMDA receptors, Schizophrenia, Amperometry, In vivo recordings

1 Introduction: D-Serine in Healthy Versus Diseased Brains

D-serine, the D-enantiomer of the amino acid serine, is present in the central nervous system (CNS) at high concentrations. In the brain, D-serine constitutes almost one-third of the total serine present (1), and is a modulatory transmitter whose major target is the co-agonist site of the *N*-methyl-D-aspartate (NMDA) receptor located on the NR1 subunit. D-serine and glycine both recognize the same site on the NMDA receptor, although with slightly different affinities. For example, in mice, D-serine has a threefold higher affinity for the co-agonist site of NMDA receptors than glycine (2).

Stéphane Marinesco and Nicholas Dale (eds.), *Microelectrode Biosensors*, Neuromethods, vol. 80,
DOI 10.1007/978-1-62703-370-1_9,

However, in humans, glycine affinity is dependent on the presence of other NMDA receptor subunits, with glycine affinity being approximately tenfold higher for NR1/NR2b receptors than NR1/NR2a receptors (3). Conversely, D-serine has a higher affinity for NR1/NR2a receptors. In addition to NMDA receptors, D-serine also binds delta 2 glutamate (gluδ2) receptors (4).

As an NMDA co-agonist, D-serine has been associated with many of the brain functions associated with NMDA receptors. For example, it has been implicated in the synaptic plasticity of the hippocampus and hypothalamus (5, 6), and is hypothesized to be released by astrocytes in response to synaptic glutamate. In the latter case, the release results in an increase in NMDA receptor activity and can induce long-term potentiation or long-term depression. Interestingly, the binding of D-serine to Gluδ2 receptors also regulates synaptic plasticity, especially long-term depression in the developing cerebellum (4). Accordingly, from a behavioral point of view, appropriate systemic doses of D-serine have the potential to improve spatial memory (7). Moreover, in aged rats, decreased levels of D-serine in the brain were found to be associated with impaired synaptic plasticity and reduced performance in spatial memory tests. Correspondingly, memory deficits could be rescued in part by administration of D-serine, or its analog, D-cycloserine, a partial agonist of the glycine/D-serine site on the NMDA receptor (8–10).

D-serine has also been implicated in brain development based on its role in the guidance of immature granule neurons along Bergmann glia in the cerebellum (11). For example, the release of both glutamate and D-serine by Bergmann glia has been shown to activate NMDA receptors located on granule neurons, and to stimulate neuronal migration mediated by chemotactic interactions.

D-serine is also expressed in the retina, and participates in the processing of light signals by retinal neuronal circuits. For example, ambient D-serine levels determine the occupancy of NMDA receptors on retinal ganglion cells, and the amplitude of light-evoked responses (12, 13). Accordingly, a transient, light-evoked release of D-serine has been hypothesized to occur, and support for this hypothesis includes the observation that an NMDA co-agonist is released in response to light, although it remains unclear whether it is D-serine or glycine (13–15).

In addition to these physiological functions, there is also strong evidence to indicate that D-serine dysfunction contributes to several neurological and psychiatric diseases. For example, D-serine has been shown to have a role in the pathophysiology of schizophrenia, with decreased concentrations of D-serine detected in the serum and cerebrospinal fluid of schizophrenic patients (16–18). Correspondingly, mRNA levels and the immunoreactivity of D-amino acid oxidase (DAAO), an enzyme that metabolizes D-serine, were found to be higher in postmortem brain samples obtained from schizophrenic patients (19). In addition, pIG72, a protein that interacts with DAAO, has been associated with schizophrenia (20–22). As a result, therapeutic strategies based on D-serine have been tested in patients. For example, oral administration of D-serine in conjunction with conventional antipsychotic medications has been found to be efficient for reducing positive, negative, and cognitive symptoms experienced by schizophrenic patients (23–25).

With NMDA receptor overactivation being a major cause of excitotoxicity, D-serine also represents a natural candidate for mediating at least some aspects of the neurological insults evidenced in neurodegenerative disease or ischemia. For example, brain slice studies have shown that decreasing levels of ambient D-serine have a neuroprotective effect during oxygen-glucose deprivation (26) and NMDA administration (27). Moreover, release of D-serine has also been associated with oxygen-glucose deprivation (28, 29). In addition, studies of serine racemase (SR) knockout mice, which are depleted of D-serine, reported reduced infarct volumes following occlusion of the middle cerebral artery (30, 31).

In models of neurodegenerative diseases such as Alzheimer's disease or amyotrophic lateral sclerosis, D-serine has also been shown to play a role. For example, in rat hippocampal cultures, the release of D-serine can be stimulated by amyloid-beta peptide (32), or by amyloid precursor protein (33). Moreover, in SR knockout mice, amyloid-beta peptide neurotoxicity has been attenuated (34). In addition, for mice and patients with amyotrophic lateral sclerosis detected in the spinal cord, Sasabe et al. also detected elevated levels of D-serine (35). In combination, these studies suggest that D-serine has an important role in the pathophysiology of neurodegenerative diseases.

Based on the possible effects mediated by D-serine in neurological and psychiatric diseases, D-serine has become an attractive candidate for drug development. For example, a drug capable of increasing extracellular levels of D-serine would be a strong candidate for treating schizophrenia. Alternatively, pharmacological tools that could lower D-serine levels could potentially prevent excitotoxicity and neurodegeneration in the brain. However, a key limiting factor to these studies of drug development is that the biochemical mechanisms that control the extracellular concentration of D-serine in the brain remain poorly understood.

2 Understanding the Regulation of D-Serine Requires In Vivo Measurements

D-serine is synthesized from L-serine by SR, a pyridoxal phosphate-dependent enzyme that is expressed exclusively in the brain (36). Although SR was initially proposed to be a glial enzyme, SR is now unequivocally associated with neurons, while astrocytes also express low levels of SR (37–40). SR activity requires the presence of ATP, Mg^{2+}, and possibly Ca^{2+} (41, 42), and increases following phosphorylation of two threonine sites (43). In addition, SR activity is modulated by interactions with at least three other proteins: glutamate receptor interacting protein (GRIP), which has a stimulatory effect (11), Golga3, which facilitates the ubiquitination and degradation of SR (44), and protein interacting with C Kinase 1 (PICK1), whose function remains unclear (45). SR has also been found to be inhibited by nitric oxide (NO) (46) and phosphatidyl inositol biphosphate (PIP2) (47). Interestingly, SR also exhibits an α,β-elimination activity which leads to the degradation of L- and D-serine to pyruvate (37, 48, 49).

Although D-serine synthesis is now well understood, there are other steps in the regulation of D-serine, including the release, reuptake, and degradation of D-serine, that remain unclear. For example, several studies have indicated that D-serine can be released by transport mechanisms involving both alanine–serine–cysteine (Asc) transporters and alanine–serine–cysteine–threonine (ASCT) transporters which mediate the heteroexchange of small neutral amino acids. Correspondingly, within these two families, Asc-1 and ASCT2 are present in the brain and accept D-serine as a substrate. Moreover, Asc-1 transporters are preferentially neuronal, Na^+ independent, and are able to transport D-serine with high affinity ($K_m = 20\ \mu M$) (50). In contrast, ASCT2 is preferentially glial, Na^+ dependent, and has a low affinity for D-serine ($K_m = 1.6$ mM) (51, 52). Therefore, although Asc-1 is the primary candidate for mediating the uptake of D-serine in vivo (53–55), a role for ASCT2 transporters cannot be excluded. Moreover, both Asc-1 and ASCT2 are heteroexchangers and have been shown to release D-serine. For example, Kim et al. (11) observed that astrocytes in culture have a low capacity for storing D-serine, leading to the hypothesis that D-serine is synthesized and released on demand. Correspondingly, D-serine release has been shown to be induced in the presence of L-serine, L-threonine, L-alanine, or L-glutamine (38, 53, 56), further suggesting a role for Asc transporters in the release of D-serine.

In addition to transporter-mediated D-serine release, studies of astrocyte cultures have suggested that D-serine may be released via Ca^{2+}-dependent exocytosis (57, 58). For example, an analysis of synaptic currents in hippocampal slices by Henneberger et al. (6) supports an exocytotic mechanism for D-serine release. However, another recent study found that D-serine is released by neurons in primary culture, in acute brain slices, and in vivo, and that this release is essentially Ca^{2+} insensitive, making it incompatible with exocytosis (38).

Finally, the mechanisms mediating D-serine degradation also remain unclear. In the cerebellum and brainstem, DAAO is highly expressed by astrocytes and efficiently degrades D-serine (56, 59, 60).

However, this enzyme is not present in the forebrain, leaving α,β-elimination by SR the only plausible mechanism for D-serine degradation (37). Moreover, the capacity for SR to mediate both D-serine synthesis and degradation remains to be elucidated.

There are many important aspects of D-serine regulation in the CNS that remain unresolved. These include (1) the mechanistic details of D-serine release that may involve heteroexchange and exocytosis, (2) the need for precise transporter subtypes responsible for D-serine reuptake to be identified, and (3) elucidation of the mechanisms of D-serine degradation that occur in the forebrain. Up to this point, the use of multiple biological preparations, including isolated proteins, cell cultures, brain slices, and in vivo models, has provided contrasting results in the study of D-serine regulation in the CNS. For example, cells in culture can develop traits characteristic of early development, resulting in the identification of mechanisms for D-serine regulation that are distinct from those activated in an adult brain. Correspondingly, Miya et al. (40) detected expression of SR in both neurons and astrocytes maintained in culture, yet in the adult brain, SR was found to be exclusively expressed in neurons. Therefore, direct in vivo measurements are needed to confirm the data obtained from cell culture studies.

3 Detection of D-Serine Using Microdialysis

The most widely used technique for monitoring molecules present in the interstitial fluid in vivo is microdialysis. This technique uses a concentric probe that is covered with a dialysis membrane and allows circulation of an artificial extracellular medium. When this probe is implanted in the CNS, molecules are able to diffuse in and out of the dialysis membrane and a sampling of the interstitial fluid can be collected. The molecules present are then analyzed using high-performance liquid chromatography (HPLC) or capillary electrophoresis (61). Advantages of using microdialysis include the ability to monitor several molecules in parallel, and the analytical power of HPLC or capillary electrophoresis that provides excellent selectivity.

Detection of D-serine using microdialysis has been demonstrated by several laboratories. For example, Hashimoto et al. used HPLC coupled to fluorometric detection to detect D-serine in rat brain microdialysates following their derivatization with Boc-L-cysteine and *o*-phthaldialdehyde (62). In this study, basal D-serine levels were found to be 5.75 μM in the prefrontal cortex, 7.25 μM in the striatum, and 0.5 μM in the cerebellum. A few years later, Ciriacks and Bowser (63) developed a D-serine detection method using capillary electrophoresis coupled to laser-induced fluorescence (CE-LIF) and *o*-phthaldialdehyde derivatization (Fig. 1). Using this method, an extremely high temporal resolution of about 3 s per sample was achieved. Furthermore, concentrations of D-serine

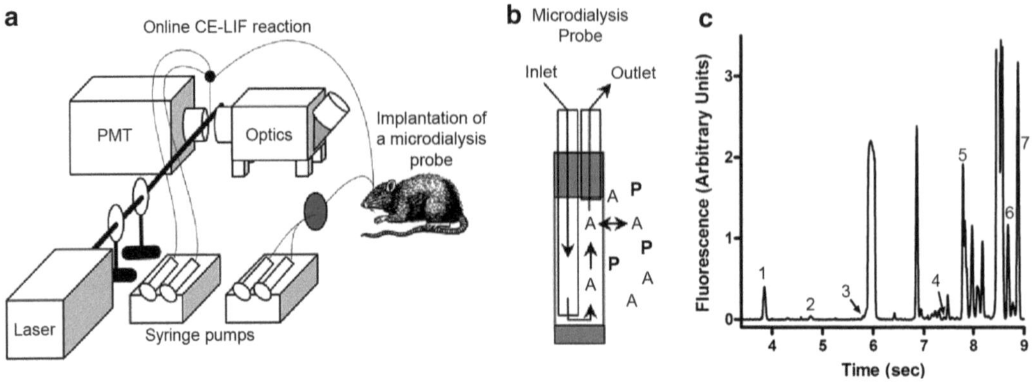

Fig. 1 Principles of D-serine detection using microdialysis coupled to capillary electrophoresis and laser-induced fluorescence (CE-LIF). (**a**) Schematic of a microdialysis-CE instrument coupled to LIF detection. (**b**) Schematic representation of the microdialysis probe used. "A" represents the analyte to be detected and "P" represents other larger molecules such as proteins or peptides. (**c**) A representative electropherogram obtained from CE-LIF analysis of dialysate collected from the striatum of an anesthetized rat. Peak identifications (1) aminoadipic acid (internal standard), (2) glutamate, (3) aspartate, (4) GABA, (5) taurine, (6) D-serine, and (7) L-serine. Adapted with permission from (63)

were detected in the cortex and the striatum (8 μM vs. 8.5 μM, respectively) (63, 64). However, the ability to use microdialysis in an in vivo model is affected by the size of the probes that are available for implantation in the brain. Currently, these probes have a diameter that ranges from 200 to 250 μm, and lengths that range from 500 to 3,000 μm. Accordingly, probe implantation is often traumatic for the brain parenchyma, and can lead to edema, changes in microcirculation, cytokine production, and gliosis (65–67). Moreover, following insertion, it has been hypothesized that a microdialysis probe becomes surrounded by a thin layer of dead tissue which can subsequently act as a diffusion barrier to impede the detection of rapid physiological variations in neurotransmitter concentrations (65). Therefore, we have sought to develop a D-serine biosensor that provides greater spatial and temporal resolution than microdialysis, and that can be implanted in the CNS without major injury.

4 Enzyme-Based D-Serine Biosensors That use D-Amino Acid Oxidase

DAAO is a flavin adenine dinucleotide (FAD)-containing enzyme that catalyzes the oxidative deamination of D-isomers of neutral and polar amino acids. It is present in most eukaryotic organisms and bacteria. It degrades D-serine, as well as other neutral D-amino acids such as D-alanine, D-phenylalanine, and D-proline (59). In the case of D-serine, the degradation product includes hydroxypyruvate as shown in reaction (9.1):

$$\text{D-serine} + O_2 \xrightarrow{\text{DAOO}} \text{hydroxypyruvate} + NH_3 + H_2O_2 \quad (1)$$

Based on the production of H_2O_2 as part of this degradation process, DAOO-based biosensors represent excellent candidates for enzymatic biosensors. Correspondingly, DAAO-based biosensors have already been developed for the analysis of D-amino acids (including D-serine) as markers of bacterial proliferation in food samples (68–72). These biosensors have primarily been applied to aspects of quality control in the food industry, and involve a DAAO purified from pig kidney (pkDAAO). However, this isoform is associated with a low specific activity, and this has been found to impact the overall sensitivity of the biosensor. Moreover, biosensors used for the analysis of food samples are too large to be inserted into the CNS. Therefore, we have sought to optimize this biosensor system in order to achieve successful electrochemical detection of D-serine in the CNS using an implantable microelectrode biosensor.

The biosensor we have developed is based on a platinum microelectrode (25 μm diameter, 100–150 μm length, see Chap. 1 for more details) that is covered with a layer of poly-*m*-phenylenediamine for increased selectivity (see Chap. 6 for more details) and a layer of DAAO purified from the yeast, *Rhodotorula gracilis* (RgDAAO) (Fig. 2). In the latter case, this isoform was selected based on its high specific activity and its capacity to bind its cofactor, FAD, with high affinity (59). Cofactor binding increases enzyme stability. Moreover, the high enzymatic activity generates biosensors with 1.5-fold higher sensitivity than biosensors coated with pkDAAO (73).

RgDAAO is able to be immobilized on a microelectrode using glutaraldehyde (73, 74). As a result, an enzyme layer can be deposited on a platinum microelectrode by dipping the electrode into a concentrated DAAO solution containing 60 mg/mL DAAO (specific activity 55 U/mg, gift from Prof. L. Pollegioni), 25 mg/ml bovine serum albumin, and 1 % glycerol in 0.01 M phosphate-buffered saline (PBS, pH 7.4). The coated microelectrode is then placed in a chamber with saturated glutaraldehyde vapors for approximately 2 min. More recently, we have developed an alternative immobilization method that uses poly(ethylene glycol) diglycidyl ether (PEGDE), a nontoxic cross-linking agent (75). In this case, 6 mg/mL PEGDE is added to the enzyme solution and the biosensor is cured at 55°C for 2 h. Biosensors prepared with PEGDE have been shown to be more sensitive than those obtained with glutaraldehyde (e.g., 212 ± 119 μA/mM/cm^2 for PEGDE vs. 87 ± 27 for glutaraldehyde). Moreover, this immobilization method does not appear to notably influence other biochemical parameters of the biosensors. Overall, the response time for these biosensors is ~3 s, consistent with the kinetics of immobilized DAAO, and the linear range of the biosensor extends up to ~200 μM for D-serine.

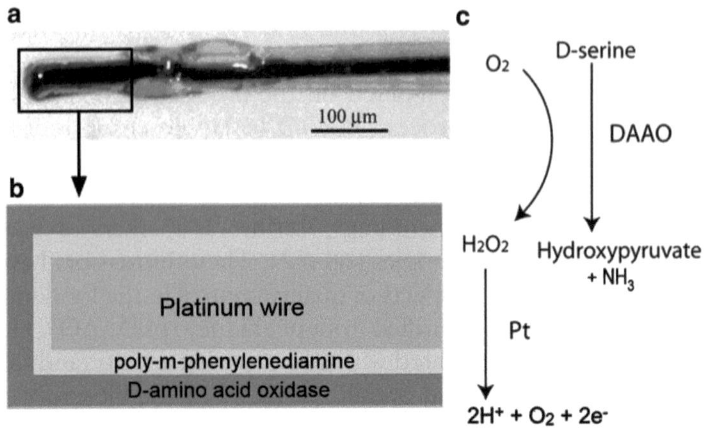

Fig. 2 Design of a D-serine biosensor. (**a**) A photomicrograph of the tip of a microbiosensor coated with a layer of D-amino acid oxidase purified from the yeast, *Rhodotorula gracilis* (RgDAAO). The enzymatic layer present appears as a translucid, *yellow* membrane, and the *box* outline is used to indicate the region referred to in *panel B*. The scale bar included represents 100 μM. (**b**) Schematic representation of the coatings used for a D-amino acid oxidase (DAAO) microbiosensor. A platinum wire is covered with a layer of poly-*m*-phenylenediamine (PPD) and then an enzymatic membrane of DAAO. (**c**) The enzymatic reaction involving DAAO that occurs for detection of D-serine by the microbiosensor: D-serine is oxidized into hydroxypyruvate by DAAO with equimolar production of hydrogen peroxide (H_2O_2) that diffuses through the PPD layer and is oxidized by the platinum wire. H_2O_2 oxidation gives rise to two electrons that are then detected by the amplifier. Reprinted with permission from (73)

In vitro, the detection limit of these biosensors is 16 nM, although in vivo, it is closer to 0.5 μM due to the high electrical noise (mostly local field potentials) present in the living brain.

Since DAAO is able to oxidize most neutral D-amino acids, and therefore is not selective for D-serine, the selectivity of our biosensor is needed to be thoroughly evaluated. As expected, DAAO biosensors generated large electrochemical responses to D-serine, as well as other D-amino acids including D-tryptophan, D-cysteine, D-alanine, D-proline, D-phenylalanine, and D-methionine. However, they exhibited very low sensitivity for acidic D-amino acids such as D-aspartate (6.5 % of the D-serine response) and D-glutamate (3.5 % of the D-serine response). Moreover, they were virtually inactive towards L-amino acids such as L-serine (Fig. 3) (73). Fortunately, however, D-serine is by far the most concentrated D-amino acid present in the brain, followed by D-aspartate (~10 % of D-serine levels) and D-alanine (<3 % of D-serine contents, Fig. 4b) (76). Therefore, based on the very low sensitivity of DAAO biosensors to D-aspartate, it was hypothesized that these biosensors would be selective for D-serine when implanted in the CNS. Pernot et al. (73) verified the selectivity of DAAO biosensors

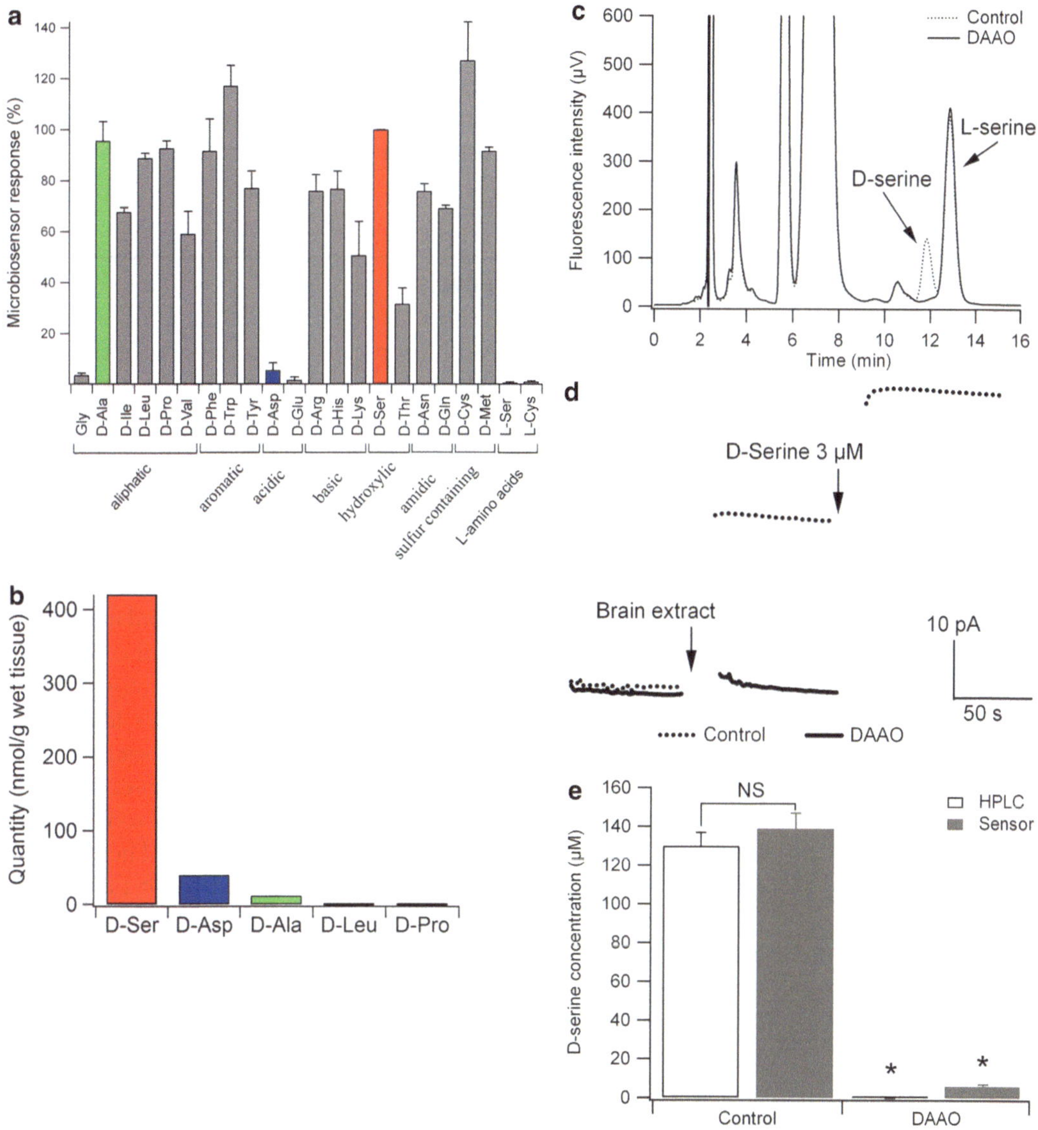

Fig. 3 Selectivity for D-serine. (**a**) The microbiosensor response recorded for each of the D-isomers of the 20 standard amino acids (10 μM), in addition to glycine, L-serine, and L-cysteine (10 μM). Data are expressed as a percentage of the response obtained with 10 μM D-serine (*red*; $n=4$) ± SD. The DAAO biosensor detected all D-amino acids that were tested, except for D-aspartate (*blue*) and D-glutamate. Glycine and L-amino acids produced negligible responses. (**b**) The nmol/g wet tissue detected for the most abundant D-amino acids present in mouse brain as reported by Hamase et al. (76). The predominant D-amino acid was found to be D-serine (*red*), followed by D-aspartate (*blue*), D-alanine (*green*), D-Leu, and D-Pro (both *gray*). (**c**) Representative chromatogram from rat brain homogenates (control) analyzed by HPLC that demonstrates the separation between D-serine (elution time = 12 min) and L-serine (= 13 min) that can be achieved. The D-serine peak completely disappeared after homogenates were incubated with RgDAAO and catalase (DAAO). (**d**) Example of the electrochemical responses recorded by a microbiosensor upon addition of brain extract (40× dilution) and 3 μM D-serine. This response almost disappeared when the brain extract was preincubated with RgDAAO and catalase. (**e**) D-serine concentrations ± SEM detected in brain extracts based on estimates provided by HPLC and the RgDAAO biosensor (sensor). Both techniques yielded comparable results, thus indicating an excellent selectivity for the detection of D-serine in brain extracts. Data are expressed as mean ± SEM ($n=6$). *NS* not significant, $^{*}p<0.05$. Data adapted with permission from (73)

by performing head-to-head comparisons between biosensors and HPLC for determining D-serine concentrations in brain extracts (Fig. 3c, d). Based on these studies, a discrepancy rate of less than 6 % was observed for the two techniques. Moreover, the data indicated that DAAO biosensors can specifically detect D-serine when implanted in the CNS.

5 Measurements of D-Serine In Vivo: Basal Levels

We then implanted DAAO biosensors in various parts of the brain, usually close to a control biosensor (devoid of DAAO) that is used to detect nonspecific variations in oxidation current. In addition, biosensors are usually given about an hour to stabilize the background currents present. For example, in the frontal cortex, differences between the DAAO biosensor and the control biosensor have typically shown a difference of 40–50 pA in basal electrochemical currents. Moreover, this difference is not detected in calibrations made using PBS or an artificial extracellular medium with an ionic composition that approximates that of rat cerebrospinal fluid. Since the control electrode and the DAAO biosensor are equally sensitive to endogenous oxidizable molecules like H_2O_2, ascorbate, etc., have similar surface areas, and since D-serine is the predominant D-amino acid in the brain, the difference in background current detected mostly reflects the ambient basal concentration of D-serine present in the brain parenchyma. Therefore, to estimate the D-serine concentration represented by this difference in background currents, biosensors are calibrated in an artificial extracellular solution at 37°C. Based on these calibration experiments, the basal extracellular D-serine concentration has been estimated at ~3.1 μM under ketamine–xylazine anesthesia, or 2.3 μM with urethane anesthesia (78).

DAAO biosensors have also been implanted in other brain structures. For example, in the hippocampus, DAAO biosensors have detected basal levels of D-serine to be ~2.8 μM, while D-serine concentrations in the cerebrospinal fluid (CSF) have been found to differ between the *cisterna magna* (1.6 μM) and the third lateral ventricle (7 μM). The reasons for these differences remain unclear, but suggest that the ventricular network of the CNS is not homogenous. We speculate that D-serine levels decline in CSF exposed to more caudal brain regions, which have low D-serine concentration and high endogenous activity of the degrading enzyme DAAO, thus increasing reuptake and degradation.

In contrast, D-serine levels in the cerebellum have been undetectable, indicating that the concentrations were less than 0.5–0.7 μM (78).

Methodological note (78): To estimate the detection limit for D-serine basal levels in vivo, *differences in basal currents detected by individual biosensors in the brain parenchyma must be accounted for. For example, our control biosensors have detected basal currents* (I_0) *of 26.6 pA, with a standard deviation* (σ_0) *of 10.9 pA. However, to reliably detect a significant basal level of* D*-serine, the current recorded by DAAO biosensors must be significantly higher than that recorded by control biosensors. Based on experiments that were repeated using 6–10 different biosensors in different animals, the standard deviation of the recorded currents* (σ_{DAAO}) *was 17.6 pA. The minimum level of* D*-serine oxidation current necessary to reach the p = 0.05 level of significance with a Student's t-test can then be estimated according to the sample size. For n values of 6 or 10, the t value necessary to reach p = 0.05 is 2.228 and 2.101, respectively. Therefore, the standard error of the two data groups is the defined as*

$$SE = \sqrt{\frac{\sigma_0^2 + \sigma_{DAAO}^2}{n}}, \tag{2}$$

with

$$t = \frac{I_{DAAO} - I_0}{SE}. \tag{3}$$

Therefore, the minimum difference in current that must be detected in order to yield a significant value for basal extracellular concentrations of D*-serine can be computed as*

$$I_{DAAO} - I_0 = t \times \sqrt{\frac{\sigma_0^2 + \sigma_{DAAO}^2}{n}}. \tag{4}$$

which yields a minimum difference in current of 18.8 pA (n = 6) or 13.8 pA (n = 10), and an in vivo *detection limit for basal* D*-serine of 0.77 μM (n = 6) or 0.56 μM (n = 10), taking into account the mean sensitivity of the biosensors at 37°C.*

Overall, these results confirm the excellent selectivity of our biosensors since D-serine was not detected in the cerebellum, a structure known to be devoid of D-serine in the adult brain. Moreover, the basal concentrations of D-serine measured using the DAAO biosensors were about half the values measured by microdialysis (63, 79), possibly due the reduced amount of damage associated with the use of microelectrode biosensors versus microdialysis probes (65).

6 Diffusion of D-Serine Through the Blood–Brain Barrier

Currently, oral administration of D-serine is used as a complement to antipsychotic treatment. Therefore, it is important to determine the kinetics of D-serine penetration through the blood–brain barrier.

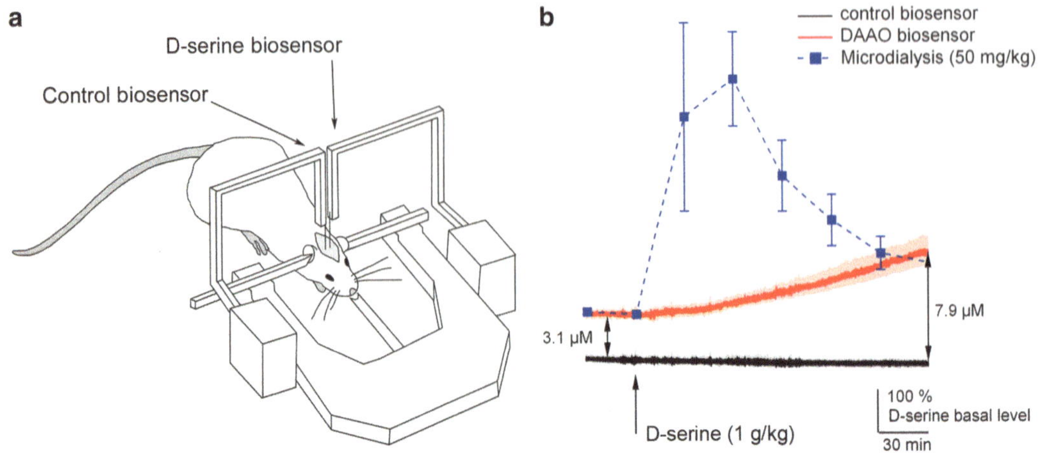

Fig. 4 (**a**) Schematic representation of the experimental design used where a control biosensor (coated with BSA) was implanted near a D-serine biosensor (coated with DAAO) in the frontal cortex of an anesthetized rat. (**b**) Following an intraperitoneal injection of D-serine (1 g/kg body weight), the electrochemical signals detected in the frontal cortex by a DAAO biosensor (indicated in *red*) and a control (BSA) microbiosensor (indicated in *black*) implanted in its vicinity are shown. By comparison, microdialysis measurements extracted from Fukushima et al. (77) (indicated in *blue*) resemble the kinetics of D-serine plasmatic concentrations. These results demonstrate that the extracellular concentration of D-serine increases slowly after the peripheral administration of D-serine, and suggest that DAAO biosensors preserve the integrity of the blood–brain barrier compared with microdialysis. Data are expressed as the mean (*solid line*) ± SEM (*shading*) of six experiments

Correspondingly, when 1 g/kg D-serine was administered intraperitoneally to anesthetized adult rats, changes in extracellular concentrations of D-serine were detected using DAAO biosensors. This peripheral administration of D-serine resulted in a steady increase in the extracellular concentration of D-serine in the frontal cortex. In our study, the increase in oxidation current detected over 3 h by a DAAO microbiosensor was 2.3–4.5 μM under urethane anesthesia, and 3.1–7.9 μM under ketamine–xylazine anesthesia (78). These increases in D-serine concentrations are consistent with previous measurements obtained using postmortem whole tissues (80). However, these data are inconsistent with an earlier report that used microdialysis to detect levels of D-serine, where a sixfold increase in D-serine levels was detected following an intraperitoneal injection of a smaller dose of D-serine (50 mg/kg) (77) (Fig. 4). Implantation of microdialysis probes in the brain is usually associated with significant tissue damage, especially to neighboring blood vessels (66, 81). Therefore, biosensors of micrometric size seem preferable for studies requiring maintenance of blood–brain barrier integrity. Correspondingly, the slow and modest increase in D-serine levels detected by our microbiosensors following a peripheral D-serine injection suggests that damage to the blood–brain barrier produced upon implantation of the electrodes in the frontal cortex is indeed minimal. Moreover, the average distance between brain capillaries in the cortex is ~15 μm (82), which suggests that no more than 10–20 capillaries are injured upon microelectrode implantation.

7 D-Serine Compartmentalization and Storage

Brain tissue can be divided into three separate compartments: blood, extracellular space, and intracellular space. Theoretical models based on ion diffusion experiments determined that these three compartments represent 5 %, 20 %, and 75 % of total brain volume, respectively (Fig. 5a) (83). We used HPLC to quantify D-serine concentrations in blood (4 ± 0.9 μM), as well as its total brain contents (188 ± 10 pmol/mg of wet tissue, Fig. 5b).

Based on a rat brain density value of 1.033 ± 0.002 that we determined experimentally, we converted the total brain D-serine concentration from 188 ± 10 pmol/mg of wet tissue to 182 ± 10 μM. We then took into account the biosensor estimations of D-serine extracellular concentration (2–3 μM D-serine), to compute the D-serine concentration in the remaining intracellular space. Our calculations yielded an estimated intracellular D-serine concentration of about 242 μM (78). At 3 h after D-serine administration (1 g/kg, i.p.), intracellular concentration was estimated to be around 400 μM, indicating that intracellular stores of D-serine nearly doubled by 3 h after administration.

Therefore, D-serine was concentrated by a factor of ~100 in the intracellular compartment, which represented 99.5 %–99.6 % of the total D-serine present in the brain (Fig. 5a). This value represents an average of all D-serine present inside brain cells, but its intracellular concentration is probably not homogeneous throughout different cell types of the forebrain. The issue of D-serine intracellular storage is currently a matter of debate. In C6 glioma cell cultures, a stable astrocytic cell line, extracellular D-serine levels in the culture medium are typically ~50 fold higher than intracellular levels (11), suggesting an ongoing release process and limited intracellular storage capacity. However, intracellular organelles immunoreactive to D-serine have been observed in vitro (58) and ex vivo (84), suggesting that intracellular stores do exist in the brain. More recently, Bergersen et al. (85) detected D-serine-immunoreactive synaptic-like vesicles in hippocampal astrocytes and estimated the D-serine intravesicular concentration at about 6 mM. It is likely that these stores are located in glial cells, especially astrocytes, radial glia, and microglia (84). However, serine racemase, the D-serine synthesizing enzyme, is preferentially expressed in neurons (40) and the possibility of neuronal D-serine storage cannot be excluded.

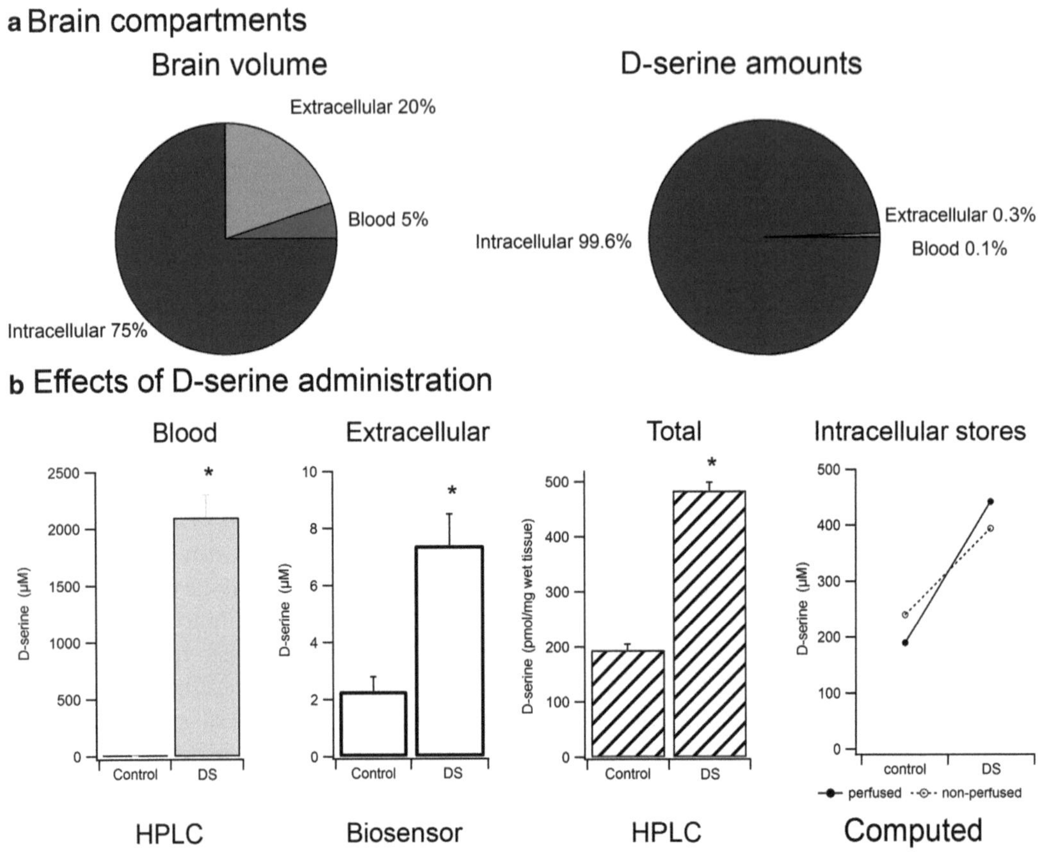

Fig. 5 (**a**) The brain can be separated into three compartments: blood, extracellular space, and intracellular space, representing 5 %, 20 %, and 75 % of the total brain volume, respectively. D-serine present in the intracellular compartment represented more than 99 % of total brain D-serine. (**b**) *Left*: D-serine concentrations in blood, extracellular space, and whole brain before and after D-serine (1 g/kg, i.p.). *Right*: Estimate of D-serine concentration in intracellular stores. DS, D-serine. Reprinted with permission from (78) Copyright© 2012, Elsevier

An essential property of chemical transmitters is their ability to be stored intracellularly and released into the interstitial medium in response to an appropriate stimulus. Our quantification of D-serine intracellular stores therefore fully supports the function of D-serine as a neuro- or gliotransmitter to modulate synaptic transmission (5, 6).

8 Conclusion

Although many questions remain regarding the extracellular regulation of D-serine concentrations in the CNS, the development of a D-serine microelectrode biosensor has provided valuable insight into potential mechanisms of D-serine modulation. Since D-serine has previously been identified as a potential new target for therapeutic intervention in cases of schizophrenia, amyotrophic lateral sclerosis, and ischemia, our biosensors may further refine the development of new drugs targeting NMDA receptor activity. Looking ahead, we believe that D-serine-sensitive biosensors will continue to have a significant impact on investigations of D-serine modulation in the CNS. Correspondingly, D-serine recordings in freely moving animals are under way, and should facilitate the direct investigation of D-serine release during various behavior and pathological states. Accordingly, such studies could further clarify the signaling involved in the release of D-serine, and thereby elucidate D-serine functions relevant to the CNS.

References

1. Hashimoto A, Nishikawa T, Oka T, Takahashi K, Hayashi T (1992) Determination of free amino acid enantiomers in rat brain and serum by high-performance liquid chromatography after derivatization with *N*-tert.-butyloxycarbonyl-L-cysteine and *o*-phthaldialdehyde. J Chromatogr 582:41–48
2. Matsui T, Sekiguchi M, Hashimoto A, Tomita U, Nishikawa T, Wada K (1995) Functional comparison of D-serine and glycine in rodents: the effect on cloned NMDA receptors and the extracellular concentration. J Neurochem 65:454–458
3. Priestley T, Laughton P, Myers J, Le Bourdelles B, Kerby J, Whiting PJ (1995) Pharmacological properties of recombinant human *N*-methyl-D-aspartate receptors comprising NR1a/NR2A and NR1a/NR2B subunit assemblies expressed in permanently transfected mouse fibroblast cells. Mol Pharmacol 48:841–848
4. Kakegawa W, Miyoshi Y, Hamase K, Matsuda S, Matsuda K, Kohda K, Emi K, Motohashi J, Konno R, Zaitsu K, Yuzaki M (2011) D-Serine regulates cerebellar LTD and motor coordination through the delta2 glutamate receptor. Nat Neurosci 14:603–611
5. Panatier A, Theodosis DT, Mothet JP, Touquet B, Pollegioni L, Poulain DA, Oliet SH (2006) Glia-derived D-serine controls NMDA receptor activity and synaptic memory. Cell 125: 775–784
6. Henneberger C, Papouin T, Oliet SH, Rusakov DA (2010) Long-term potentiation depends on release of D-serine from astrocytes. Nature 463:232–236
7. Zhang Z, Gong N, Wang W, Xu L, Xu TL (2008) Bell-shaped D-serine actions on hippocampal long-term depression and spatial memory retrieval. Cereb Cortex 18:2391–2401
8. Mothet JP, Rouaud E, Sinet PM, Potier B, Jouvenceau A, Dutar P, Videau C, Epelbaum J, Billard JM (2006) A critical role for the glial-derived neuromodulator D-serine in the age-related deficits of cellular mechanisms of learning and memory. Aging Cell 5:267–274
9. Junjaud G, Rouaud E, Turpin F, Mothet JP, Billard JM (2006) Age-related effects of the neuromodulator D-serine on neurotransmission and synaptic potentiation in the CA1 hippocampal area of the rat. J Neurochem 98: 1159–1166
10. Aura J, Riekkinen P Jr (2000) Pre-training blocks the improving effect of tetrahydroaminoacridine and D-cycloserine on spatial navigation performance in aged rats. Eur J Pharmacol 390:313–318
11. Kim PM, Aizawa H, Kim PS, Huang AS, Wickramasinghe SR, Kashani AH, Barrow RK, Huganir RL, Ghosh A, Snyder SH (2005) Serine racemase: activation by glutamate neurotransmission via glutamate receptor interacting protein and mediation of neuronal migration. Proc Natl Acad Sci USA 102:2105–2110
12. Stevens ER, Esguerra M, Kim PM, Newman EA, Snyder SH, Zahs KR, Miller RF (2003) D-serine and serine racemase are present in the vertebrate retina and contribute to the physiological activation of NMDA receptors. Proc Natl Acad Sci USA 100:6789–6794
13. Sullivan SJ, Miller RF (2011) AMPA receptor mediated D-serine release from retinal glial cells. J Neurochem 115:1681–1689
14. Kalbaugh TL, Zhang J, Diamond JS (2009) Coagonist release modulates NMDA receptor subtype contributions at synaptic inputs to retinal ganglion cells. J Neurosci 29:1469–1479
15. Stevens ER, Gustafson EC, Sullivan SJ, Esguerra M, Miller RF (2010) Light-evoked NMDA receptor-mediated currents are reduced by blocking D-serine synthesis in the salamander retina. Neuroreport 21:239–244

16. Hashimoto K, Fukushima T, Shimizu E, Komatsu N, Watanabe H, Shinoda N, Nakazato M, Kumakiri C, Okada S, Hasegawa H, Imai K, Iyo M (2003) Decreased serum levels of D-serine in patients with schizophrenia: evidence in support of the N-methyl-D-aspartate receptor hypofunction hypothesis of schizophrenia. Arch Gen Psychiatry 60:572–576
17. Hashimoto K, Engberg G, Shimizu E, Nordin C, Lindstrom LH, Iyo M (2005) Reduced D-serine to total serine ratio in the cerebrospinal fluid of drug naive schizophrenic patients. Prog Neuropsychopharmacol Biol Psychiatry 29:767–769
18. Bendikov I, Nadri C, Amar S, Panizzutti R, De Miranda J, Wolosker H, Agam G (2007) A CSF and postmortem brain study of D-serine metabolic parameters in schizophrenia. Schizophr Res 90:41–51
19. Verrall L, Walker M, Rawlings N, Benzel I, Kew JN, Harrison PJ, Burnet PW (2007) D-Amino acid oxidase and serine racemase in human brain: normal distribution and altered expression in schizophrenia. Eur J Neurosci 26:1657–1669
20. Chumakov I, Blumenfeld M, Guerassimenko O, Cavarec L, Palicio M, Abderrahim H, Bougueleret L, Barry C, Tanaka H, La Rosa P, Puech A, Tahri N, Cohen-Akenine A, Delabrosse S, Lissarrague S, Picard FP, Maurice K, Essioux L, Millasseau P, Grel P, Debailleul V, Simon AM, Caterina D, Dufaure I, Malekzadeh K, Belova M, Luan JJ, Bouillot M, Sambucy JL, Primas G, Saumier M, Boubkiri N, Martin-Saumier S, Nasroune M, Peixoto H, Delaye A, Pinchot V, Bastucci M, Guillou S, Chevillon M, Sainz-Fuertes R, Meguenni S, Aurich-Costa J, Cherif D, Gimalac A, Van Duijn C, Gauvreau D, Ouellette G, Fortier I, Raelson J, Sherbatich T, Riazanskaia N, Rogaev E, Raeymaekers P, Aerssens J, Konings F, Luyten W, Macciardi F, Sham PC, Straub RE, Weinberger DR, Cohen N, Cohen D (2002) Genetic and physiological data implicating the new human gene G72 and the gene for D-amino acid oxidase in schizophrenia. Proc Natl Acad Sci USA 99: 13675–13680
21. Sacchi S, Bernasconi M, Martineau M, Mothet JP, Ruzzene M, Pilone MS, Pollegioni L, Molla G (2008) pLG72 modulates intracellular D-serine levels through its interaction with D-amino acid oxidase: effect on schizophrenia susceptibility. J Biol Chem 283:22244–22256
22. Detera-Wadleigh SD, McMahon FJ (2006) G72/G30 in schizophrenia and bipolar disorder: review and meta-analysis. Biol Psychiatry 60:106–114
23. Tsai G, Yang P, Chung LC, Lange N, Coyle JT (1998) D-serine added to antipsychotics for the treatment of schizophrenia. Biol Psychiatry 44:1081–1089
24. Tsai GE, Yang P, Chung LC, Tsai IC, Tsai CW, Coyle JT (1999) D-serine added to clozapine for the treatment of schizophrenia. Am J Psychiatry 156:1822–1825
25. Heresco-Levy U, Javitt DC, Ebstein R, Vass A, Lichtenberg P, Bar G, Catinari S, Ermilov M (2005) D-serine efficacy as add-on pharmacotherapy to risperidone and olanzapine for treatment-refractory schizophrenia. Biol Psychiatry 57:577–585
26. Katsuki H, Nonaka M, Shirakawa H, Kume T, Akaike A (2004) Endogenous D-serine is involved in induction of neuronal death by *N*-methyl-D-aspartate and simulated ischemia in rat cerebrocortical slices. J Pharmacol Exp Ther 311:836–844
27. Shleper M, Kartvelishvily E, Wolosker H (2005) D-serine is the dominant endogenous coagonist for NMDA receptor neurotoxicity in organotypic hippocampal slices. J Neurosci 25:9413–9417
28. Kirschner DL, Wilson AL, Drew KL, Green TK (2009) Simultaneous efflux of endogenous D-ser and L-glu from single acute hippocampus slices during oxygen glucose deprivation. J Neurosci Res 87:2812–2820
29. Katsuki H, Watanabe Y, Fujimoto S, Kume T, Akaike A (2007) Contribution of endogenous glycine and D-serine to excitotoxic and ischemic cell death in rat cerebrocortical slice cultures. Life Sci 81:740–749
30. Mustafa AK, Ahmad AS, Zeynalov E, Gazi SK, Sikka G, Ehmsen JT, Barrow RK, Coyle JT, Snyder SH, Dore S (2010) Serine racemase deletion protects against cerebral ischemia and excitotoxicity. J Neurosci 30:1413–1416
31. Mustafa AK, Ahmad AS, Zeynalov E, Gazi SK, Sikka G, Ehmsen JT, Barrow RK, Coyle JT, Snyder SH, Dore S (2011) Serine racemase deletion protects against cerebral ischemia and excitotoxicity. J Neurosci 30:1413–1416
32. Wu SZ, Bodles AM, Porter MM, Griffin WS, Basile AS, Barger SW (2004) Induction of serine racemase expression and D-serine release from microglia by amyloid beta-peptide. J Neuroinflammation 1:2
33. Wu S, Basile AS, Barger SW (2007) Induction of serine racemase expression and D-serine release from microglia by secreted amyloid precursor protein (sAPP). Curr Alzheimer Res 4:243–251
34. Inoue R, Hashimoto K, Harai T, Mori H (2008) NMDA- and beta-amyloid1-42-induced neurotoxicity is attenuated in serine racemase knock-out mice. J Neurosci 28:14486–14491
35. Sasabe J, Chiba T, Yamada M, Okamoto K, Nishimoto I, Matsuoka M, Aiso S (2007)

D-serine is a key determinant of glutamate toxicity in amyotrophic lateral sclerosis. EMBO J 26:4149–4159
36. Wolosker H, Blackshaw S, Snyder SH (1999) Serine racemase: a glial enzyme synthesizing D-serine to regulate glutamate-N-methyl-D-aspartate neurotransmission. Proc Natl Acad Sci USA 96:13409–13414
37. Wolosker H (2011) Serine racemase and the serine shuttle between neurons and astrocytes. Biochim Biophys Acta 1814(11):1558–1566
38. Rosenberg D, Kartvelishvily E, Shleper M, Klinker CM, Bowser MT, Wolosker H (2010) Neuronal release of D-serine: a physiological pathway controlling extracellular D-serine concentration. FASEB J 24:2951–2961
39. Yoshikawa M, Takayasu N, Hashimoto A, Sato Y, Tamaki R, Tsukamoto H, Kobayashi H, Noda S (2007) The serine racemase mRNA is predominantly expressed in rat brain neurons. Arch Histol Cytol 70:127–134
40. Miya K, Inoue R, Takata Y, Abe M, Natsume R, Sakimura K, Hongou K, Miyawaki T, Mori H (2008) Serine racemase is predominantly localized in neurons in mouse brain. J Comp Neurol 510:641–654
41. De Miranda J, Panizzutti R, Foltyn VN, Wolosker H (2002) Cofactors of serine racemase that physiologically stimulate the synthesis of the *N*-methyl-D-aspartate (NMDA) receptor coagonist D-serine. Proc Natl Acad Sci USA 99:14542–14547
42. Neidle A, Dunlop DS (2002) Allosteric regulation of mouse brain serine racemase. Neurochem Res 27:1719–1724
43. Foltyn VN, Zehl M, Dikopoltsev E, Jensen ON, Wolosker H (2010) Phosphorylation of mouse serine racemase regulates D-serine synthesis. FEBS Lett 584:2937–2941
44. Dumin E, Bendikov I, Foltyn VN, Misumi Y, Ikehara Y, Kartvelishvily E, Wolosker H (2006) Modulation of D-serine levels via ubiquitin-dependent proteasomal degradation of serine racemase. J Biol Chem 281:20291–20302
45. Fujii K, Maeda K, Hikida T, Mustafa AK, Balkissoon R, Xia J, Yamada T, Ozeki Y, Kawahara R, Okawa M, Huganir RL, Ujike H, Snyder SH, Sawa A (2006) Serine racemase binds to PICK1: potential relevance to schizophrenia. Mol Psychiatry 11:150–157
46. Mustafa AK, Kumar M, Selvakumar B, Ho GP, Ehmsen JT, Barrow RK, Amzel LM, Snyder SH (2007) Nitric oxide *S*-nitrosylates serine racemase, mediating feedback inhibition of D-serine formation. Proc Natl Acad Sci USA 104:2950–2955
47. Mustafa AK, van Rossum DB, Patterson RL, Maag D, Ehmsen JT, Gazi SK, Chakraborty A, Barrow RK, Amzel LM, Snyder SH (2009) Glutamatergic regulation of serine racemase via reversal of PIP2 inhibition. Proc Natl Acad Sci USA 106:2921–2926
48. Foltyn VN, Bendikov I, De Miranda J, Panizzutti R, Dumin E, Shleper M, Li P, Toney MD, Kartvelishvily E, Wolosker H (2005) Serine racemase modulates intracellular D-serine levels through an alpha, beta-elimination activity. J Biol Chem 280:1754–1763
49. Panizzutti R, De Miranda J, Ribeiro CS, Engelender S, Wolosker H (2001) A new strategy to decrease *N*-methyl-D-aspartate (NMDA) receptor coactivation: inhibition of D-serine synthesis by converting serine racemase into an eliminase. Proc Natl Acad Sci USA 98:5294–5299
50. Rutter AR, Fradley RL, Garrett EM, Chapman KL, Lawrence JM, Rosahl TW, Patel S (2007) Evidence from gene knockout studies implicates Asc-1 as the primary transporter mediating d-serine reuptake in the mouse CNS. Eur J Neurosci 25:1757–1766
51. Gliddon CM, Shao Z, LeMaistre JL, Anderson CM (2009) Cellular distribution of the neutral amino acid transporter subtype ASCT2 in mouse brain. J Neurochem 108:372–383
52. Dun Y, Mysona B, Itagaki S, Martin-Studdard A, Ganapathy V, Smith SB (2007) Functional and molecular analysis of D-serine transport in retinal Muller cells. Exp Eye Res 84:191–199
53. Ribeiro CS, Reis M, Panizzutti R, de Miranda J, Wolosker H (2002) Glial transport of the neuromodulator D-serine. Brain Res 929:202–209
54. O'Brien KB, Bowser MT (2006) Measuring D-serine efflux from mouse cortical brain slices using online microdialysis-capillary electrophoresis. Electrophoresis 27:1949–1956
55. Rosenberg D, Kartvelishvily E, Shleper M, Klinker CM, Bowser MT, Wolosker H (2010) Neuronal release of D-serine: a physiological pathway controlling extracellular D-serine concentration. FASEB J 24(8):2951–2961
56. Schell MJ, Molliver ME, Snyder SH (1995) D-serine, an endogenous synaptic modulator: localization to astrocytes and glutamate-stimulated release. Proc Natl Acad Sci USA 92:3948–3952
57. Mothet JP, Pollegioni L, Ouanounou G, Martineau M, Fossier P, Baux G (2005) Glutamate receptor activation triggers a calcium-dependent and SNARE protein-dependent release of the gliotransmitter D-serine. Proc Natl Acad Sci USA 102:5606–5611
58. Martineau M, Galli T, Baux G, Mothet JP (2008) Confocal imaging and tracking of the exocytotic routes for D-serine-mediated gliotransmission. Glia 56:1271–1284
59. Pollegioni L, Piubelli L, Sacchi S, Pilone MS, Molla G (2007) Physiological functions of

D-amino acid oxidases: from yeast to humans. Cell Mol Life Sci 64:1373–1394
60. Horiike K, Tojo H, Arai R, Nozaki M, Maeda T (1994) D-amino-acid oxidase is confined to the lower brain stem and cerebellum in rat brain: regional differentiation of astrocytes. Brain Res 652:297–303
61. Benveniste H, Huttemeier PC (1990) Microdialysis–theory and application. Prog Neurobiol 35:195–215
62. Hashimoto A, Oka T, Nishikawa T (1995) Extracellular concentration of endogenous free D-serine in the rat brain as revealed by in vivo microdialysis. Neuroscience 66:635–643
63. Ciriacks CM, Bowser MT (2004) Monitoring D-serine dynamics in the rat brain using online microdialysis-capillary electrophoresis. Anal Chem 76:6582–6587
64. Ciriacks CM, Bowser MT (2006) Measuring the effect of glutamate receptor agonists on extracellular D-serine concentrations in the rat striatum using online microdialysis-capillary electrophoresis. Neurosci Lett 393:200–205
65. Borland LM, Shi G, Yang H, Michael AC (2005) Voltammetric study of extracellular dopamine near microdialysis probes acutely implanted in the striatum of the anesthetized rat. J Neurosci Methods 146:149–158
66. Khan AC, Michael AC (2003) Invasive consequences of using micro-electrodes and microdialysis probes in the brain. Trends Anal Chem 22:6
67. Benveniste H, Diemer NH (1987) Cellular reactions to implantation of a microdialysis tube in the rat hippocampus. Acta Neuropathol 74:234–238
68. Johanson E, Marko-Varga G, Gorton L (1993) Study of a reagent- and mediator-less biosensor for D-amino acids based on co-immobilized D-amino acid oxidase and peroxidase in carbon paste electrodes. J Biomater Appl 8:146–173
69. Inaba Y, Mizukami K, Hamada-Sato N, Kobayashi T, Imada C, Watanabe E (2003) Development of a D-alanine sensor for the monitoring of a fermentation using the improved selectivity by the combination of D-amino acid oxidase and pyruvate oxidase. Biosens Bioelectron 19:423–431
70. Stefan RI, Nejem RM, van Staden JF, Aboul-Enein HY (2004) New amperometric biosensors based on diamond paste for the assay of L- and D-pipecolic acids in serum samples. Prep Biochem Biotechnol 34:135–143
71. Dominguez R, Serra B, Reviejo AJ, Pingarron JM (2001) Chiral analysis of amino acids using electrochemical composite bienzyme biosensors. Anal Biochem 298:275–282
72. Wcislo M, Compagnone D, Trojanowicz M (2007) Enantioselective screen-printed amperometric biosensor for the determination of D-amino acids. Bioelectrochemistry 71:91–98
73. Pernot P, Mothet JP, Schuvailo O, Soldatkin A, Pollegioni L, Pilone M, Adeline MT, Cespuglio R, Marinesco S (2008) Characterization of a yeast D-amino acid oxidase microbiosensor for D-serine detection in the central nervous system. Anal Chem 80:9
74. Zain ZM, O'Neill RD, Lowry JP, Pierce KW, Tricklebank M, Dewa A, Ab Ghani S (2010) Development of an implantable D-serine biosensor for in vivo monitoring using mammalian D-amino acid oxidase on a poly (*o*-phenylenediamine) and nafion-modified platinum-iridium disk electrode. Biosens Bioelectron 25:1454–1459
75. Vasylieva N, Barnych B, Meiller A, Maucler C, Pollegioni L, Lin JS, Barbier D, Marinesco S (2011) Covalent enzyme immobilization by poly(ethylene glycol) diglycidyl ether (PEGDE) for microelectrode biosensor preparation. Biosens Bioelectron 26:3993–4000
76. Hamase K, Konno R, Morikawa A, Zaitsu K (2005) Sensitive determination of D-amino acids in mammals and the effect of D-amino-acid oxidase activity on their amounts. Biol Pharm Bull 28:1578–1584
77. Fukushima T, Kawai J, Imai K, Toyo'oka T (2004) Simultaneous determination of D- and L-serine in rat brain microdialysis sample using a column-switching HPLC with fluorimetric detection. Biomed Chromatogr 18:813–819
78. Pernot P, Maucler C, Tholance Y, Vasylieva N, Debilly G, Pollegioni L, Cespuglio R, Marinesco S (2012) D-Serine diffusion through the blood-brain barrier: effect on D-serine compartmentalization and storage. Neurochem Int 60:837–845
79. Hashimoto A, Oka T, Nishikawa T (1995) Extracellular concentration of endogenous free D-serine in the rat brain as revealed by in vivo microdialysis. Neuroscience 66:635–643
80. Hashimoto A, Chiba S (2004) Effect of systemic administration of D-serine on the levels of D- and L-serine in several brain areas and periphery of rat. Eur J Pharmacol 495:153–158
81. Morgan ME, Singhal D, Anderson BD (1996) Quantitative assessment of blood-brain barrier damage during microdialysis. J Pharmacol Exp Ther 277:1167–1176
82. Tsai PS, Kaufhold JP, Blinder P, Friedman B, Drew PJ, Karten HJ, Lyden PD, Kleinfeld D (2009) Correlations of neuronal and microvascular densities in murine cortex revealed by direct counting and colocalization of nuclei and vessels. J Neurosci 29:14553–14570
83. Nicholson C, Sykova E (1998) Extracellular space structure revealed by diffusion analysis. Trends Neurosci 21:207–215

84. Williams SM, Diaz CM, Macnab LT, Sullivan RK, Pow DV (2006) Immunocytochemical analysis of D-serine distribution in the mammalian brain reveals novel anatomical compartmentalizations in glia and neurons. Glia 53:401–411

85. Bergersen LH, Morland C, Ormel L, Rinholm JE, Larsson M, Wold JF, Roe AT, Stranna A, Santello M, Bouvier D, Ottersen OP, Volterra A, Gundersen V (2011) Immunogold detection of L-glutamate and D-serine in small synaptic-like microvesicles in adult hippocampal astrocytes. Cereb Cortex 22(7):1690–1697

Chapter 10

Measurement of Purine Release with Microelectrode Biosensors

Nicholas Dale

Abstract

The purines, adenosine triphosphate (ATP) and adenosine, are now recognized as extracellular signaling agents in both the brain and peripheral organs. The mechanisms of purinergic signaling can differ from those of conventional neurotransmission especially in the variety of release mechanisms and routes of production of extracellular ATP and adenosine. Direct real time measurements of these purinergic agents have therefore been of great value in elucidating the functional roles of ATP and adenosine in several diverse contexts. This review introduces the concept of microelectrode biosensors for ATP and adenosine and explains how they operate, how they can best be used, and gives examples of new mechanistic insight into the mechanosensory control of respiration, synaptic physiology, and glucosensing in the hypothalamus.

Key words Biosensor, Real-time measurement, Chemosensory mechanisms, ATP, Adenosine

1 Introduction

1.1 An Overview of Purinergic Signaling

The purines, ATP, adenosine diphosphate (ADP) and adenosine, are fundamentally linked to cellular metabolism because ATP acts as the universal store of chemical energy that powers cellular and molecular processes. However, the purines are also important extracellular signaling molecules. The role of adenosine as a modulator acting on the heart was described in the 1920s (1). It took far longer for the roles of ATP, firstly as a peripheral neurotransmitter (2, 3) and then as a centrally acting neurotransmitter (4) to be recognized. By now the purines are universally accepted as important extracellular signaling molecules in virtually every organ of the body including the brain (5–7).

1.1.1 ATP

ATP acts at both ionotropic (P2X) (8) and metabotropic receptors (P2Y) (9). The P2X receptors are so far unique amongst ligand-gated ion channels in being trimeric (10). There are seven subtypes each having somewhat different properties, and some of which can

Stéphane Marinesco and Nicholas Dale (eds.), *Microelectrode Biosensors*, Neuromethods, vol. 80,
DOI 10.1007/978-1-62703-370-1_10, © Springer Science+Business Media, LLC 2013

combine to form heterotrimeric receptors (10). The P2X receptors are only activated by ATP and certain ATP analogues; uridine triphosphate (UTP) and ADP are ineffective agonists. A number of P2X receptors, particularly the P2X7 receptor, have an unusual property known as pore dilation: in the face of prolonged agonist activation, the channel enters a high conductance state that allows passage of molecules of up to a few hundred Daltons (11–13). The mechanisms of pore dilation are controversial: some authors contend that accessory proteins, namely pannexin-1, are required (14–17), while others posit that it is an intrinsic property of the channel itself (13, 18, 19). P2X7 receptors have attracted considerable interest as they have been characterized as "death channels" owing to their ability to undergo pore dilation, permeabilize cells, and ultimately trigger apoptosis.

Although there are 14 P2Y subtypes (each being a typical G-protein coupled receptor), only a subset of these are considered to be ATP/ADP receptors (P2Y1, P2Y2, P2Y4, P2Y6, P2Y11, P2Y12, and P2Y13) (9). Some receptors can be activated by ATP alone, some by ATP or ADP and others by ADP alone. For example the P2Y1 receptor can be activated by ATP and ADP, couples to G_q and G_o, and can activate the IP_3 pathway leading to mobilization of intracellular Ca^{2+}. Some P2Y receptors also couple to, and modulate, specific ion channels such as N-type Ca^{2+} channels and M-type K^+ channels. Overall, the P2Y receptors mediate a range of modulatory effects very similar to those of other G-protein coupled neurotransmitter receptors.

Although receptors for ATP are widespread in the central nervous system, the source of ATP that activates them in many cases remains unclear. ATP can indeed be released at synapses both peripherally and centrally (4, 20). However in the CNS it seems that it acts mainly as a cotransmitter (21, 22), and there are very few synapses where it acts as the principal transmitter. ATP is also released by glial cells (23, 24) for the most part via exocytosis (25, 26), but see (27).

Unlike many other neurotransmitters, it appears that ATP can be released by several other mechanisms in addition to exocytosis. There is compelling evidence for channel-mediated mechanisms of ATP release. For example ATP release has been documented via gap junction hemichannels (both pannexins and connexins) (28–31), volume activated channels (32, 33) and even the P2X7 receptor itself following pore dilation (34).

1.1.2 Adenosine

Adenosine is perhaps the most pervasive modulator in the central nervous system. It has many physiological roles including acting as a sleep promoting substance (somnogen) (35, 36) and as the brain's endogenous neuroprotective agent [37)—it helps to inhibit glutamate release during periods of brain hypoxia and ischaemia and acts to dampen brain excitability and prevent epileptic seizures.

The most widely used psychoactive drug is caffeine, a mild stimulant, which at the concentrations ingested from coffee acts mostly as an antagonist of adenosine receptors. There are no ionotropic channels for adenosine. It acts instead at a set of four G-protein coupled receptors (A_1, A_{2a}, A_{2b}, and A_3) sometimes collectively referred to as P1 receptors (38). These receptors mediate a range of effects very similar to those of other G-protein coupled receptor systems.

The routes for adenosine release and the cellular sources seem to be even more mysterious than those for ATP. For example there is abundant evidence that adenosine can arise from the breakdown of previously released ATP via the actions of specialized enzymes termed ectonucleotidases (39). Indeed one source of the adenosine that acts as a somnogen may be from ATP released from astrocytes (40).

There is also evidence that adenosine can be released in an activity-dependent manner (41–43), although whether this represents direct adenosine release or production via the prior release of ATP and extremely rapid conversion to adenosine remains, at least for the time being, open to debate (42).

Equilibrative nucleoside transporters (44) are abundant in the CNS and effectively equalize adenosine concentrations in both directions across the plasma membrane. This means that, somewhat counter intuitively, one of the main determinants of extracellular adenosine is the intracellular enzyme adenosine kinase (45). This enzyme converts adenosine to AMP, thus reducing the intracellular concentration of adenosine and hence maintain an inward equilibrium for the transport of adenosine from the extracellular space.

Under pathological conditions, especially when ATP production drops and intracellular adenosine levels rise, adenosine can indeed be released directly (46–50). The mechanisms by which this occurs (cellular sources molecules and transporters involved) have still not been settled.

1.1.3 Other Agents

The diadenosine polyphosphates can act at certain P2Y receptors. These molecules can have three, four or even more phosphates linking the two adenosine moieties and are found in brain tissue, and there is some evidence that they may be released (51–53). While not strictly a purine, UTP is also an agonist at some P2Y receptors and has the potential to be an endogenous agonist. UDP-glucose, a potential endogenous ligand for which there is evidence of release, is an effective agonist acting at P2Y14 receptors (54, 55) and at the recently "deorphanized" receptor, GPR17 (56).

1.1.4 Ectonucleotidases

The extracellular enzymes that control the metabolism of ATP in the extracellular space do much more than terminate the actions of ATP; they also initiate the actions of downstream signaling molecules such as ADP and adenosine. It should also be noted that these enzymes will hydrolyze most nucleotides; however, for the sake of simplicity, I refer only to their actions on adenine nucleotides.

There are four possible gene families that contribute to this conversion of ATP (57). The first is the NTPDase family (also known as CD39) that has eight members, four of which are found on the plasma membrane, while the others are found only in intracellular compartments (NTPDase4 and 7), while others have the potential to be secreted (39, 58–62). The NTPDases are integral membrane proteins. NTPDases1–4 and 8 have two transmembrane segments with a large extracellular loop, whereas NTPDases5 and 6 lack the second transmembrane segment and have a cleavage site allowing the potential secretion of the extracellular domain into the extracellular space.

All of the members of the NTPDase family exhibit distinct patterns of expression suggesting somewhat different functional roles. The members of the NTPDase family differ in their substrate affinities and enzymatic reactions. For example NTPDase1 will metabolize both ATP and ADP with equal efficiency, whereas NTPDase2 acts only on ATP. Thus, expression of NTPDase1 results in accumulation of AMP, whereas expression of NTPDase2 will result in accumulation of ADP. Interestingly NTPDase2 and P2Y1 receptors can act in concert in the functional development of the eye (63).

The NPPase family comprises seven members. These enzymes have a single transmembrane domain. These enzymes again exhibit differing expression patterns and enzymatic activities (60, 64–66). NPPases1–3 are able to hydrolyze both ATP and ADP.

The alkaline phosphatase family, linked to the membrane via a GPI anchor, have very wide substrate specificity (ATP, ADP, AMP and even pyrophosphate) (60). Alkaline phosphatases have particularly been associated with neural progenitor cells and their differentiation (67–69).

The final gene is that of the ecto 5′-nucleotidase (also known as CD73) that converts AMP to adenosine (60, 70). This enzyme is linked to the membrane via a GPI anchor and is a key enzyme for the production of adenosine from AMP in the brain. It also has the curious property that it can be inhibited by ATP and ADP (71–73), which may be of significance in providing temporal separation between the release of ATP and its conversion to adenosine (73, 74).

The regulation of the conversion of ATP to adenosine will depend on the characteristics and distribution of the ectonucleotidases present in the extracellular space. As these enzymes will not necessarily be arranged in discrete anatomical structures, the appearance of adenosine in the extracellular space may be widespread and diffuse. Furthermore, several of the enzymes are secretable, and those enzymes linked via a GPI anchor to the membrane also have the possibility of secretion into the extracellular space.

2 Real-Time Measurements of Purine Release

Conventional synaptic transmission involves vesicular release of transmitter from defined presynaptic structures (the bouton). The transmitter enters the synaptic cleft (about 50 nm in width) where it diffuses on a millisecond timescale to activate the receptors on the closely apposed postsynaptic membrane. Transmitter release is time-locked to and dependent on the presynaptic action potential resulting in the predictable occurrence of postsynaptic currents at a characteristic latency. For most purposes, many investigators have therefore regarded the postsynaptic currents as an indirect but sufficient measure of transmitter release.

While capturing the essence of synaptic transmission, we now appreciate that this conventional description is an over-simplification. For example transmitters such as GABA (75) and glutamate (76) can spill out of the synaptic cleft to have more diffuse and widespread actions than this classical model suggests. Under pathological conditions, neurotransmitters such as glutamate (77) can be released by the reversal of Na^+-dependent concentrative transporters. From the above discussion it should be clear that the purines, ATP and adenosine, depart from the paradigm of conventional neurotransmission even more comprehensively. ATP can be released at synapses, but there is abundant evidence for its release from astrocytes and by non-exocytotic mechanisms. Adenosine too arises from via a number of processes such as equilibrative transport, production in the extracellular space and possibly direct activity-dependent release.

Given that much of purinergic signaling occurs outside of traditional synaptic structures and that there may not be any convenient electrophysiological correlates of purine release that can be measured, a method for directly measuring real time release of ATP and adenosine has great potential value in the analysis of this extracellular signaling system. My group recognized this need many years ago and invented and applied the first biosensors to the study of purinergic signaling (46, 74, 78).

2.1 Microelectrode Biosensors for ATP and Adenosine

We have developed electrochemical biosensors based around oxidases to provide a real time signal for both ATP and adenosine and downstream purines. The basic principle of these biosensors is that an enzymatic cascade recognizes the purine of interest and produces a product (H_2O_2) proportional to analyte concentration. This product can then be detected electrochemically at the microelectrode surface. We have developed two enzymatic cascades to achieve sensitive detection of ATP and adenosine.

For ATP, this relies upon the enzymes glycerol kinase and glycerol-3-phosphate oxidase (Fig. 1a). If glycerol is provided at a level that saturates glycerol kinase (>0.5 mM), the reaction becomes solely

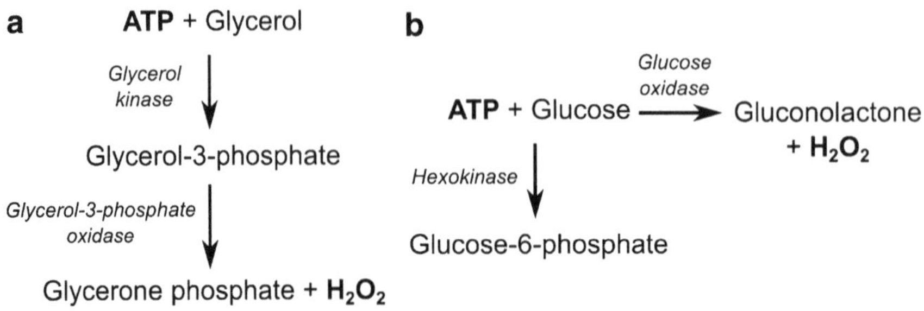

Fig. 1 *Enzyme cascades for electrochemical detection of ATP.* The two cascades result in a change in production of H_2O_2 that is proportional to the analyte concentration. For the cascade in (**a**) the production of H_2O_2 is directly proportional to the concentration of ATP, assuming a saturating concentration of glycerol. By contrast the cascade in (**b**) the production of H_2O_2 is inversely proportional to ATP concentration as ATP causes a *loss* of H_2O_2 as it drives the hexokinase reaction at the expense of gluconolactone and H_2O_2

dependent on ambient ATP concentrations (79). The product of the first step, glycerol-3-phosphate, is the substrate for the second, peroxide-producing, enzyme that gives an electrochemical signal proportional to ATP concentration. An alternative ATP detection cascade has been reported: hexokinase and glucose oxidase (80). Here the hexokinase converts glucose to glucose-6-phosphate, which is not a substrate for glucose oxidase (Fig. 1b). Thus, the signal is manifested as a loss of the glucose oxidation current. This can work well in vitro (81, 82), but this biosensor necessarily has to operate at glucose concentrations around the K_m for glucose oxidase, and is therefore also glucose sensitive. Thus in vivo, where local glucose concentrations may well change with time, this detection cascade cannot be used with confidence because it will likely report a combination of changes in ATP and glucose concentrations.

For adenosine, we use a linear cascade of three enzymes: adenosine deaminase, purine nucleoside phosphorylase and xanthine oxidase to produce an electrochemical signal proportional to adenosine concentration (Fig. 2) (74, 78).

2.2 Direct Electrochemistry: An Alternative Method for Detection of Adenosine

It is possible to detect adenosine by direct electrochemical means utilizing fast scan cyclic voltammetry. Adenosine will oxidize at quite positive potentials, but with appropriately treated carbon electrodes (carbon fiber (83, 84) or boron-doped diamond (85, 86)) sensitive detection with relatively rapid temporal resolution is possible. Venton and coworkers have demonstrated the selectivity of these sensors for adenosine versus other purine (83) and used them in brain slices to demonstrate adenosine release (87, 88). Their results largely replicate the findings we have made with enzymatic microelectrode biosensors (41, 43) giving confidence that both techniques accurately capture the dynamics of adenosine in the extracellular space.

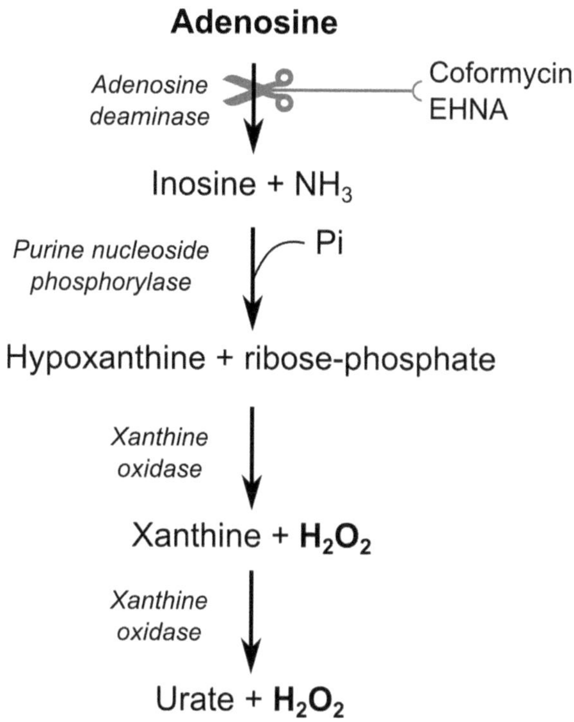

Fig. 2 *Enzyme cascade for the detection of adenosine and downstream purines.* The contribution of adenosine to the biosensor signal can be assessed through the use of specific blockers of adenosine deaminase (coformycin or EHNA)

2.3 Aptamers: Bringing ADP and Other Purines into the Measurement Fold?

As ADP is produced in the extracellular space from ATP and can act at several types of P2Y receptors (P2Y1, P2Y12, and P2Y13), it is of considerable interest to measure the release/production of this nucleotide. One way is to include a regenerating enzyme that converts ADP to ATP (e.g., creatine kinase) into the enzymatic cascade to make the biosensor sensitive to ADP as well as ATP (79). However, the responses of these biosensors tend to be very slow, and the signal cannot discriminate between ATP and ADP.

At present there is no very good solution to this problem, but an exciting potential alternative would be to exploit ADP-selective aptamers in a reagent free design biosensor (see Chap. 15 for more details on aptamer design). Aptamer-based biosensors might also offer ways of detecting other important signaling agents such as UDP-glucose and the diadenosine polyphosphates that appear to be physiologically functional but about which very little is known in terms of release.

2.4 Multienzyme Cascades

As multienzyme biosensors are sensitive to the substrates of all enzymes present in the cascade, the signals measured in tissue with a multienzyme biosensor could arise from any of these substrates. Measurement protocols must therefore take this into account if it is desired to identify the particular analyte or analytes being measured. One way to do this is to compare the signal derived from biosensors having the complete measurement cascade, with reference sensors that lack the first enzyme (46–48, 89). This has for example allowed the determination of the relative contributions of adenosine, inosine and hypoxanthine to biosensor signals recorded in brain slices. Alternatively, a specific blocker can be used for the first enzyme in the cascade with the aim of demonstrating loss of the signal. For adenosine biosensors we have used blockers of adenosine deaminase: coformycin (46, 74) and EHNA (41, 43) (Fig. 2). This manipulation can demonstrate the contribution of adenosine as opposed to the downstream metabolites to the signals recorded by the multienzyme adenosine biosensors.

2.5 Interference Rejection

A further requirement for the practical application of biosensing to physiological measurements is that the biosensors should be selective for the compound(s) of interest. While this can be conferred by the enzymes involved, electrochemical biosensors have a further complication. In general, these biosensors operate at rather positive potentials necessary for the oxidation of H_2O_2. Unfortunately at these potentials many other compounds, some of which are common in physiological systems (e.g., ascorbate, urate, dopamine) will also oxidize and generate a current unrelated to the specific analyte—so-called electrochemical interference.

There are several ways around this (89). One is to use rather complex electrode designs which involve electrochemical mediators so that the operating potential can be reduced to the point where very few compounds are capable of being oxidized (or reduced) to give a non-specific current. This solution may be very useful for diagnostic applications where a high premium is placed on selectivity. However, the use of electrochemical mediators often greatly slows the response time of the biosensor. This is a significant disadvantage for applications to physiological signaling which can occur on very rapid timescales.

A common alternative is the use of a permselective barrier layer that excludes interferences, but still permits the detection of the enzymatically produced H_2O_2. Some of these layers can be formed by electrochemical deposition—phenylenediamine and related compounds (see Chap. 6 for more details on phenylenediamine). Others such as nafion are applied by dip-coating (see Chap. 2 for more details on Nafion). In both cases the enzymatic layer is then formed on top of the permselective layer. For the purine biosensors that we make, we use a layer of phenylene diamine deposited by cyclic voltammetry. We then use sol–gel electrodeposition methods to deposit the enzymatic layer. Although this procedure does some damage to the inner permselectivity layer, it is possible to repair this damage by further depositing phenylene diamine following addition of the sol–gel enzymatic layer.

Because all the layers involved can be very thin, this latter method has the advantage that it does not slow biosensor response time. No matter what strategy is used to enhance the selectivity of electrochemical biosensors, no method gives an absolute guarantee of perfect selectivity. For example mediators may react directly with an interference to give a spurious signal; permselective layers may not effectively exclude very low molecular weight interferences (e.g., NO). Therefore, any use of biosensors should be accompanied by well performed controls to check the veracity of the observed signal. These issues have been considered in detail previously (89). In brief, we use a "null" sensor that lacks enzymes as a control for electroactive interferences—an idea similar to that of "self referencing" described in Chap. 2. The null sensor will not react with the specific analyte but will detect any interference. Subtraction of the null from the specific biosensor will therefore give a selective signal (31, 41, 48, 78, 79, 90–94).

3 Insights from Purine Measurements

A very useful and powerful approach to employing biosensors as part of an experimental program is to make predictions from the biosensor recordings that can subsequently be tested by independent experiment. This has the advantage of providing independent confirmation of the findings derived from the biosensors. An obvious way to achieve this is by use of selective antagonists to block the actions of purines predicted on the basis of their release. I give some examples of how we have used biosensors to significantly advance understanding of purinergic signaling. These examples also demonstrate integration of biosensor measurements with other physiological, electrophysiological and imaging methods both in vivo and in vitro. For a fuller account of the application of microelectrode biosensors to purinergic signaling, we refer the reader to (95).

3.1 Mechanosensory Afferent Control of Breathing: Breur–Hering Reflex

The Breur–Hering reflex is a key negative feedback mechano-reflex that limits the duration and extent of inspiration, thus preventing over-inflation of the lungs. Activation of slowly adapting pulmonary stretch receptors (SARs) during lung inflation initiates this reflex (96, 97). The SARs directly contact second order relay neurones or pump cells located in a discrete region of the dorsal medullary nucleus of the solitary tract (NTS) (97–99). These pump cells then contact the brain stem respiratory networks (99, 100) to evoke rapid suppression of inspiratory activity. Several studies suggest that the action of SARs on the pump cells is at least partly mediated by the action of glutamate at AMPA type receptors (98, 99, 101, 102). As many neurons in the NTS possess P2X receptors (103–106), and there is a precedent for ATP acting as a cotransmitter with glutamate, we used ATP biosensors to test whether there might be a contribution of ATP signaling to transmission at the central synapses of the SARs (92).

We placed ATP and null microelectrode biosensors into the NTS at the location where the SARs terminate in anesthetized artificially ventilated rats (Fig. 3a). The ATP sensors detected a signal that was phase locked to changes in tracheal pressure (a measure of lung inflation) (92) (Fig. 3b). There was no corresponding signal on the null biosensors suggesting that this signal was specifically related to ATP release in the NTS. This raised three possibilities (1) ATP release was related to inspiratory activity in the central networks; (2) ATP release originated from the SARs themselves; or (3) ATP release was dependent on SARs activity but not released from them directly.

The first possibility was easy to eliminate, as over-ventilation reduced respiratory drive and hence inspiratory activity as measured on the phrenic nerve. It was possible to completely block phrenic nerve activity by this means without altering the rhythmic ATP release phase-locked to lung inflation (92) (Fig. 3c). The SARs travel from the lungs to the brain stem via the vagal nerve. This offered a way of testing the second possibility (dependence of ATP release on SARs activity) by temporarily blocking nerve conduction in the vagal nerve through application of a local anesthetic. This manipulation completely and reversibly blocked the rhythmic release of ATP, demonstrating the dependence of this release on activity within the SARs (92) (Fig. 3d). The final possibility (that ATP release was downstream of activity in the SARs, but not released from them directly) was eliminated in two ways each designed to test whether neurons postsynaptic to the SARs might be involved. Firstly we applied a glutamate receptor antagonist to the dorsal surface of the brain stem—this greatly reduced phrenic nerve activity, indicating efficacy of the treatment, but had no effect on ATP release. Secondly we used a massive dose of the inhibitory transmitter GABA to effectively shunt neurons in the brain stem networks by profoundly inhibiting them. This treatment greatly reduced inspiratory activity but had no effect on ATP release (92). Our data therefore indicate that additional neurons downstream from the SARs are very unlikely to be involved and that ATP is most probably released from the SARs themselves.

As previous evidence suggested that glutamate was released from the SARs to mediate their effects, we confirmed this prior finding by utilizing glutamate biosensors to demonstrate glutamate release from the region of the NTS where the SARs afferents terminate. By performing a series of experiments for glutamate release analogous to those described above for ATP release, we confirmed that SARs also release glutamate (92). Our biosensor measurements therefore made a clear prediction: that the activation of pump cells by the SARs should depend not just on glutamate but

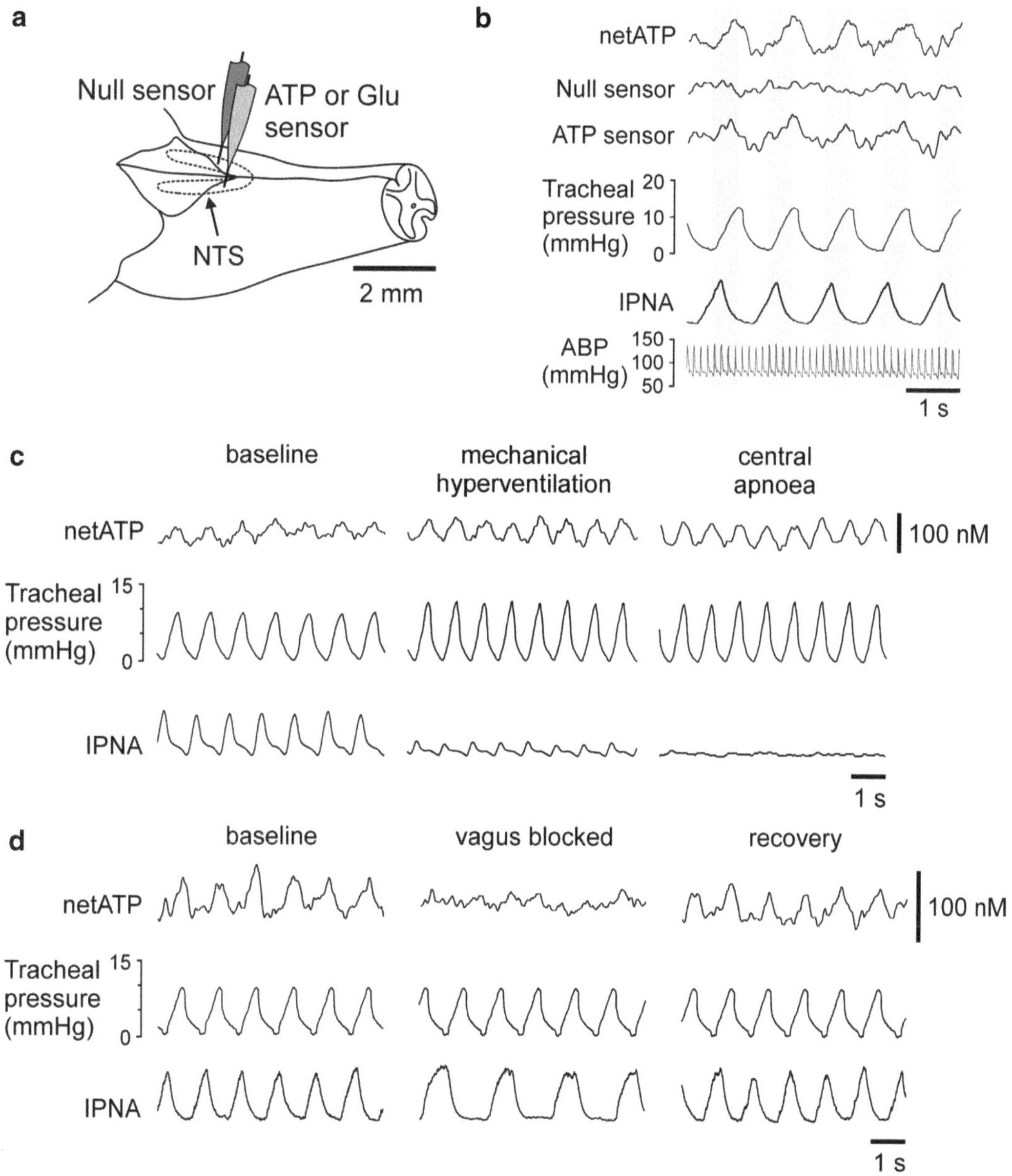

Fig. 3 *Rhythmic ATP release from the afferent terminals in the NTS of the slowly adapting stretch receptors of the lung*. (**a**) Diagram showing the brain stem of the rat and the placement of the biosensors to target the NTS. (**b**) Example records showing the ATP and null biosensor recordings, along with tracheal pressure (a measure of lung inflation), the integrated phrenic nerve recording (IPNA) and arterial blood pressure (ABP). Note that fluctuations resulting from blood pressure changes appear on both the null and ATP biosensor recordings but are largely absent from the difference trace (ATP—null, netATP). (**c**) Demonstration that ATP release depends on lung inflation rather than inspiratory activity as mechanical over-ventilation increases the amplitude of ATP release while abolishing inspiratory activity recorded on IPNA. (**d**) Rhythmic ATP release depends on electrical activity in the vagus nerve. Application of a local anesthetic to the vagus nerve reversibly abolished ATP release. Reproduced with permission from (92)

also on ATP. By making recordings from pump cells we showed that (1) they could be excited by ATP; (2) their activation during lung inflation was reduced by ATP receptor antagonists or glutamate receptor antagonists applied separately; and (3) the combination of ATP and glutamate receptor antagonists was more effective at preventing the firing of pump cells than either antagonist applied alone (92).

The interplay between physiological experimentation and biosensor measurements in this example has allowed new insight into the signaling from the SARs to their follower cells. Our studies have demonstrated that there is unexpected complexity and demonstrated that two transmitters are involved in relaying the excitation from SARs. Whether these transmitters are simultaneously released at the same synapses or there are subsets of SARs some of which use glutamate and others which use ATP remains to be determined.

3.2 The Properties of Activity-Dependent Adenosine Release

Adenosine can be released in an activity-dependent manner under normal physiological conditions, e.g., following a brief train of spikes (41–43). While it is generally accepted that under pathological conditions (e.g., metabolic stress), adenosine can be released directly via transporters; the direct release of adenosine is not thought to occur normally. Instead the prevailing consensus is that adenosine arises instead from the hydrolysis of previously released ATP mediated by many extracellular enzymes capable of mediating this conversion (39, 59–62). Thus, the several examples of activity-dependent adenosine release (42) have largely been viewed as arising from exocytotic release of ATP and subsequent conversion to adenosine.

Adenosine biosensors have sufficient sensitivity and temporal resolution to measure spike-evoked adenosine release in the cerebellum (41). The biosensor recordings reveal that adenosine release has the characteristics of exocytosis—TTX and Ca^{2+} dependence (41). Furthermore, adenosine release depends on spike width and can exhibit short-term plasticity more typically connected with conventional synaptic potentials. For example at very short intervals (50 ms), adenosine release undergoes paired pulse facilitation, while at longer intervals (30–120 s) it undergoes paired pulse depression (43) (Fig. 4). These data are even more suggestive of an exocytotic release mechanism, which can undergo depletion and rather slow restocking (43). Interestingly the activity-dependent adenosine release can be modulated by a number of different transmitters acting via G-protein coupled receptors including adenosine A_1, mGluR4 and $GABA_B$ receptors (41, 107).

It remains possible that this activity-dependent adenosine release conforms to the current orthodoxy of prior synaptic ATP release and conversion to adenosine. However, under the same conditions where adenosine release is observed, no corresponding

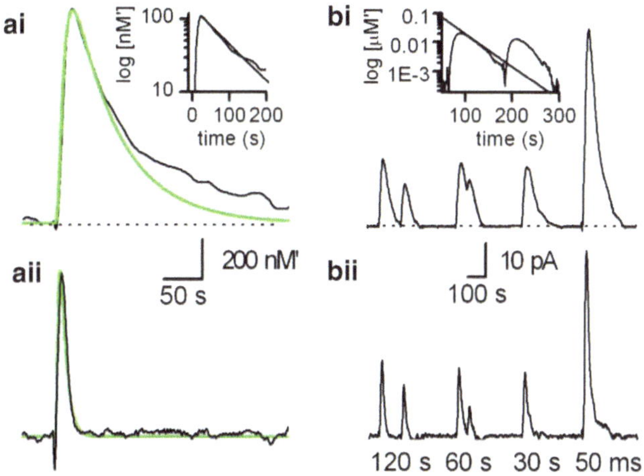

Fig. 4 *Spike evoked adenosine release displays short-term plasticity.* (**a**) Adenosine release evoked by a single action potential in the parallel fibers of a cerebellar slice. (**a1**) raw trace, *inset* shows the decay of the signal can be fitted by a single exponential process; *green line* is a minimal theoretical model of the data (rapid initial rise of adenosine followed by exponential decay, details in (43)); and (**a2**) deconvolution of the signal to remove the slowest component of the decay due to diffusion; *green line* is deconvolution of the minimal theoretical model. (**b**) Demonstration of short-term plasticity of adenosine release during paired stimuli at different interpulse intervals. (**b1**) raw records at intervals indicated below, *inset* showing that the decay can again be fitted by a single exponential process. (**b2**) The deconvolved records. At intervals of 30 s or longer, adenosine release undergoes paired pulse depression. At very short intervals (50 ms) adenosine release exhibits paired pulse facilitation. Reproduced with permission from (43)

"upstream" ATP release has been resolved with ATP biosensors, even when the adenosine release has been greatly enhanced by the use of 4AP to broaden the presynaptic spikes (41, 107). The data at present would suggest that if this does indeed occur, the conversion must be localized extremely close to the sites of adenosine release and must be very efficient. The current lack of potent and selective inhibitors of the enzymes that break ATP down has not allowed rigorous testing of the two competing mechanisms to date. Nevertheless use of biosensor recordings has revealed a richness to activity-dependent adenosine release, irrespective of underlying mechanism, that was unknown before the application of these methods.

3.3 Glucosensing in the Hypothalamus

The brain controls body weight, both through the conscious desire for food and attendant pleasures and at a subconscious or autonomic level. The first evidence for this autonomic control of appetite and body was the demonstration that lesions of the ventromedial hypothalamus of rats led to over-eating and obesity (108), whereas more lateral hypothalamic lesions led to a dramatic drop in food intake (109). Neuronal nuclei within the hypothalamus are in effect multivalent chemosensors and integrators of information that control: appetite and food intake; the burning of energy and the deposition of fat (three factors that collectively determine energy balance). Physiological control of body weight requires the monitoring and integration of many signals that circulate in the body and provide information from the periphery (stomach, pancreas, fat stores, etc.) to the brain (110). These key signals include not only circulating metabolites such as glucose, free fatty acids and amino acids but also specifically secreted hormones such as leptin, grehlin and insulin.

While hypothalamic neurons are clearly part of these control mechanisms, a group of enigmatic glia-like cells called tanycytes exist at the border of the third ventricle (Fig. 5a). These cells have a soma that contacts the ventricular fluid, and a single process that penetrates into the brain parenchyma. They are found from the very ventral part of the ventricle wall at the median eminence (so-called β tanycytes) and progress dorsally to about midway up the ventricle wall (α tanycytes). The functions of the more dorsal tanycytes are intriguing as their processes project into the arcuate nucleus and ventromedial hypothalamic nucleus, two key areas associated with energy balance.

Several authors have hypothesized a role for tanycytes as glucosensors, as they posses the glucose transporter involved in this role in pancreatic β cells, along with K-ATP channels (KIR 6.1) and associated SUR subunit as well as glucokinase. This has led to the idea that tanycytes sense glucose via the same mechanism as that of pancreatic β cells. However, the mere presence of these components does not mean that tanycytes are indeed glucosensors. A demonstration that tanycytes respond to variations in glucose concentration over a physiological range would greatly strengthen this hypothesis. There is some indirect correlative evidence that tanycytes may have a physiological role in responses to glucoprivation. Injection of alloxan (an inhibitor of glucokinase) into the third ventricle kills a large number of GFAP-positive cells, some of which are tanycytes. These cells eventually regenerate. The glucoprivic response disappears and recovers with a time course similar to that of the disappearance and reappearance of the GFAP positive cells including the tanycytes (111).

We came to this field with the hypothesis that tanycytes may sense glucose via an ATP receptor-mediated mechanism (94, 112). We used a combination of Ca^{2+} imaging and ATP biosensing to tackle the problem. When we selectively stimulated the tanycyte cell bodies with glucose by focal applications (an approach that mimics their selective sensing of the cerebrospinal fluid), we were then able to see rapid Ca^{2+} responses to alterations of bathing glucose over physiological concentration ranges (3–8 mM) (94). Interestingly, even non-metabolizable analogues of glucose were able to evoke these responses. ATP biosensors revealed localized glucose-dependent ATP release from the tanycyte layer (Fig. 5b),

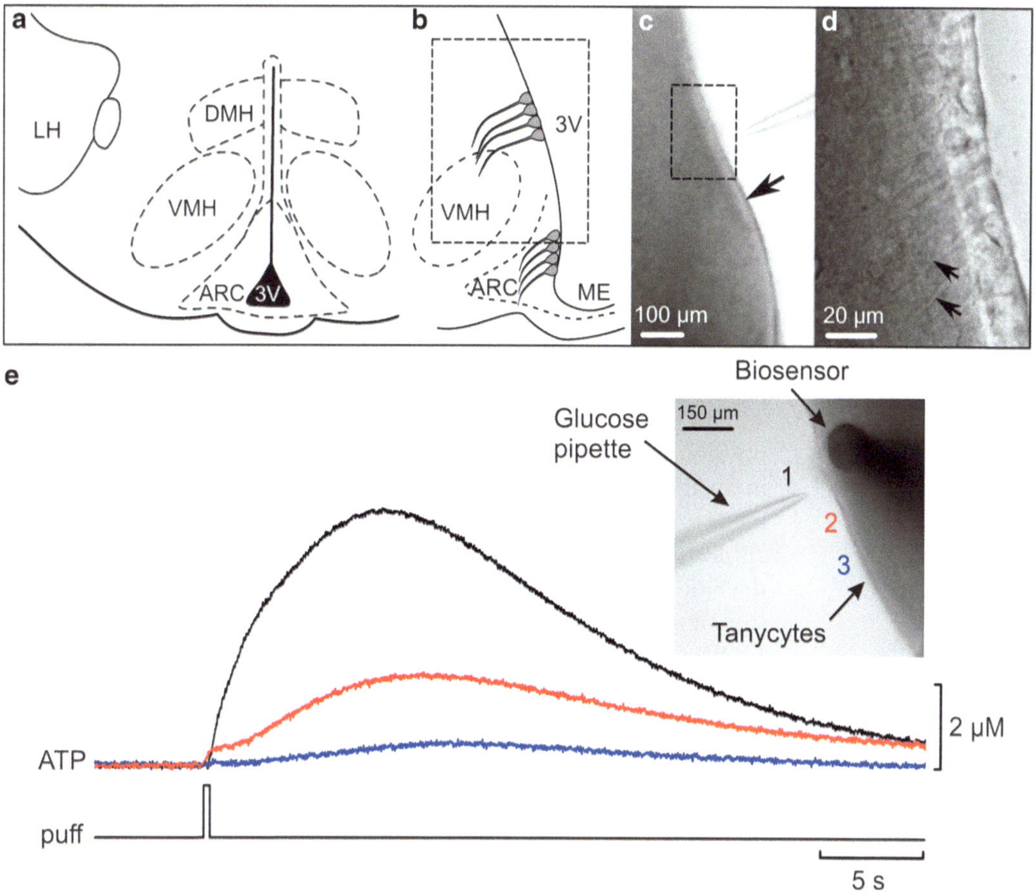

Fig. 5 *Glucose-evoked ATP release from hypothamalic tanycytes.* (**a**) Diagram showing the major features of the hypothalamus in a coronal section. *VMH* ventromedial hypothalamic nucleus, *DMH* dorsomedial hypothalamic nucleus, *ARC* arcuate nucleus, *LH* lateral hypothalamus, *3V* third ventricle. (**b**) More detailed diagram of the third ventricle and adjacent areas showing the location of the tanycytes and their processes projecting into the VMH and ARC. *ME* median eminence. (**c**) Photomicrograph of an acute hypothalamic slice, the *arrow points* to the tanycyte layer at the border of the slice. (**d**) Higher magnification picture of the area indicated by the *dashed rectangle* in (**c**), *arrows* indicate the processes of tanycytes. (**e**) Glucose evoked ATP release from tanycytes. The *inset* shows the recording arrangement of the biosensor placed over the tanycyte layer and the glucose puffing pipette. A brief puff ejected sufficient glucose from the pipette to activate the tanycytes. More ATP release was seen when glucose was puffed nearer to the biosensor, indicating that a localized signal is generated by release of ATP from tanycytes. Reproduced with permission from (94)

and the glucose-evoked Ca^{2+}-signaling in tanycytes depended on the activation of P2Y1 receptors (94).

Our data are important as they overturn a common assumption in the field: that tanycytes sense glucose by the same mechanism as pancreatic β cells. This latter mechanism requires the phosphorylation of glucose to glucose-6-phosphate. This then enters the Krebs cycle thereby generating ATP and hence altering the ATP:ADP ratio ultimately resulting in closure of the K-ATP channel and consequent depolarization. Since tanycytes respond to non-metabolizable analogues of glucose, our data suggest that this model does not apply to tanycytes. The ATP-receptor-dependent detection of glucose that we have described is unique, and the precise transduction mechanism of glucosensing remains to be established (94, 112).

4 Concluding Remarks

Microelectrode biosensors have been applied to the study of purinergic signaling in a number of different contexts, both in vitro and in vivo for nearly 10 years. Their use and integration into physiological experiments has contributed to substantial advances in a variety of different fields where signaling by ATP and adenosine is important. I hope this brief review serves to stimulate further application of these methods, and also helps to inspire further developments, such as new biosensors for related purinergic signaling agents such as ADP, the diadenosine polyphosphates and UDP-glucose.

Acknowledgements

I thank the Wellcome Trust and MRC for providing funding to make possible some of the work described in this review, and the many collaborators who have helped to apply biosensors for the purines to a range of physiological questions.

References

1. Drury AN, Szent-Gyorgyi A (1929) The physiological activity of adenine compounds with especial reference to their action upon the mammalian heart. J Physiol 68:213–237
2. Su C, Bevan JA, Burnstock G (1971) (3H) adenosine triphosphate: release during stimulation of enteric nerves. Science 173:336–338
3. Burnstock G (1972) Purinergic nerves. Phar macol Rev 24:509–581
4. Edwards FA, Gibb AJ, Colquhoun D (1992) ATP receptor-mediated synaptic currents in the central nervous system. Nature 359:144–147
5. Burnstock G (1997) The past, present and future of purine nucleotides as signalling molecules. Neuropharmacology 36:1127–1139
6. Burnstock G (2006) Historical review: ATP as a neurotransmitter. Trends Pharmacol Sci 27:166–176
7. Burnstock G, Fredholm BB, Verkhratsky A (2011) Adenosine and ATP receptors in the brain. Curr Top Med Chem 11:973–1011
8. Khakh BS (2001) Molecular physiology of P2X receptors and ATP signalling at synapses. Nat Rev Neurosci 2:165–174
9. Abbracchio MP, Burnstock G, Boeynaems JM, Barnard EA, Boyer JL, Kennedy C, Knight GE, Fumagalli M, Gachet C, Jacobson KA, Weisman GA (2006) International Union of Pharmacology LVIII: update on the P2Y G protein-coupled nucleotide receptors: from molecular mechanisms and pathophysiology to therapy. Pharmacol Rev 58:281–341
10. Kawate T, Michel JC, Birdsong WT, Gouaux E (2009) Crystal structure of the ATP-gated P2X(4) ion channel in the closed state. Nature 460:592–598

11. Surprenant A, Rassendren F, Kawashima E, North RA, Buell G (1996) The cytolytic P2Z receptor for extracellular ATP identified as a P2X receptor (P2X7). Science 272:735–738
12. Khakh BS, Bao XR, Labarca C, Lester HA (1999) Neuronal P2X transmitter-gated cation channels change their ion selectivity in seconds. Nat Neurosci 2:322–330
13. Chaumont S, Khakh BS (2008) Patch-clamp coordinated spectroscopy shows P2X2 receptor permeability dynamics require cytosolic domain rearrangements but not Panx-1 channels. Proc Natl Acad Sci USA 105:12063–12068
14. Pelegrin P, Surprenant A (2006) Pannexin-1 mediates large pore formation and interleukin-1beta release by the ATP-gated P2X7 receptor. EMBO J 25:5071–5082
15. Locovei S, Wang J, Dahl G (2006) Activation of pannexin 1 channels by ATP through P2Y receptors and by cytoplasmic calcium. FEBS Lett 580:239–244
16. Locovei S, Scemes E, Qiu F, Spray DC, Dahl G (2007) Pannexin1 is part of the pore forming unit of the P2X(7) receptor death complex. FEBS Lett 581:483–488
17. Iglesias R, Locovei S, Roque AP, Alberto AP, Dahl G, Spray DC, Scemes E (2008) P2X7 receptor-Pannexin1 complex: pharmacology and signaling. Am J Physiol Cell Physiol 295: C752–C760
18. Yan Z, Li S, Liang Z, Tomic M, Stojilkovic SS (2008) The P2X7 receptor channel pore dilates under physiological ion conditions. J Gen Physiol 132:563–573
19. Yan Z, Khadra A, Li S, Tomic M, Sherman A, Stojilkovic SS (2010) Experimental characterization and mathematical modeling of P2X7 receptor channel gating. J Neurosci 30: 14213–14224
20. Pankratov Y, Lalo U, Verkhratsky A, North RA (2006) Vesicular release of ATP at central synapses. Pflugers Archiv 452:589–597
21. Pankratov Y, Castro E, Miras-Portugal MT, Krishtal O (1998) A purinergic component of the excitatory postsynaptic current mediated by P2X receptors in the CA1 neurons of the rat hippocampus. Eur J Neurosci 10: 3898–3902
22. Jo YH, Schlichter R (1999) Synaptic corelease of ATP and GABA in cultured spinal neurons. Nat Neurosci 2:241–245
23. Wang Z, Haydon PG, Yeung ES (2000) Direct observation of calcium-independent intercellular ATP signaling in astrocytes. Anal Chem 72:2001–2007
24. Fields RD, Stevens B (2000) ATP: an extracellular signaling molecule between neurons and glia. Trends Neurosci 23:625–633
25. Pascual O, Casper KB, Kubera C, Zhang J, Revilla-Sanchez R, Sul JY, Takano H, Moss SJ, McCarthy K, Haydon PG (2005) Astrocytic purinergic signaling coordinates synaptic networks. Science 310:113–116
26. Bowser DN, Khakh BS (2007) Vesicular ATP is the predominant cause of intercellular calcium waves in astrocytes. J Gen Physiol 129: 485–491
27. Hamilton NB, Attwell D (2010) Do astrocytes really exocytose neurotransmitters? Nat Rev Neurosci 11:227–238
28. Bao L, Locovei S, Dahl G (2004) Pannexin membrane channels are mechanosensitive conduits for ATP. FEBS Lett 572:65–68
29. Kang J, Kang N, Lovatt D, Torres A, Zhao Z, Lin J, Nedergaard M (2008) Connexin 43 hemichannels are permeable to ATP. J Neurosci 28:4702–4711
30. Huckstepp RT, Eason R, Sachdev A, Dale N (2010) CO2-dependent opening of connexin 26 and related beta connexins. J Physiol 588: 3921–3931
31. Huckstepp RT, id Bihi R, Eason R, Spyer KM, Dicke N, Willecke K, Marina N, Gourine AV, Dale N (2010) Connexin hemichannel-mediated CO2-dependent release of ATP in the medulla oblongata contributes to central respiratory chemosensitivity. J Physiol 588: 3901–3920
32. Fields RD, Ni Y (2010) Nonsynaptic communication through ATP release from volume-activated anion channels in axons. Sci Signal 3:ra73
33. Sabirov RZ, Okada Y (2005) ATP release via anion channels. Purinergic signal 1: 311–328
34. Pellegatti P, Falzoni S, Pinton P, Rizzuto R, Di Virgilio F (2005) A novel recombinant plasma membrane-targeted luciferase reveals a new pathway for ATP secretion. Mol Biol Cell 16:3659–3665
35. Radulovacki M, Virus RM, Djuricic-Nedelson M, Green RD (1984) Adenosine analogs and sleep in rats. J Pharmacol Exp Ther 228: 268–274
36. Porkka-Heiskanen T, Strecker RE, Thakkar M, Bjorkum AA, Greene RW, McCarley RW (1997) Adenosine: a mediator of the sleep-inducing effects of prolonged wakefulness. Science 276:1265–1268
37. Rudolphi KA, Schubert P, Parkinson FE, Fredholm BB (1992) Neuroprotective role of

adenosine in cerebral-ischemia. Trends Pharmacol Sci 13:439–445

38. Fredholm BB, IJzerman AP, Jacobson KA, Klotz KN, Linden J (2001) International Union of Pharmacology. XXV. Nomenclature and classification of adenosine receptors. Pharmacol Rev 53:527–552
39. Zimmermann H (2000) Extracellular metabolism of ATP and other nucleotides. Naunyn Schmiedebergs Arch Pharmacol 362: 299–309
40. Halassa MM, Florian C, Fellin T, Munoz JR, Lee SY, Abel T, Haydon PG, Frank MG (2009) Astrocytic modulation of sleep homeostasis and cognitive consequences of sleep loss. Neuron 61:213–219
41. Wall MJ, Dale N (2007) Auto-inhibition of rat parallel fibre-Purkinje cell synapses by activity-dependent adenosine release. J Physiol 581:553–565
42. Wall M, Dale N (2008) Activity-dependent release of adenosine: a critical re-evaluation of mechanism. Curr Neuropharmacol 6: 329–337
43. Klyuch BP, Richardson MJ, Dale N, Wall MJ (2011) The dynamics of single spike-evoked adenosine release in the cerebellum. J Physiol 589:283–295
44. Baldwin SA, Beal PR, Yao SY, King AE, Cass CE, Young JD (2004) The equilibrative nucleoside transporter family, SLC29. Pflugers Arch 447:735–743
45. Boison D (2006) Adenosine kinase, epilepsy and stroke: mechanisms and therapies. Trends Pharmacol Sci 27:652–658
46. Dale N, Pearson T, Frenguelli BG (2000) Direct measurement of adenosine release during hypoxia in the CA1 region of the rat hippocampal slice. J Physiol 526:143–155
47. Pearson T, Nuritova F, Caldwell D, Dale N, Frenguelli BG (2001) A depletable pool of adenosine in area CA1 of the rat hippocampus. J Neurosci 21:2298–2307
48. Frenguelli BG, Llaudet E, Dale N (2003) High-resolution real-time recording with microelectrode biosensors reveals novel aspects of adenosine release during hypoxia in rat hippocampal slices. J Neurochem 86: 1506–1515
49. Frenguelli BG, Wigmore G, Llaudet E, Dale N (2007) Temporal and mechanistic dissociation of ATP and adenosine release during ischaemia in the mammalian hippocampus. J Neurochem 101:1400–1413
50. Martin ED, Fernandez M, Perea G, Pascual O, Haydon PG, Araque A, Cena V (2007) Adenosine released by astrocytes contributes to hypoxia-induced modulation of synaptic transmission. Glia 55:36–45
51. Pintor J, Puche JA, Gualix J, Hoyle CH, Miras-Portugal MT (1997) Diadenosine polyphosphates evoke Ca^{2+} transients in guinea-pig brain via receptors distinct from those for ATP. J Physiol 504(Pt 2):327–335
52. Klishin A, Lozovaya N, Pintor J, Miras-Portugal MT, Krishtal O (1994) Possible functional role of diadenosine polyphosphates: negative feedback for excitation in hippocampus. Neuroscience 58:235–236
53. Pintor J, Diaz-Rey MA, Torres M, Miras-Portugal MT (1992) Presence of diadenosine polyphosphates–Ap4A and Ap5A–in rat brain synaptic terminals. Ca^{2+} dependent release evoked by 4-aminopyridine and veratridine. Neurosci Lett 136:141–144
54. Gao ZG, Ding Y, Jacobson KA (2010) UDP-glucose acting at P2Y14 receptors is a mediator of mast cell degranulation. Biochem Pharmacol 79:873–879
55. Bassil AK, Bourdu S, Townson KA, Wheeldon A, Jarvie EM, Zebda N, Abuin A, Grau E, Livi GP, Punter L, Latcham J, Grimes AM, Hurp DP, Downham KM, Sanger GJ, Winchester WJ, Morrison AD, Moore GB (2009) UDP-glucose modulates gastric function through P2Y14 receptor-dependent and -independent mechanisms. Am J Physiol Gastrointest Liver Physiol 296:G923–G930
56. Ciana P, Fumagalli M, Trincavelli ML, Verderio C, Rosa P, Lecca D, Ferrario S, Parravicini C, Capra V, Gelosa P, Guerrini U, Belcredito S, Cimino M, Sironi L, Tremoli E, Rovati GE, Martini C, Abbracchio MP (2006) The orphan receptor GPR17 identified as a new dual uracil nucleotides/cysteinyl-leukotrienes receptor. EMBO J 25:4615–4627
57. Langer D, Hammer K, Koszalka P, Schrader J, Robson S, Zimmermann H (2008) Distribution of ectonucleotidases in the rodent brain revisited. Cell Tissue Res 334: 199–217
58. Masse K, Eason R, Bhamra S, Dale N, Jones EA (2006) Comparative genomic and expression analysis of the conserved NTPDase gene family in Xenopus. Genomics 87:366–381
59. Robson SC, Sevigny J, Zimmermann H (2006) The E-NTPDase family of ectonucleotidases: structure function relationships and pathophysiological significance. Purinergic Signal 2:409–430
60. Yegutkin GG (2008) Nucleotide- and nucleoside-converting ectoenzymes: important modulators of purinergic signalling cascade. Biochim Biophys Acta 1783:673–694

61. Zimmermann H (1999) Two novel families of ectonucleotidases: molecular structures, catalytic properties and a search for function. Trends Pharmacol Sci 20:231–236
62. Zimmermann H, Braun N (1999) Ecto-nucleotidases–molecular structures, catalytic properties, and functional roles in the nervous system. Prog Brain Res 120:371–385
63. Masse K, Bhamra S, Eason R, Dale N, Jones EA (2007) Purine-mediated signalling triggers eye development. Nature 449:1058–1062
64. Stefan C, Jansen S, Bollen M (2005) NPP-type ectophosphodiesterases: unity in diversity. Trends Biochem Sci 30:542–550
65. Stefan C, Jansen S, Bollen M (2006) Modulation of purinergic signaling by NPP-type ectophosphodiesterases. Purinergic signal 2:361–370
66. Masse K, Bhamra S, Allsop G, Dale N, Jones EA (2010) Ectophosphodiesterase/nucleotide phosphohydrolase (Enpp) nucleotidases: cloning, conservation and developmental restriction. Int J Dev Biol 54:181–193
67. Diez-Zaera M, Diaz-Hernandez JI, Hernandez-Alvarez E, Zimmermann H, Diaz-Hernandez M, Miras-Portugal MT (2011) Tissue-nonspecific alkaline phosphatase promotes axonal growth of hippocampal neurons. Mol Biol Cell 22:1014–1024
68. Kermer V, Ritter M, Albuquerque B, Leib C, Stanke M, Zimmermann H (2010) Knockdown of tissue nonspecific alkaline phosphatase impairs neural stem cell proliferation and differentiation. Neurosci Lett 485:208–211
69. Langer D, Ikehara Y, Takebayashi H, Hawkes R, Zimmermann H (2007) The ectonucleotidases alkaline phosphatase and nucleoside triphosphate diphosphohydrolase 2 are associated with subsets of progenitor cell populations in the mouse embryonic, postnatal and adult neurogenic zones. Neuroscience 150: 863–879
70. Zimmermann H (1992) 5 -Nucleotidase: molecular structure and functional aspects. Biochem J 285:345–365
71. Gordon EL, Pearson JD, Slakey LL (1986) The hydrolysis of extracellular adenine nucleotides by cultured endothelial cells from pig aorta. Feed-forward inhibition of adenosine production at the cell surface. J Biol Chem 261:15496–15507
72. Slakey LL, Cosimini K, Earls JP, Thomas C, Gordon EL (1986) Simulation of extracellular nucleotide hydrolysis and determination of kinetic constants for the ectonucleotidases. J Biol Chem 261:15505–15507
73. James S, Richardson PJ (1993) Production of adenosine from extracellular ATP at the striatal cholinergic synapse. J Neurochem 60: 219–227
74. Dale N (1998) Delayed production of adenosine underlies temporal modulation of swimming in frog embryo. J Physiol 511:265–272
75. Wall MJ, Usowicz MM (1997) Development of action potential-dependent and independent spontaneous GABAA receptor-mediated currents in granule cells of postnatal rat cerebellum. Eur J Neurosci 9:533–548
76. Kullmann DM, Asztely F (1998) Extrasynaptic glutamate spillover in the hippocampus: evidence and implications. Trends Neurosci 21:8–14
77. Rossi DJ, Oshima T, Attwell D (2000) Glutamate release in severe brain ischaemia is mainly by reversed uptake. Nature 403: 316–321
78. Llaudet E, Botting NP, Crayston JA, Dale N (2003) A three-enzyme microelectrode sensor for detecting purine release from central nervous system. Biosens Bioelectron 18:43–52
79. Llaudet E, Hatz S, Droniou M, Dale N (2005) Microelectrode biosensor for real-time measurement of ATP in biological tissue. Anal Chem 77:3267–3273
80. Kueng A, Kranz C, Mizaikoff B (2004) Amperometric ATP biosensor based on polymer entrapped enzymes. Biosens Bioelectron 19:1301–1307
81. Masson JF, Kranz C, Mizaikoff B, Gauda EB (2008) Amperometric ATP microbiosensors for the analysis of chemosensitivity at rat carotid bodies. Anal Chem 80:3991–3998
82. Patel BA, Rogers M, Wieder T, O'Hare D, Boutelle MG (2011) ATP microelectrode biosensor for stable long-term in vitro monitoring from gastrointestinal tissue. Biosens Bioelectron 26:2890–2896
83. Swamy BE, Venton BJ (2007) Subsecond detection of physiological adenosine concentrations using fast-scan cyclic voltammetry. Anal Chem 79:744–750
84. Cechova S, Venton BJ (2008) Transient adenosine efflux in the rat caudate-putamen. J Neurochem 105:1253–1263
85. Dong H, Wang S, Galligan JJ, Swain GM (2011) Boron-doped diamond nano/microelectrodes for biosensing and in vitro measurements. Front Biosci (Schol Ed) 3:518–540
86. Fortin E, Chane-Tune J, Mailley P, Szunerits S, Marcus B, Petit JP, Mermoux M, Vieil E (2004) Nucleosides and ODN electrochemical

detection onto boron doped diamond electrodes. Bioelectrochemistry 63:303–306

87. Pajski ML, Venton BJ (2010) Adenosine release evoked by short electrical stimulations in striatal brain slices is primarily activity dependent. ACS Chem Neurosci 1:775–787
88. Cechova S, Elsobky AM, Venton BJ (2010) A1 receptors self-regulate adenosine release in the striatum: evidence of autoreceptor characteristics. Neuroscience 171:1006–1015
89. Dale N, Hatz S, Tian F, Llaudet E (2005) Listening to the brain: microelectrode biosensors for neurochemicals. Trends Biotechnol 23:420–428
90. Gourine AV, Llaudet E, Dale N, Spyer KM (2005) ATP is a mediator of chemosensory transduction in the central nervous system. Nature 436:108–111
91. Gourine AV, Dale N, Llaudet E, Poputnikov DM, Spyer KM, Gourine VN (2007) Release of ATP in the central nervous system during systemic inflammation: real-time measurement in the hypothalamus of conscious rabbits. J Physiol 585:305–316
92. Gourine AV, Dale N, Korsak A, Llaudet E, Tian F, Huckstepp R, Spyer KM (2008) Release of ATP and glutamate in the nucleus tractus solitarii mediate pulmonary stretch receptor (Breuer-Hering) reflex pathway. J Physiol 586:3963–3978
93. Tian F, Gourine AV, Huckstepp RT, Dale N (2009) A microelectrode biosensor for real time monitoring of L-glutamate release. Anal Chim Acta 645:86–91
94. Frayling C, Britton R, Dale N (2011) ATP-mediated glucosensing by hypothalamic tanycytes. J Physiol 589:2275–2286
95. Dale N, Frenguelli BG (2012) Measurement of purine release with microelectrode biosensors. Purinergic Signal 8:27–40
96. Head H (1889) On the reglation of respiration: Part I. Experimental. J Physiol 10 (1–152):153
97. Widdicombe J (2001) Airway receptors. Respir Physiol 125:3–15
98. Miyazaki M, Tanaka I, Ezure K (1999) Excitatory and inhibitory synaptic inputs shape the discharge pattern of pump neurons of the nucleus tractus solitarii in the rat. Exp Brain Res 129:191–200
99. Kubin L, Alheid GF, Zuperku EJ, McCrimmon DR (2006) Central pathways of pulmonary and lower airway vagal afferents. J Appl Physiol 101:618–627
100. Ezure K, Tanaka I, Saito Y, Otake K (2002) Axonal projections of pulmonary slowly adapting receptor relay neurons in the rat. J Comp Neurol 446:81–94
101. Bonham AC, McCrimmon DR (1990) Neurones in a discrete region of the nucleus tractus solitarius are required for the Breuer-Hering reflex in rat. J Physiol 427:261–280
102. Zhang W, Mifflin SW (1995) Excitatory amino-acid receptors contribute to carotid sinus and vagus nerve evoked excitation of neurons in the nucleus of the tractus solitarius. J Auton Nerv Syst 55:50–56
103. Kanjhan R, Housley GD, Burton LD, Christie DL, Kippenberger A, Thorne PR, Luo L, Ryan AF (1999) Distribution of the P2X2 receptor subunit of the ATP-gated ion channels in the rat central nervous system. J Comp Neurol 407:11–32
104. Yao ST, Barden JA, Finkelstein DI, Bennett MR, Lawrence AJ (2000) Comparative study on the distribution patterns of P2X(1)-P2X(6) receptor immunoreactivity in the brainstem of the rat and the common marmoset (Callithrix jacchus): association with catecholamine cell groups. J Comp Neurol 427:485–507
105. Yao ST, Barden JA, Lawrence AJ (2001) On the immunohistochemical distribution of ionotropic P2X receptors in the nucleus tractus solitarius of the rat. Neuroscience 108:673–685
106. Yao ST, Gourine AV, Spyer KM, Barden JA, Lawrence AJ (2003) Localisation of P2X2 receptor subunit immunoreactivity on nitric oxide synthase expressing neurones in the brain stem and hypothalamus of the rat: a fluorescence immunohistochemical study. Neuroscience 121:411–419
107. Klyuch BP, Dale N, Wall MJ (2012) Receptor-mediated modulation of activity-dependent adenosine release in rat cerebellum. Neuropharmacology 62:815–824
108. Hetherington AW, Ranson SW (1940) Hypothalamic lesions and adiposity in the rat. Anat Rec 78:149–172
109. Anand BK, Brobeck JR (1951) Hypothalamic control of food intake in rats and cats. Yale J Biol Med 24:123–140
110. Morton GJ, Cummings DE, Baskin DG, Barsh GS, Schwartz MW (2006) Central nervous system control of food intake and body weight. Nature 443:289–295
111. Sanders NM, Dunn-Meynell AA, Levin BE (2004) Third ventricular alloxan reversibly impairs glucose counterregulatory responses. Diabetes 53:1230–1236
112. Dale N (2011) Purinergic signaling in hypothalamic tanycytes: potential roles in chemosensing. Semin Cell Dev Biol 22: 237–244

Chapter 11

Glucose and Lactate Monitoring Across the Rat Sleep–Wake Cycle

Raymond Cespuglio, Larissa Netchiporouk, and Nataliya Shram

Abstract

Glucose biosensors were prepared by immobilizing glucose oxidase on carbon fiber microelectrodes (CFMEs) either by cross-linking glutaraldehyde vapors or via enzyme entrapment in films of *m*-phenylene-diamine or resorcinol. The enzymatic layer was then covered with a membrane made of Nafion or cellulose acetate. The biosensors were tested using differential normal pulse voltammetry (DNPV) to detect signals. The calibration curves for glucose were linear from 0.3 to 6.5 mM. The DNPV response was essentially insensitive to potentially interfering molecules. Glucose concentrations in plasma and cerebrospinal fluid (CSF) corresponded with those measured by standard procedures. In freely moving rats, cortical extracellular glucose concentration averaged 0.59 ± 0.3 mM. Using combined polysomnographic and DNPV assessment, glucose levels in the cortex were determined during various phases of the sleep–wake cycle. Compared to the waking state (W, 100 %), spontaneous variations were observed during active waking triggered by a water puff (AW, −32 %), slow-wave sleep (SWS, +13 %), and paradoxical sleep (PS, −11 %). To prepare lactate biosensors, CFMEs were coated with lactate oxidase. These sensors, combined with DNPV measurements, showed a linear response to lactate at concentrations ranging from 0.1 to 2.0 mM. Lactate measurement was insensitive to common potentially interfering molecules. In anesthetized rats, cortical lactate concentration averaged 0.41 ± 0.02 mM. Lactate levels in samples of brain tissue, plasma, and CSF corresponded to those measured by standard procedures. In freely moving rats, lactate changes were also related to sleep–wake state. Compared to W (100 %), lactate level increased during AW (+53 %) and decreased during SWS (−16.2 %). However, during PS, lactate level increased (+8.5 %) compared to that noted during SWS. Finally, long-term lactate monitoring revealed the existence of a circadian influence on lactate production rate.

Key words Biosensor, Glucose, Lactate, Sleep, Wake, Rat

1 Background and Historical Overview of Energy Metabolism in Mammals

Interest in energy metabolism of the brain has increased noticeably over the last two decades. Although the brain represents only 2 % of total body weight, it is responsible for about 20 % of total oxygen consumption and 25 % of glucose utilization. Glucose metabolism contributes, via production of intermediate metabolites such as lactate and pyruvate, to the production of ATP (1–3), the major source of energy in the mammalian brain. Anaerobic metabolism,

Stéphane Marinesco and Nicholas Dale (eds.), *Microelectrode Biosensors*, Neuromethods, vol. 80,
DOI 10.1007/978-1-62703-370-1_11, © Springer Science+Business Media, LLC 2013

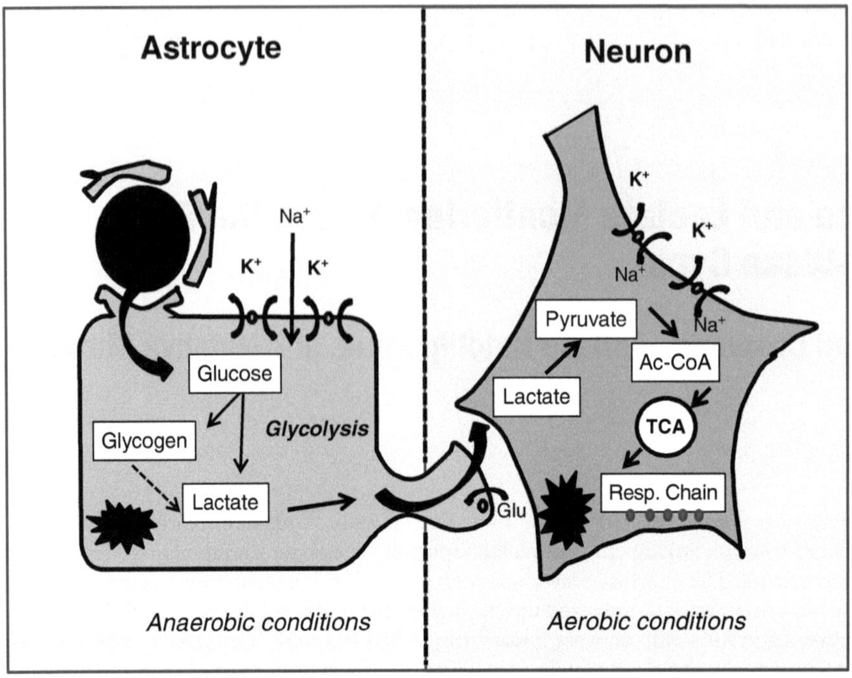

Fig. 1 Energy metabolic pathways in homeotherms. Glucose biosynthesized from simpler precursors or derived from breakdown of a polysaccharide (glycogen) is the main energy source for synthesis of adenosine triphosphate (ATP) in endotherms. Glucose is metabolized to pyruvate under anaerobic conditions via glycolysis. Two moles of ATP are formed per mole of glucose. Under anaerobic conditions, pyruvate can be subjected to lactic acid fermentation and transformed into lactate. Under aerobic conditions (when the Krebs cycle and the respiratory chain are active), pyruvate oxidized to CO_2 and H_2O contributes to the production of 34 additional moles of ATP. Total degradation of 1 mole of glucose therefore produces 36 moles of ATP. Lipids, proteins, and amino acids may also be involved in ATP production, entering the phosphorylative oxidation pathway in the form of acetyl-coenzyme-A (acetyl CoA). Glutamate (Glu), binding principally to NMDA receptors (intrinsic to astrocytes), may stimulate glycolysis to enhance astrocytic production of the basic neuronal fuel, lactate. This view is supported by our results (Fig. 4d). *Abbreviations. Resp. Chain* respiratory chain, *TCA* tricarboxylic acid cycle

via glycolysis, produces two moles of ATP per mole of glucose, whereas, under aerobic conditions (when oxidative phosphorylation is active), 34 additional moles of ATP are produced (3–5) (Fig. 1). In terms of processes whereby glucose is metabolized, metabolites such as lactate and pyruvate do not only enter the tricarboxylic acid cycle, but can rather be released from tissue and removed via the blood. Glucose can also be indirectly incorporated into lipids, proteins, and glycogen, or may serve as a precursor of certain neurotransmitters such as γ-aminobutyric acid (GABA), glutamate, and acetylcholine (3). The processes active throughout the sleep–wake cycle of mammals include energy production, storage, and consumption (5, 6). Advances in the field of sleep-energy interdependence have been constantly associated with development of new technical procedures or tools facilitating the monitoring of key parameters of energy metabolism, either after death or directly in vivo. Some valuable technical procedures include deoxyglucose autoradiography (7, 8), positron emission tomography (PET) (9), various spectroscopic methods (10), ATP determination using nuclear magnetic resonance (NMR) featuring [^{31}P] spectroscopy (11), ATP determination using a luciferin–luciferase-based assay (12), and brain glycogen assay (13, 14). We have made a concerted effort to develop biosensors capable of detecting glucose directly in vivo, in non-anesthetized animals (glucose is the key substrate in energy metabolism) (15). Also, we have developed sensors that measure lactate (16, 17) and ATP (18) levels. These sensors have been employed in combination with polysomnography to precisely define the relationship between sleep–wake state and energy metabolism. In chronically implanted freely moving animals, we can now combine the use of glucose and lactate biosensors with polysomnography (19, 20). To date, our approach remains entirely original, with no similar procedure reported.

In this chapter we describe our glucose and lactate biosensors, the main steps in biosensor production, the tests employed to verify specificity, the use of biosensors in combination with polysomnography, and some of our most relevant results.

2 Glucose and Lactate Biosensors

2.1 Principle of Detection

Our assay employs a widely used enzymatic method. Glucose undergoes enzyme-catalyzed oxidation as follows:

2.1.1 Glucose Biosensor

$$\beta\text{-D-glucose} + \text{GOx} / \text{FAD f} \rightarrow$$
$$(\text{a})\,\text{glucone-}\beta\text{-lactone} + \text{GOx} / \text{FADH}_2$$
$$(\text{b})\,\text{GOx} / \text{FADH}_2 + \text{O}_2 \rightarrow \text{GOx} / \text{FAD} + \text{H}_2\text{O}_2$$

where GOx (glucose oxidase) is an oxidoreductase and FAD is the oxidized form of the prosthetic flavin adenine dinucleotide group. In common with other approaches, amperometric oxidation of the hydrogen peroxide (H_2O_2) produced is simultaneously performed; GOx is immobilized at or near the electrode surface.

2.1.2 Lactate biosensor

The enzymatic principle described above was also used in construction of lactate biosensor. Lactate undergoes enzyme-catalyzed oxidation as follows:

$$\text{lactate} + \text{O}_2 + \text{LOx} \rightarrow (\text{a})\,\text{pyruvate} + \text{H}_2\text{O}_2$$
$$(\text{b})\,\text{H}_2\text{O}_2 \rightarrow \text{O}_2 + 2\text{H} + 2\text{e}^-$$

where LOx (lactate oxidase) is an oxidoreductase. The measurement of the oxidation current ($2e^-$) produced in reaction (b) allows determination of lactate concentration in the vicinity of the electrode surface.

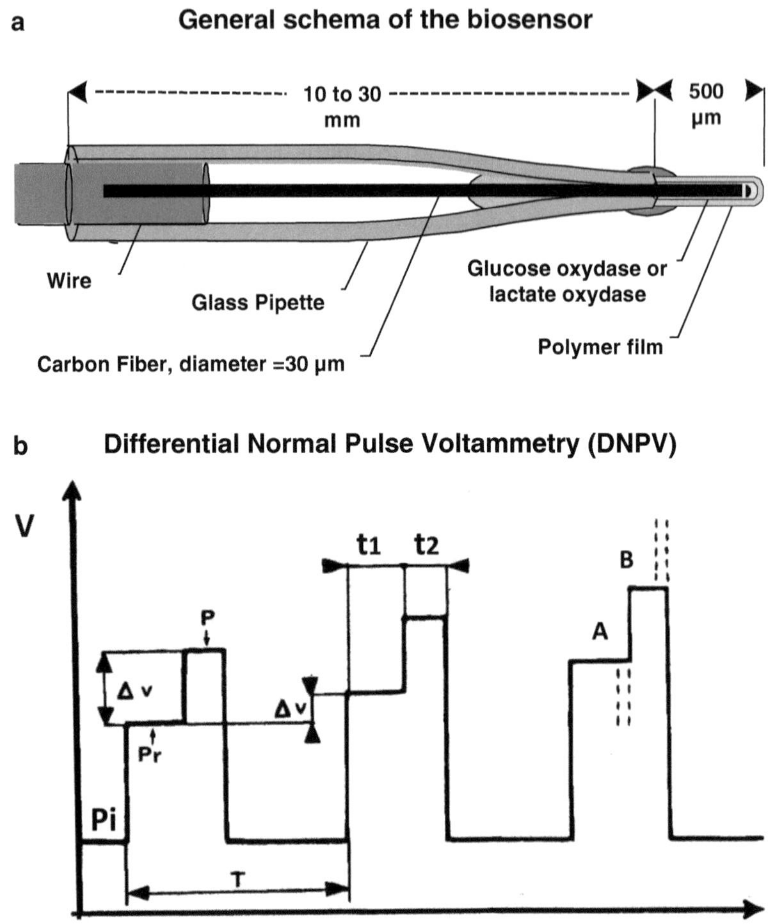

Fig. 2 (**a**) General overview of the carbon fiber-based biosensor. This biosensor is employed to monitor glucose and lactate levels. The active component is the carbon fiber (500 μm) emerging from the pipette. This part is layered with specific enzymes (glucose oxidase or lactate oxidase) and additional polymer films. *Abbreviations*: *Wire* copper wire, *μm* micrometer, *mm* millimeter. (**b**) Differential normal pulse voltammetry (DNPV). This method employs successive pre-pulses (increasing in amplitude) and pulses (constant in amplitude). The current is sampled just at the end of the pre-pulse (Pr in A) and at the pulse (P in B). The difference (differential) between these two measurements is recorded. The advantage of DNPV compared to other methods is that, between each double pulse (Pr + P), the electrode is held at the initial potential (Pi). Thus, background noise is decreased. *Abbreviations*: *T* period of each double pulse (about 400 ms); *t1*, *t2* pre-pulse and pulse durations; Δv pulse amplitude and pre-pulse increment

2.2 Preparation

2.2.1 Glucose Biosensor

This biosensor, previously described (15) (Fig. 2a), employs a single carbon fiber (35 μm in diameter; type AVCO; Lowell, MA) that is inserted into a glass micropipette and pushed until the fiber emerges from the drawn tip. At the emergence point, the fiber is sealed with epoxy resin and cut to a length of 500 μm. At the back of the biosensor, a contact is made using silver wire. This basic structure, used also in the construction of lactate and other biosensors, is electrochemically pretreated in PBS via successive cycling (+2.4 V for 10 s; +1.3 V for 3 s) using triangular pulses delivered at 70 Hz, prior to deposition of an enzyme-containing film. Such pretreatment is performed using a PRT 20-2X potentiostat (Tacussel, Villeurbanne, France) together with a pilot voltage generator (type GSTP, Tacussel). Such treatment enhances detection sensitivity and allows a well-defined half-wave potential for hydrogen peroxide oxidation to be obtained. Next, the sensor is functionalized by deposition of glucose oxidase (GOx) (Fig. 2a) on the surface of the carbon fiber. The experimental procedure involves formation of a thick cross-linked enzyme-containing membrane. The microelectrodes are dipped into a solution of GOx (7.5×10^6 UA/ml) in PBS with 10 % (v/v) glycerol. They are dried for 10 min, exposed for 90 min to saturated glutaraldehyde vapor, and dried again for 15 min. An additional membrane is finally deposited by dip-coating with cellulose acetate (a 10 % [w/v] solution in a 1:1 [v/v] mixture of acetone and cyclohexanone) and the sensors are next dried for at least 10 min. The response to glucose is linear in the range 0.3–6.5 mM (Fig. 3a,b).

2.2.2 Lactate Biosensor

This biosensor, prepared as described previously (17), employs the basic structure developed for glucose biosensor construction (above) (Fig. 2a). This basic sensor is functionalized via lactate oxidase (LOx) deposition onto the surface of the carbon fiber. The enzyme immobilization solution is prepared by dissolving 10 mg of LOx, 5 mg of BSA, and 5 μl of glycerol in 40 μl of 3 % (w/v) hydroxyethylcellulose (HEC) in 0.1 M phosphate buffer, pH 7.0. The active part of the electrodes is dipped into 6 μl of this mixture, to which 1.2 μl of 5 % (w/v) glutaraldehyde is added. Next, electrodes are dried for 1 h and dip-coated to form a cellulose acetate (CA) membrane (2 % [w/w] CA in a 1:1 [v/v] mixture of acetone and cyclohexanone) and allowed to dry for 2 h.

2.3 Measurements

2.3.1 Glucose Biosensor

Measurement employs the DNPV method which is now well known and widely used. This method involves application of successive double pulses (a pre-pulse and pulse). The pre-pulse amplitude increases in successive steps. The electrical current measured is the current difference between the end of the pulse and that of the pre-pulse. The signal obtained is defined as an oxidation peak (Fig. 2b). This method, unlike potentiostatic procedures, permits differentiation of oxidation currents yielded by various electroactive species. However, compared with potentiostatic methods, the measurement time is longer. In both in vitro and in vivo experiments (15), a silver chlorinated wire (Ag/AgCl) is used as a reference electrode. The Ag/AgCl wire is introduced into a micropipette filled with NaCl (3 M) and the tip of the micropipette (containing a bundle of glass wool) placed in touch with the dura of the brain.

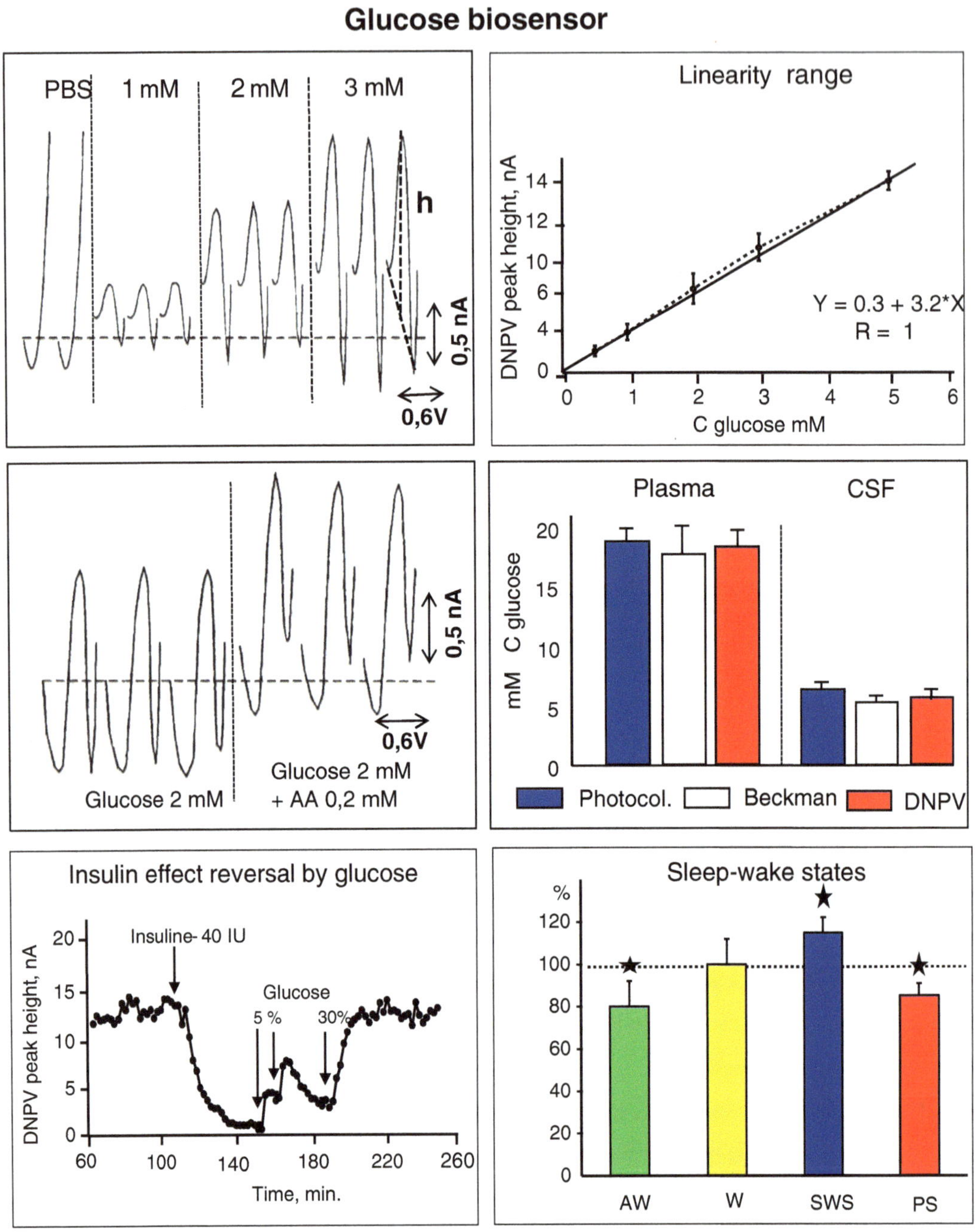

Fig. 3 Glucose biosensor. (**a**) Differential normal pulse voltammograms recorded using solutions of different glucose concentrations (0–3 mM). The signals yielded by H_2O_2 oxidation (0.8 V) are shown. (**b**) Calibration curve showing that the response is linear from 0.3 to 6.5 mM glucose. Data are means ± SEMs, $n = 5$. (**c**) Appearance of the DNPV peaks recorded using a glucose solution compared with signals recorded in the presence of ascorbic acid (AA). The latter compound influences the form of the oxidation signal and the level of background noise. However, signal height is not affected because AA is oxidized at a lower potential (0.1 V) than is H_2O_2 (0.8 V). (**d**) Bar graphs showing glucose concentrations (±SEMs) in rat plasma and CSF (cerebrospinal fluid, drawn from the *cisterna magna*). Values were determined using a photocolorimetric method

The electrode system is successively immersed, using an electrode holder, in plastic cells containing glucose solutions of various concentrations. Potentially interfering compounds may be added, depending on the experimental protocol. Glucose concentrations can be also determined in plasma and CSF; 150 μl samples are conveniently collected for this purpose. The height of the DNPV peak is measured from the top to the baseline, identified by drawing a line connecting the two minima of the voltammetric signal (Fig. 3a). Electrochemical pretreatment of microelectrodes and differential normal pulse voltammetry (DNPV) measurements are achieved using a pulsed voltammetric unit (Biopulse; Solea, Villeurbanne, France) and a chart recorder. Experimentally, the following DNPV parameters have been established: 10–30 mV for pulse magnitude and 60–100 ms for pulse and pre-pulse duration. To prevent depletion of hydrogen peroxide in the area around the working electrode, the interval between successive pulses should be no less than 0.3 s. After establishing DNPV parameters, measurements are performed every 2 min at a scan rate of 10 V/s (a step of 4 mV per interval of 0.4 s). The effect of oxygen on the response of electrodes covered with an additional CA membrane was studied by O_2 bubbling of calibration glucose solutions. An extension of the initial dynamic range of sensor response was obtained, up to glucose concentrations of 15–20 mM, although the initial slope of the calibration curve remains unchanged. To further improve the performance of our biosensor, GOx was trapped in *m*-phenylenediamine (PPD) or resorcinol films. A disadvantage of such biosensors is that the DNPV peak for glucose does not appear immediately after film preparation. It becomes evident only after 2 days of storage in PBS or after DNPV potential

Fig. 3 (continued) (Photocol.), a standard Beckman glucose analyzer (Beckman), and the glucose biosensor operating in Differential Normal Pulse Voltammetry (DNPV) mode. The plasma and CSF samples were diluted 1:1 (v/v) in buffer. Averages ± SEMs ($n = 6$) are shown. (**e**) Changes in glucose concentration in the rat brain (measured using DNPV) when insulin (40 IU) and glucose (5 % and 30 %, both w/v) were successively administered intraperitoneally to anesthetized rats. (**f**) Changes in the extracellular level of glucose across the sleep–wake cycle. Voltammetric measurements were performed in freely moving animals equipped with polysomnographic electrodes. The mean value obtained during the spontaneous waking state (W) was normalized to 100 %, whereas those obtained during active wakefulness (AW), slow-wave sleep (SWS), and paradoxical sleep (PS or REM sleep) are expressed as percentages relative to W. Changes in glucose levels observed during AW ($n = 42$), SWS ($n = 400$), and REM ($n = 89$) were significant relative to W ($n = 543$). *Abbreviations*: *n* sample number, *Sleep–wake states*: *W* the electroencephalogram (EEG) exhibits fast electrical waves (theta, 5–8 Hz; beta-1, 14.5–18.6 Hz; beta-2, 19–30 Hz) and sustained activity of the neck muscles; *AW* the EEG exhibits electrical activity with waves analogous to those of W; the activity of the neck muscles is, however, more significant than during the W state; *SWS* the EEG exhibits spindles (sigma, 10–14 Hz) and slow waves (delta, 0.5–4 Hz); *REM sleep or PS* the EEG is as described for W, but the amplitude of the theta rhythm is more significant. *Statistics*: in the sleep–wake experiments, samples (*n*) were obtained from eight rats; *stars*: ANOVA followed by a LSD test, $p < 0.05$ (using "state" as independent variable) sweeping (which takes one-to-several hours for either GOx/polyresorcinol or GOx/PPD films). However, when the sensitivity (assuming that GOx is immobilized in a thin insulating layer) and the extended dynamic range are considered, such sensors are nonetheless valuable (15).

2.3.2 Lactate Biosensor

Measurements are conducted using a chlorinated silver wire as the reference electrode (Ag/AgCl), a platinum wire as auxiliary electrode (Pt), and a carbon fiber lactate sensor as the working electrode (WE). The electrodes are connected to a Biopulse voltammetric unit (Tacussel), used in the DNPV mode, and to a chart recorder. DNPV measurements are collected using the following parameters: a potential sweep from 0.45 to 1.25 V with a scan rate of 10 mV/s (the potential step was at 5 mV per 0.5 s); a pre-pulse duration of 60–80 ms; and a measuring pulse of 20 mV in amplitude and 80 ms in duration. The measurement period is 2 min and the time lapse between two successive scans is thus 40 s. Under such conditions, the sensor yields a well-resolved DNPV peak appearing at 0.8–1.0 V *versus* Ag/AgCl. Commonly, the linearity range of such sensors is 0.1–2.0 mM lactate (Fig. 4a, b). The effect of oxygen on sensor response was studied by calibrating the sensor in solutions after bubbling with O_2. O_2-dependence was not significant until a lactate concentration of 1 mM. Such dependence increased with higher lactate concentrations. In O_2-saturated solutions, the linearity of the sensor response extended to 3–4 mM lactate. These estimations indicate that, despite the relatively low basal pO_2 level in the cerebral milieu (about 35–90 Torr), the sensor can detect low levels of lactate. When it is considered that

Fig. 4 (continued) cisterna magna). Values were determined by a photocolorimetric method (Photocol.) and via use of the lactate biosensor with data collected in the Differential Normal Pulse Voltammetry (DNPV) mode. Values (±SEMs) from brain tissue ($n=7$), plasma ($n=7$), and CSF ($n=6$) were calculated after samples were diluted 1:9 (v/v) in buffer. No significant difference between Photocol. and DNPV measurements is apparent. (**e**) An example of in vivo changes in lactate concentration (in the cortex and extracellular space) following intraperitoneal injection of NMDA (5 mg/kg, $n=6$). The baseline level prior to substance administration was taken as 100 %. (**f**) Changes in the extracellular level of lactate across the sleep–wake cycle. Voltammetric measurements were performed in freely moving animals equipped with polysomnographic electrodes. The mean value obtained during the spontaneous waking state (W, $n=668$) was normalized to 100 %. Values obtained during active waking (AW, $n=18$), slow-wave sleep (SWS, $n=717$), and paradoxical sleep (PS or REM sleep, $n=60$) are expressed as percentages relative to W values. Changes were seen during PS compared to SWS. During AW (compared to W) the lactate level increased significantly (+53 %). During SWS (compared to W) the cortical level of lactate fell significantly (−16 %). During PS, lactate level was somewhat lower, but the difference was not significant (−9 %). However, when lactate level was compared between PS and SWS states, a significant increase was apparent (+8.5 %). *Abbreviations*: *n* sample, *V* volt; *nA* nano-Ampere, *NMDA* *N*-methyl-D-aspartic acid. *Sleep–wake states*: For a definition, see legend to Fig. 3. *Statistics*: Changes induced by NMDA were measured in six rats; *Star*: ANOVA followed by the LSD test, $p<0.02$ (DNPV values were considered to be the independent variable); Samples for sleep–wake states samples were obtained from eight rats; ANOVA followed by the LSD test, $p<0.05$ ("state" as the independent variable)

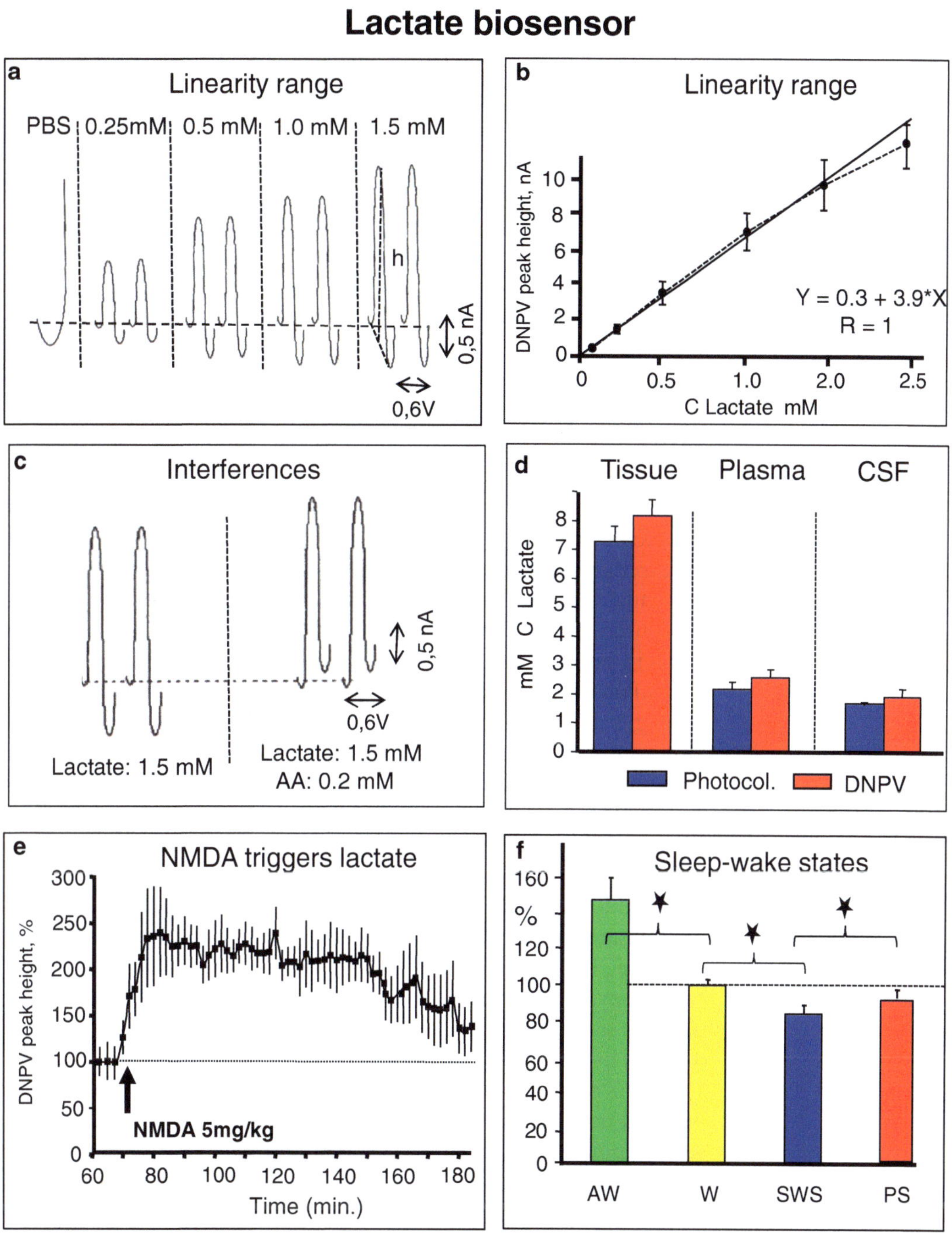

Fig. 4 Lactate biosensor. (**a**) Differential normal pulse voltammograms obtained in vitro using a lactate biosensor. For each concentration of lactate, the first two signals are shown. The height of each signal (h), reflecting oxidation of H_2O_2 (0.8 V), is shown. (**b**) Calibration curve showing that the response is linear in the range 0.1–2.0 mM lactate. Data are means ± SEMs, *n* = 6. (**c**) Differential normal pulse voltammograms obtained using solutions of lactate with or without ascorbic acid (AA). In the presence of AA, both the form of the oxidation signal and the background noise are affected. This does not change signal height. (**d**) Bar graphs showing lactate concentrations (±SEMs) in rat brain tissue (tissue) plasma and CSF (cerebrospinal fluid, drawn from the

lactate concentration in the rat brain is in the range 0.35–1.12 mM, in vivo brain monitoring is thus possible. The response of the sensor to 1.5 mM lactate was also examined in the 6–9 pH range. Under physiological conditions (pH 7.4 and buffer concentration of about 40 mM), the sensor worked optimally (17).

2.4 Specificity

2.4.1 Glucose Biosensor

The effects of potentially interfering species were first examined. We noted that, using a GOx-modified microelectrode, the baseline shifted when 0.2 mM of ascorbic acid (AA) was added to the glucose solution, whereas both the shape and amplitude of the signal remained unchanged (Fig. 3c). The same observation was made when acetaminophen was added to a physiological concentration (0.2 mM). The fact that sensitivity is not affected by potentially interfering species with oxidation potentials far more negative than that of H_2O_2 can be attributed to the mode of measurement employed. The measured signal is the first derivative of the polarographic curve, assuming that the chosen DNPV parameters are correct for extraction of capacitive currents. However, it is important to carefully discriminate between molecular species oxidizable at a potential close to that of hydrogen peroxide. When in vivo brain measurements are conducted, peptides (present in large concentrations in cerebral extracellular fluid) may pose problems. To evaluate the influence of peptides on the microbiosensor response to glucose (using 2 mM solutions for GOx-modified electrodes with or without additional membranes; 5 mM for the GOx/PPD and GOx/polyresorcinol electrodes), measurements were performed using pure glucose solutions or after addition of post-hypophyseal extract (CHOAY). The peptides in this extract have an oxidation potential of about +700 mV *versus* an Ag/AgCl reference and also cause a noticeable increase in the DNPV baseline. Addition of a Nafion membrane considerably restricts the contribution of peptides to the baseline, but the response to glucose is decreased. Only coating with enzymatic membranes adequately overcomes the interference. In such situations, the addition of a CA membrane separates signals elicited by peptides or hydrogen peroxide.

The reliability of our DNPV measurements of glucose was verified using brains of anesthetized (chloral hydrate, 400 mg/kg, i.p. administration) Sprague–Dawley rats. Injection of insulin, well known to interfere with glucose metabolism (21), caused a decrease in the DNPV signal, whereas a secondary injection of glucose sufficed to restore the signal (Fig. 3d). When sensors were inserted into the cortex, stable and reproducible DNPV peaks were obtained after a delay 20–40 min. The basal glucose level was monitored for at least 2 h before insulin injection. After insulin administration, the average time to observation of a 50 % decrease in peak height was about 30 min. Subsequently, the effect of the drug could be reversed by injection of a bolus of glucose; 2 ml of a hypertonic solution afforded baseline recovery in about 1 h (Fig. 3d). Thus, the specificity of the DNPV signal for glucose was clearly established. All biosensors were calibrated before and after implantation. A loss of sensitivity (of up to 20 %) occurred after continuous in vivo use for about 8 h. This was taken into account when baseline glucose levels in extracellular brain compartments were estimated (15).

We explored the reliability of our DNPV measurements by comparing the derived glucose concentrations with those obtained using standard procedures. To this end, plasma and CSF samples, drawn from rats anesthetized as described above, were analyzed using the biosensor and two commercial glucose analyzers (the Beckman Glucose Analyzer 2 that employs the electrochemical glucose oxidase–H_2O_2 principle and the Ciba Corning CL 7200 glucose analyzer running a colorimetric test employing glucose oxidase and a chromogen). Plasma and CSF samples were diluted 1:2 v/v with PBS. The results showed that glucose concentrations determined using the glucose biosensor were in good agreement with those obtained by standard procedures (Fig. 3e) (15).

2.4.2 Lactate Biosensor

As with glucose biosensors, we explored, in vitro, whether easily oxidizable species, such as AA, uric acid, and peptides, could affect the biosensor response (Fig. 4c). To minimize such difficulties, lactate biosensors were also coated with CA membranes. We also monitored lactate levels in the extracellular space of the somatosensory cortex. In animals anesthetized with pentobarbital (50 mg/kg, i.p.), the lactate level was 0.41 ± 0.02 mM. To confirm the ability of our biosensor to monitor lactate levels in vivo, we showed that injection of *N*-methyl-D-aspartate (NMDA, 5 mg/1 ml NaCl [0.9 % w/v]; i.p.), an ionotrophic glutamate receptor agonist, induced an immediate rise (+150 %) in the lactate signal (Fig. 4d). Earlier reports suggest that this rise is most likely attributable to stimulation of glycolysis in astrocytes (via the actions of NMDA receptors and lactate carriers) (22). Thus, glucose uptake rises and lactate production increases. Finally, to further validate our measurements, lactate concentrations in CSF, plasma, and brain tissue were determined ex vivo, either using our lactate biosensor or a conventional clinical analyzer. Prior to DNPV detection, each sample was diluted ninefold with a 50 mM phosphate buffer and measurements were performed using at least three electrodes. The results of voltammetric lactate detection and those of the standard methods were in good agreement (Fig. 4b) (17).

3 Glucose and Lactate Changes During the Sleep–Wake Cycle

3.1 Glucose

Extracellular glucose levels were monitored in the somatosensory cortex of freely moving (non-anesthetized) rats equipped with polysomnographic electrodes that allowed sleep–wake states to be distinguished. Electrode implantation was achieved under chloral hydrate anesthesia, as described above. Three extradural stainless steel screws (EEG electrodes) and three multistranded iron wires (EMG electrodes) were implanted using a standard procedure (23). For voltammetric measurements, two microreference electrodes (Ag/AgCl, 3 M) and an auxiliary electrode (tungsten) were placed in contact with the dura mater. Finally, a microlift system (using a dedicated electrode holder from Unimécanique, Paris, France) was fixed over a hole drilled in the skull. This enabled insertion of a glucose-specific working microelectrode into the somatosensory cortex. When voltammetric recordings were discontinued, a dummy electrode (a micropipette filled with epoxy resin) was inserted into the microlift system to prevent healing of the dura. Because it was particularly important to avoid blood vessel damage, the glucose electrode implantation site differed slightly for each animal (P: ±1.7 to ±2.3; L: 2.6 to 3.6; V: 0.5 to 1 from *bregma*; see the atlas of Paxinos and Watson, 1997). Next, electrode leads were soldered to miniature contact pins and cemented to the skull with dental acrylic resin. The animals were allowed to recover from surgery for at least 1 week; the dummy electrodes were removed and cleaned daily using an antibiotic-containing solution. The recording chamber was a sound-attenuated home cage with a 12-h light–dark cycle. Rats were individually housed in large Plexiglas bowls with free access to food and water. A gold slip ring commutator (Air Precision, Paris, France), linked to a flexible connecting cable, allowed each animal to move freely. During experimental sessions, the dummy electrode was removed from the microlift and replaced by a glucose biosensor (15). DNPV measurements were performed using a standard pulsed-voltammetry device (Biopulse, Tacussel) and a paper chart recorder (Tacussel). Before insertion into the rat brain, sensors were calibrated in vitro in a phosphate-buffered saline solution (50 mM KH_2PO_4–NaOH; 150 mM NaCl; pH 7.4) containing glucose at various concentrations. The sensitivity and linear range were determined for each sensor, using the initial slope of the curve and the equation [hDNPV = f(Cglucose)], where Cglucose was the glucose concentration in solution and hDNPV the height of the corresponding DNPV signal. After in vivo use, all sensors were tested again in the same manner. Sleep–wake states were assessed using a clinical paper chart polygraph (ECEM, Paris, France). Polysomnographic recordings and voltammetric measurements were conducted alternately. A DNPV scan, 90 s in duration, was performed every 3 min. In this way, the second half of each measurement interval was devoted to EEG–EMG recording. A recording session typically lasted 8 h.

The basal extracellular glucose concentration in conscious rats averaged 0.59 ± 0.3 mM, and increased by up to 180 % of this level under chloral hydrate anesthesia. During the sleep–wake cycle, the existence of significant spontaneous variations was noted in the mean glucose level during the active waking triggered by a water puff (AW) (−32 % after 15 min), during SWS (+18 % after 12 min), and during PS (−11 %), compared with the level of the spontaneous waking state (100 %) (Fig. 3f). Voltammetric techniques employing enzymatic biosensors permit thus direct glucose measurements in freely moving animals. Overall, PS is an energy-dependent state and SWS an energy-saving state (5, 19).

3.2 Lactate

The equipment used and the methods employed were similar to those described above. Measurements were conducted on somatosensory cortical fluid of freely moving (non-anesthetized) rats. Briefly, electrode implantation was performed in male Wistar rats (IFFA-Credo, L'Arbresle, France) anesthetized with sodium pentobarbital (60 mg/kg, i.p.). To permit EEG recording, three stainless steel screws were inserted into the skull (in the frontal and occipital cortices). For EMG monitoring, three multi-wire steel electrodes ($\varphi = 150\ \mu m$) were introduced into the neck muscles. An electrode holder, allowing permanent replacement of the lactate sensor (Unimécanique), together with auxiliary and reference electrodes, was positioned as described above for experiments involving the glucose biosensor. Recordings were achieved using a potential sweep of 10 mV/s, covering the 550–1,250 mV window (measurement duration: 70 s, repeated every 3 min). Immediately before and after use, sensors were calibrated in standard solutions of lactate. Polysomnographic recordings were alternated with voltammetric measurements. At the end of each experimental session, data were scored and a given sleep–wake state was linked to each voltammetric signal (reflecting the instantaneous extracellular lactate level). This was done only when sleep–wake status did not change during a DNPV measurement (for example, wakefulness—DNPV scan—wakefulness). This criterion is particularly important when short PS episodes are to be analyzed (20).

The mean lactate concentration in the extracellular space of the somatosensory cortex varied significantly from the light (0.46 mM) to the dark period (1.26 mM). The observed changes were associated with the sleep–wake states. During the AW triggered by a water puff, lactate level increased significantly (+53 %) compared to that seen during the spontaneous waking (W, 100 %) (Fig. 4f). During SWS, lactate level decreased significantly (−16.2 %) (Fig. 4d). During PS, lactate level remained low (−9.0 %), but this variation was not statistically significant. However, when PS change was compared to SWS variation, a significant increase in lactate level was evident (+8.5 %) (Fig. 4f). Long-term monitoring also permitted identification of a circadian influence on lactate production; the lowest and highest values were obtained during the light and dark periods, respectively. The acrophase of the circadian change (+89 %) occurred during the dark period, about 3 h after lights-off (20).

It has been suggested that, during wakefulness, astrocyte metabolism allows transformation of blood-borne glucose to lactate. The increase in lactate observed during PS may be used to enhance oxidative phosphorylation, to supply the high level of ATP needed during PS (20).

4 Importance of the Biosensors and Problems in Practice

The biosensors described above are of great importance to gain a better understanding, under in vivo conditions, of the energy metabolism of the brain. They constitute unique tools that allow direct assessment of the changes occurring in the key compounds of energy metabolism in non-anesthetized freely moving animals. Both the selectivity and sensitivity of the sensing layer still require improvements. The present difficulties can be classified into two categories. First, the performance of enzyme-bearing electrodes depends on the enzyme immobilization method used, the thickness and stability of the entrapping membrane, and the activity and stability of the enzyme. Second, specificity of measurements taken under working conditions is very important. Amperometric detection of glucose in the living brain suffers from direct faradic interference from AA, uric acid, and molecular species with oxidation potentials much lower than that of H_2O_2. Several approaches may allow this difficulty to be overcome; they may be summarized as follows:

1. It is possible to cover the enzymatic layer with an additional membrane that can exclude interfering molecules on the basis of charge or size. An interesting variant of this approach is immobilization of GOx in permselective electropolymerized films that allow the passage of H_2O_2 but prevent interfering species from reaching the electrode surface. Apart from reducing interferences, additional membranes have been used as diffusional barriers for glucose, thus reducing the "oxygen deficit" (reaction 2). However, although up to four additional membranes have sometimes been employed, the in vivo performance of such amperometric sensors remains poorly reliable.
2. Incorporation of an enzyme, a redox reagent, or a ligand in the outer layer of the sensor to inactivate incoming interfering species has been suggested. Although this method has found *some in vitro applications, its use in vivo remains difficult.*
3. Lowering of the potential applied to the electrode may be helpful; this could be achieved by incorporation of a mediator shuttling electrons between the reduced form of the enzyme and the electrode; by co-immobilization of GOx with horseradish peroxidase (HRP), to detect H_2O_2 at a working potential of 0 mV *versus* an SCE (standard calomel electrode) via direct electron transport between HRP and the electrode surface; and by optimization of the applied potential to decrease the current produced by oxidation of interfering species. Unfortunately, the construction of a multicomponent system also creates problems with sensor performance.

4. Use of a pre-oxidizing cell maintained at a potential sufficiently high to destroy the interfering compounds. This method can only be used with the dialysis technique, the main disadvantages of which are the low time resolution (from 2 to 10 min) and the relatively large size of the probes.

Taking into consideration the above difficulties, we employed carbon fiber electrodes and voltammetric pulsed techniques for current detection. Both the good specificity and the small size of the sensors allowed us to investigate glucose and lactate concentrations directly in precise brain areas of living animals also equipped to permit polysomnography. Because of the noise on polysomnographic recordings produced by voltammetric pulsed scanning, the measurements were performed alternately. The digitization of recordings and the ability to reject DNPV-generated noise via active filtering now open the way for simultaneous voltammetric–polysomnographic measurements.

5 Conclusion

Differential normal pulse voltammetry, using modified carbon fiber microelectrodes, is a reliable method for direct and specific monitoring of glucose and lactate levels in the brain of non-anesthetized freely moving animals. Carbon fibers are minimally invasive, strong enough, and conveniently biocompatible. They meet the requirements imposed by the use of a pulsed technique. DNPV offers better selectivity in terms of analysis of enzymatically produced hydrogen peroxide, compared to conventional amperometry. Potentially interfering species (peptides) with oxidation potentials close to that of H_2O_2 can be effectively eliminated. Good agreement is evident between glucose and lactate concentrations determined by standard procedures and the DNPV technique. Finally, our approach permits the study of brain metabolism under conditions of sleep–wake alternation or various other behavioral situations.

Acknowledgements

This work was supported by the University of Lyon and the Neuroscience Center of Lyon. The English form was checked by English Manager Sciences Editing.

References

1. Magistretti PJ, Pellerin L, Martin JL, Bloom FE (1995) Brain energy metabolism: an integrated cellular perspective. In: Bloom FE, Kupfer DJ (eds) Psychopharmacology: the fourth generation of progress. Raven, New York, pp 657–670
2. Magistretti PJ, Pellerin L, Rothman DL, Shulman RG (1999) Energy on demand. Science 283:496–497
3. Magistretti PJ, Allaman I (2007) Glycogen: a Trojan horse for neurons. Nat Neurosci 10:1341–1342

4. Cespuglio R, Colas D, Gautier-Sauvigné S (2005) Energy processes underlying the sleep-wake cycle. In: Parmeggiani PL, Velluti R (eds) The physiological nature of sleep. Imperial College Press, London, pp 3–22
5. Erecinska M, Silver IA (1989) ATP and brain function. J Cereb Blood Flow Metab 9:2–19
6. Benington JH, Heller C (1995) Restoration of brain energy metabolism as the function of sleep. Prog Neurobiol 45:347–360
7. Lydic R, Baghdoyan HA, Hibbard L, Bonyak EV, De Joseph MR, Hawkins RA (1991) Regional brain glucose metabolism is altered during rapid eye movements in cat: a preliminary study. J Comp Neurol 304:517–529
8. Sokoloff L, Reivich M, Kennedy C, Des Rosiers MH, Patlak CS, Pettigrew KD, Sakurada O, Shinohara M (1977) The [14C]deoxyglucose method for the measurement of local cerebral glucose utilization: theory, procedure, and normal values in the conscious and anaesthetised albino rat. J Neurochem 28:897–916
9. Phelps ME, Huang SC, Hoffman EJ, Selin C, Sokoloff L, Kuhl DE (1979) Tomographic measurement of local cerebral glucose metabolic rate in humans with (F-18) 2-Fluoro-2-deoxy-D-glucose: validation of method. Ann Neurol 6:371–388
10. Mottin S, Laporte P, Jouvet M, Cespuglio R (1997) Determination of NADH in the rat brain during sleep-wake states with an optic fibre sensor and time-resolved fluorescence procedures. Neuroscience 79:683–693
11. Chahboune H, Mahdjoub R, Desgoutte P, Rousset C, Briguet A, Cespuglio R (2008) Effects of chloramphenicol on brain energy metabolism using 31P spectroscopy: influences on sleep-wake states in rat. J Neurochem 106: 1552–62
12. Dworak M, McCarley RW, Kim T, Kalinchuk AV, Basheer R (2010) Sleep and brain energy levels: ATP changes during sleep. J Neurosci 30:9008–9016
13. Kong J, Shepel PN, Holden CP, Mackiewicz M, Pack AI, Geiger JD (2002) Brain glycogen decreases with increased periods of wakefulness: implications for homeostatic drive to sleep. J Neurosci 22:5581–5587
14. Morgenthaler FD, Lanz BR, Petit JM, Frenkel H, Magistretti PJ, Gruetter R (2009) Alteration of brain glycogen turnover in the conscious rat after 5 h of prolonged wakefulness. Neurochem Int 55:45–51
15. Netchiporouk LI, Shram NF, Jaffrezic-Renault N, Martelet C, Cespuglio R (1996) In vivo brain glucose measurements: differential normal pulse voltammetry with enzyme-modified carbon fiber microelectrodes. Anal Chem 68:4358–4364
16. Shram NF, Netchiporouk LI, Martelet C, Jaffrezic-Renault N, Cespuglio R (1997) Brain glucose: voltammetric determination in normal and hyperglycaemic rats using a glucose microsensor. Neuroreport 8:1109–1112
17. Shram NF, Netchiporouk LI, Martelet C, Jaffrezic-Renault N, Bonnet C, Cespuglio R (1998) In vivo voltammetric detection of rat brain lactate with carbon fiber microelectrodes coated with lactate oxidase. Anal Chem 13:2618–2622
18. Soldatkin O, Schuvailo O, Marinesco S, Cespuglio R, Soldatkin A (2009) Microbiosensor based on glucose oxidase and hexokinase co-immobilised on platinum microelectrode for selective ATP detection. Talanta 78:1023–1028
19. Netchiporouk L, Shram N, Cespuglio R (2001) Brain extracellular glucose assessed by voltammetry throughout the rat sleep-wake cycle. Eur J Neurosci 13:1429–1434
20. Shram N, Netchiporouk L, Cespuglio R (2002) Lactate in the brain of the freely moving rat: voltammetric monitoring of the changes related to the sleep-wake states. Eur J Neurosci 16:461–466
21. Schwartz MW, Figlewicz DP, Baskin DG, Woods SC, Porte D Jr (1992) Insulin in the brain: a hormonal regulator of energy balance. Endocrinol Rev 13:387–414
22. Schipke CG, Ohlemeyer C, Matyash M, Nolte C, Kettenmann H, Kirchhof F (2001) Astrocytes of the mouse neocortex express functional N-methyl-D-aspartate receptors. FASEB J 15:1270–1272
23. Burlet S, Leger L, Cespuglio R (1999) Nitric oxide and sleep in the rat: a puzzling relationship. Neuroscience 92:627–639

Chapter 12

Forebrain Cholinergic Systems and Cognition: New Insights Based on Rapid Detection of Choline Spikes Using Enzyme-Based Biosensors

Vinay Parikh and Martin Sarter

Abstract

Changes in acetylcholine (ACh) release in the forebrain represent a critical step in the activation of larger neuronal circuits mediating cognitive functions. Therefore, the development of methods which allow the assessment of ACh release at a high temporal and spatial resolution is of immense significance for behavioral and cognitive neuroscience research. In this chapter, we review evidence from in vivo studies utilizing choline-sensitive microelectrodes for the detection of rapid changes in extracellular choline levels as a spatially discrete and sensitive measure of cortical cholinergic transmission. We then summarize empirical data from our lab based on studies involving electrochemical recordings and *real-time* monitoring of ACh release in animals performing a cued-appetitive response task. Phasic cholinergic activation in the prefrontal cortex, demonstrated by second-based increases in cholinergic transmission during behaviorally significant cues triggering reward approach, suggests that cholinergic transients encode specific information pertaining to the detection of attention-demanding cues. These findings indicate that cortical cholinergic transmission mediates defined cognitive operations which are contrary to the traditional description of cortical ACh as a slowly acting neuromodulator influencing different arousal states.

Key words Acetylcholine, Choline-sensitive microelectrodes, Amperometry, Attention

1 Introduction

The cholinergic system is one of the most important modulator neurotransmitter system implicated in diverse behavioral and cognitive processes. Basal forebrain cholinergic neurons projecting to entire cortical mantle constitute the most rostral component of brain cholinergic system. Substantial evidence supports the hypothesis that basal forebrain cortical cholinergic inputs mediate fundamental aspects of attentional information processing (1–6). Therefore, there is a considerable interest in the dynamics of cortical acetylcholine (ACh) release and cholinergic regulation of attentional processes and capacities.

Stéphane Marinesco and Nicholas Dale (eds.), *Microelectrode Biosensors*, Neuromethods, vol. 80,
DOI 10.1007/978-1-62703-370-1_12,

The past 50 years have witnessed an exponential increase in the understanding of cholinergic system by measuring the changes in extracellular ACh levels in discrete brain areas. Earlier studies involving the assessment of ACh efflux using the push–pull cannula and cortical cups suggested that slowly changing levels of cortical ACh mediate different states of arousal and wakefulness (7–10). A major problem with these techniques was that they were mostly limited to experimentation in anesthetized animal preparations. Furthermore, the collection and preservation of ACh-containing media and the analysis of ACh in these media entailed extremely challenging methods. The development of in vivo microdialysis methods coupled with advanced high-performance liquid chromatography was a major breakthrough as this technique exhibited impressive sensitivity to detect neurotransmitter release and this was readily applied to study brain–behavior relationships (11). Microdialysis studies measuring ACh efflux from task-performing animals revealed performance-associated increases in cortical ACh release in cognitive tasks that produces demands on attentional processes (3, 12–14). However, the limited temporal resolution associated with HPLC analytics that required several minutes of sampling time, and lower spatial resolution attributed to larger sampling areas, posed potential limitations to understand the dynamics of ACh release during behavioral manipulations using microdialysis. Therefore, future questions pursuing the relationship between cortical cholinergic transmission and specific cognitive operations, and regional variations for such cholinergic regulation, require a method with higher temporal and spatial resolution.

Evolution of electrochemical approaches like voltammetry and amperometry made it possible to study the kinetics and dynamics of rapid neurochemical events in *real time* and at a higher spatial and temporal resolution (15–21). Enzyme-based sensors configured for oxidizing non-electroactive neurochemicals such as choline, glutamate, glucose, and lactate allow for detection of such molecules (22–27). The purpose of this chapter is to provide an overview of the use of enzyme-based microelectrodes and amperometric methods to detect changes in cholinergic transmission, and how these techniques could serve as a valuable tool for neuroscientists to study the regulation and cognitive functions of forebrain cholinergic systems. We will specifically focus on the use of choline-sensitive microelectrodes to measure rapid changes in extracellular choline as a potential marker for cortical cholinergic transmission. To that end, we will summarize some empirical data from our laboratory gathered over the past several years that indicate phasic cholinergic transmission in the prefrontal cortex (PFC) encodes specific cognitive operations. These findings not only describe how forebrain cholinergic circuits process cognitive information but also question the traditional view concerning the regulation and function of neuromodulator systems. Finally, we conclude with a discussion on the future applications of choline-sensitive electrodes to understand the neurobiology of cognitive symptoms of neuropsychiatric and age-related neurodegenerative disorders, and to screen procholinergic drugs for the treatment of cognitive disorders.

2 Enzyme-Based Biosensors to Detect Changes in Cholinergic Transmission

2.1 Choline-Sensitive Microelectrodes

Considerable progress has been made to develop enzyme-based microelectrodes for the detection of non-electroactive chemicals in the brain including glutamate, acetylcholine, choline, and glucose (22, 25, 26, 28–31). These microelectrodes involve enzyme-associated oxidation of non-electroactive analyte to hydrogen peroxide (H_2O_2), which is then oxidized at a fixed potential generating measurable current. Several groups have developed choline-sensitive microelectrodes to detect changes in ACh release by measuring variations in extracellular choline levels. Wightman and group utilized carbon fiber electrodes to detect in vivo choline concentration using an organic conducting salt that mediates an electron transport chain between the enzyme choline oxidase (CO) and the electrode (32). Michael and coworkers developed a redox polymer layer with horseradish peroxidase and choline oxidase to measure choline using carbon fibers (27, 33). Although very selective for detecting choline, these carbon fiber microelectrodes do exhibit some limitations. The detection limit of these microelectrodes ranged from 5 to 10 μM, which may not be sufficient to detect smaller changes in extracellular choline as a consequence of the hydrolysis of behaviorally induced synaptic ACh release. Furthermore, relatively slower response times (tens of seconds) and clearance (several minutes) rates of choline-derived signals recorded from enzyme-based carbon fiber electrodes suggested a limited ability to capture the spatiotemporal kinetics of ACh-derived choline and transport-mediated clearance. Gerhardt and group used a different approach that involved the use of ceramic-based platinum microelectrode arrays to detect rapid choline measures in brain (34). The platinum recording sites (50 × 100 μm) lithographed onto ceramic wafers were coated with CO to electrochemically measure choline. The development, construction, and properties of these platinum microelectrodes are discussed earlier (22, 25, 26). A major advantage of using these electrodes was the presence of multiple platinum recording channels on the microelectrode; selective coating of adjacent channels with enzyme or control protein (BSA) allows for self-referencing against electroactive interferents, such as ascorbic acid (AA) or dopamine (DA), to improve signal-to-noise ratios and detection limits. These choline-sensitive microelectrodes were significantly improved over previous designs and exhibited a sensitivity of 0.5 μM and response time <1 s.

2.2 Rapid Changes in Extracellular Choline Levels as an In Vivo Measure of Cholinergic Transmission

As ceramic-based multisite, choline-sensitive microelectrodes provided the resolution to address questions on the relationship between ACh release and specific behavioral responses in cognitive tasks known to recruit the forebrain cholinergic system, we first investigated the usefulness of these electrodes to measure amperometrically rapid changes in extracellular choline levels as a marker for the assessment of cortical cholinergic transmission. In this study (35), we tested the ability of choline-sensitive microelectrodes to measure increases in choline resulting from the hydrolysis of newly released ACh by endogenous acetylcholinesterase (AChE).

The geometry of a typical microelectrode array and the measurement scheme are illustrated in Fig. 1a. The electrode consists of 2 pairs of platinum recording sites. Each site is 15 μm wide and 333 μm long, and the distance between the two pairs is approximately 100 μm. Prior to enzyme immobilization, the microelectrodes were dipped in nafion and the coating was dried at 170°C for 4–5 min. These steps were repeated 3–4 times. Nafion is a charged polymer that limits the access of potentially interfering analytes (such as ascorbate, uric acid, and DOPAC), but not H_2O_2, from reaching the platinum recording site. Alternatively, *meta*-phenylenediamine (*m*-PD) could also be polymerized at +0.5 V on

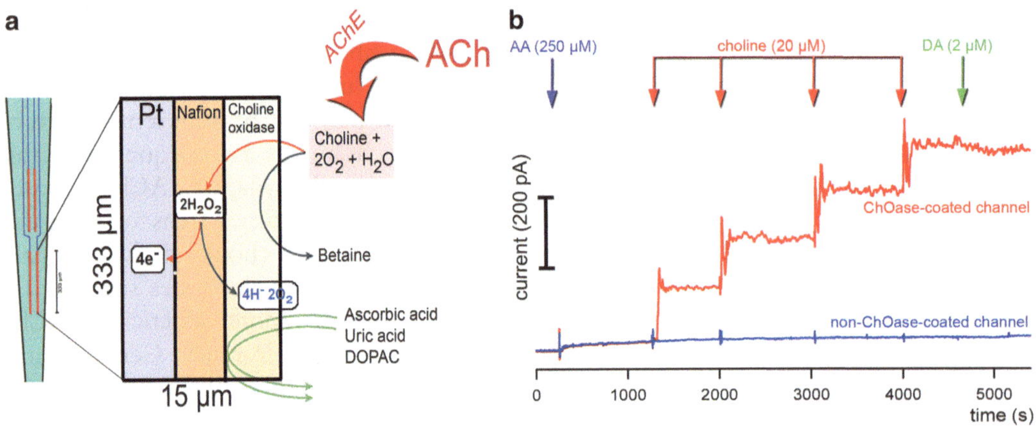

Fig. 1 (**a**) Schematic diagram depicting the choline-sensitive microelectrode and major steps underlying the measurement of extracellular choline concentration. Arrangement and dimensions of the four recording areas (marked in *red*) on the ceramic-based microelectrode are shown. Individual recording sites were 333 x 15 μm (see scale for recording site length on the *right*). The lower pair of recording sites was typically coated with choline oxidase (ChOase), while the upper pair was coated with BSA–glutaraldehyde matrix and used for self-referencing. Choline is generated by hydrolysis of newly released acetylcholine (ACh) by acetylcholinesterase (AChE). Nafion serves to repel anions from the surface of the platinum electrode. Hydrogen peroxide (H_2O_2) is generated as a result of choline oxidation and measured using fixed potential amperometry. (**b**) In vitro calibration of a ChOase-coated microelectrode. Raw current traces recorded from a ChOase-coated (*red*) and non-ChOase-coated (*blue*) channel are shown. Currents in response to the successive addition of AA, choline, and DA indicate the sensitivity of the ChOase-coated channel to choline but not to AA and DA. The figure was modified from Parikh et al. (2008) Glutamatergic contributions to nicotinic acetylcholine receptor agonist-evoked cholinergic transients in the prefrontal cortex, *J Neurosci* 28, 3768–80

platinum recording sites instead of nafion to enhance choline selectivity by blocking potentially electroactive interferents (36, 37). The enzyme CO—cross-linked with bovine serum albumin (BSA) and glutaraldehyde—was immobilized on a pair of platinum recording channels. The other pair of channels, the sentinel control sites, was similarly coated except that they did not contain the enzyme. As discussed above, CO converts the non-electroactive choline into a reporter molecule (H_2O_2) that is oxidized on the surface of platinum recording sites to which a fixed potential (+0.7 V) is applied, generating current. Cross talk between enzyme-coated and sentinel channels may appear due to the placement of very thick coat of enzyme closer to the non-CO channel and the diffusion of H_2O_2 on the latter. Therefore, over-coating of enzyme channels should be avoided.

The electrodes were calibrated in vitro using a potentiostat for choline sensitivity and selectivity prior to implantation in animals. Calibrations were performed by applying a constant voltage of +0.7 V to the platinum recording sites versus Ag/AgCl reference electrode in a beaker containing a stirred solution of 0.05 M PBS maintained at 37°C. Amperometric currents were digitized at a frequency of 5 Hz following gradual additions of choline (20–80 μM). Electrode response to the addition of AA (500 μM) and DA (2 μM) was recorded to determine the specificity of choline-sensitive sites (Fig. 1b). To be used for subsequent in vivo experiments, electrodes were typically required to meet the following characteristics: slope (sensitivity) for choline >3 pA/μM, LOD < 0.3 μM, selectivity for choline vs. AA > 80:1, detection of increasing choline concentration (20–80 μM) $R^2 > 0.98$.

Amperometric recordings of cholinergic signals were made from the frontoparietal cortex of urethane-anesthetized male rats. A micropipette (15–20 μm internal tip diameter) was attached to the ceramic platform of the microelectrode array and positioned such that the tip was equidistant between the two pairs of channels (50–100 μm from the platform). The microelectrode/micropipette assembly was lowered into the frontoparietal cortex using a microdrive, and an Ag/AgCl reference electrode was implanted at a site remote to the recording region. Drug solutions were pressure ejected from the micropipettes using a picospritzer and the ejected volumes were monitored using a stereomicroscope fitted with a reticule. The resulting choline or sentinel microelectrode signals were recorded from all four platinum recording sites. Normalized currents recorded from non-CO-coated sites were subtracted from normalized currents recorded via CO-coated sites ("self-referencing") based on DA response from in vitro calibration to yield currents that selectively reflect choline concentrations. Self-referencing was utilized to remove interference signals including background noise.

Local application of ACh (200 nL; 10 and 100 mM) produced concentration-dependent increases in choline-related signal amplitudes. These signals were rapidly removed (within 10 s) from the extracellular space, suggesting a rapid choline transporter (CHT)-mediated clearance mechanism (38). Furthermore, co-administration of neostigmine, an AChE inhibitor, with ACh but not choline eliminated ACh-associated change in oxidizing current. We also determined the ability of two pharmacological stimuli, known to increase cortical ACh release, to increase the choline signal through stimulation of *endogenous* ACh. Local depolarization with potassium (70 mM; 66 and 200 nL) and blockade of muscarinic autoreceptors with scopolamine markedly increased extracellular choline signals which exhibited rapid rise times (1–2 s) and clearance. Choline signal amplitudes were again inhibited by locally applying neostigmine. Collectively, these findings suggested that ACh-associated increases in oxidizing current on CO-coated recording sites required the hydrolysis of either exogenously applied or endogenously generated ACh into choline by tissue AChE. Moreover, choline-sensitive electrodes are sensitive enough to detect endogenously generated ACh in vivo.

We also demonstrated that blockade of voltage-gated sodium channels with tetrodotoxin robustly inhibited choline signals produced by local application of scopolamine, confirming the ability of these electrodes to detect impulse-dependent release of ACh. Finally, in a separate study (38), we showed that removal of cholinergic projections to the recording region by infusions of 192-IgG saporin into the basal forebrain attenuated potassium-evoked cholinergic signals (Fig. 2a–d). These data confirmed that the demonstration of depolarization-dependent generation of choline signals in vivo depends on the presence and integrity of cholinergic terminals.

2.3 Sensitivity Issues

As discussed above, choline-sensitive microelectrodes are sufficiently sensitive to detect depolarization-evoked ACh release. However, their sensitivity to detect basal ACh is not known. Our attempts to determine the contribution of neuronal ACh release under baseline conditions to changes in baseline choline concentration following local application of TTX were unsuccessful (Parikh and Sarter; unpublished data). This may be attributed to higher baseline extracellular choline levels that fluctuate between 3 and 5 μM and are efficiently buffered with plasma choline levels. Furthermore, removal of cholinergic inputs from the recording field did not affect basal extracellular choline levels detected by subtracting background currents on non-CO channels from CO-coated channels. These findings indicated extreme stability of basal extracellular choline levels (38). As the detection limits of these electrodes to detect choline range from 0.1 to 0.3 μM, and due to the ubiquitous presence of extra-potent metabolizing enzyme AChE in the extrasynaptic space that keeps basal ACh levels incredibly low, it also

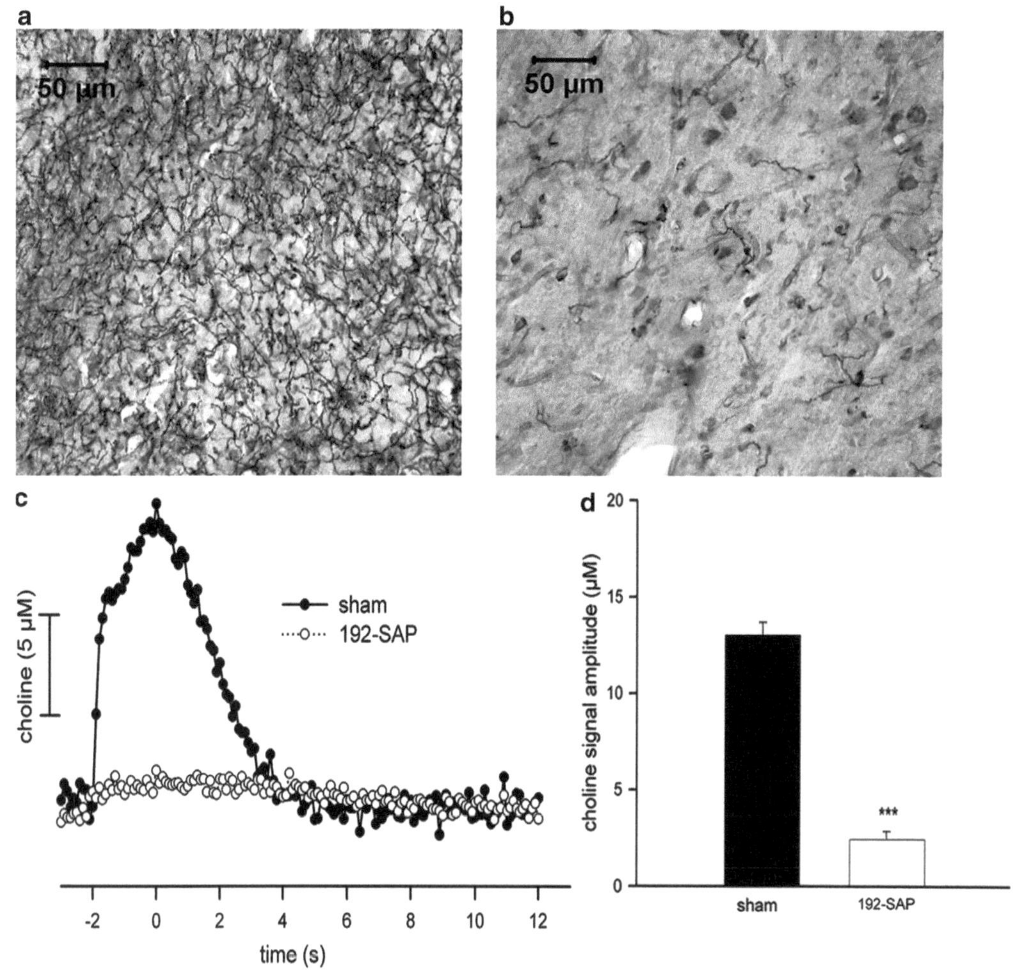

Fig. 2 Effects of the loss of cholinergic inputs to the recording region on KCl-induced choline signals. Cholinergic deafferentation was produced by bilateral stereotaxic infusions of 192-IgG saporin (192-SAP), a selective cholino-immunotoxin, into the nucleus basalis of Meynert and substantia innominata region of rat basal forebrain (BF). Sham lesions were produced by infusing mouse saporin into the BF. Amperometric recordings were conducted 3 weeks after sham or lesion surgeries. AChE-positive fiber densities in the recording region from a sham (**a**) and a 192-IgG saporin (192-SAP)-lesioned animal (**b**). Comparison between cholinergic fiber densities indicated a highly significant, >80% reduction in the density of cholinergic inputs to the recording region in the lesioned animal. (**c**) Example of a KCl-induced choline signal in a sham (*filled circles*) and a 192-SAP-lesioned (*open circles*) animal. As a result of the extensive decrease of cholinergic terminals, KCl-induced signal amplitudes were significantly decreased in the lesioned (*white bars*) as compared to sham (*black bars*) animals (**d**; $P < 0.001$ vs. sham). The figure was modified from Parikh, V. and Sarter, M. (2006) Cortical choline transporter function measured in vivo using choline-sensitive microelectrodes: clearance of endogenous and exogenous choline and effects of removal of cholinergic terminals, *J Neurochem 97*, 488–503

cannot be ruled out that our inability to measure ambient levels of ACh arose from lower sensitivity of these electrodes. It is important to note that measures of ACh release obtained with microdialysis indicated ACh levels in the lower nanomolar to micromolar range depending upon whether the dialysate contained neostigmine. Moreover, local tissue damage produced with the insertion of microdialysis probe and reactive gliosis may also account for variations in resting ACh levels. Ceramic-based microelectrode arrays produce minimal tissue damage and microglial activation following acute (35) or chronic use (39) due to smaller size of the platinum recording sites and thickness of the ceramic wafer. Therefore, efforts to directly measure ACh by coating the electrodes with both AChE and CO have gained interest (40, 41). As a result of the relatively high concentration of AChE at the recording site, extracellular ACh that escapes metabolism through tissue AChE would be hydrolyzed on the recording sites generating more peroxide-mediated current than CO-coated site.

Recently, we tested the possibility that co-immobilization of AChE with CO on platinum recording sites increases the level of detection of depolarization-evoked ACh release in the prefrontal cortex of rats (42). Currents were recorded from AChE+CO- and CO-coated sites following potassium depolarization from five recording regions spanning the dorso-ventral extension of the prelimbic region of the cortex. The amplitudes of evoked choline signals did not significantly differ between the two coating sites, suggesting that ACh spillover into the extracellular space of the cortex is extremely limited. These findings apply to the regulation of ACh by AChE in the cortex and must be viewed cautiously for other brain regions where AChE is differentially regulated. Although the issue that ACh can escape hydrolysis by AChE and reaches the extracellular space (volume transmission) or remains confined to synaptic space (wiring transmission) will remain debated (43), our studies suggest that choline-sensitive microelectrodes are sufficiently sensitive and generate valid and sufficiently sensitive measures of synaptic ACh release in cortex.

3 Real-Time Monitoring of ACh Release in Attention Task-Performing Animals

The persistent attentional impairments that result from selective removal of the cortical cholinergic input system indicate that the integrity of the basal forebrain cholinergic system is necessary for performance in tasks that tax attentional processes (2–5, 44). Furthermore, increases in cortical acetylcholine (ACh) release were observed in attentional task-performing animals, suggesting a necessary role of this neuromodulator for attentional performance (12–14, 45–47). However, the precise cognitive operations supported by changes in cortical cholinergic activity have remained unknown. Traditional methods used for monitoring ACh release in performing animals, particularly the widely used microdialysis method, have generated important evidence concerning the role of this neurotransmitter system in attentional functions at a relatively low temporal resolution. These experiments typically required 5–10 min of sample collection to detect ACh concentration in the dialysate using HPLC, over tens of trials, to generate a single data point. Although numerous control experiments confirmed that increases in cortical ACh release mediate attentional processes, the low temporal resolution of neurotransmitter release data limited the demonstration of the specific task events or the behavioral or cognitive operations that are responsible for the increases in cortical cholinergic activity. Consequently, conclusions based on microdialysis data were consistent with the conventional characteristics of ACh as a slowly acting cortex-wide neuromodulator optimizing input processing by differentially regulating arousal states. However, the presence of ACh metabolizing enzyme AChE, and ionotropic nicotinic aceylcholine receptors (nAChRs) that mediate faster cholinergic signaling, suggests that functions of forebrain cholinergic system are not sufficiently described by such notions.

With the advent of enzyme-based biosensors and amperometric techniques, the measurement of ACh levels on a faster timescale is possible. Cortical cholinergic inputs, particularly to the PFC, have been hypothesized to mediate cue detection and facilitate attentional performance. Detection refers to a set of cognitive processes that involve the integration of a stimulus into ongoing behavior, including a shift in attention from ongoing activities towards stimulus-evoked behavior, response rule processing and preparation, outcome expectation, and the timing of such an outcome. To test the hypothesis that cholinergic activity in the PFC controls signal detection, we employed ceramic-based multisite microelectrode arrays for the electrochemical measurement of ACh release at a sub second resolution.

3.1 Choline-Sensitive Microelectrodes for Awake Animal Recordings

Ceramic-based microelectrodes featuring four platinum recording sites (Fig. 1a) were differentially coated with CO and BSA coated (above) and modified for amperometric recordings in *awake* freely moving task-performing animals (see Fig. 3 for steps involved in the construction of these microelectrodes). These modifications were required to connect the implanted electrode to the miniaturized rat hat preamplifier (39). The electrode paddle was modified by soldering one end of four enamel-coated magnet wires (30 ga) to the terminals on the electrode panel and the other end to gold-pin connectors. The soldering joints were fixed with epoxy and allowed to dry overnight. The electrode paddle beyond the soldering joints was cut using an electronic pliers cutter. Reference electrodes were constructed by soldering Ag/AgCl reference electrodes prepared from 0.008 silver wire to gold-pin connectors. The gold pins were connected to the silver wire and the magnet wires were inserted into a miniature 9-pin plastic strip connector. The connector was glued to the microelectrode using epoxy (for details on the design and construction of enzyme-coated microelectrodes for

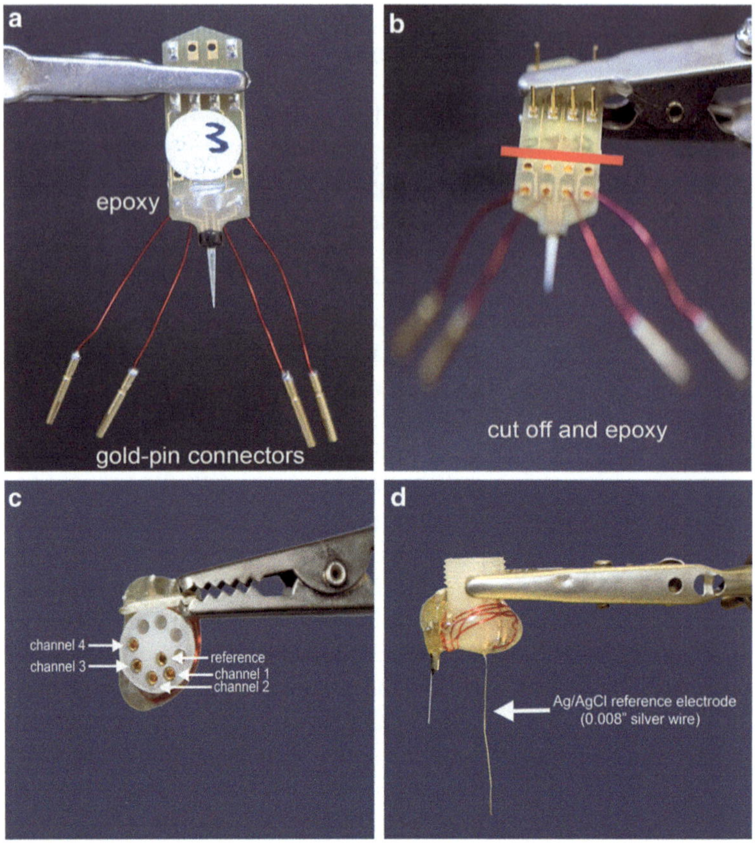

Fig. 3 Photomicrographs depicting the steps involved in the construction of microelectrode assembly to record cholinergic transmission from freely moving awake animals. (**a**) Enzyme-coated ceramic microelectrodes were modified by soldering one end of four enamel-coated magnet wires (30 ga) to the terminals on the ceramic wafer and the other end to gold-pin connectors. The soldered joints were epoxied and allowed to dry. (**b**) Backside view of the electrode depicting the position to cut the paddle. Reference electrode was constructed by soldering Ag/AgCl wire to gold-pin connector. The gold pins containing the wires connected to the electrode and reference were inserted into the 9-pin plastic strip connector. (**c**, **d**) Fully assembled choline-sensitive microelectrode depicting miniaturized connector glued to the modified electrode paddle using epoxy. Channels 1 and 2 depict gold pins for CO-coated channels, while channels 3 and 4 are gold pins for non-enzyme (BSA) channels. Reference represents gold pin for Ag/AgCl electrode

recordings in freely moving animals, see (39)). Finally, an infusion cannula was attached to the electrode such that the tip of the cannula was placed between the two pairs of platinum recording sites at a distance of ~150–200 μm. CO-coated microelectrodes were calibrated using a potentiostat for choline sensitivity and selectivity prior to implantation in animals as described above.

3.2 Behavioral Apparatus and Task Design

Amperometric recordings in performing animals are associated with substantial technical and experimental problems, including interference by static energy sources. Therefore, our first set of experiments necessitated a relatively simple test of the hypothesis that phasic cholinergic signals are associated with, and are necessary for, shifts from endogenously generated behavior and associational processing to the detection of signals and signal-evoked cognitive and behavioral processes. These experiments utilized a relatively simple cued-appetitive response task (48). Behavioral training and testing in this task were carried out using a plywood-constructed chamber, with ventilation ports on top and bottom and on both sides. Reward ports were located on the left and right side walls. The cue light was presented via water clear White T1 ¾ LED attached to the ceiling. Video tracking of the animals' behavior and task events was accomplished by installing a CCTV bullet camera at the ceiling and connecting it to a monitor and a video recorder. For amperometric recordings in performing animals, the chamber was shielded with copper wire screen to avoid interference from static energy. Furthermore, the chamber was equipped with a removable top to center a tether port, swivel, low noise commutator, and counter weight to allow animals to move freely in the chamber during recording sessions.

Partially food-restricted rats were trained to predict the cue, a ceiling light on for 1 s with subsequent reward (food pellet) delivery (6 ± 2 s later) at one of two reward ports (randomly selected). This task involved relatively longer intertrial intervals (ITI; 90 ± 30 s) to foster animals' disengagement from task (Fig. 4a). Behaviorally, the cue generated a distinct shift from ongoing behavior (typically grooming) toward the monitoring of the reward ports, followed by port approach and reward retrieval in response to reward delivery. Following training, such a cue-evoked shift in behavior ("detected cue") occurred in the majority, but not all trials. In 30–40 % of the trials the cue failed to evoke such behavior and animals typically continued grooming ("missed cue"). Importantly, in such trials, reward was also delivered, and because of the salient stimuli associated with reward delivery, animals approached the ports and retrieved the reward, albeit with longer latencies when compared with animals which detected the cue and therefore were already monitoring the ports.

Choline-sensitive electrodes were implanted in the PFC of trained animals. Following recovery, animals were placed in the chamber and the microelectrode assembly was connected to the potentiostat through a shielded cable and a low-impedance commutator via a miniature potentiostat headstage (Fig. 4b). Recording sessions took place after the rats had returned to the criterion performance following 3–5 training sessions post-surgery. Amperometric recordings were conducted at 2 Hz by applying a fixed potential of +0.7 V to the microelectrode. The recording events were

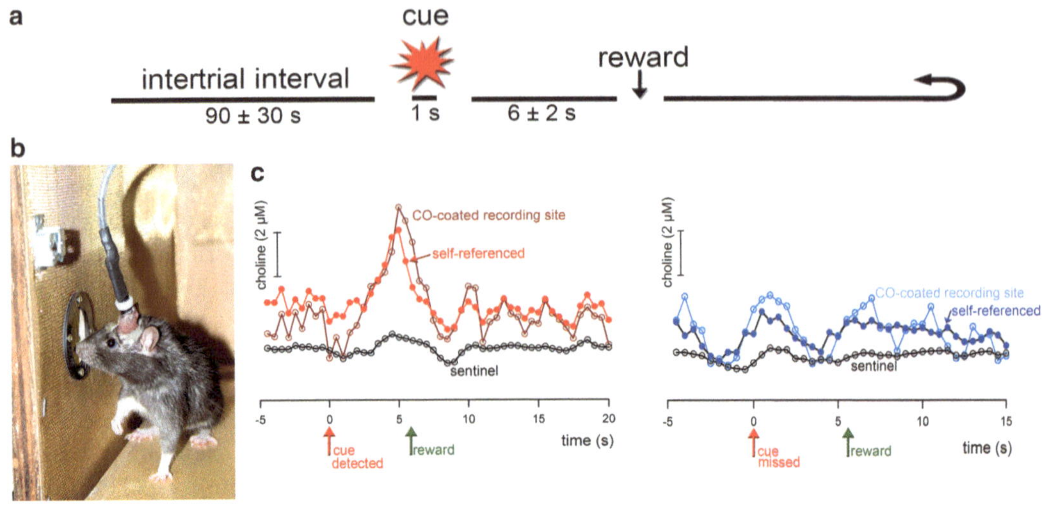

Fig. 4 (**a**) Illustration of the main events constituting the cued-appetitive response task. Following ITI (90 ± 30 s), a cue was presented followed by reward delivery at one of the two random ports (randomly selected) 6 ± 2 s later. Trials involving cue detection were classified as such based on cue-elicited disengagement from ongoing behavior and monitoring of the food ports. A missed cue was defined as such based on the absence of a cue-evoked shift in behavior. (**b**) A task-performing animal undergoing electrochemical recording of cholinergic transmission from the medial prefrontal cortex. (**c**) Representative traces from a CO-coated (*dark red*) and a non-CO-coated ("sentinel"; *black*) recording channel during a trial involving cue detection; amperometric recordings were made at 2 Hz. Furthermore, the self-referenced trace (*red*; *filled circles*) is shown (*arrows* on the abscissa depict the time of cue and reward presentation). (**d**) Raw traces (CO coated: *blue*; sentinel: *black*) and self-referenced (*dark blue*) traces from a trial during which the cue was missed. Phasic increases in cholinergic activity were not observed in such trials. The figure was modified from Parikh et al. (2007). Prefrontal acetylcholine release controls cue detection on multiple timescales, *Neuron* 56, 141–54

time-locked and synchronized by marking all task events via TTL pulses. In vivo sensitivity of electrodes to choline was determined by infusing choline (5–20 nmol) via guide cannula at the end of the recording sessions (48).

3.3 Phasic Cholinergic Transients and Cue Detection

Amperometric recordings of cholinergic activity in the medial PFC of performing rats indicated phasic cholinergic signals evoked by cues that were detected but not in trials involving missed cues. Representative raw traces from CO-coated and control platinum recording sites and self-referenced traces from a typical detected and a missed cue are shown (Fig. 4c, d). It should be noted that other task events like port approach, reward delivery, and reward retrieval did not generate phasic cholinergic signals. The onset of the cholinergic signal, defined as a 25% increase of pre-cue cholinergic activity, was highly correlated with the onset of the behavioral shift. Removal of cholinergic inputs to the recording region by locally infusing cholino-immunotoxin 192-IgG saporin completely abolished cue-evoked cholinergic signals in detected trials, confirming that signals recorded from CO-coated sites indeed required the presence of cholinergic terminals. Collectively, these findings suggested that transient or "phasic" increases in prefrontal cholinergic activity are evoked by attended cues. In trials involving missed cues, the delivery of reward did trigger port approach and reward retrieval. However, the absence of cholinergic transients in those trials supported the evidence that cholinergic transients mediate cue-evoked cognitive operations.

Additional experiments indicated that variation in the time interval between cue and reward delivery caused variation of the timing of the peak amplitude of cue-evoked cholinergic signals (47, 48). This was an important observation as it indicated that cholinergic transients do not merely reflect sensory encoding of the cue. If that was the case, variations of cue–reward intervals should not affect the timing of the cholinergic transients. The variation of the timing of cue-evoked cholinergic transients indicates that phasic ACh release in the PFC is associated with a cognitive operation (cue detection), the timing of which is a function of cue–reward intervals.

We also tested the hypothesis that cholinergic control of cue detection represents a specific function of the PFC by recording cholinergic activity from the motor cortex (forepaw region) of task-performing animals. This non-associational control region was selected to exclude the possibility that cholinergic inputs to other associational regions, particularly the posterior parietal cortex, directly or indirectly complement the cognitive functions of cholinergic inputs to the PFC. Neither detected nor missed cues, nor any other task-related event, evoked reliable and robust cholinergic transients in the motor cortex. Additionally, and as would be expected, removal of cholinergic inputs to motor cortex did not affect the animals' performance of the cued-appetitive response task. It is noteworthy that bilateral removal of cholinergic inputs to the medial PFC did decrease the rate of cue detection (48). Thus, the functional role of cholinergic transients in PFC and motor cortex appears to be clearly dissociated. The putative functions of cholinergic transients in the motor cortex remain unknown but may contribute to the learning of fine motor skills (49, 50).

3.4 Tonic Cholinergic Transmission

In addition to cue-evoked cholinergic transients, we also observed minute-based increases in fluctuations in cholinergic activity from animals performing the task (48) (Fig. 5). Such session-based slower tonic changes in cholinergic transmission were not observed in trained animals that were placed into test chambers but that were not allowed to perform. Therefore, tonic cholinergic activity was not evoked by the performance context and associated cognitive and motivational states and expectations. Rather, tonic changes represent, necessarily, a correlate of performance. Moreover, such tonic changes were observed in medial PFC and motor cortex (Fig. 5a, b) and therefore are hypothesized to manifest cortex-wide.

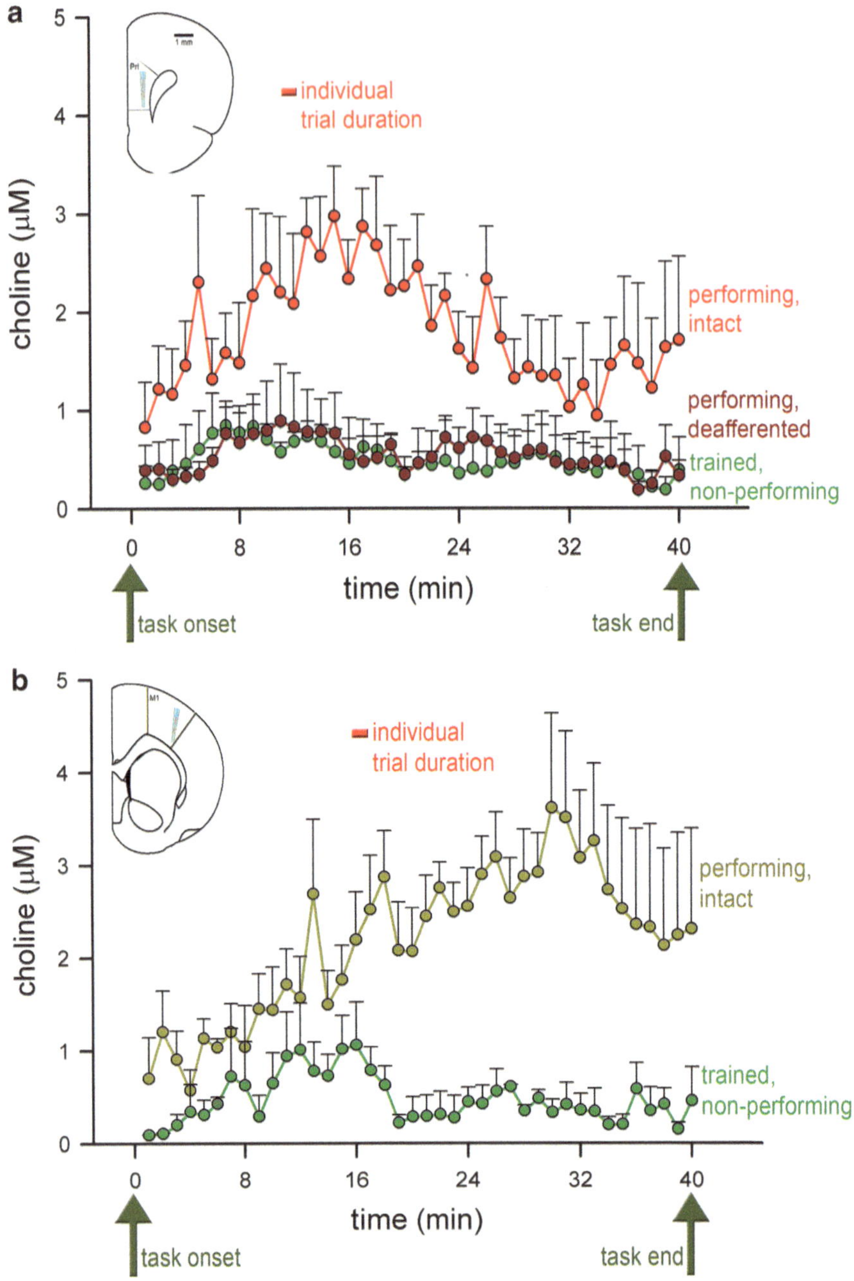

Fig. 5 Performance session-related tonic changes in cholinergic activity in the medial PFC and motor cortex. These data were extracted from amperometric recordings (2 Hz) by boxcar-filtering self-referenced traces and expressing minute-based data points as a change from average signal levels recorded during a 3-min pre-session baseline. (**a**) Session-related changes in cholinergic activity in the PFC of intact (*red trace*), trained but not performing (animals placed into chambers but task not turned on; *green trace*), and task-performing animals after unilateral removal of cholinergic inputs to the recording region (*dark red trace*). Note the bar indicating the duration of a single trial relative to the abscissa depicting the entire 40-min session. (**b**) Session-related changes in cholinergic activity in the motor cortex of intact (*light green trace*) and trained but non-performing animals (*dark green trace*). In intact rats, session-related changes in cholinergic activity occurred in both regions,

Absence of tonic cholinergic activity following removal of cholinergic inputs from the recording region confirmed that these choline measures reflected ACh release (Fig. 5a).

The pattern of these session-related changes in cholinergic activity resembled performance-associated ACh release measured by microdialysis in animals performing operant attention tasks (12, 45). Therefore, we also measured ACh release in the medial PFC and motor cortex in animals performing the cued-appetitive response task and treated the amperometric and microdialysis data for analysis so that their temporal resolution was identical and that statistical comparisons could be conducted (48, 51). This analysis substantiated that there was no difference between tonic changes recorded using choline-selective microelectrodes and microdialysis. Furthermore, there were no regional differences. In other words, both methods recorded the same type of cholinergic activity in mPFC as well as motor cortex. Absence of session-related changes in ACh release measured using microdialysis and electrochemistry clearly assists in the interpretation of data on the component of ACh release measured by the former method. In vivo microdialysis studies measure performance-related changes in tonic cholinergic activity, on a scale of minutes (but see (43)).

4 Functional Contributions: Phasic Versus Tonic ACh

The functional contributions of tonic changes in cholinergic activity to performance in relation to phasic ACh release are not clear. On the one hand, we observed a significant positive correlation between tonic signal levels and the amplitudes of cue-evoked cholinergic transients (48). These findings led us to speculate that the level of tonic cholinergic activity modulates the general efficacy of cortical information processing. The amplitude of the cholinergic transients correlated with response latency, suggesting that, indirectly, tonic levels likewise contribute to cue-evoked shifting in attention, processing of response rules, and, generally, cue-evoked behavioral operations. On the other hand, lesions of the cholinergic input to the motor cortex did not affect the animals' behavior (48). It is possible that these session-related, cortex-wide tonic changes in cholinergic activity act in concert with other ascending arousal systems to foster and maintain a general readiness for cortical

Fig. 5 (continued) but they did not differ in magnitude. The different time course of minute-based changes in cholinergic activity in the two regions was reflected by a significant interaction between time and region. Increases in cholinergic activity were not observed in trained animals placed in the test chamber but not allowed to perform. Likewise, following unilateral cholinergic deafferentation of the recording region, session-related increases in cholinergic activity were not observed. The figure was modified from Parikh et al. (2007). Prefrontal acetylcholine release controls cue detection on multiple timescales, *Neuron* 56, 141–54 information. Therefore, the removal of individual contributions of neuronal systems to tonic cholinergic activity may have very limited functional consequences. This view also corresponds with the observation that in order to disrupt attention performance by cortical cholinergic deafferentation, relatively extensive removal of cholinergic inputs from greater parts of the cortex is required to produce robust and persistent impairments in performance (52). Contrary to this, restricted cholinergic deafferentation of the medial PFC produced profound impairments in attentional performance when visual stimuli were presented at a higher event rate (2). We do not yet know what the critical role of medial PFC cholinergic inputs in conditions of high attentional load manifests with respect to phasic or tonic cholinergic activity. Microdialysis experiments revealed that PFC cholinergic activity reflects demands on attentional effort, as opposed to task performance levels (45, 53). However, the interactions between tonic and phasic cholinergic activity in the PFC and elsewhere in the cortex in the mediation of increases in attentional effort remain to be studied.

Absence of changes in session-related tonic cholinergic signal levels between medial PFC and motor cortex (above) suggests that this measure does not appear to be influenced by phasic signals that occurred only in medial PFC. Therefore, tonic cholinergic activity may not be necessarily confounded by phasic activity and these two modes of cholinergic transmission are dissociated (51, 54). Future studies exploring the relative contributions of phasic versus tonic cholinergic changes to attentional capacities require experimental approaches that involve the individual manipulation of cholinergic signals on different timescales.

5 Regulation of Extracellular Choline by Choline Transporters and Electrochemical Detection of Choline Signals

Recycling of choline into the cholinergic synapses is required to sustain ACh synthesis and release. Cholinergic terminals recover choline following ACh hydrolysis by AChE from the extracellular space through a hemicholinium-3 (HC-3)-sensitive high-affinity choline transporter (CHT). Therefore, the transport of choline into presynaptic cholinergic terminals via CHTs is considered as the rate-limiting step for ACh synthesis (55, 56). The capacity for choline transport is regulated by the density of CHTs located in synaptic plasma membrane. Rapid changes in CHT capacity as a result of pharmacological or behavioral manipulations could influence cholinergic transmission (57, 58). Such changes in CHT capacity are caused by altered trafficking of CHTs between intracellular reserve pools and presynaptic plasma membrane (59, 60).

Previous studies primarily based on in vitro synaptosomal uptake assays demonstrated a close relationship between cholinergic neuronal activity and CHT capacity ((61, 62) but see (63, 64)). Therefore, prolonged activation of cholinergic neurons may change the dynamics of extracellular choline flux over time due to increased density of CHTs on presynaptic terminals and enhanced choline uptake. As electrochemical detection of cholinergic transmission using choline-sensitive electrodes relies on increases in extracellular choline from ACh hydrolysis, assessment of ACh release in situations that involve sustained cholinergic firing, e.g., behavioral tasks, may pose some limitation as choline measures could change with time. However, our findings from studies designed to assess CHT function in vivo using these electrodes are not suggestive of a tight coupling between CHT-mediated choline uptake and ACh synthesis (38). We demonstrated that both exogenously applied choline and endogenously generated hydrolyzed choline are cleared by CHTs from the extracellular space in an identical fashion. CHT-driven choline clearance in the absence of cholinergic neuronal activity supports the view that CHTs are partially substrate regulated. Moreover, phasic cholinergic signals evoked by physiological stimulation, e.g., local depolarization of cholinergic terminals or behavior, may rarely exceed 10–15 μM, and thus appear to stay within the limits of CHT capacity (60). Thus, changes in plasma CHT density as a consequence of neuronal activity-driven CHT trafficking do not seem to influence the amplitudes of phasic cholinergic signals assessed using choline-sensitive microelectrodes over time. However, the regulation of CHTs by tonic cholinergic activity and its impact on minute-based changes in choline signals are not known and need to be further investigated.

6 Conclusions

Amperometric techniques employing enzyme-based choline-sensitive microelectrodes are uniquely suited to detect in vivo cholinergic transmission at high temporal and spatial resolution. These electrochemical approaches have now allowed us to understand the role of multiple and dynamic modes of cholinergic transmission in cognition. Moreover, these methodologies could be useful in understanding the cellular regulation of cortical ACh release and attentional functions. In this context, we have recently shown that local application of brief pulses of nAChR agonists generates phasic cholinergic signals in the prefrontal cortex that are similar to transients observed in attention task-performing animals. nAChR-mediated generation of cholinergic transients involves activation of thalamocortical glutamatergic inputs and stimulation of ionotropic glutamate receptors (36, 37, 65). As nAChR agonists are known to enhance cognitive abilities, activation of these receptors may predict enhancement of detection processes and facilitation of attentional functions by augmenting temporally orchestrated cue-evoked cholinergic activity via glutamatergic regulation (65, 66).

The findings that prefrontal cholinergic transients encode specific cognitive operations have major implications in understanding the role of cholinergic dysfunction in the manifestation of cognitive symptoms of neuropsychiatric disorders and age-related dementias. Dysregulated phasic cholinergic transients could disrupt attentional abilities of patients suffering from schizophrenia (67). Abnormalities in the orchestration of phasic cholinergic signaling may precede global and structural decline in cholinergic function and consequently the loss of cholinergic neurons in Alzheimer's disease (68).

As cholinergic signaling constitutes a necessary component of the neuropharmacological mechanisms mediating drug-induced cognition enhancement (69), and procholinergic drugs modulate attentional performance, choline biosensors could potentially be used as a screening tool to identify and characterize procholinergic drugs for the treatment of cognitive disorders. Such a screening strategy would be based on characterizing drugs that specifically amplify performance-evoked phasic cholinergic transients, e.g., nAChR agonists. Similarly, drugs that influence molecular mechanisms to enhance the capacity of cholinergic synapses to increase phasic ACh release, e.g., CHT-mediated choline uptake mechanisms (59, 70), may exert beneficial effects on attentional functions and can be easily screened. Therefore, choline-sensitive microelectrode-based screening approaches that monitor drug-induced modulation of prefrontal cholinergic transients in task-performing animals should be able to predict clinical efficacy of putative cognition enhancers.

Acknowledgements

The authors' research was supported by PHS grants MH080426, MH080322, and AG092592.

References

1. Chiba AA, Bucci DJ, Holland PC, Gallagher M (1995) Basal forebrain cholinergic lesions disrupt increments but not decrements in conditioned stimulus processing. J Neurosci 15:7315–7322
2. Dalley JW, Theobald DE, Bouger P, Chudasama Y, Cardinal RN, Robbins TW (2004) Cortical cholinergic function and deficits in visual attentional performance in rats following 192 IgG-saporin-induced lesions of the medial prefrontal cortex. Cereb Cortex 14:922–932
3. McGaughy J, Dalley JW, Morrison CH, Everitt BJ, Robbins TW (2002) Selective behavioral and neurochemical effects of cholinergic lesions produced by intrabasalis infusions of 192 IgG-saporin on attentional performance in a five-choice serial reaction time task. J Neurosci 22:1905–1913
4. McGaughy J, Everitt BJ, Robbins TW, Sarter M (2000) The role of cortical cholinergic afferent projections in cognition: impact of new selective immunotoxins. Behav Brain Res 115:251–263
5. McGaughy J, Kaiser T, Sarter M (1996) Behavioral vigilance following infusions of 192 IgG-saporin into the basal forebrain: selectivity

of the behavioral impairment and relation to cortical AChE-positive fiber density. Behav Neurosci 110:247–265
6. Muir JL, Dunnett SB, Robbins TW, Everitt BJ (1992) Attentional functions of the forebrain cholinergic systems: effects of intraventricular hemicholinium, physostigmine, basal forebrain lesions and intracortical grafts on a multiple-choice serial reaction time task. Exp Brain Res 89:611–622
7. Phillis JW (1968) Acetylcholine release from the cerebral cortex: its role in cortical arousal. Brain Res 7:378–389
8. Bartolini A, Pepeu G (1967) Investigations into the acetylcholine output from the cerebral cortex of the cat in the presence of hyoscine. Br J Pharmacol Chemother 31:66–73
9. Szerb JC (1967) Cortical acetylcholine release and electroencephalographic arousal. J Physiol 192:329–343
10. Pepeu G, Mantegazzini P (1964) Midbrain hemisection: effect on cortical acetylcholine in the cat. Science 145:1069–1070
11. Westerink BH, Cremers TI (2007) Handbook of microdialysis: methods, applications and perspectives, vol 16. Academic, London
12. Arnold HM, Burk JA, Hodgson EM, Sarter M, Bruno JP (2002) Differential cortical acetylcholine release in rats performing a sustained attention task versus behavioral control tasks that do not explicitly tax attention. Neuroscience 114:451–460
13. Dalley JW, McGaughy J, O'Connell MT, Cardinal RN, Levita L, Robbins TW (2001) Distinct changes in cortical acetylcholine and noradrenaline efflux during contingent and noncontingent performance of a visual attentional task. J Neurosci 21:4908–4914
14. Passetti F, Dalley JW, O'Connell MT, Everitt BJ, Robbins TW (2000) Increased acetylcholine release in the rat medial prefrontal cortex during performance of a visual attentional task. Eur J Neurosci 12:3051–3058
15. Cahill PS, Walker QD, Finnegan JM, Mickelson GE, Travis ER, Wightman RM (1996) Microelectrodes for the measurement of catecholamines in biological systems. Anal Chem 68:3180–3186
16. Michael DJ, Wightman RM (1999) Electrochemical monitoring of biogenic amine neurotransmission in real time. J Pharm Biomed Anal 19:33–46
17. Wightman RM, Robinson DL (2002) Transient changes in mesolimbic dopamine and their association with 'reward'. J Neurochem 82:721–735
18. Phillips PE, Stuber GD, Heien ML, Wightman RM, Carelli RM (2003) Subsecond dopamine release promotes cocaine seeking. Nature 422:614–618
19. Daws LC, Toney GM, Gerhardt GA, Frazer A (1998) In vivo chronoamperometric measures of extracellular serotonin clearance in rat dorsal hippocampus: contribution of serotonin and norepinephrine transporters. J Pharmacol Exp Ther 286:967–976
20. Zahniser NR, Larson GA, Gerhardt GA (1999) In vivo dopamine clearance rate in rat striatum: regulation by extracellular dopamine concentration and dopamine transporter inhibitors. J Pharmacol Exp Ther 289:266–277
21. Staal RG, Mosharov EV, Sulzer D (2004) Dopamine neurons release transmitter via a flickering fusion pore. Nat Neurosci 7:341–346
22. Burmeister JJ, Pomerleau F, Palmer M, Day BK, Huettl P, Gerhardt GA (2002) Improved ceramic-based multisite microelectrode for rapid measurements of L-glutamate in the CNS. J Neurosci Methods 119:163–171
23. Pomerleau F, Day BK, Huettl P, Burmeister JJ, Gerhardt GA (2003) Real time in vivo measures of L-glutamate in the rat central nervous system using ceramic-based multisite microelectrode arrays. Ann NY Acad Sci 1003:454–457
24. Burmeister JJ, Palmer M, Gerhardt GA (2005) L-lactate measures in brain tissue with ceramic-based multisite microelectrodes. Biosens Bioelectron 20:1772–1779
25. Burmeister JJ, Gerhardt GA (2001) Self-referencing ceramic-based multisite microelectrodes for the detection and elimination of interferences from the measurement of L-glutamate and other analytes. Anal Chem 73:1037–1042
26. Burmeister JJ, Moxon K, Gerhardt GA (2000) Ceramic-based multisite microelectrodes for electrochemical recordings. Anal Chem 72:187–192
27. Garguilo MG, Michael AC (1996) Amperometric microsensors for monitoring choline in the extracellular fluid of brain. J Neurosci Methods 70:73–82
28. Oldenziel WH, Dijkstra G, Cremers TI, Westerink BH (2006) In vivo monitoring of extracellular glutamate in the brain with a microsensor. Brain Res 1118:34–42
29. Oldenziel WH, van der Zeyden M, Dijkstra G, Ghijsen WE, Karst H, Cremers TI, Westerink BH (2007) Monitoring extracellular glutamate in hippocampal slices with a microsensor. J Neurosci Methods 160:37–44
30. Day BK, Pomerleau F, Burmeister JJ, Huettl P, Gerhardt GA (2006) Microelectrode array studies of basal and potassium-evoked release of L-glutamate in the anesthetized rat brain. J Neurochem 96:1626–1635
31. Burmeister JJ, Pomerleau F, Huettl P, Gash CR, Werner CE, Bruno JP, Gerhardt GA (2008) Ceramic-based multisite microelectrode

arrays for simultaneous measures of choline and acetylcholine in CNS. Biosens Bioelectron 23:1382–1389
32. Xin Q, Wightman RM (1997) Transport of choline in rat brain slices. Brain Res 776:126–132
33. Garguilo MG, Michael AC (1994) Quantitation of choline in the extracellular fluid of brain tissue with amperometric microsensors. Anal Chem 66:2621–2629
34. Burmeister JJ, Palmer M, Gerhardt GA (2003) Ceramic-based multisite microelectrode array for rapid choline measures in brain tissue. Anal Chim Acta 481:65–74
35. Parikh V, Pomerleau F, Huettl P, Gerhardt GA, Sarter M, Bruno JP (2004) Rapid assessment of in vivo cholinergic transmission by amperometric detection of changes in extracellular choline levels. Eur J Neurosci 20:1545–1554
36. Parikh V, Man K, Decker MW, Sarter M (2008) Glutamatergic contributions to nicotinic acetylcholine receptor agonist-evoked cholinergic transients in the prefrontal cortex. J Neurosci 28:3769–3780
37. Parikh V, Ji J, Decker MW, Sarter M (2010) Prefrontal beta2 subunit-containing and alpha7 nicotinic acetylcholine receptors differentially control glutamatergic and cholinergic signaling. J Neurosci 30:3518–3530
38. Parikh V, Sarter M (2006) Cortical choline transporter function measured in vivo using choline-sensitive microelectrodes: clearance of endogenous and exogenous choline and effects of removal of cholinergic terminals. J Neurochem 97:488–503
39. Rutherford EC, Pomerleau F, Huettl P, Stromberg I, Gerhardt GA (2007) Chronic second-by-second measures of L-glutamate in the central nervous system of freely moving rats. J Neurochem 102:712–722
40. Mitchell KM (2004) Acetylcholine and choline amperometric enzyme sensors characterized in vitro and in vivo. Anal Chem 76:1098–1106
41. Bruno JP, Gash C, Martin B, Zmarowski A, Pomerleau F, Burmeister J, Huettl P, Gerhardt GA (2006) Second-by-second measurement of acetylcholine release in prefrontal cortex. Eur J Neurosci 24:2749–2757
42. Giuliano C, Parikh V, Ward JR, Chiamulera C, Sarter M (2008) Increases in cholinergic neurotransmission measured by using choline-sensitive microelectrodes: enhanced detection by hydrolysis of acetylcholine on recording sites? Neurochem Int 52:1343–1350
43. Sarter M, Parikh V, Howe WM (2009) Phasic acetylcholine release and the volume transmission hypothesis: time to move on. Nat Rev Neurosci 10:383–390
44. Sarter M, Hasselmo ME, Bruno JP, Givens B (2005) Unraveling the attentional functions of cortical cholinergic inputs: interactions between signal-driven and cognitive modulation of signal detection. Brain Res Brain Res Rev 48:98–111
45. Kozak R, Bruno JP, Sarter M (2006) Augmented prefrontal acetylcholine release during challenged attentional performance. Cereb Cortex 16:9–17
46. Himmelheber AM, Sarter M, Bruno JP (1997) Operant performance and cortical acetylcholine release: role of response rate, reward density, and non-contingent stimuli. Brain Res Cogn Brain Res 6:23–36
47. Himmelheber AM, Sarter M, Bruno JP (2000) Increases in cortical acetylcholine release during sustained attention performance in rats. Brain Res Cogn Brain Res 9:313–325
48. Parikh V, Kozak R, Martinez V, Sarter M (2007) Prefrontal acetylcholine release controls cue detection on multiple timescales. Neuron 56:141–154
49. Conner JM, Chiba AA, Tuszynski MH (2005) The basal forebrain cholinergic system is essential for cortical plasticity and functional recovery following brain injury. Neuron 46: 173–179
50. Conner JM, Culberson A, Packowski C, Chiba AA, Tuszynski MH (2003) Lesions of the Basal forebrain cholinergic system impair task acquisition and abolish cortical plasticity associated with motor skill learning. Neuron 38:819–829
51. Parikh V, Sarter M (2008) Cholinergic mediation of attention: contributions of phasic and tonic increases in prefrontal cholinergic activity. Ann NY Acad Sci 1129:225–235
52. McGaughy J, Sarter M (1998) Sustained attention performance in rats with intracortical infusions of 192 IgG-saporin-induced cortical cholinergic deafferentation: effects of physostigmine and FG 7142. Behav Neurosci 112:1519–1525
53. Sarter M, Gehring WJ, Kozak R (2006) More attention must be paid: the neurobiology of attentional effort. Brain Res Rev 51:145–160
54. Hasselmo ME, Sarter M (2011) Modes and models of forebrain cholinergic neuromodulation of cognition. Neuropsychopharmacology 36:52–73
55. Blusztajn JK, Wurtman RJ (1983) Choline and cholinergic neurons. Science 221:614–620
56. Lockman PR, Allen DD (2002) The transport of choline. Drug Devel Ind Pharm 28:749–771
57. Ferguson SM, Bazalakova M, Savchenko V, Tapia JC, Wright J, Blakely RD (2004) Lethal impairment of cholinergic neurotransmission in hemicholinium-3-sensitive choline transporter

knockout mice. Proc Natl Acad Sci USA 101:8762–8767

58. Apparsundaram S, Martinez V, Parikh V, Kozak R, Sarter M (2005) Increased capacity and density of choline transporters situated in synaptic membranes of the right medial prefrontal cortex of attentional task-performing rats. J Neurosci 25:3851–3856
59. Ferguson SM, Savchenko V, Apparsundaram S, Zwick M, Wright J, Heilman CJ, Yi H, Levey AI, Blakely RD (2003) Vesicular localization and activity-dependent trafficking of presynaptic choline transporters. J Neurosci 23:9697–9709
60. Ferguson SM, Blakely RD (2004) The choline transporter resurfaces: new roles for synaptic vesicles? Mol Interv 4:22–37
61. Atweh S, Simon JR, Kuhar MJ (1975) Utilization of sodium-dependent high affinity choline uptake in vitro as a measure of the activity of cholinergic neurons in vivo. Life Sci 17:1535–1544
62. Simon JR, Atweh S, Kuhar MJ (1976) Sodium-dependent high affinity choline uptake: a regulatory step in the synthesis of acetylcholine. J Neurochem 26:909–922
63. Collier B, Ilson D (1977) The effect of preganglionic nerve stimulation on the accumulation of certain analogues of choline by a sympathetic ganglion. J Physiol 264:489–509
64. Kessler PD, Marchbanks RM (1979) Choline transport is not coupled to acetylcholine synthesis. Nature 279:542–544
65. Sarter M, Parikh V, Howe WM (2009) nAChR agonist-induced cognition enhancement: integration of cognitive and neuronal mechanisms. Biochem Pharmacol 78:658–667
66. Howe WM, Ji J, Parikh V, Williams S, Mocaer E, Trocme-Thibierge C, Sarter M (2010) Enhancement of attentional performance by selective stimulation of alpha4beta2(*) nAChRs: underlying cholinergic mechanisms. Neuropsychopharmacology 35:1391–1401
67. Sarter M, Lustig C, Taylor SF (2012) Cholinergic contributions to the cognitive symptoms of schizophrenia and the viability of cholinergic treatments. Neuropharmacology 62(3):1544–53
68. Mesulam M, Shaw P, Mash D, Weintraub S (2004) Cholinergic nucleus basalis tauopathy emerges early in the aging-MCI-AD continuum. Ann Neurol 55:815–828
69. Potter AS, Newhouse PA, Bucci DJ (2006) Central nicotinic cholinergic systems: a role in the cognitive dysfunction in attention-deficit/hyperactivity disorder? Behav Brain Res 175:201–211
70. Sarter M, Parikh V (2005) Choline transporters, cholinergic transmission and cognition. Nat Rev Neurosci 6:48–56

Part III

The Future?

Chapter 13

Carbon Nanotubes-Based Microelectrode (Bio)sensors

Paloma Yáñez-Sedeño, Lourdes Agüí, and José M. Pingarrón

Abstract

The last developments of sensors and biosensors involving microelectrodes modified with carbon nanotubes to detect species with biochemical or biological relevance are reviewed in this chapter. Apart from the different electrode configurations employed, special consideration is given to monitoring of neurotransmitters and the nitric oxide radical. The use of implantable microsensors for glucose detection in cells or tissues, the direct electrochemistry of proteins and some few examples of immunosensors and DNA sensors making use of this kind of modified microelectrodes are also treated. Finally we consider the future perspectives in this field.

Key words Carbon nanotubes, Microelectrodes, Biosensors, Neurotransmitters, Proteins

1 Introduction

Carbon nanotubes (CNTs) have attracted a great interest in different areas of science mainly due to their excellent electric and mechanical properties. In the particular case of electrochemical (bio)sensors, the characteristics of this nanomaterial are profited to enhance the sensitivity of the detection and to promote the electron transfer reactions which is of high relevance when biomolecules are involved. CNTs provide large active surface area which gives rise to enhanced electroanalytical signals and, subsequently, enhanced sensitivity. Moreover, CNTs exhibit remarkable electrocatalytic effects towards numerous electrochemical reactions of biochemical significance which allows using less extreme detection potential thereby minimizing the impact of potential interferences from electroactive substances. This electrocatalytic ability is attributed to the enhanced electron transfer rate occurring at the edge-plane-like graphite sites at the CNT ends (1), and it has been demonstrated that this electrocatalytic activity is dependent on the CNTs fabrication and dispersion procedures. Besides these properties, the presence of CNTs has demonstrated to minimize the problems

Stéphane Marinesco and Nicholas Dale (eds.), *Microelectrode Biosensors*, Neuromethods, vol. 80,
DOI 10.1007/978-1-62703-370-1_13, © Springer Science+Business Media, LLC 2013

associated with the electrode surface fouling produced in the electrochemical oxidation of species typically passivant such as NADH. These characteristics have led to claiming CNTs as an ideal substrate to develop biosensors. In fact, till present, a huge amount of scientific articles regarding the preparation of biosensors using CNTs-modified electrodes have been reported. Most of them use conventional electrochemical devices and techniques. However, the development of in vivo monitoring methods requires the use of small size electrodes and high rate electrochemical techniques. Therefore, a great interest has arisen in last years on the use of microelectrodes modified with CNTs, thus coupling the practical advantages of the nanomaterial commented above with the particular characteristics of microelectrodes. In principle, both single-walled (SWCNTs) and multi-walled (MWCNTs) carbon nanotubes can be used for implementing biosensing devices, although SWCNTs can be metallic or semiconducting depending on their structure, while MWCNTs are conducting materials. However, the electronic properties of perfect MWCNTs are rather similar to those of perfect SWCNTs, because the coupling between the cylinders is weak in MWCNTs (2). Furthermore, chemical modification of CNTs surface is usually needed to generate biocompatibility of the nanomaterial and to modulate the position of attached biomolecules, which enables control on their architecture and functionality. Several strategies are available to generate relatively stable CNTs suspensions including oxidative acid treatments and surfactant- and polymer-assisted dispersion. For example, a very popular methodology is the generation of carboxylic groups on the ends and walls of CNTs by refluxing and sonication in a concentrated H_2SO_4/HNO_3 mixture. The defects caused by this treatment in the CNTs structure provide sites for covalent coupling of biomaterials through the formation of amide and ester bonds. Also, wrapping of CNTs with biopolymers can provide control over the interfacial properties, thus allowing their assembly into architectures suitable for biosensing applications.

This chapter is devoted to make a short review of the methods developed using (bio)sensors based on microelectrodes modified with CNTs for the detection of species with biochemical or biological relevance. Special emphasis is made on in vivo monitoring methods, which are mainly applied to the detection of neurotransmitters, but also some examples of methods that have not been yet used in vivo, but are devoted to the analysis of substances especially relevant from the biochemical point of view, have been considered.

2 Types of Configurations

Most of the electrochemical designs described in the literature make use of microelectrodes prepared with a single carbon fiber. This kind of microelectrodes has been employed for different bioelectrochemical applications since they require small volumes of sample and provide a high signal-to-noise ratio (3). Modification of these microelectrodes is very simple. Good results are obtained with many configurations by simply immersing the active surface of the electrode in a nanotubes suspension (4), or by dropping a small volume of that suspension on the microelectrode surface (5), and allowing evaporation of the solvent to proceed. Besides carbon fibers, Pt microelectrodes chemically etched to obtain a small cavity in which nanotubes are introduced have been also proposed (6), as well as metal wires of W (7) or stainless steel (8) which are modified by procedures such as carbon method deposition (CVD) or electrophoretical deposition (EPD). Also, microelectrodes patterned with insulating layers which were modified with carbon nanotubes using different methods were proposed to be employed as sensors in microfluidics (9).

3 Sensors for Neurotransmitters Monitoring

In vivo neurotransmitters monitoring has been mainly feasible thanks to the use of microelectrodes (10). Since 1976, the small electrode size, the facility of construction, and the fast electrochemical response achieved, together with the high sensitivity of modern electroanalytical techniques, allowed a better understanding of the neurochemical systems (11). Nowadays, the use of electrodes modified with CNTs has improved largely the conditions for the detection of some particularly relevant biological molecules such as the neurotransmitters dopamine, epinephrine, and norepinephrine. The presence of the nanotubes induces a strong electrocatalytic activity towards the redox processes of these compounds. This effect enhances the selectivity detection making possible that species such as ascorbic acid do not interfere. For instance, the electrochemical responses of uric acid, ascorbic acid, and norepinephrine can be observed separately at these modified electrodes (12).

Dopamine is one of the most studied compounds on electrodes modified with CNTs due to its good electrochemical response and biological significance. A practically reversible electrochemical behavior corresponding to the oxidation to dopamine-*o*-quinone and the further reduction of this compound in the reverse scan is observed by cyclic voltammetry. Carbon fiber microelectrodes modified with CNTs simply by immersion in a dispersion of multi-walled carbon nanotubes with Nafion (MWCNTs/CFMEs) showed an excellent electrochemical behavior allowing dopamine to be determined in the presence of ascorbate at physiological pH (4). Moreover, the high conductivity and area-to-volume ratio of CNTs gave rise to a good sensitivity. These facts together with the inherent permselectivity of Nafion make these electrode materials very appropriate for this specific application. MWCNTs/CFMEs were also used for in vivo determinations. For instance, a recent

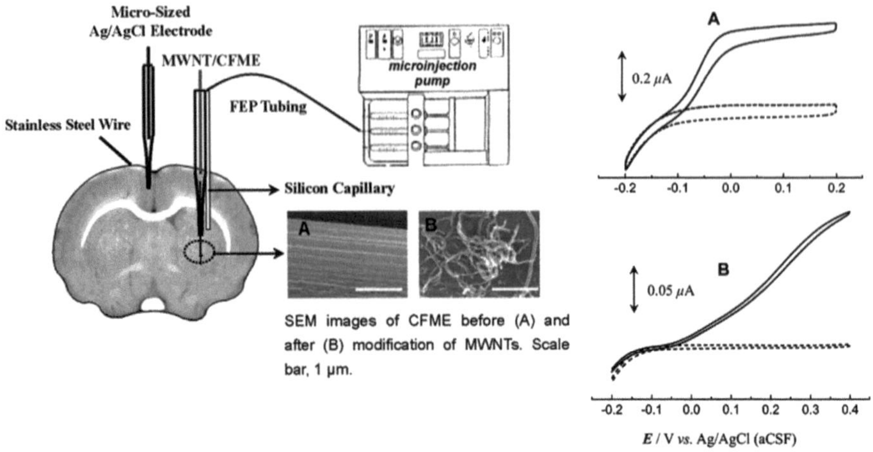

Fig. 1 Schematic illustration of the experimental setup for in vivo voltammetric measurement of striatum ascorbic acid (AA) with a MWCNTs/CFME and for exogenous infusion of standard solutions of AA and AAox into rat striatum through a silicon capillary and FEP tubing pumped with a microinjection pump. : CVs at the MWCNTs modified (**a**) and (**b**) bare carbon fiber microelectrodes in artificial cerebrospinal fluid (pH 7.4) in the absence (·····) and presence (——) of 0.50 mM ascorbic acid. Scan rate, 10 mV/s. Reproduced from (13) with permission from ACS

application to detect ascorbic acid in brain tissue used carbon fiber cylindrical microelectrodes, 500–1,000 μm in length, modified by immersion of the microelectrode in a drop of a MWCNT suspension in *N*,*N*-dimethylformamide. The modified electrodes exhibited a strong electrocatalytic effect towards the oxidation of ascorbic acid as well as a high stability, characteristics that allowed the monitoring of this substance in rat brain with no interference from other present electroactive species such as 3,4-dihydroxyphenylacetic acid, uric acid, or 5-hydroxytryptamine (5-HT). Differential pulse voltammetry was employed for the in vivo observation of homeostatic regulation of striatum ascorbic acid with exogenous infusion of this acid into the brain (Fig. 1). The basal level of striatum AA was determined to be 0.20 ± 0.05 mM ($n = 3$) (13).

Joint determination of dopamine and serotonin is of interest because of the interaction that is produced between both neurotransmitters (14), for example in the cases of schizophrenia (15), and the effect that some antidepressant medicines and abuse drugs such as cocaine exert on both substances (16). Similarly to dopamine, serotonin is also electroactive but its detection is more difficult due to the strong electrode fouling caused by its oxidation products. Nevertheless, the in vivo monitoring of these neurotransmitters has been described by using high scan rate cyclic voltammetry at carbon fiber microelectrodes modified with single-walled carbon nanotubes (SWCNTs). The changes in the concentration of both neurotransmitters were detected simultaneously in the striatum of anesthetized rats upon administration of 5-hydroxytryptamine, a precursor of the serotonin synthesis. The main advantages with respect to the use of bare microelectrodes for this application were the higher sensitivity and the minimization of the electrode fouling as a consequence of serotonin oxidation. So, 90% of the original response for the flow detection of 1 μM serotonin stayed after 25 injections whereas only 60% was maintained at the bare microelectrode.

Serotonin release in the digestive system of intact zebrafish embryos was also monitored by differential pulse voltammetry with implanted CNTs/CFMEs. The use of the modified microelectrodes decreased notably the fouling of the bare carbon fiber and, furthermore, produced an 11% increase in the slope value of serotonin calibration graphs. This allowed serotonin levels of up to 29.9 nM to be measured in vivo under normal physiological conditions. Moreover, the sensor was able to quantify pharmacological alterations of serotonin with PCPA (4-chloro-DL-phenylanaline methyl ester hydrochloride) and reserpine (17).

SWCNTs-based multielectrode arrays (MEA) were also constructed for electrophysiological recordings. Carbon nanotubes were directly deposited onto four, 40 μm diameter Pt electrodes arranged in a quasi matrix distribution, and the device was used to record electrophysiological activity in cultured retinal explants (18).

A recent trend in this area is the preparation of microelectrodes modified with a paste composed of CNTs and ionic liquids (ILs). At these modified microelectrodes, it is possible to observe glucose oxidase direct electron transfer exhibiting a reversible behavior even at high potential scan rates. Also, the CNTs/ILs/CFMEs show an excellent electrochemical performance for the electrocatalytic oxidation of dopamine, ascorbic acid, and NADH (19). The methodology implied the use of the IL 1-butyl-3-methylimidazolium hexafluorophosphate (bmim)PF_6, and the microelectrode was modified by spreading carefully the IL/MWCNTs mixture on the electrode surface. The strong electrocatalytic effect observed allowed the simultaneous detection of dopamine and ascorbic acid using square wave voltammetry with a 200 mV separation between the corresponding oxidation peaks.

On the other hand, indium tin oxide (ITO) electrodes, which possess a high electrical conductivity and good photo-penetrability, have widely demonstrated their usefulness for the detection of biomolecules. In this context, the modification of ITO electrodes with CNTs in a single microfluidic chip allowed the fabrication of an ultra-sensitive electrochemical sensor, which was applied to the detection of dopamine from single rat pheochromocytoma (PC 12) cells. Figure 2 displays a cross-sectional scheme of the microfluidic electrochemical sensor containing the CNTs/ITO microelectrode 4 μm in diameter. Modification with carbon nanotubes was performed from a CNTs suspension in sodium dodecyl sulfate (SDS) spin-coated onto the ITO surface. Monitoring of the single-cell dopamine release was accomplished by depositing an

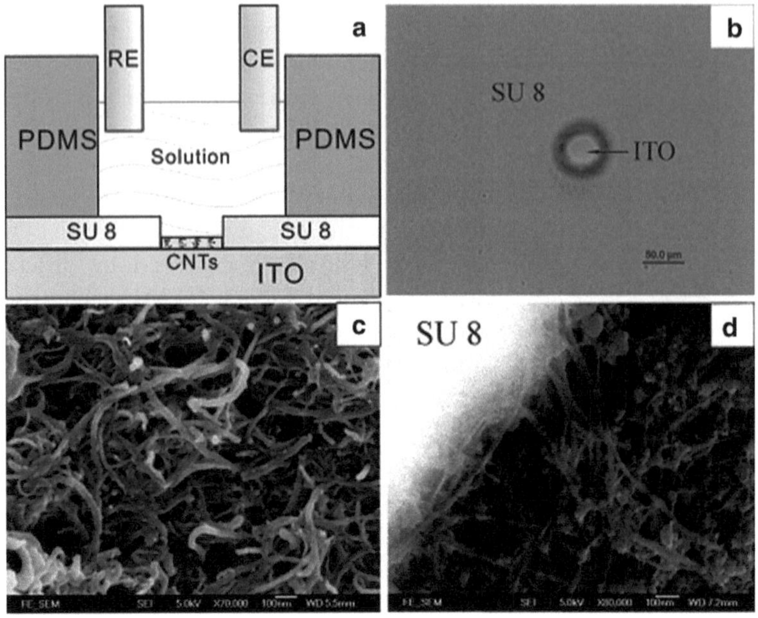

Fig. 2 (**a**) Cross-sectional scheme of a microfluidic electrochemical sensor involving CNTs/ITO microelectrodes. Ag/AgCl acted as reference electrode (RE) and Pt wire as counter electrode (CE). (**b**) Optical photograph of CNTs/ITO microelectrode patterned with SU-8 insulating layer. (**c**) SEM image of CNTs used for ITO surface modification. (**d**) SEM image of the patterned CNTs/ITO microelectrode. Reproduced from (7) with permission from Elsevier

aliquot of PC 12 cells in the PDMS chamber positioned over the modified ITO microelectrode. When the cell was adhered to the electrode, potassium ion was injected with a micropipette and dopamine release was detected amperometrically using a potential of 0.3 V vs. Ag/AgCl. The chip demonstrated to provide high sensitivity, the modification with CNTs improving it by two to three orders of magnitude. The obtained results validated the developed configuration for monitoring of biological exocytotic phenomena (9).

Stainless steel microelectrodes were also modified with SWCNTs using electrophoretical deposition (EPD), and they were applied to the detection of epinephrine in the presence of ascorbic acid. The modification procedure was carried out in an electrochemical cell by applying a DC voltage of 7 V/cm between the stainless steel microelectrode and the reference electrode immersed in the nanotube suspension. SWCNTs were purified and treated with concentrated KOH solutions in order to produce electron–donor groups as –OH- on the nanotube walls. The electrostatic barrier formed improved the selectivity of the microsensor by repelling negatively charged molecules such as ascorbic acid or uric acid, and attracting positively charged compounds such as epinephrine.

Therefore, the modified microelectrodes exhibited a high sensitivity towards the electrochemical detection of epinephrine due to the significant enhancement of the corresponding electrochemical oxidation response. A wide linear range (2–100 μM) and a detection limit of 2 μM for epinephrine, a response time of 6 s, an operational stability of 13 h under continuous working conditions and a long-term stability of 1 month are the main analytical characteristics of these stainless steel-modified microelectrodes (8).

Carbon vapor deposition was employed to incorporate SWCNTs onto W wires, which were previously modified with Fe nanoparticles. Using this procedure, microelectrodes with oriented bundles of carbon nanotubes that showed a good performance towards neurotransmitters were obtained. These devices offered a combination of wide potential window, low background current, and high stability. The limits of detection obtained by cyclic voltammetry for dopamine, epinephrine, and serotonin were 7 nM, 32 nM, and 50 nM, respectively (7).

Carbon fiber nanoelectrodes (CFNEs) modified with SWCNTs were also fabricated. The tip of the carbon fiber was etched slowly in the bottom of a flame to form CFNEs with tip dimensions from 100 to 300 nm. The so prepared nanoelectrode was subsequently immersed in a suspension of SWCNTs in SDS. The SWCNTs/CFNEs exhibited a high sensitivity with low detection limits for dopamine, epinephrine, and norepinephrine in spite of the nanometer scale of the modified electrode. This is attributed to the large available area of the modified electrode and the high conductivity of SWCNTs (20).

4 Detection of Nitric Oxide

Nitric oxide (NO) plays various relevant functions in living organisms. It has an important role in various physiological processes affecting neurotransmission and immune response, as well as participating as a mediator in various antimicrobial and antitumor activities (21, 22). The detection of NO in biological systems is not easy due to its instability, since NO, which is a free radical, reacts quickly with oxygen, peroxides, metals, and other species. This makes that the physiological concentrations are in the nanomolar range. Microelectrodes are probably the unique analytical tools to accomplish this kind of determination, and various configurations have been applied to monitoring NO release from cells in different tissues. This suitability is mainly due to the fact that microelectrodes provide a rapid and sensitive response, and their small size allows implantation in the cell proximity. The electrochemical detection of NO is based on its oxidation reaction because the reduction reaction usually suffers from interference by oxygen. Nevertheless, the anodic response can also be affected by the presence of other substances in the biological medium such as nitrite anion, ascorbic acid, or dopamine among others. In order to increase the sensitivity and selectivity of the detection, microelectrodes modified with different materials, mainly polymers, were used. In particular, Nafion avoids the interference of different compounds in anionic form at the working pH. Microelectrodes modification with CNTs is a novel strategy insufficiently explored yet for this kind of detection. Few examples of application of CNTs-modified microelectrodes for NO detection are available in the literature, although it is known that the presence of CNTs provides high stability and strong catalytic effect towards the electrochemical oxidation of nitric oxide (23). So, a carbon fiber cylindrical microelectrode modified with MWCNTs dispersed in dihexadecyl hydrogen phosphate (DHP), and coated later with Nafion, was used for the NO detection in fish liver mitochondria. Figure 3 displays SEM images of the bare and modified carbon fiber microelectrode as well as an amperogram showing the incremental electrochemical response produced by L-Arg addition corresponding to the release of NO from mitochondria (5). The stability of the electrode for in vivo experiments was high, being possible to be used continuously for 8 h. Furthermore, dopamine, ascorbic acid, and nitrite did not interfere when were present at concentrations slightly higher than

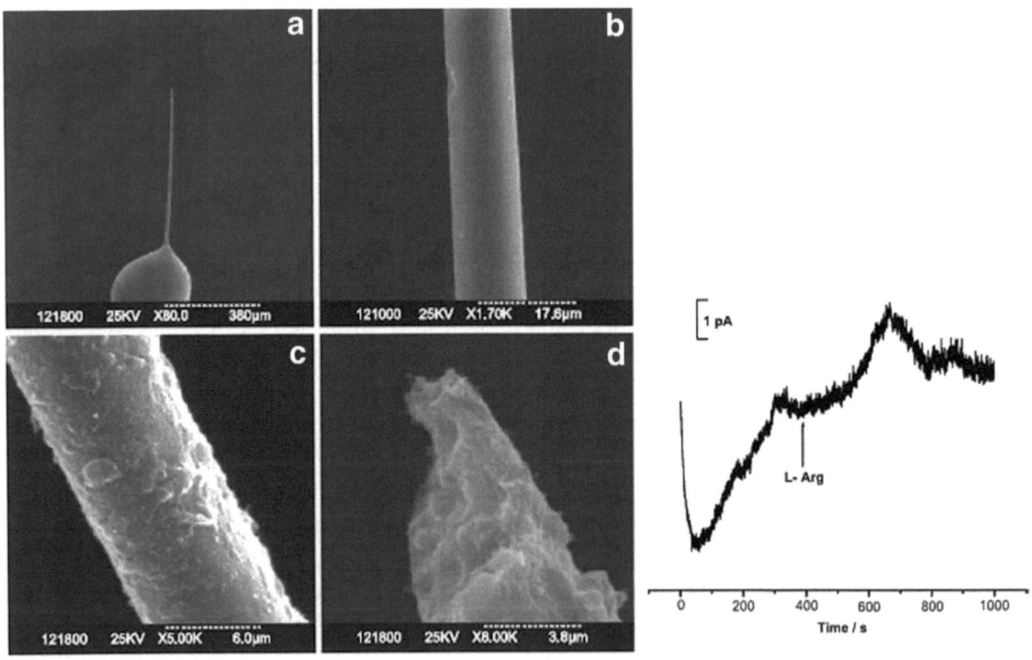

Fig. 3 (**A**) SEM images of the general view of CFME (**a**); the midsection bare CFME (**b**); the midsection of MWCNTs/CFME (**c**); the tip and cross section of MWCNTs/CFME (**d**); (**B**) Amperogram showing NO release from fish liver mitochondria without L-Arg and adding 0.1 mM L-Arg measured with the Nafion/MWCNTs/CFME; applied potential, 800 mV. Reproduced from (3) with permission from Elsevier

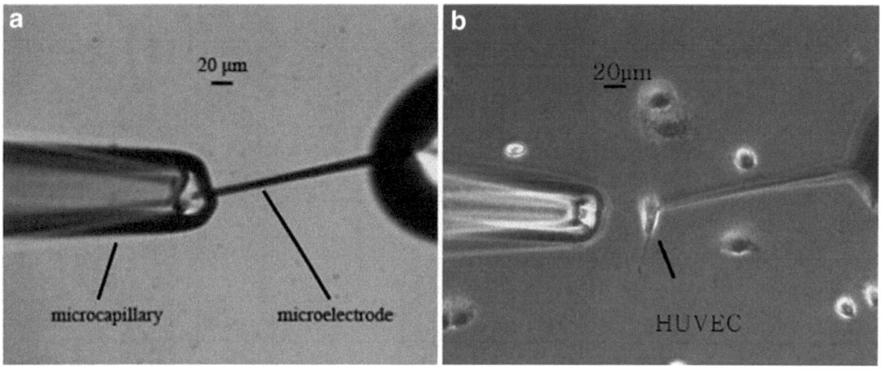

Fig. 4 (**a**) Calibration system to calibrate the carbon fiber microdisk electrode modified with SWCNTs and Nafion with NO solution of different concentrations. (**b**) Detection system in single cell experiments: the outlet of the microcapillary rightly faces the cell to inject stimulus towards it. Reproduced from (24) with permission from Elsevier

that of NO. Dopamine at a concentration slightly lower did not interfere either.

Another sensitive method for nitric oxide was also developed by employing an electrochemical microsensor consisting of a carbon fiber microdisk electrode modified with SWCNTs and Nafion (24). The method was applied to the determination of NO release from single isolated human umbilical vein endothelial cells (HUVECs). The presence of SWCNTs caused a great improvement in sensitivity, achieving a detection limit of 4.3 nM. Figure 4 shows the microcapillary and the microelectrode used both placed in aerated PBS solution (1) to calibrate with NO solution of different concentrations and (2) in a single cell experiment injecting stimulus towards the microcapillary.

SWCNTs and room temperature ionic liquid (RTIL) were used to prepare a gel microelectrode which was applied to study the NO oxidation. The gel composite was prepared by mixing carboxylated SWCNTs with the ionic liquid 1-hexyl-3-methylimidazolium hexafluorophosphate ($HMIMPF_6$) and inserted into the microcavity resulting from the etching of a 50 μm platinum wire. The external microdisk surface and the thin layer effect produced by the porous electrode material both contribute to the good performance of the microelectrode giving a linear concentration relationship ranging from 100 to 100 μM (25).

5 Enzyme Biosensors

Glucose monitoring in cells or tissues is a matter of concern in different biochemical investigations. The use of implantable microsensors makes it possible to face this significant analytical challenge. However, the responses obtained from miniaturized

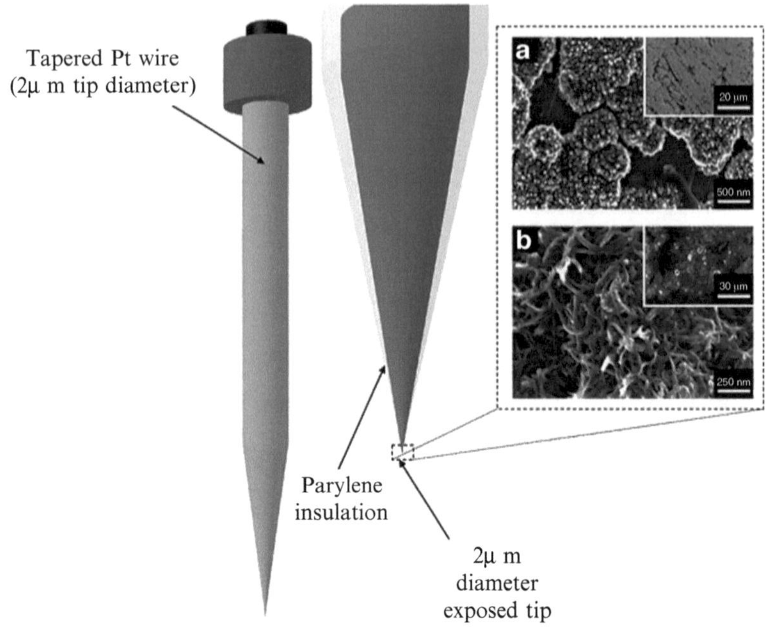

Fig. 5 Scheme of the glucose oxidase biosensor using a Pt microelectrode and parylene insulation. Inset (**a**) amorphous Pt black nanoclusters electrodeposited on the surface of the Pt microelectrode, (**b**) MWNT/Nafion/GOx matrix deposited on the Pt black microelectrode tip. Reproduced from (26) with permission from Elsevier

sensors used in physiological studies show often a relatively high noise level that becomes worse by the drift in concentration produced by the metabolic activity of cells. With the aim of minimizing this problem, microsensors able to filter the electronic noise and to provide directly the measurement of the analyte flow through the cell membrane or the tissue have been designed recently. These microsensors are called self-referencing. An interesting example of this kind of microsensors is the one based on the deposition of platinum black, MWCNTs, Nafion, and the enzyme glucose oxidase on a Pt/Ir microelectrode (see Fig. 5). This biosensor exhibited a high sensitivity and selectivity and allowed quantification of the physiological glucose flux in tumourogenic breast endothelial cells and pharmacological glucose transport inhibition in pancreatic cells (26).

6 Direct Electrochemistry of Proteins

The direct electrochemistry of enzymes at electrode surfaces has attracted more and more attention in the last years due to the significance of these processes to investigate electron transfer mechanisms in live organisms which are responsible for the energy transformations and metabolic actions (6). However, it should be mentioned that direct electron transfer between the redox center of enzymes and electrodes is usually prevented because the centers are buried into the complex proteic structure which impedes adequate contact with the electrode surface. The use of electrodes modified with CNTs has facilitated the solution to this problem. Regarding the specific case of microelectrodes, some configurations of this type have been reported. For instance, direct electrochemistry of horseradish peroxidase (HRP) was observed at a Pt microelectrode (PtME) modified with the HRP/CNTs mixture (6). The 76 μm Pt microelectrode was etched to form a cavity of 10 μm deep. A mixture of CNTs powder and HRP was prepared, and the etched tip was ground together with the dried mixture on a glass slide, until the micro cavity was filled. Cyclic voltammograms of HRP/CNTs/PtME showed a pair of nearly reversible peaks which corresponded to the Fe^{3+}/Fe^{2+} couple in HRP. The observation of direct electron transfer was attributed to the special nano-structure of CNTs that can act as a molecular wire lead electrons to the redox centers of HRP. The same authors also demonstrated the existence of direct electron transfer between hemoglobin (Hb) and CNTs powder-modified microelectrodes prepared by mixing CNTs with acetate buffer and Hb (27). Direct electrochemistry of microperoxidase (MP-11) was also found at Pt microelectrodes modified with MWCNTs. In this case, a 10-μm radius Pt wire was coated by dropping a suspension of MWCNTs in DMF and the modified microelectrode was immersed in MP-11 solution. The redox active center in MP-11 is an iron protoporphyrin IX not shielded by a polypeptide. Thus, electron transfer can easily be achieved under appropriate conditions (28). MP-11/MWCNTs/Pt microelectrodes showed an intense electrocatalytic activity towards oxygen reduction.

7 Immunosensors

Gold microelectrodes with a surface area of 9×10^{-4} cm^2 were used to develop impedimetric immunosensors for detecting deep venous thrombosis biomarker (D-dimer), an essential element in the diagnostics of deep vein thrombosis and pulmonary embolism. The microelectrodes were fabricated using silicon technology and modified with an 11-amino-1-undecanethiol self-assembled monolayer where carboxylated single-walled carbon nanotubes (SWCNT-COOH) were covalently attached. Then, the antibody (anti-D-dimer) was immobilized by covalent bonding on the functionalized SWCNTs. The interaction of D-dimer with the biofunctionalized surface produced a decrease in the electron transfer resistance which was attributed to the increase of the negative charge of the immunocomplex which would induce an increase in CNT conductivity. The impedimetric microimmunosensors allowed a detection limit of 0.1 pg/mL (0.53 fM) to be obtained, which is four thousand times lower that that obtained using a polypyrrole-modified microelectrode (29).

PSA is an androgen-regulated serine protease existing in numerous forms. Total PSA (T-PSA) level significantly increases in serum during prostatic diseases, whose diagnosis has been dramatically improved by the monitoring of this antigen. A label-free electrochemical immunosensor using microelectrode arrays modified with SWCNTs were prepared for the determination of T-PSA. The method was based on the measurement of the intrinsic oxidation signal of the electroactive amino acid residues, tyrosine and tryptophan, contained in T-PSA, which increases with the interaction of T-PSA and T-PSA-mAb. The detection limit achieved using differential pulse voltammetry was 0.25 ng/mL (30).

8 DNA Sensors

A platinum interdigitated microelectrode was fabricated using a conventional photolithographic method with a finger width of 10 μm and a gap of 10 μm. Carboxylated MWCNTs were dispersed in ethylenediamine to form amine groups and the MWCNTs dispersion was drop-coated on the Pt interdigitated electrodes. Finally, the DNA probe was immobilized by means of covalent binding. The modified bioelectrode was used for direct and label-free detection of influenza virus (type A). The hybridization of the DNA probe and the target DNA were detected by measuring changes in the conductance on the surface of sensors leading to the change in the output signal of the system. The sensor response originated from conductivity changes from single- to double-stranded DNA molecules, which are inter-linking the CNTs cushion on top of the microelectrode structure. This DNA sensor was able to detect as low as 0.5 nM of the target DNA samples (31).

9 Conclusions and Future Remarks

The ability that microelectrodes modified with CNTs possess for the sensitive and selective detection of different molecules in complex environments together with the possibility of applying fast electrochemical techniques can be profited for a better understanding of some biochemical processes in living organisms. In particular, significant advances have been made in the field of the neurotransmission and the effects of pharmacological substances on neurotransmitters release. Possibly, it would be necessary to develop microelectrodes with smaller dimensions to achieve a better adaptation to the studied systems. A promising example belongs to the so-called nanoneedle electrodes, where, for instance, SWCNTs extending from the end of the electrode acts as nanoneedle capable of electrochemical measurements (32). Furthermore, these capabilities should be extended to the in vivo detection of other molecules. In this context, the miniaturization of the glucose biosensor and other enzyme-based biosensors to be implanted for continuous monitoring of relevant analytes is still a challenge that needs to be overcome. Related to this, an important subject to be studied in the next years is the possible toxicity and adverse effects of the implantable devices based on electrodes modified with CNTs.

Acknowledgements

Financial support from the Spanish Ministerio de Ciencia e Innovación (Projects DPS2008-07005-C02-01, CTQ2012-34238 and CTQ2012-35041) and Comunidad de Madrid S2009/PPQ-1642, programme AVANSENS, is gratefully acknowledged.

References

1. Liu G, Riechers SL, Mellen MC, Lin Y (2005) Sensitive electrochemical detection of enzymatically generated thiocholine at carbon nanotube modified glassy carbon electrode. Electrochem Commun 7:1163
2. Sun X, Li C-Z (2007) Fundamental aspects and applications of nanotubes and nanowires for biosensors. In: Knopf GK, Bassi AS (eds) Smart biosensor technology. CRC, Boca Raton, FL
3. Amatore C, Maisonhaute E (2005) When voltammetry reaches nanoseconds. Anal Chem 77:303A–311A
4. Ho evar SB, Wang J, Deo RP, Musameh M, Ogorevc B (2005) Carbon nanotube modified microelectrode for enhanced voltammetric detection of dopamine in the presence of ascorbate. Electroanalysis 17:417–422
5. Wang Y, Li Q, Hu S (2005) A multiwall carbon nanotubes film-modified carbon fiber ultramicroelectrode for the determination of nitric oxide radical in liver mitochondria. Bioelectrochemistry 65:135–142
6. Zhao YD, Zhang WD, Chen H, Luo QM, Li SFY (2002) Direct electrochemistry of horseradish peroxidase at carbon nanotube powder microelectrode. Sens Actuators B 87:168–172
7. Valentini F, Orlanducci S, Tamburri E, Terranova ML, Curulli A, Palleschi G (2005) Single-walled carbon nanotubes on tungsten wires: a new class of microelectrochemical sensors. Electroanalysis 17:28–37
8. Valentini F, Palleschi G, Lopez Morales E, Orlanducci S, Tamburri E, Terranova ML (2007) Functionalized single-walled carbon nanotubes modified microsensors for the selective response of epinephrine in presence of ascorbic acid. Electroanalysis 19:859–869
9. Shi B-X, Wang Y, Zhang K, Lam T-L, Chan HL-W (2011) Monitoring of dopamine release in single cell using ultrasensitive ITO microsensors modified with carbon nanotubes. Biosens Bioelectron 26:2917–2921
10. Troyer KP, Heien MLAV, Venton BJ, Wightman RM (2002) Neurochemistry and electroanalytical probes. Curr Opin Chem Biol 6:696–703
11. Adams RN (1976) Probing brain chemistry with electroanalytical techniques. Anal Chem 48:1126A–1138A
12. Wang J, Li M, Shi Z, Li N, Gu Z (2002) Investigation of the electrocatalytic behavior of single-wall carbon nanotube films on an Au electrode. Microchem J 73:325–333
13. Zhang M, Liu K, Xiang L, Lin Y, Su L, Mao L (2007) Carbon nanotube-modified carbon fiber microelectrodes for *in vivo* voltammetric measurement of ascorbic acid in rat brain. Anal Chem 79:6559–6565
14. Kumara Swamy BE, Jill Venton B (2007) Carbon nanotube-modified microelectrodes for simultaneous detection of dopamine and serotonin in vivo. Analyst 132:876–884
15. Kapur S, Remington G (1996) Serotonin-dopamine interaction and its relevance to schizophrenia. Am J Psychiatry 153:466–476
16. Elliott JM, Beveridge TJR (2005) Psychostimulants and monoamine transporters: upsetting the balance. Curr Opin Pharmacol 5:94–100

17. Njagi J, Ball M, Best M, Wallace KN, Andreescu S (2010) Electrochemical quantification of serotonin in the live embryonic zebrafish intestine. Anal Chem 82:1822–1830
18. Gabriel G, Gómez R, Bongard M, Benito N, Fernández E, Villa R (2009) Easily made single-walled carbon nanotube surface microelectrodes for neuronal applications. Biosens Bioelectron 24:1942–1948
19. Liu Y, Zou X, Dong S (2006) Electrochemical characteristics of facile prepared carbon nanotubes-ionic liquid gel modified microelectrode and application in biochemistry. Electrochem Commun 8:1429–1434
20. Chen RS, Huang WH, Tong H, Wang ZL, Cheng JK (2003) Carbon fiber nanoelectrodes modified by single-walled carbon nanotubes. Anal Chem 75:6341–6345
21. Bedoui F, Villeneuve N (2003) Electrochemical nitric oxide sensors for biological samples—principle, selected examples and applications. Electroanalysis 15:5–18
22. Lincoln J, Hoyle CH, Burnstock G (1997) Nitric oxide in health and disease. Cambridge University Press, Cambridge, p 83
23. Wu F-H, Zhao G-C, Wei X-W (2002) Electrocatalytic oxidation of nitric oxide at multi-walled carbon nanotubes modified electrode. Electrochem Commun 4:690–694
24. Du F, Huang W, Shi Y, Wang Z, Cheng J (2008) Real-time monitoring of NO release from single cells using carbon fiber microdisk electrodes modified with single-walled carbon nanotubes. Biosens Bioelectron 24:415–421
25. Li CM, Zang J, Zhan D, Chen W, Sun CQ, Teo AL, Chua YT, Lee VS, Moochhala SM (2006) Electrochemical detection of nitric oxide on a SWCNT/RTIL composite gel microelectrode. Electroanalysis 18:713–718
26. McLamore ES, Shia J, Jaroch D, Claussen JC, Uchida A, Jiang Y, Zhang W, Donkin SS, Banks MK, Buhmane KK, Teegardene D, Rickusa JL, Porterfield DM (2011) A self referencing platinum nanoparticle decorated enzyme-based microbiosensor for real time measurement of physiological glucose transport. Biosens Bioelectron 26:2237–2245
27. Zhao Y-D, Bi Y-H, Zhang W-D, Luo Q-M (2005) The interface behavior of hemoglobin at carbon nanotube and the detection for H_2O_2. Talanta 65:489–494
28. Wang M, Zhao F, Liu Y, Dong S (2005) Direct electrochemistry of microperoxidase at Pt microelectrodes modified with carbon nanotubes. Biosens Bioelectron 21:159–166
29. Bourigua S, Hnaien M, Bessueille F, Lagarde F, Dzyadevych S, Maaref A, Bausells J, Errachid A, Jaffrezic Renault N (2010) Impedimetric immunosensor based on SWCNT-COOH modified gold microelectrodes for label-free detection of deep venous thrombosis biomarker. Biosens Bioelectron 26:1278–1282
30. Okuno J, Maehashi K, Kerman K, Takamura Y, Matsumoto K, Tamiya E (2007) Label-free immunosensor for prostate-specific antigen based on single-walled carbon nanotube array-modified microelectrodes. Biosens Bioelectron 22:2377–2381
31. Tam PD, Hieu NV, Chien ND, Le AT, Tuan MA (2009) DNA sensor development based on multi-wall carbon nanotubes for label-free influenza virus (type A) detection. J Immunol Methods 350:118–124
32. Jacobs CB, Peairs MJ, Venton BJ (2010) Carbon nanotube based electrochemical sensors for biomolecules. Anal Chim Acta 662: 105–127

Chapter 14

Label-Free Affinity Biosensors Based on Electrochemical Impedance Spectroscopy

Nicole Jaffrezic-Renault

Abstract

There is an increasing interest in Electrochemical Impedance Spectroscopy (EIS) as the transducing method for label-free biosensors. This technique allows the direct detection of the affinity complex formation through the variation of the capacitance or that of the charge transfer resistance at the biofunctionalized electrode/biological liquid interface.

This chapter will present the physical principles of EIS, the different designs of the micro/nanoelectrodes, the new ways for the elaboration of the recognition biofilm, the performances of detection of the different biotargets, and the perspectives for the in situ detection.

Key words Impedimetric biosensors, Affinity biosensors, Enzymatic biosensors, Cell sensors, Charge transfer resistance, Capacitance, Interdigitated electrodes

1 Introduction

In analytical science, the term biosensor is used when the biochemical recognition event is integrated with the signal transduction into a single device. Thus, an electrochemical biosensor represents some kind of "bioelectronic device" which is based on the interaction of biomolecules with a conductive (or semiconductive) support and the electronic transduction of a specific biological reaction associated with the biological function. The biological components might be proteins, including enzymes, receptors, and antibodies, nucleic acids, or low-molecular-weight molecules such as cofactors, inhibitors, or antigens. More recently, artificial recognition elements, such as molecular imprinted polymers (MIPs), peptide nucleic acids (PNA), aptamers, abzymes, and so on, have been developed for sensor applications. The detection process involves the formation of a recognition complex between the sensing biomolecule and the analyte at the interface of the electronic transducer

Stéphane Marinesco and Nicholas Dale (eds.), *Microelectrode Biosensors*, Neuromethods, vol. 80,
DOI 10.1007/978-1-62703-370-1_14, © Springer Science+Business Media, LLC 2013

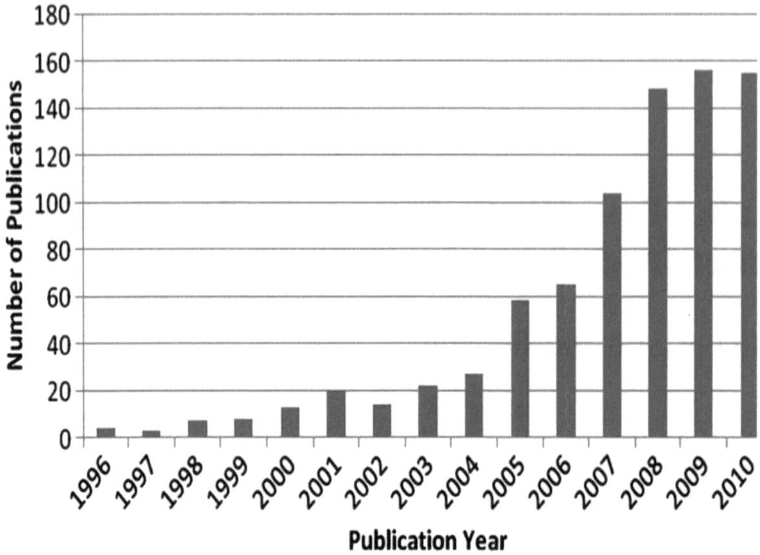

Fig. 1 Number of publications on impedance spectroscopy and biosensors since 1996, corresponding to bibliographic searches using the keyword combination "impedance spectroscopy AND biosensor" with ISI Web of Science

which directly or indirectly alters the electrical properties of the recognition surface. Electrochemical biosensors for direct and often label-free measurement of various analytes are especially attractive because of their fast performance and less demanding instrumentation. In this context, impedance spectroscopy is perfectly qualified for the readout of biosensors since it provides detailed information on the resistive and capacitive properties of the recognition surface. Because of the affordable availability of impedance instrumentation nowadays, a trend towards the development of impedimetric biosensors has currently been observed. This can be seen from the increasing number of publications on this topic over the past few years (cf. Fig. 1).

Impedance spectroscopy has been applied in bioanalytics to monitor both the fabrication of biosensors and the biosensor recognition process. Applications have been demonstrated for all types of (bio)analytes such as proteins, nucleic acids, whole cells, microorganisms, antibodies, and antigens. A detailed and complete description of the different approaches is difficult; thus, the present chapter tries to give an overview on the various strategies that have been attempted to design impedimetric biosensors without claiming to be complete. The presentation is focused on general concepts and some selected examples and discusses novel designs which were published.

2 Principles

The impedance Z of a system is determined by applying a voltage perturbation of small amplitude and detecting the current response. From the definition impedance is the quotient of the voltage–time function $V(t)$ and the resulting current–time function $I(t)$:

$$Z = \frac{V(t)}{I(t)} = \frac{V_0}{I_0}\frac{\sin(2\pi ft)}{\sin(2\pi ft + \varphi)} = \frac{1}{\Upsilon}. \tag{1}$$

In analogy to the conductance definition in a dc circuit one can define the admittance Υ as the reciprocal impedance or complex conductance. The impedance is a complex value since the current can not only differ in the amplitude (as in the case for a pure ohmic resistance) but can also show a phase shift ϕ compared to the voltage–time function (as in the case of capacitive or inductive resistances). Thus, the value can be described either by the modulus $|Z|$ and the phase shift ϕ or alternatively by the real part Z_R and the imaginary part Z_I of the impedance. This is illustrated in Fig. 2. Therefore the result of an impedance measurement can be illustrated in two different ways, the Bode plot using $|Z|$ and ϕ as a function of log f or the Nyquist plot using Z_R and Z_I.

The name impedance "spectroscopy" is derived from the fact that usually not only a single frequency is measured but impedance is determined at different frequencies. Thus, an impedance spectrum is obtained allowing the characterization of surfaces, layers, or membranes as well as exchange and diffusion processes. For this purpose the impedance spectrum is often analyzed by means of an equivalent circuit. The circuit, commonly consisting of resistances and capacitances, represents the different physicochemical properties of the system under investigation (1–3). Alternatively the description of the system can be performed on the basis of transfer functions, which are derived from the basic laws of the processes involved such as electrokinetics, diffusion, partition, etc.

However, it is not only possible to describe a system, but the measurement can be also used for analytical purposes. In this case the change of one impedance element, a resistance or a capacitance, depending on the solution composition is evaluated. In some cases it is also suitable to correlate the overall impedance to a concentration change. This can result in a more simple measurement since here it is often sufficient to determine the impedance at one selected frequency or within a limited frequency window (where the relative changes are largest).

2.1 Impedance Elements and Simple Circuits

In electrochemical impedance spectroscopy, where the electrolyte solution is one component of the system to be investigated, usually four elements are used for the description of the impedance behavior:

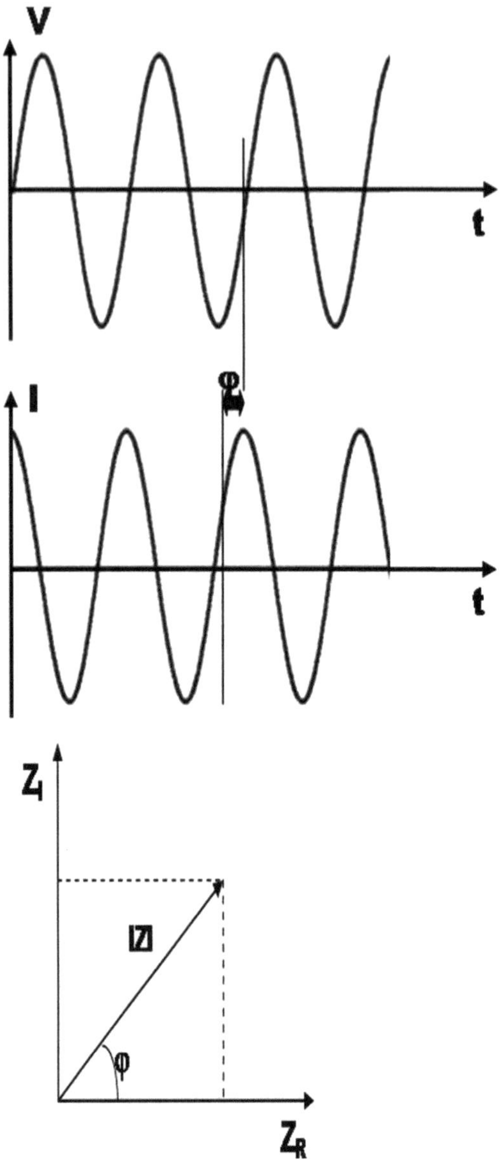

Fig. 2 Impedance is a complex value and defined as quotient of the voltage (time) and current (time) function. It can be expressed as modulus |*Z*| and phase angle ϕ or can be given by the real part (Z_R) and the imaginary part (Z_I) of impedance

ohmic resistance, capacitance, constant phase element, and Warburg impedance. These elements and their definitions are summarized in Table 1.

An ohmic resistance *R* appears in the impedance spectrum when charge carriers are transported within the bulk of a material, through thin layers or across interfaces. An ohmic resistance causes no phase shift between the ac voltage and the ac current and the impedance value is independent of the frequency used for

Table 1
Impedance definition, frequency dependence, and phase shift of different elements most often used for the description of the (bio)electrochemical systems

Impedance element	Definition	Phase angle	Frequency dependence
R	$Z = R$	0°	No
C	$Z = \frac{I}{j\omega C}$	90°	Yes
CPE	$Z = \frac{1}{Aj\omega^{\alpha}}$	0–90°	Yes
W (semi-finite)	$Z = \frac{\sigma}{\sqrt{\omega}}(1-j)$ $\sigma = \frac{RT}{n^2F^2\sqrt{2}}\left(\frac{I}{\sqrt{D_O c_O}} + \frac{1}{\sqrt{D_R c_R}}\right)$	45°	Yes

ω angular frequency
D diffusion coefficient
A has the numerical value of the admittance $(1/|Z|)$ at $\omega = 1$ rad/s
σ Warburg coefficient
c_O, c_R concentration of oxidized and reduced species

the measurement. This means that the impedance is constant over the whole frequency range.

A capacitance C appears when charge carriers are separated at interfaces (e.g., electrochemical double layer) or at membranes or films. In a simplified model this can be interpreted as a plate capacitor where the two "plates" are separated by a nonconducting medium. A capacitance results in a phase shift of −90° of the current–time function compared to the applied voltage function and the impedance is strongly dependent on the frequency of measurement (as also seen from the impedance definition in Table 1). Particularly at low frequencies the capacitive impedance becomes very high.

Besides these two "basic" elements, so-called distributed impedance elements have often to be included for an appropriate description of the impedance behavior of a real system. A constant phase element (CPE) is used when an ideal capacitive behavior with a phase shift of −90° is not observed. In electrical engineering, this corresponds to the behavior of a "loss capacitor." The physicochemical background can be of different origin but is very often the heterogeneity of the surface (e.g., microroughness) or the heterogeneity of the bulk material (e.g., water content in a polymeric membrane). The degree of frequency dispersion is indicated by the α-value: When α is 1, a pure capacitive behavior is observed, with decreasing α, the properties show also resistive behavior and when α is 0 a pure ohmic resistance appears.

A second distributed impedance element often used in equivalent circuits is the Warburg impedance *W*. Here, the system is under diffusion control; this means that the diffusion of substances is limiting the current. This impedance element is normally relevant at low frequencies and results in a linear behavior with a slope of 45° in the Nyquist plot. From this impedance behavior the diffusion coefficient of the relevant species can be calculated (see also Table 1). However, the so-called semi-infinite Warburg impedance can only be observed when the region available for diffusion is not limiting.

The individual impedance elements can be connected in two ways: in series and in parallel. A series circuit, including one resistance and one capacitance, means that the overall impedance is the sum of the individual elements. Since the capacitive impedance is frequency dependent, the whole impedance is also dependent on the frequency,

$$Z = R + \frac{I}{j\omega C}, \tag{2}$$

j being the imaginary unit ($j^2 = -1$).

In a parallel circuit, including one resistance and one capacitance, the current has two ways to flow through the system and will preferably choose the way of low impedance. Here the overall impedance is the sum of the reciprocal impedance of the individual elements (or, in other words, the reciprocal impedance is the sum of the individual admittances).

$$\frac{1}{Z} = \frac{1}{R} + j\omega C. \tag{3}$$

2.2 Equivalent Circuits

Equivalent circuits are used in order to approximate the experimental impedance data with the above-mentioned ideal or distributed impedance elements, both arranged in series and/or in parallel. Many electrochemical systems have been analyzed according to this procedure. For an individual system under investigation one can probably find models in the literature for the description of the impedance behavior—at least as a starting point for the analysis.

2.2.1 Electrode in Electrolyte Contact

When an electrode is immersed into an electrolyte solution the ac current has to flow through the solution and has to pass the interface between the electron- and ion-conducting phase. The equivalent circuit for this situation is shown in Fig. 3 and is often referred to as Randles circuit. According to the electrochemical theory a current flow is possible by charging and discharging the double layer at the interface or by oxidizing/reducing substances from the electrode or in solution. Subsequently, the model of the interface is a parallel circuit of the double layer capacitance C_{dl} and the charge transfer resistance R_{ct}. At low frequencies where the

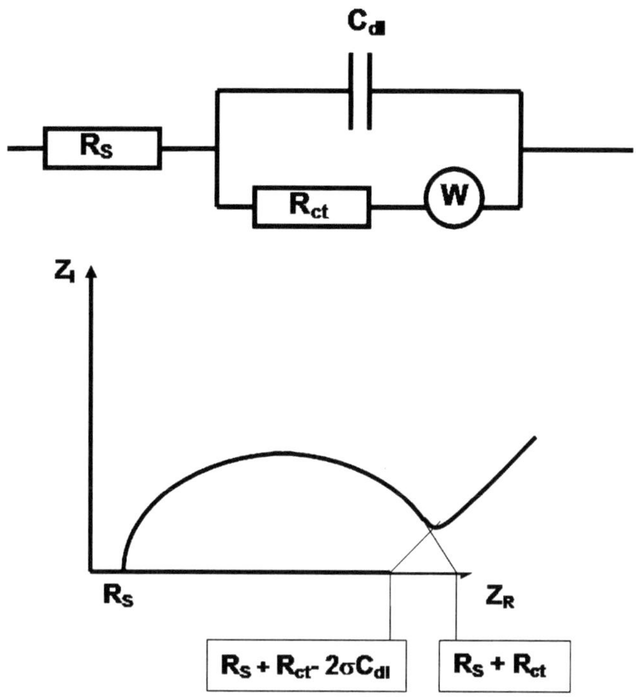

Fig. 3 Randles equivalent circuit for an electrode in electrolyte contact. The double layer capacitance C_{dl} results from the possibility of charge storage in the double layer at the interface. The charge transfer resistance R_{ct} denotes the possibility of current flow by redox reactions at the interface and the Warburg impedance results from the current limitation by diffusion from the bulk of the solution to the interface. R_s is the solution resistance given by the ion concentration and the cell geometry. From the Nyquist plot R_s and R_{ct} can be easily determined, but also the σ value from the Warburg impedance can be evaluated (see Table 1)

charge transfer might be faster than the diffusion of the substances to the electrode, a Warburg impedance needs to be introduced in series to R_{ct} for an appropriate impedance description. In series to this parallel circuit the solution resistance R_s is added. In the Nyquist plot shown in Fig. 3 the values for R_s and R_{ct} can be easily determined. From the frequency at the maximum of the semicircle the double layer capacitance can be calculated ($\omega = 2\Pi f = 1/R_{ct}C_{dl}$). The product of R_{ct} and C_{dl} is often termed as time constant τ of the electrochemical process. The 45° line indicating Warburg limited behavior can be extrapolated to the real axis. The intercept is equal to $R_s + R_{ct} - 2\sigma C_{dl}$ from which σ and subsequently diffusion coefficients can be calculated (see Table 1).

For analytical applications, however, the equivalent circuit is often simplified by neglecting the Warburg impedance. This can be done by choosing a frequency range where no 45° line in the Nyquist plot is observed and the interfacial or bulk impedance is dominating.

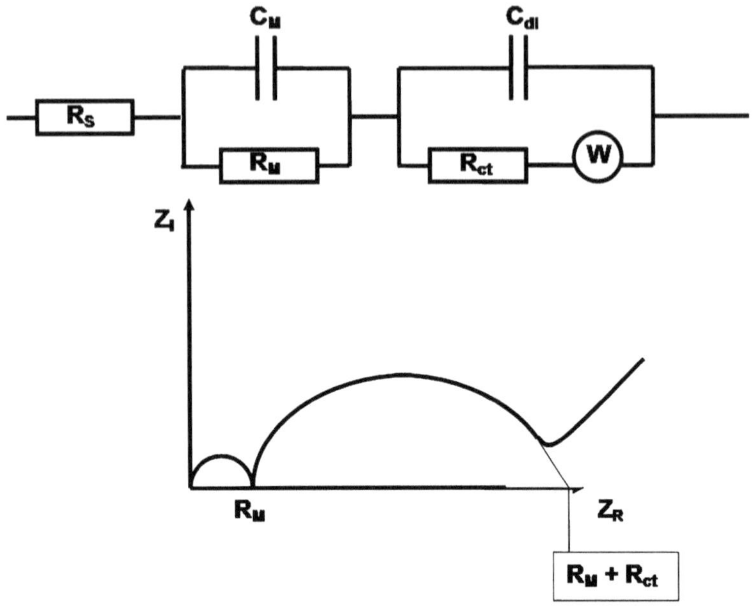

Fig. 4 The figure presents first the diagram of the equivalent circuit of an electrode covered with a polymer membrane and illustrates schematically the principle impedance spectrum of a polymer membrane in electrolyte contact. The *first semicircle* describes the impedance of the bulk of the membrane (first RC circuit), the *second semicircle* represents the membrane-solution interface, and the 45° line shows the infinite Warburg impedance. It has to be noted that depending on the nature of the membrane not all impedance contributions can always be seen. One impedance may dominate over the others

2.2.2 Polymer Membranes in Electrolyte Contact

Polymeric membranes are often applied for the construction of sensors as immobilization matrix for the recognition element or as filters improving the selectivity of the sensor by rejecting disturbing substances. In general, the impedance behavior of such membranes can be described by a series circuit of two parallel RC combinations describing the bulk properties of the membrane and the properties of the membrane/electrolyte interface and a Warburg impedance for the diffusion limitation (4). Under the assumption of three well-separated impedance contributions, Fig. 4 illustrates the behavior in the Nyquist plot. From the first RC term the membrane resistance and the geometrical membrane capacitance can be determined. These values are governed by the chemical nature of the polymer, the pore size and distribution, the water content, the electrolyte uptake, and the membrane thickness. From the second RC term (second semicircle in the Nyquist plot) the values for the exchange resistance and the double layer capacitance of the interface can be determined. Both values are also influenced by the above-mentioned membrane parameters with the exception of the membrane thickness. At lower frequencies the Warburg impedance dominates the impedance spectrum. However, this general behavior will not be necessarily observed for all membranes. The values for the different impedance values might differ significantly, so that for example only one semicircle occurs in the impedance spectrum when the bulk impedance of the membrane is strongly dominating. This is an important aspect in impedance spectroscopy: In a series of impedance contributions the largest impedance determines the overall impedance behavior to a large extent so that minor components cannot be well characterized. In the literature one can find examples concentrating on the one hand on analysis of the bulk properties and on the other hand on ion exchange at the interface—mainly because of membrane applications in separation (e.g., (5, 6)), sensor construction (7–9), and fuel cell electrodes (10–12) or applications in chemical ion sensors (ion-selective electrodes) (13, 14).

2.3 Impedance Measurements

The impedance can be measured in the frequency domain. A perturbing signal is applied to the system and the response is recorded. In the frequency domain, an ac excitation voltage of small amplitude (usually within the range 5–50 mV) is applied and the corresponding ac current will be detected as a function of the frequency of the applied voltage. Since the impedance is a complex value the modulus/phase shift or the real/imaginary part couple has to be determined. Nowadays this can be easily performed with a frequency response analyzer (FRA). Alternatively a lock-in amplifier can be used. To obtain the impedance in the frequency domain from this kind of measurement a Fourier transformation is necessary. Another technique which can be termed as multisine technique applies different frequencies at the same time to the system. In this case impedance data can be obtained rather fast; however, also here a Fourier transformation is necessary to transfer the data into the normal format of the frequency domain.

In general some requirements have to be fulfilled in order to get a reliable impedimetric description of the system:

1. The response of the system should only originate from the perturbation applied. This means other conditions have to be controlled to be constant (e.g., temperature).
2. The system should react linearly to the imposed frequency. This means that the impedance should be independent of the amplitude of the perturbation applied which has to be verified experimentally.
3. The system has to be stable during the measurement and should not change because of the perturbation applied. (This can be tested by repeated measurements. For changeable systems the faster multisine technique might be advantageous.)
4. The impedance has to be a continuous and finite-valued function in the frequency range of the measurement.

These are important features which have to be controlled during the impedance analysis of a system. More accurately, the measured impedance data should follow the Kramers–Kroning relation which allows the calculation of the imaginary part of the impedance $Z_I = f(\omega)$ function from the real part of the impedance $Z_R = f(\omega)$ function (1).

The frequency range which has to be analyzed varies with the system under investigation. However, the range in general is from several tens of kilohertz, where the capacitive impedance becomes rather small and often the impedance is determined by the solution resistance, to 10^{-1} or 10^{-3} Hz, where often the diffusion is limiting. Commercial instruments easily cover a range from 10^{-5} to 10^{6} Hz. Few frequencies per decade are analyzed, and around ten measurements are performed for each frequency; obtained mean value is then registered.

For electrochemical impedance spectroscopy it is often necessary to control the situation at the working electrode (1) In some cases impedance can be measured at the so-called open circuit potential (OCP). This means that an equilibrium situation is reached at the electrode depending on the solution composition. However, this situation has to be stable during the measurement as mentioned above. Here a simple two-electrode arrangement can often be used. (2) Alternatively, a dc potential has to be fixed at the working electrode in order to define a situation at the electrode or to enforce a process to occur (e.g., an oxidation process). This is a typical potentiostatic mode where a three-electrode arrangement is necessary. However, ac voltage and current appear only in the working/counter electrode circuit. (3) Another possibility is to define a current to flow over the working electrode. This can be achieved in a galvanostatic mode again with a three-electrode arrangement.

Modern impedance analyzers also contain data evaluation software and fitting programs which allow the calculation of transfer functions according to an equivalent circuit. These equivalent circuits are based on a physicochemical model of the system under investigation, as described, e.g., in Sect. 14.2.2. As a starting point for the description of an individual system one can find models in the literature or one can choose an equivalent circuit from the implemented software. In order to verify the validity of the chosen circuit not only does the fit quality in relation to the experimental curve need to be evaluated, but also a variation of experimental parameters such as electrolyte or analyte concentration, layer thickness, or electrode area. From such an analysis of the impedance data the frequency range necessary to determine a parameter of interest can also be deduced. For analytical application, it is also possible not only to reduce the frequency range, but to limit the measurement to a few or even one selected frequency. Here the impedance parameter of interest (e.g., resistance or capacitance as a function of a concentration to be determined) can be obtained from a fit of the measured impedance data by a verified equivalent circuit or the overall impedance can be directly correlated to a concentration.

In the following, the impedimetric biosensors will be divided into two categories. The first approach uses the electric properties at the surface of a single working electrode ("one sensor electrode" approach). The second method is based on two shortly spaced electrodes ("in-plane" geometry). In the latter case, the recognition reaction can also occur in the gap between the two electrodes.

3 Biosensor Applications

3.1 Impedimetric Detection Using a Single Sensor Electrode

Biosensors of this type follow the idea that the impedance at a single working electrode should determine the overall impedance. Thus after modification with the recognition element, this electrode becomes the "sensing" electrode. This approach is sometimes also referred to as "interfacial" measurement. Figure 5 shows the schematic structure of such an impedimetric biosensor. Different options exist for the evaluation of the recognition event at the sensor surface; besides the simple detection of the overall impedance at selected frequencies, one can analyze the changes in electrode capacitance or resistance.

3.1.1 Capacitance and Resistance Measurements

When the biosensor electrode is immersed in an aqueous solution an electrochemical double layer is formed. The capacitance C_{dl} depends on all compounds present at the interface, which are predominantly solvent molecules, ions, immobilized biomolecules, and optional films which promote the immobilization and detection processes. A capacitance change is induced when the dielectric constant or the thickness of the double layer on the transducer surface changes. Both options have been a matter of examination (15, 16). When constructing a capacitive biosensor, the electrode surface is usually covered with an additional insulating layer to reduce Faradaic currents (i.e., to increase the charge transfer resistance in parallel to the capacity). The biorecognition element is immobilized on top of this layer (17).

Capacitance measurements in the absence of Faradaic currents also allow for the investigation of the influence of the electrical field on the biological recognition event by applying different dc working potentials (18). The interaction between the recognition element and the analyte as well as their orientation might be influenced by the applied dc potential, particularly when the components of the interaction are charged.

The alternative approach for impedimetric detection of analyte binding is the determination of the charge transfer resistance R_{ct}. These Faradaic impedance measurements are usually performed in

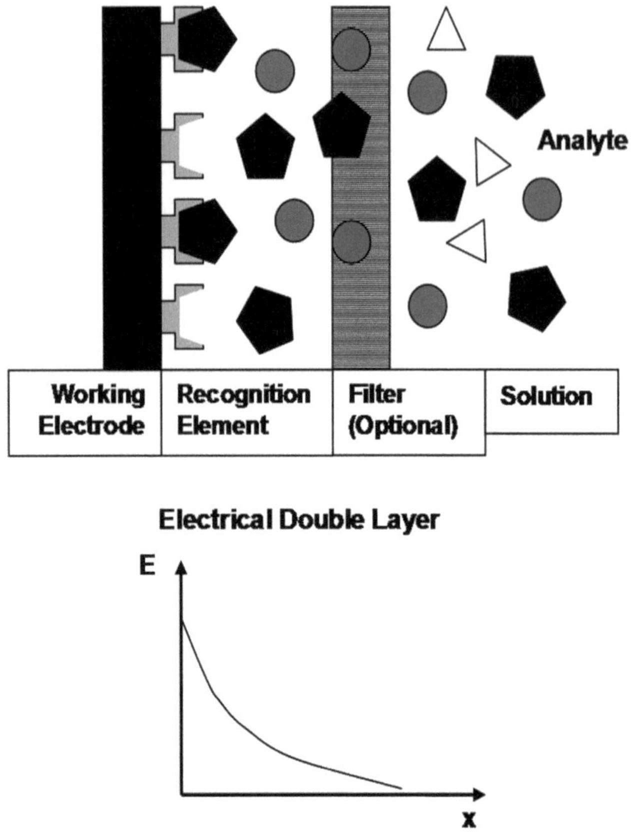

Fig. 5 Scheme of impedimetric biosensors using a modified working electrode as sensor immobilization occurs on the electrode

the presence of a redox couple in solution and rely on the change in the barrier for the redox conversion by the formation of the recognition complex itself or a subsequent complex. The ferri-/ferrocyanide system is often chosen for this purpose. For sensitive detection, impedance spectra are generally recorded for dc potentials equal to the open circuit potential (OCP) of the employed redox couple. Although an appropriate redox probe is necessary, the Faradaic impedance measurement is considered to be more sensitive since the changes in R_{ct} are usually larger than the respective capacitance changes.

Although very often used, the presence of a redox couple is not absolutely necessary. The measurement in an inert electrolyte solution results usually in a much higher impedance, but as long as the impedance changes upon the biorecognition process are reproducible as well this kind of measurement can result in reliable data. Particularly interesting is this approach when the conductivity of the recognition layer is changed by the specific binding event (as shown for example with proteins embedded in a lipid layer) (19).

3.1.2 Sensor and Sensor Preparation

Typical transducer elements are carbon and metal electrodes. Preferentially gold is chosen, because it allows the application of thiol chemistry for immobilization (20–23). Additionally, thin oxide layers on the electrode surface like indium tin oxide (ITO) or platinum/platinum hydroxide electrodes provide hydroxyl groups for silanization chemistry (24). Metal oxides of platinum (25, 26), tantalum (27, 28), and titanium (29) have also been used for capacitance-based detection.

Self-assembled monolayers (SAM) are frequently used to promote the immobilization of the biorecognition element. Such films are formed spontaneously by chemisorption and self-organization from solution (21, 30–32). Long-chain thiols can effectively block the electrode surface from unwanted reactions and thus are often used in capacitive sensors. Since the SAM films do not have specific recognition properties, the biocompound is normally coupled to the surface with functionalized SAMs (33). Alternatively the SAM layer has to be structured in order to create recognition sites. This can be termed as a kind of surface imprinting (34, 35). An alternative method to obtain well-ordered recognition layers is the transfer of a lipid bilayer (from liposomes) or a Langmuir–Blodgett (LB) film onto the electrode surface (19, 36, 37). Other immobilization protocols rely on the preparation of thin polymer films, electropolymers, or bioaffinity layers (e.g., avidin films capable for binding biotinylated biomolecules). More recently polyelectrolyte films have been introduced for biomolecule immobilization. This technique is based on the layer-by-layer adsorption of oppositely charged polymers on the transducer surface and is particularly suited for the preparation of multilayers (38, 39). The introduction of nanomaterials in the biolayer is a highly increasing new field and will be treated separately.

3.1.3 Impedimetric Immuno- and DNA Sensing

As already implied, impedimetric analyte detection is frequently applied for immunosensing. This originates from the fact that classical immunoassays (ELISA) are widespread and used in clinical diagnostics, food quality control, and environmental analysis and for the detection of pathogens, toxins, and explosives or drugs. Even though immunosensing is an established technique, background for the sensorial developments with impedimetric detection are the demands of point-of-care diagnostics with the need of rather simple equipment and the potential for a label-free analysis.

For immunosensing two immobilization strategies are possible. The sensor chip can either be modified with antibodies binding the respective antigen molecule or the antigen itself is immobilized, which binds then the complementary antibody. In both cases the binding event results in a change of the electrical surface properties, although in the latter case larger changes can be expected because of the high-molecular weight and the low dielectric constant of antibodies.

Initially capacitive sensors were preferably used for the detection of immune reactions (21, 22, 40, 41). Charge transfer can be diminished, e.g., by long-chain SAM layers. In combination with a potential-jump method very sensitive measurements have been reported (42). In order to reduce the influence of nonspecific binding, which would influence the capacitance in a similar way compared to the specific interaction, a differential mode of measurement was proposed (26). Antibodies can be incorporated into Langmuir–Blodgett films while retaining their binding properties. This can result in very ordered recognition structures (36, 43).

Resistance-based sensors have also been developed (44–46). For example, the human mammary tumor-associated glycoprotein was detected with specific antibodies immobilized on gold by spontaneous adsorption. The binding of the complementary antigen resulted in a change of the charge transfer resistance R_{ct} (45). The prior modification of the electrode surface by tethered nanomaterials such as carbon nanotubes leads to a large improvement of the analytical characteristics of the immunosensors (46) in terms of the decrease of the detection limit and of the enlargement of the dynamic range. Another system used ultrathin platinum layers and evaluated the conductivity changes based on an impedance model analysis (47). The change in ion conductivity of a lipid bilayer with incorporated ion channels can also be used for the antibody–antigen binding detection (48).

A special approach for the immobilization can be seen in the use of conducting polymers such as polypyrrole. This can help in amplifying the response since the conductivity of the polymeric network is strongly influenced by conformational changes induced by binding events (49–51). Biotinylated polypyrrole films were used for the immobilization of biotinylated antibodies via avidin (52). Antigen binding was connected with an increase in the charge transfer resistance. The respective detection limit was 10 pg/mL of human IgG.

Besides immunosensing, the impedimetric detection of nucleic acids is also a matter of research. ssDNA and dsDNA have been discriminated using a hanging mercury drop electrode (53, 54). Desorption of denatured ssDNA was accompanied by a larger dielectric loss than desorption of native dsDNA. This was explained with the higher flexibility of the ssDNA compared to the rigid dsDNA double strand.

For a sensor application more interesting are investigations on gold electrodes. Gold is a very conductive material, quite resistive to corrosion. Moreover it can be easily functionalized using thiolized molecules, leading to a semi-covalent bond. Several sensor systems have been reported for a voltammetric analysis of the hybridization (55–58). But here the addition or the coupling of a redox indicator is necessary. The presence of ss- or dsDNA on the electrode surface was found to result in different molecular structures which can be evaluated by different spectroscopic and microscopic methods (59). This provides the basis for impedimetric detection of oligonucleotide concentrations and single base pair mismatches (60). In analogy to the immune detection using conductive polymers, probe DNA can also be linked to conducting polymer films to evaluate the change in interface properties after target DNA binding. Here an increase in resistance was observed which could be used for a sensorial detection of suitable sensitivity (61–64). Amplification of the response by the use of nanoparticles coupled to an oligonucleotide has been reported (65). As an alternative to polypyrrole a quinone-containing polymer was developed (66) or polythiophene was used (67). The impedance technique can be used not only for the detection of DNA–DNA binding but also for the analysis of DNA–analyte interactions. Several small organic molecules but also large proteins show a specific interaction with DNA duplexes. This can be used for an analytical determination of these substances. Examples are cisplatin (68) or mythramycin, a.o. (69), but also specific DNA-binding proteins (70).

Molecular imprinted polymers are a rather new group within the recognition elements. They provide a surface which acts as a "negative" of the analyte and thus can recognize it within a mixture. Because of its chemical nature it can be reproducibly fabricated in different formats and is rather long-term stable in different environments. Besides its application in chromatographic separation, interesting examples have been reported for applications in sensors. For an impedimetric transduction of the binding event, particularly thin surface films are suited. Thus, capacitive sensors have been developed for organic molecules (34, 71) but also for whole cells and viruses (72, 73).

3.1.4 Impedimetric Enzyme- and Cell-Based Sensors

In the area of enzyme sensors amperometric electrodes are by far the most attractive transducers. This is because of the ease of use and their wide dynamic range. However, the change in redox conversion of a redox-active compound by the turnover of an enzyme (in the presence of the respective substrate) can also be detected by following the charge transfer resistance. This has been demonstrated for glucose detection using glucose oxidase and parabenzoquinone as mediator (74) and for urea detection using urease on insulated silicon electrodes (75). With EIS glucose sensor, a wide linear relation with more than two orders of magnitude was found between $1/R_{ct}$ and glucose concentration. Since the DC potentials applied in the EIS measurements (e.g., +0.280 V) are less anodic than that applied in amperometric measurements (e.g., +0.600 V), more accurate data, with less interfering effects, can be obtained in the EIS method.

In contrast to the transduction process itself impedance measurements have found increasing applications in optimizing biosensors mainly by following the different steps of surface modification during sensor preparation. This includes formation of SAMs, Langmuir–Blodgett films, adsorption layers, and so on, but also the binding of the recognition element to the transducer surface (e.g., (76–78)). In order to get rid of unwanted nonspecific binding of proteins, polyoxyethylene (POE) chains should be grafted for covering the surface between probe biomolecules.

Impedance spectroscopy was used to study proton transport across the membrane. Impedance spectra could be correlated to the concentration of inhibitor and substrate of the enzyme. Other examples are the detection of paraoxon (79), potato glycoalkaloids (80), cyanide (81), quinacrine (82), or the bacterial toxin streptolysin (83).

In analytics not only substrates and inhibitors of enzymes have been determined, but also the enzyme activity itself has often been analyzed. Here an interesting approach was introduced by using degradable polymer films. These polymers can be coated onto an electrode and impedimetrically characterized. The degradation by a biocatalytic reaction can occur directly by the enzyme acting on the polymer chains or by products of the enzymatic conversion. The diminished film thickness is easily followed by impedance measurements. This has been shown for urea (84), glucose (85), and enzymes such as chymotrypsin or lipase (86). This scheme can be nicely combined with immunoassays using enzymes as labels (84).

Recent developments are cell-based assays for the detection of specific analytes or for the screening of large libraries of substances for their biological effect. This relies on the concept of effect monitoring. Here, parameters are needed that report the response of the whole biological system with respect to the stimulus under investigation. Impedance is one of the suitable parameters because it can be used for the indirect detection of metabolic activity, cell adhesion on surfaces, and response to potential drugs and for cytotoxicity tests (87, 88). First examples have been reported analyzing the status of a cell culture (89–92). A simplified equivalent circuit describing the impedance of a cell layer as a parallel circuit of intracellular and extracellular pathways has been proposed. It has to be mentioned that cells can also be cultivated directly on the working electrode with an appropriate surface modification (93).

Impedance is not only suited to give information on the status of surface-fixed cells, but can also be used for the detection of the number of cells in solution. This has been intensively investigated in microbiology and was used for the detection, quantification, and identification of bacteria using specific antibodies (94) or bacteriophages (95), and resulted also in micromachined devices for cell counting (96) or cell differentiation (97).

3.2 Impedimetric Sensors Using Interdigitated Electrodes

Instead of using the surface of a single working electrode an in-plane geometry can be applied, which makes use of two closely spaced electrodes. Two electrodes are prepared on an insulating substrate in a fingerlike structure leaving a small nonconductive gap between them. These electrodes are termed as interdigitated electrodes (IDEs) and can be formed from different metals such as palladium (98), titanium (99), platinum (100), and gold (101). The geometry takes advantage of the changed conditions for the current flow which occurs mainly very near to the surface and thus shows a much higher sensitivity towards surface changes compared to the conventional design. The biorecognition element can be immobilized on both electrodes, in the gap between the electrodes or on top of both. This electrode design has an additional advantage since by preparing the electrodes on a chip the whole electrochemical setup is already integrated and no external electrodes are needed. The basic scheme is shown in Fig. 6.

Immune detection using IDEs takes advantage of the change in resistance close to the surface due to protein accumulation and the change in the dielectric properties due to the presence of substances with a lower dielectric constant compared to water. Thus, a label-free detection becomes feasible (102, 103). This kind of measurement can also be applied to the detection of DNA hybridization without

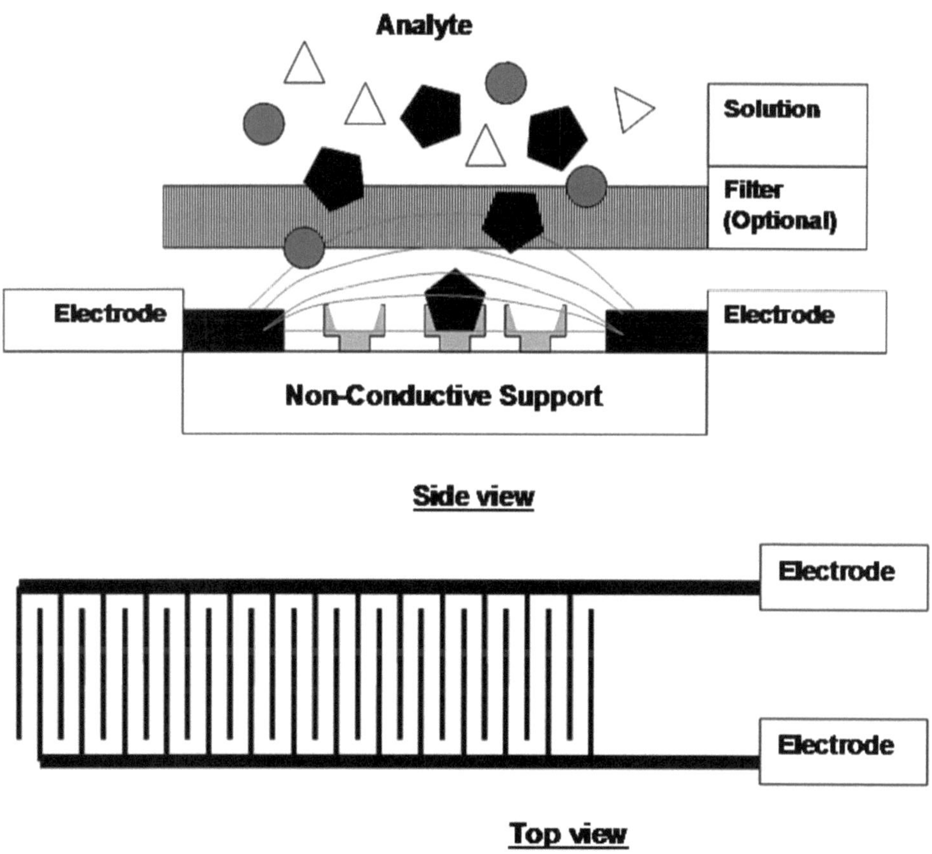

Fig. 6 Schemes of impedimetric biosensors using an in-plane arrangement of two electrodes (interdigitated electrodes IDE) between and/or on the electrodes

the need of a label. This relies mainly on the changed charge accumulation after double strand formation (100, 101, 104).

Receptor–ligand binding is often accompanied by structural changes of the complex. This can be also used for a sensitive ligand detection particularly when a receptor is used, which is a chemically gated transport protein such as the nicotinic acetylcholine receptor. Incorporated in a lipid membrane, acetylcholine but also receptor antagonists can be measured by capacitance changes (105).

IDE-based sensors have found application in the detection of enzyme substrates and of enzyme inhibitors, especially for environmental applications (106). They also found application in enzyme detection. For example, an enzyme-based impedimetric sensor was developed for the detection of collagenase. Impedance changes were induced by the proteolytic digestion of gelatine-coated interdigitated electrodes (107). In this case, enzyme degradation of the layer produced a rapid increase in the impedance due to a changed thickness of the layer. The change in impedance with protease digestion was correlated to the solubilization of the gelatine layer from the sensor surface. Other examples based on the enzyme-catalyzed polymer transformation, explained in Sect. 14.3.1.4, are sensors for urea (108, 109) or creatinine (109).

A highly sensitive response of the sensor requires strong electrical fields due to short distances of the electrodes (110). Typical distances between the electrode fingers are in the range of 1–10 μm and are limited by the fabrication procedure. However, electrode widths from 500 nm to less than 200 nm have meanwhile been achieved. A theoretical analysis demonstrates that 80% of the current between the electrodes with a spacing of about 250 nm will flow in a layer not higher than 250 nm above the surface. This may illustrate that a reproducible production of interdigitated electrodes with dimensions in the nanometer range will significantly increase the sensitivity for a direct binding analysis.

4 Perspectives for In Vivo Implantation

In conclusion, it can be stated that impedimetric detection can achieve the necessary analytical sensitivity for practical applications in medicine, gene analysis, food industry, or environmental analysis, brought by the design of nanoIDEs and the use of nanomaterials. A lot of commercial setups for electrochemical measurements are equipped with a module allowing impedance measurements in the frequency range applicable for biosensor application (100 mHz to 100 kHz). New cheaper setups are now equipped with a module for impedance measurements with a narrower range (100 mHz to 1 kHz), which is very often quite convenient for biosensor applications. With most of these setups, measurements on microelectrodes with a surface of a few hundred μm^2 can be performed without any limitation. For surface areas smaller than 10 μm^2, a low current module should be used.

Label-free and direct binding detection is very attractive and could lead to bioanalytical systems for real-time detection (111). The measurement device should then be simplified (e.g., single or dual frequency measurement) and the impedimetric biosensor should be integrated in a microfluidic cartridge for implementation of a washing procedure for canceling the nonspecific effects. Based on this principle, a point-of-care device was conceived for diagnosis of deep vein thrombosis (112).

Implantable biosensors based on impedance measurements are still in their infancy. A preliminary study based on EIS has been proposed to measure the electrical impedance of cells and tissues associated with the neointimal growth that characterizes in-stent restenosis in coronary artery stents. The advantages of such a device would be that monitoring could be noninvasively and easily carried out, allowing more routine follow-ups and the early identification and management of any device complications (113).

References

1. Macdonald JR (1987) Impedance spectroscopy. Wiley, New York
2. Rubinstein I (1995) Physical electrochemistry. Dekker, New York
3. Krause S (2003) Encyclopedia of electrochemistry, vol 3. Instrumentation and electroanalytical chemistry. Wiley-VCH, Weinheim
4. Buck RP, Ciani S (1975) Electroanalytical chemistry of membranes. Crit Rev Anal Chem 5:323–420
5. Malmgreen-Hansen B, Sorensen TS, Jensen JB, Henneberg M (1989) Electric impedance of cellulose acetate membranes and a composite membrane at different salt concentrations. J Colloid Interface Sci 130:359–385
6. Canas A, Ariza MJ, Benavente J (2001) Characterization of active and porous sublayers of a composite reverse osmosis membrane by impedance spectroscopy, streaming and membrane potentials, salt diffusion and X-ray photoelectron spectroscopy measurements. J Memb Sci 183: 135–146
7. Lisdat F, Moritz W (1994) Investigation of polymer membranes for diffusion restriction in miniaturised reference sensors. Thin Solid Films 248:126–133
8. Wiegand G, Arribas-Layton N, Hillebrandt H, Sackmann E, Wagner P (2002) Electrical properties of supported lipid bilayer membranes. J Phys Chem B 106:4245–4254
9. Pejcic B, de Marco R (2006) Impedance spectroscopy: over 35 years of electrochemical sensor optimization. Electrochim Acta 51:6217–6229
10. Wagner N (2002) Characterization of membrane electrode assemblies in polymer electrolyte fuel cells using a.c. impedance spectroscopy. J Appl Electrochem 32:859–863
11. Himanen O, Hottinen T (2006) Characterization of membrane–electrode assembly with hydrogen–hydrogen cell and ac-impedance spectroscopy: Part II. Modeling. Electrochim Acta 52:581–588
12. Agel E, Bouet J, Fauvarque JF (2001) Characterization and use of anionic membranes for alkaline fuel cells. J Power Sources 101:267–274
13. Nagarale RK, Gohil GS, Shahi VK (2006) Recent developments on ion-exchange membranes and electro-membrane processes. Adv Colloid Interface Sci 119:97–130
14. Touzi H, Chevalier Y, Kalfat R, Ben Ouada H, Zarrouk H, Chapel JP, Jaffrezic-Renault N (2006) Elaboration and electrical characterization of silicone-based anion-exchange materials. Mater Sci Eng C 26:462–471
15. Berggren C, Bjarnason B, Johansson G (2001) Capacitive biosensors. Electroanalysis 13:173–180
16. Saby C, Jaffrezic-Renault N, Martelet C, Colin B, Charles MH, Delair T, Mandrand B

(1993) Immobilization of antibodies onto a capacitance silicon-based transducer. Sens Actuators B 16:458–462

17. Hays HCW, Millner PA, Prodromidis MI (2006) Development of capacitance based immunosensors on mixed self-assembled monolayers. Sens Actuators B 114: 1064–1070
18. Bart M, Stigter ECA, Stapert HR, de Jong GJ, van Bennekom WP (2005) On the response of a label-free interferon-γ immunosensor utilizing electrochemical impedance spectroscopy. Biosens Bioelectron 21:49–59
19. Hou Y, Jaffrezic-Renault N, Martelet C, Tlili C, Zhang A, Pernollet JC, Briand L, Gomila G, Errachid A, Samitier J, Salvagnac L, Torbiéro B, Temple-Boyer P (2005) Study of Langmuir and Langmuir-Blodgett films of odorant-binding protein/amphiphile for odorant biosensor. Langmuir 21:4058–4065
20. Knichel M, Heiduschka P, Beck W, Jung G, Gopel W (1995) Utilization of a self-assembled peptide monolayer for an impedimetric immunosensor. Sens Actuators B 28:85
21. Rickert R, Gopel W, Beck W, Jung G, Heiduschka P (1996) A 'mixed' self-assembled monolayer for an impedimetric immunosensor. Biosens Bioelectron 11:757–768
22. Taira H, Nakano K, Maeda M, Takagi M (1993) Electrode modification by long-chain, dialkyl disulfide reagent having terminal dinitrophenyl group and its application to impedimetric immunosensors. Anal Sci 9:199–206
23. Ameur S, Martelet C, Jaffrezic-Renault N, Chovelon JM (2000) Sensitive immunodetection through impedance measurements onto gold functionalized electrodes. Appl Biochem Biotechnol 89:161–170
24. Sadik OA, Xu H, Gheorghiu E, Andreescu D, Balut C, Gheorghiu M, Bratu D (2002) Differential impedance spectroscopy for monitoring protein immobilization and antibody–antigen reactions. Anal Chem 74:3142–3150
25. Ameur S, Maupas H, Martelet C, Jaffrezic-Renault N, Ben Ouada H, Cosnier S, Labbe P (1997) Impedimetric measurements on polarized functionalized platinum electrodes: application to direct immunosensing. Mater Sci Eng C 5:111–119
26. Maupas H, Soldatkin AP, Martelet C, Jaffrezic-Renault N, Mandrand B (1997) Direct immunosensing using differential electrochemical measurements of impedimetric variations. J Electroanal Chem 421:165–171
27. Gebbert A, Alvarezicaza M, Peters H, Jager V, Bilitewski U, Schmid RD (1994) On-line monitoring of monoclonal antibody production with regenerable flow-injection immuno systems. J Biotechnol 32:213–220
28. Gebbert A, Alvarezicaza M, Stocklein W, Schmid RD (1992) Real-time monitoring of immunochemical interactions with a tantalum capacitance flow-through cell. Anal Chem 64:997–1003
29. Varlan AR, Suls J, Sansen W, Veelaert D, de Loof A (1997) Capacitive sensor for the allatostatin direct immunoassay. Sens Actuators B 44:334–340
30. Bain CD, Troughton EB, Tao YT, Evall J, Whitesides GM, Nuzzo RG (1989) Formation of monolayer films by the spontaneous assembly of organic thiols from solution onto gold. J Am Chem Soc 111:321–335
31. Anderson JL, Bowden EF, Pickup PG (1996) Dynamic electrochemistry: methodology and application. Anal Chem 68:R379–R444
32. Mirsky VM, Riepl M, Wolfbeis OS (1997) Capacitive monitoring of protein immobilization and antigen–antibody reactions on monomolecular alkylthiol films on gold electrodes. Biosens Bioelectron 12:977–989
33. Wink T, Zuilen SJ, Bult A, van Bennekom WP (1997) Self-assembled monolayers for biosensors. Analyst 122:43R–50R
34. Prodromidis MI, Hirsch T, Mirsky VM, Wolfbeis OS (2003) Enantioselective artificial receptors formed by the spreader-bar technique. Electroanalysis 15:1795–1798
35. Mirsky VM (2002) New electroanalytical applications of self-assembled monolayers. Trends Anal Chem 21:439–450
36. Barraud A, Perrot H, Billard V, Martelet C, Therasse J (1993) Study of immunoglobulin G thin layers obtained by the Langmuir-Blodgett method: application to immunosensors. Biosens Bioelectron 8:39–48
37. Wang LG, Li YH, Tien HT (1995) Electrochemical transduction of an immunological reaction via s-BLMs. Bioelectrochem Bioenerg 36:145–147
38. Cassier T, Lowack K, Decher G (1998) Layer-by-layer assembled protein/polymer hybrid films: nanoconstruction via specific recognition. Supramol Sci 5:309–315
39. Calvo EJ, Battaglini F, Danilowicz C, Wolosiuk A, Otero M (2000) Layer-by-layer electrostatic deposition of biomolecules on surfaces for molecular recognition, redox mediation and signal generation. Faraday Discuss 116:47–65
40. Berggren C, Bjarnason B, Johansson G (1998) An immunological Interleukine-6 capacitive biosensor using perturbation with a

potentiostatic step. Biosens Bioelectron 13: 1061–1068
41. Bataillard P, Gardies F, Jaffrezic-Renault N, Martelet C, Colin B, Mandrand B (1988) Direct detection of immunospecies by capacitance measurements. Anal Chem 60: 2374–2379
42. Berggren C, Johansson G (1997) Capacitance measurements of antibody–antigen interactions in a flow system. Anal Chem 69: 3651–3657
43. Hou YX, Tlili C, Jaffrezic-Renault N, Zhang AD, Martelet C, Ponsonnet L, Errachid A, Samitier J, Bausells J (2004) Study of mixed Langmuir–Blodgett films of immunoglobulin G/amphiphile and their application for immunosensor engineering. Biosens Bioelectron 20:1126–1133
44. Hleli S, Martelet C, Abdelghani A, Burais N, Jaffrezic-Renault N (2006) Atrazine analysis using an impedimetric immunosensor based on mixed biotinylated self-assembled monolayer. Sens Actuators B 113:711–717
45. Jie M, Ming CY, Jing D, Cheng LS, Na LH, Jun F, Xiang CY (1999) An electrochemical impedance immunoanalytical method for detecting immunological interaction of human mammary tumor associated glycoprotein and its monoclonal antibody. Electrochem Commun 1:425–428
46. Bourigua S, Hnaien M, Bessueille F, Lagarde F, Dzyadevych S, Maaref A, Bausells J, Errachid A, Jaffrezic-Renault N (2010) Impedimetric immunosensor based on SWCNT-COOH modified gold microelectrodes for label-free detection of deep venous thrombosis biomarker. Biosens Bioelectron 26:1278–1282
47. Pak SC, Penrose W, Hesketh PJ (2001) An ultrathin platinum film sensor to measure biomolecular binding. Biosens Bioelectron 16:371–379
48. Woodhouse GE, King LG, Wieczorek L, Cornell BA (1998) Kinetics of the competitive response of receptors immobilised to ion-channels which have been incorporated into a tethered bilayer. Faraday Discuss 111:247–258
49. Sargent A, Loi T, Gal S, Sadik OA (1999) The electrochemistry of antibody-modified conducting polymer electrodes. J Electroanal Chem 470:144–156
50. Ouerghi O, Senillou A, Jaffrezic-Renault N, Martelet C, Ben Ouada H, Cosnier S (2001) Gold electrode functionalized by electropolymerization of a cyano N-substituted pyrrole: application to an impedimetric immunosensor. J Electroanal Chem 501:62–69
51. Hafaid I, Chebil S, Korri-Youssoufi H, Bessueille F, Errachid A, Sassi Z, Ali Z, Abdelghani A, Jaffrezic-Renault N (2010) Effect of electrical conditions on an impedimetric immunosensor based on a modified conducting polypyrrole. Sens Actuators B 144:323–331
52. Ouerghi O, Touhami A, Jaffrezic-Renault N, Martelet C, Ben Ouada H, Cosnier S (2002) Impedimetric immunosensor using avidin–biotin for antibody immobilization. Bioelectrochemistry 56:131–133
53. Hason S, Dvorak J, Jelen F, Vetterl V (2002) Interaction of DNA with echinomycin at the mercury electrode surface as detected by impedance and chronopotentiometric measurements. Talanta 56:905–913
54. Strasak L, Dvorak J, Hason S, Vetterl V (2002) Electrochemical impedance spectroscopy of polynucleotide adsorption. Bioelectrochemistry 56:37–41
55. Kelley SO, Boon EM, Barton JK, Jackson NM, Hill MG (1999) Single-base mismatch detection based on charge transduction through DNA. Nucleic Acids Res 27:4830–4837
56. Fojta M, Havran L, Billova S, Kostecka P, Masarik M, Kizek R (2003) Two-surface strategy in electrochemical DNA hybridization assays: detection of osmium-labeled target DNA at carbon electrodes. Electroanalysis 15:431–440
57. Hashimoto K, Ito K, Ishimore Y (1994) Novel DNA sensor for electrochemical gene detection. Anal Chim Acta 286:219–224
58. Pänke O, Kirbs A, Lisdat F (2007) Voltammetric detection of single base-pair mismatches and quantification of label-free target ssDNA using a competitive binding assay. Biosens Bioelectron 22:2656–2662
59. Zhao YD, Pang DW, Hu S, Wang ZL, Cheng JK, Qi YP, Dai HP, Mao BW, Tian ZQ, Luo J, Lin ZH (1999) DNA-modified electrodes. Part 3: Spectroscopic characterization of DNA-modified gold electrodes. Anal Chim Acta 388:93–101
60. Kafka J, Pänke O, Abendroth B, Lisdat F (2008) A label-free DNA sensor based on impedance spectroscopy. Electrochim Acta 53:7467–7474
61. Korri-Youssoufi H, Garnier F, Srivtava P, Godillot P, Yassar A (1997) Toward bioelectronics: specific DNA recognition based on an oligonucleotide-functionalized polypyrrole. J Am Chem Soc 119:7388–7389
62. Li CM, Sun CQ, Song S, Choong VE, Maracas G, Zhang XJ (2005) Impedance labelless detection-based polypyrrole DNA biosensor. Front Biosci 10:180–186

63. Wang J, Jiang M, Fortes A, Mukherjee B (1999) New label-free DNA recognition based on doping nucleic-acid probes within conducting polymer films. Anal Chim Acta 402:7–12
64. Tlili C, Korri-Youssoufi H, Ponsonnet L, Martelet C, Jaffrezic-Renault N (2005) Electrochemical impedance probing of DNA hybridisation on oligonucleotide functionalised polypyrrole. Talanta 68:131–137
65. Peng H, Soeller C, Cannell MB, Bowmaker GA, Cooney RP, Travas-Seijdic J (2006) Electrochemical detection of DNA hybridization amplified by nanoparticles. Biosens Bioelectron 21:1727–1736
66. Piro B, Haccoun J, Pham MC, Tran LD, Rubin A, Perrot H, Gabrielli C (2005) Study of the DNA hybridization transduction behavior of a quinone-containing electroactive polymer by cyclic voltammetry and electrochemical impedance spectroscopy. J Electroanal Chem 577:155–165
67. Gautier C, Cougnon C, Pilard JF, Casse N (2006) Label-free detection of DNA hybridization based on EIS investigation of conducting properties of functionalized polythiophene matrix. J Electroanal Chem 587:276–283
68. Yan F, Sadik OA (2001) Enzyme-modulated cleavage of dsDNA for supramolecular design of biosensors. Anal Chem 73:5272–5280
69. Li CZ, Liu YL, Luong JHT (2005) Impedance sensing of DNA binding drugs using gold substrates modified with gold nanoparticles. Anal Chem 77:478–485
70. Li CZ, Long YT, Lee JS, Kraatz HB (2004) Protein–DNA interaction: impedance study of MutS binding to a DNA mismatch. Chem Commun 5:574–575
71. Liu XY, Li CY, Wang CF, Li T, Hu SS (2006) The preparation of molecularly imprinted poly(o-phenylenediamine) membranes for the specific O, O-dimethyl-α-hydroxylphenyl phosphonate sensor and its characterization by AC impedance and cyclic voltammetry. J Appl Polym Sci 101:2222–2227
72. Hayden O, Lieberzeit PA, Blaas D, Dickert FL (2006) Artificial antibodies for bioanalyte detection—sensing viruses and proteins. Adv Funct Mater 16:1269–1278
73. Dickert FL, Hayden O, Lieberzeit PA, Haderspoeck C, Bindeus R, Palfinger C, Wirl B (2003) Nano- and micro-structuring of sensor materials—from molecule to cell detection. Synth Metal 138:65–69
74. Shervedani RK, Mehrjardi AH, Zamiri N (2006) A novel method for glucose determination based on electrochemical impedance spectroscopy using glucose oxidase self-assembled biosensor. Bioelectrochemistry 69:201–208
75. Barhoumi H, Maaref A, Rammah M, Martelet C, Jaffrezic N, Mousty C, Vial S, Forano C (2006) Urea biosensor based on Zn3Al–Urease layered double hydroxides nanohybrid coated on insulated silicon structures. Mater Sci Eng C 26:328–333
76. Ye JS, Ottova A, Tien HT, Sheu FS (2003) Nanostructured platinum-lipid bilayer composite as biosensor. Bioelectrochemistry 59:65–72
77. Jeuken LJC, Connell SD, Henderson PJF, Gennis RB, Evans SD, Bushby RJ (2006) Redox enzymes in tethered membranes. J Am Chem Soc 128:1711–1716
78. Naumann R, Schmidt EK, Jonczyk A, Fendler K, Kadenbach B, Liebermann T, Offenhausser A, Knoll W (1999) The peptide-tethered lipid membrane as a biomimetic system to incorporate cytochrome c oxidase in a functionally active form. Biosens Bioelectron 14:651–662
79. Schasfoort RBM, Niedziela T (1994) Detection of inhibitory compounds of acetylcholine esterase with a novel ion responding impedance sensor (IRIS). Sens Actuators B 18:175–177
80. Benilova IV, Soldatkin AP, Martelet C, Jaffrezic-Renault N (2006) Nonfaradaic impedance probing of potato glycoalkaloids interaction with butyrylcholinesterase immobilized onto gold electrode. Electroanalysis 18(19–20):1950–1956
81. Bouyahia N, Hamlaoui ML, Hnaien M, Lagarde F, Jaffrezic-Renault N (2011) Impedance spectroscopy and conductometric biosensing for probing catalase reaction with cyanide as ligand and inhibitor. Bioelectrochemistry 80:155–161
82. Tong YH, Han XJ, Song YH, Jiang JG, Wang EK (2003) Characterization and property of DNA incorporated bilayer lipid membranes. Biophys Chem 105:1–9
83. Wilkop T, Xu DK, Cheng Q (2007) Characterization of pore formation by streptolysin O on supported lipid membranes by impedance spectroscopy and surface plasmon resonance spectroscopy. Langmuir 23:1403–1409
84. McNeil CJ, Athey D, Ball M, Ho WO, Krause S, Armstrong RD, Wright JD, Rawson K (1995) Electrochemical sensors based on impedance measurement of enzyme-catalyzed polymer dissolution: theory and applications. Anal Chem 67:3928–3935
85. Gooding JJ, Hall EA (1996) Membrane properties of acrylate bulk polymers for

biosensor applications. Biosens Bioelectron 11:1031–1040
86. Sumner C, Krause S, Sabot A, Turner K, McNeil CJ (2001) Biosensor based on enzyme-catalysed degradation of thin polymer films. Biosens Bioelectron 16:709–714
87. Hug TS (2003) Biophysical methods for monitoring cell-substrate interactions in drug discovery. Assay Drug Dev Technol 1:479–488
88. Yotter RA, Wilson DM (2004) Sensor technologies for monitoring metabolic activity in single cells. Part II: Nonoptical methods and applications. IEEE Sens J 4:412–429
89. Yeon JH, Park JK (2005) Cytotoxicity test based on electrochemical impedance measurement of HepG2 cultured in microfabricated cell chip. Anal Biochem 341:308–315
90. Xiao C, Luong JHT (2003) On-line monitoring of cell growth and cytotoxicity using electric cell-substrate impedance sensing (ECIS). Biotechnol Prog 19:1000–1005
91. K'Owino IO, Sadik OA (2005) Impedance spectroscopy: a powerful tool for rapid biomolecular screening and cell culture monitoring. Electroanalysis 17:2101–2113
92. Kyle AH, Chan CTO, Minchinton AI (1999) Characterization of three-dimensional tissue cultures using electrical impedance spectroscopy. Biophys J 76:2640–2648
93. Bouafsoun A, Helali S, Othmane A, Kerkeni A, Prigent AF, Jaffrezic-Renault N, Bessueille F, Leonard D, Ponsonnet L (2007) Evaluation of endothelial cell adhesion onto different protein/gold electrodes by EIS. Macromol Biosci 7:599–610
94. Maalouf R, Fournier-Wirth C, Coste J, Chebib H, Saikali Y, Vittori O, Errachid A, Cloarec JP, Martelet C, Jaffrezic-Renault N (2007) Label-free detection of bacteria by electrochemical impedance spectroscopy: comparison to surface plasmon resonance. Anal Chem 79:4879–4886
95. Shabani A, Zourob M, Allain B, Marquette CA, Laurence MF, Mandeville R (2008) Bacteriophage-modified microarrays for the direct impedimetric detection of bacteria. Anal Chem 80:9475–9482
96. Jonsson M, Welch K, Hamp S, Stromme M (2006) Bacteria counting with impedance spectroscopy in a micro probe station. J Phys Chem B 110:10165–10169
97. Cheung K, Gawad S, Renaud P (2005) Impedance spectroscopy flow cytometry: on-chip label-free cell differentiation. Cytometry A 65A:124–132
98. van Gerwen P, Laureyn W, Laureys W, Huyberechts G, de Beeck MO, Baert K, Suls J, Sansen W, Jacobs P, Hermans L, Mertens R (1998) Nanoscaled interdigitated electrode arrays for biochemical sensors. Sens Actuators B 49:73–80
99. Laureyn W, Nelis D, van Gerwen P, Baert K, Hermans L, Magnee R, Pireaux JJ, Maes G (2000) Nanoscaled interdigitated titanium electrodes for impedimetric biosensing. Sens Actuators B 68:360–370
100. Gheorghe M, Guiseppi-Elie A (2003) Electrical frequency dependent characterization of DNA hybridization. Biosens Bioelectron 19:95–102
101. Dharuman V, Grunwald I, Nebling E, Albers J, Blohm L, Hintsche R (2005) Label-free impedance detection of oligonucleotide hybridisation on interdigitated ultramicroelectrodes using electrochemical redox probes. Biosens Bioelectron 21:645–654
102. Taylor RF, Marenchic IG, Spencer RH (1991) Antibody- and receptor-based biosensors for detection and process control. Anal Chim Acta 249:67–70
103. Hnaien M, Hassen WM, Abdelghani A, Fournier-Wirth C, Coste J, Bessueille F, Leonard D, Jaffrezic-Renault N (2008) A conductometric immunosensor based on functionalized magnetite nanoparticles for *E. coli* detection. Electrochem Commun 10:1152–1154
104. Hang TC, Elie AG (2004) Frequency dependent and surface characterization of DNA immobilization and hybridization. Biosens Bioelectron 19:1537–1548
105. Eldefrawi ME, Sherby SM, Andreou AG, Mansour NA, Annau Z, Blum NA, Valdes JJ (1988) Acetylcholine receptor-based biosensor. Anal Lett 21:1665–1680
106. Jaffrezic-Renault N, Dzyadevych SV (2008) Conductometric microbiosensors for environmental monitoring. Sensors 8:2569–2588
107. Saum AEG, Cumming RH, Rowell FJ (1998) Use of substrate coated electrodes and AC impedance spectroscopy for the detection of enzyme activity. Biosens Bioelectron 13:511–518
108. Sheppard NF, Mears DJ, Guiseppi-Elie A (1996) Model of an immobilized enzyme conductimetric urea biosensor. Biosens Bioelectron 11:967–979
109. Ho WO, Krause S, McNeil CJ, Pritchard JA, Armstrong RD, Athey D, Rawson K (1999) Electrochemical sensor for measurement of urea and creatinine in serum based on ac impedance measurement of enzyme-catalyzed polymer transformation. Anal Chem 71:1940–1946

110. Montelius L, Tegenfeldt JO, Ling TGI (1995) Fabrication and characterization of a nanosensor for admittance spectroscopy of biomolecules. J Vac Sci Technol A 13: 1755–1760
111. Jaffrezic-Renault N, Errachid A (2010) Analytical microsystems for biomedical and environmental applications. In: Mascini M, Torbicz W, Pijanowska DG (eds) Lecture notes of the ICB seminar "Micro- and nano-systems in biochemical diagnosis. Principles and applications". Polish Academy of Sciences, Warsaw, pp 40–52
112. http://www.diagnosingdvt.com
113. Shedden L, Kennedy S, Connolly P, Wadsworth R (2010) Towards a self-reporting coronary artery stent – measuring neointimal growth associated with in-stent restenosis using impedance techniques. Biosens Bioelectron 26:661–666

Chapter 15

Aptamer-Based Electrochemical Biosensors for the Detection of Small Molecules and Plasma Proteins

Cassie Ho and Hua-Zhong Yu

Abstract

The ability to detect the presence of certain molecular analytes inside the human body is vital to the ability of a medical professional to assess a patient's health. The use of sensitive and selective biosensors would be a tremendous asset to the field of medical screening and diagnosis. The fabrication of aptamer-based electrochemical biosensors as an emerging technology has made the detection of both small and macromolecular analytes easier, faster, and more suited for the ongoing transition from fundamental analytical science to the early detection of diseases. Aptamers are synthetic oligonucleotides that have undergone iterative rounds of in vitro selection for binding with high affinity to specific analytes of choice; a sensitive yet simple method to utilize aptamers as recognition entities for the development of biosensors is to transduce the signal electrochemically. In this chapter, we will summarize the state-of-the-art research progresses on aptamer-based electrochemical biosensors for the detection of both plasma proteins and small molecules. The targeted portable electronic devices, as they can be operated by the patients during their day-to-day lives, have the potential to revolutionize the delivery of medical treatment.

Key words Aptamers, Biosensors, Electrochemistry, Biomarkers, Diagnosis

1 Introduction

1.1 Sensors

Sensors are devices composed of a detection element that respond to a stimulus by inciting the transduction element to give a detectable signal. The use of sensitive and selective sensors is very important to the field of medicine. The ability to detect the presence of certain molecular analytes inside the human body (namely, biomarkers) is essential to the assessment of a patient's health. Enzyme-link immunosorbent assays (ELISA) are currently the gold standard for biomarker detection in health care. However, ELISA assays require a multistep protocol, which necessitates the use of highly trained clinicians to process and interpret the results. In principle, the use of biosensors can eliminate many of the preparation and operation steps. Biosensors are, as the name suggests, sensors that contain a biological element for sensing; they take advantage of selective and

Stéphane Marinesco and Nicholas Dale (eds.), *Microelectrode Biosensors*, Neuromethods, vol. 80,
DOI 10.1007/978-1-62703-370-1_15,

sensitive biological recognition reactions to detect the target analytes. An ideal biosensor should require little to no sample manipulation prior to detection and should be designed so that the results are easily interpreted. One of the great advantages of using biosensors over conventional laboratory processes is their ability to deliver portable, point-of-care diagnosis. In third-world countries and other places that lack a strong healthcare infrastructure, biosensors have the ability to greatly improve medical diagnosis and can provide more timely treatments as required.

Many different types of signal transduction protocols are commonly used in biosensors; however, the two that lend itself best to adaptation for application are optical and electrochemical methods. That is because of the simple transformation process of a target detection event into a measureable signal, as well as the low theoretical limit of detection (LoD) of these types of sensors. Nonetheless, electrochemical sensors have a great potential for miniaturization; its capacity for direct electrical signal input and for label-free detection makes it very appealing for medical applications. Another significant advantage electrochemical sensors have over their optical counterparts is their compatibility with the semiconductor and microelectronics industry which will allow for a quick adaptation of laboratory prototypes for commercialization and mass production. Many commercial enterprises already have the microelectronics required for the development of clinical biosensors on the market (1). The ubiquitous electrochemical glucose monitors are perhaps the best example of a successful commercial electrochemical biosensor (2, 3).

1.2 Aptamers

Aptamers are oligonucleotides that can bind to their target ligands with high specificity and affinity. They were first reported independently by Szostak (4) and Gold (5) back in 1990. Since then, aptamers have been adapted in numerous sensor applications ranging from the detection of food and environment toxins (6) to the identification of live parasites that infect a large number of people every year (7). Although the binding mechanism of an aptamer to the target molecule is not clear in every case, it is suggested that aptamers bind by folding around their target and enveloping them (8) and, as a result, the name aptamer is derived from the Latin word *aptus* (meaning "to fit"). Like antibodies, aptamers have the ability to recognize unique epitopes of protein targets via sequence-specific hydrogen bonding between the bases and amino acid side chains (9). The secondary structures of aptamers tend to be very important to the binding affinity and specificity (10, 11); this is why aptamers are commonly comprised of single-stranded RNA or DNA (12). The G-quartet secondary structure, in particular, is a recurring theme in many different aptamers (13–15). For example, the thrombin-binding aptamer selected by Bock et al. (16) has been shown by both NMR and X-ray crystallography to fold into a

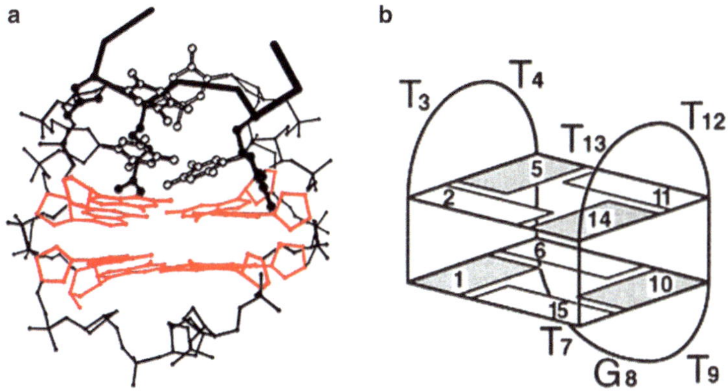

c

1 2 3 4 5 6 7 8 9 10 11 12 13 14 15

5′-GGTTGGTGTGGTTGG-3′

Fig. 1 NMR structure and sequence of the anti-thrombin aptamer. (**a**) The two G-quartets are in *red* and the thymine bridges are *bolded*. (**b**) Structural model of the aptamer; positions correlate to the sequence in (**c**). Figures were adapted with modifications from (17)

G-quartet structure, both in the presence of stabilizing salts and upon target binding (Fig. 1) (17–20).

Compared to antibodies, aptamers have many advantages that make them more suitable for use as sensing elements in biosensors. Aptamers are less toxic and less immunogenic than antibodies and will clear the body in a relatively short time frame (21). They also do not require an immune response in host animals to produce; aptamers are commonly produced chemically in an oligonucleotide synthesizer in large quantities. This allows aptamers to be produced against toxic ligands that cannot survive in a host animal. Typically, aptamers have sub-micromolar affinities to their targets that are comparable to antibodies (22). Although aptamers are selected for target binding in vitro, they are also capable of binding to their targets in vivo with good affinity and, as a result, many aptamers have been utilized for drug delivery and even as drugs themselves (23). Aptamers are also readily modified, which allows for different methods of attachment onto supportive matrices/substrates. In most types of sensor systems, the support matrix plays a large role in signal transduction; therefore the immobilization chemistry can have a large effect on the sensitivity of the sensor. For the purpose of electrochemical detection the support is a solid electrode, commonly gold, although other electrode materials, such as glassy carbon and indium-tin oxide (ITO), are also used. In particular, the properties of DNA immobilized on gold electrodes are very well characterized (24–30). Another advantage that aptamers have over antibodies for electrochemical biosensor applications is the ability to regenerate the electrode surface post target detection (31).

After incubating the electrode in a target-free buffer with chaotropic salts to denature the aptamer and release the bound target (32, 33), the aptamers on the sensor can be refolded into an active conformation to be reused. This particular characteristic of aptamers is simply not possible for most antibodies (34, 35), which also lays the cornerstone of aptamer affinity chromatography (36, 37).

Aptamers are generated via an in vitro selection method commonly known as SELEX (*s*ystematic *e*volution of *l*igands by *ex*ponential enrichment) (4, 5). Various modifications to the original SELEX procedure have been reported, such as capillary electrophoresis SELEX (38), atomic force microscopy SELEX (39), and automated SELEX (40). More comprehensive reviews on the many types of in vitro aptamer selection can be found in the litterature (41–43).

The selection process begins with the solid-phase synthesis of a library of completely random oligonucleotides, either single-stranded RNA or single-stranded DNA, which typically contains 10^{13}–10^{15} different sequences (40–80 nucleotides long) (42). Due to the importance of secondary structures to target binding, libraries containing alternating regions of purine and pyrimidine designed to increase the probability of stem–loop structures have been proposed (44). The random oligo region is flanked with two primer regions usually between 15 and 18 nucleotides in length that allows for the PCR amplification of the successful target binding sequence in the subsequent steps; for an RNA sequence it will be necessary to first transcribe into the equivalent DNA sequence prior to amplification.

In the conventional SELEX method, the target is immobilized onto a solid support, typically silica gel in a column, via biotin–streptavidin interactions. The library is then incubated in the column to allow the candidate sequences to interact with the target. The column is then washed with buffer in order to separate out the non-binding sequences. The binding sequences are then eluted out from the column under highly chaotropic conditions and then amplified using PCR. The successful sequences are then returned to the column and another iteration of the cycle begins (Fig. 2). Usually 10–20 cycles are enough to generate aptamers with consensus sequences. The incubation times in the selection column are typically on the order of a few minutes, but if an aptamer with faster binding kinetics is desired then the incubation time can be decreased during the last few rounds in order to place an increased selection pressure on the library. Sometimes counter-selection is performed using a secondary target with a similar structure so as to ensure that the successful aptamer is selective only for the desired target. In counter-selection, alternating cycles are carried out where sequences that bind to the secondary target are removed from the selection pool. Sequence minimization is then performed to determine the smallest possible binding sequence. One of the limitations

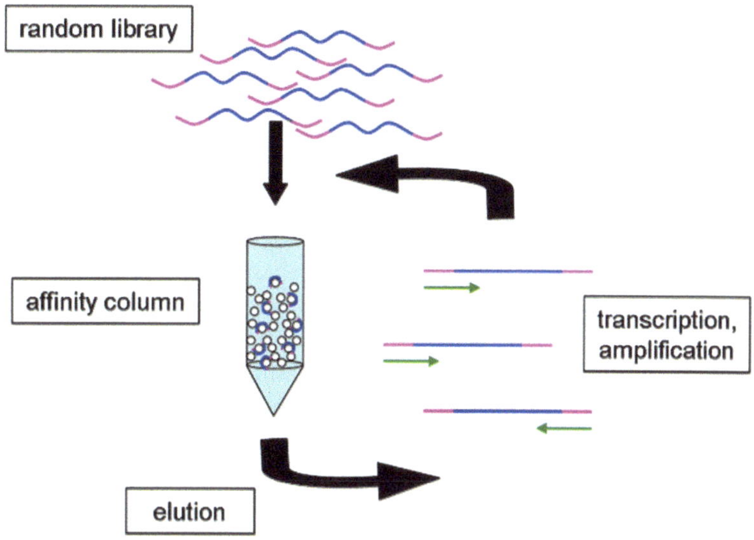

Fig. 2 General schematic of the SELEX process. A library consisting of random sequences of 40–80 nucleotides (in *blue*) flanked by PCR primer regions is introduced into an affinity column. The target is immobilized onto the solid-phase material to capture aptamer candidates which are then eluted out for PCR amplification. Aptamer candidates are then cycled through the selection process again. Usually 15–20 iterative cycles are sufficient to produce a consensus aptamer sequence

with in vitro selection is that optimal binding will usually occur only under the selection conditions (so-called "selection pressures"). This can affect the downstream application of the aptamer, so the selection conditions would typically emulate physiological environments.

The process of creating an aptamer-based electrochemical biosensor first requires the selection of an aptamer for the desired target following the procedure described above. Due to the flexible nature of the target in the selection process, it is entirely possible to select for any target, even targets whose structure is entirely unknown. In fact, in vitro selection targeting whole cells has been used to select for aptamers that bind to previously unknown protein targets on the outer membranes of those cells (11, 45). The versatility of the SELEX protocol for selecting aptamers that target a wide range of biomarker molecules is a large part of the appeal of using aptamers as the sensing element in biosensors.

2 Aptamer-Based Electrochemical Biosensors

It is possible to classify electrochemical biosensors into two broad categories based on the nature of the recognition mechanism which can then be subdivided into more specific categories. Labeling methods require the use of a redox active compound, attached either to the aptamer probe or the analyte itself, to signal a molecular recognition event and include the dual binding site-based and the conformational change-based detection schemes. Label-free methods, on the other hand, detect binding events through a change in the physical or chemical properties of the electrode itself, with or without the use of solution-diffused redox markers. These include the conductivity change-based detection scheme and the surface property-based detection schemes. The main advantage of label-free methods is that they tend to be less complex than labeled systems; however, they can also be less sensitive to the target binding. It is important to note that the majority of aptamer-based electrochemical sensors are created with DNA aptamers rather than their RNA counterparts. Although RNA aptamers tend to have more complex secondary structures compared to DNA aptamers, the resulting improvement in binding ability is offset by its instability toward hydrolysis (due to the hydroxyl group on its 2′ carbon) (46). DNA aptamers are more stable compared to unmodified RNA aptamers and, as a result, they are more amenable to sensor applications and can potentially be stored for a long period of time prior to use. Although modified RNA selection libraries can also be used to produce aptamers with increased stability (47, 48), DNA aptamers are less sensitive to nuclease degradation in biological solutions (46). It is possible to regenerate and reuse a DNA aptamer-based biosensor; it would be much more challenging with a RNA aptamer-based biosensor.

2.1 Dual Binding Site-Based Detection Schemes

The basic construction of the dual binding site-based electrochemical sensor (also called a "sandwich-type" scheme) uses the same concept as an enzyme-link immunosorbent assay (ELISA), which relies on the availability of multiple binding sites on the same analyte molecule. In this detection scheme, an unlabeled aptamer (probe aptamer) is first immobilized onto the electrode substrate and then the target is allowed to bind onto the electrode (Fig. 3). After that, a second labeled aptamer (signaling aptamer) is bound onto the target allowing for electrochemical detection.

There are two common types of labels for aptamer-based electrochemical sensors. Enzymatic labels catalyze redox reactions in the reaction buffer that generate an electrochemical current, whereas small molecule redox active labels take part directly in the redox process to generate a current. Usually in the sandwich-type assay the enzymatic labels are cross-linked to streptavidin or avidin and then allowed to interact with the biotinylated aptamer to form the aptamer–enzyme conjugate used in electrochemical signaling. The aptamers are typically covalently bound to biotin via standard phosphoramidite chemistry (35) for which commercial reaction kits are available (49, 50). It is also possible to attach biotin to the aptamers via the use of biotinylated primers in the PCR amplification

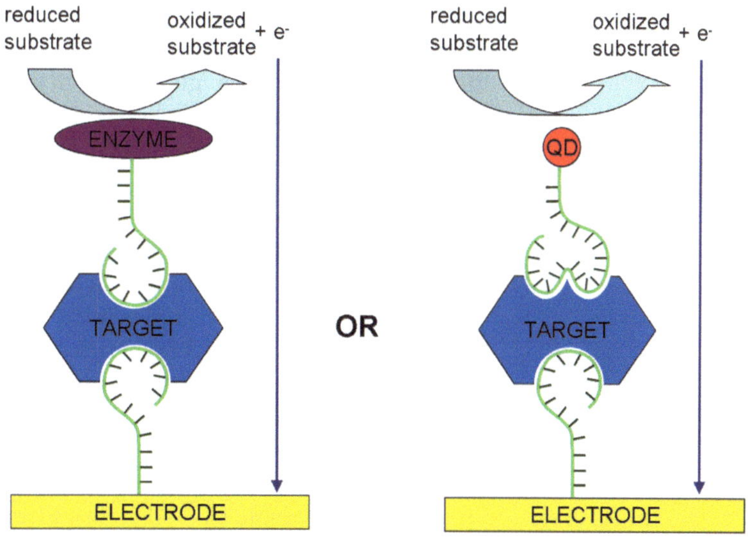

Fig. 3 General detection scheme of dual binding site-based detection schemes to create aptamer-based electrochemical biosensors. For targets with more than one binding site for a particular aptamer, it is possible to modify the aptamer with an enzyme or a quantum dot and use it as part of the electrochemical signaling mechanism. For targets with only one binding site for a particular aptamer, the selection of a second aptamer that binds on another region in the target is required

step. Small molecule redox labels can be covalently tethered to the 5′ or 3′ end of the aptamer strand or they can be non-covalently bound via base intercalation or via charge interactions (51).

For the dual binding site-based detection scheme to work there must be multiple binding sites on the target available for the same aptamer or the signaling and probe aptamers must bind at different sites on the target. Often the signaling aptamer is enzyme-modified instead of being directly labeled with a redox active marker. It is also possible to modify signaling aptamers with quantum dots and other types of nanoparticles, which will either catalyze a redox reaction (52) or be dissolved and quantified in subsequent electrochemical measurements (53, 54).

A dual binding site-based sensor design with two different thrombin-binding aptamers that recognized different epitopes of the protein has been demonstrated by Ikebukuro et al. (55). The signaling aptamer was modified with glucose dehydrogenase (GDH) which catalyzes the oxidation of glucose in the sample buffer to gluconolactone (56). The thiolated 29-nucleotide aptamer (probe) was immobilized onto a gold electrode; the 15-nucleotide aptamer (signaling) was modified with biotin, and was then coupled to an avidin-modified GDH. The protein target was incubated on the probe aptamer-modified electrode for 30 min at ambient conditions. The GDH-modified signaling aptamers were then allowed to bind to the thrombin (at a different binding site) and then the electrode was washed to remove nonspecifically bound aptamers. Coulometry was performed to quantify the number of bound thrombin in a glucose containing buffer spiked with a charge carrier, 1-methoxyphenazine methosulphate (m-PMS), to facilitate charge transfer between the enzyme and the electrode. The LoD for this sensor was on the order of 1 μM, which was far from optimal considering the nanomolar (nM) binding affinity of the thrombin-binding aptamers. It was determined that this may have been a result of the low activity of the GDH, and subsequently, Ikebukuro et al. redesigned their thrombin sensor with a new catalytic enzyme and exchanged the positions of the 15 nucleotides and 29 nucleotides aptamers in the sensor system. Pyrroquinoline quinone (PQQ) glucose dehydrogenase (GDH) conjugate was chosen as the new enzyme due to its high activity and its insensitivity to oxygen which is a major advantage for its use in a biological sample (57) (Fig. 4a). Electrochemical measurements

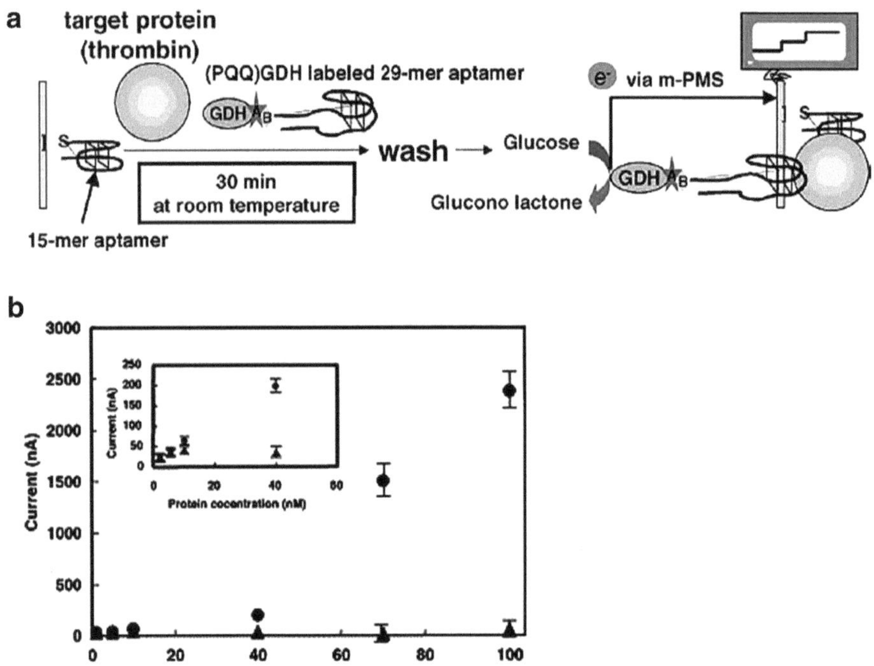

Fig. 4 (**a**) Detection scheme of the dual binding site-based electrochemical thrombin sensor using a pyrroquinoline quinone glucose dehydrogenase ((PQQ)GDH) enzymatic label. (PQQ)GDH and streptavidin were first cross-linked in the presence of glutaraldehyde and then bound to the biotinylated 29 nucleotides thrombin-binding aptamer to form the (PQQ)GDH–aptamer signaling complex. The thiolated 15 nucleotides aptamer was immobilized onto a gold electrode and allowed to bind to the target thrombin in varying concentrations. The electrode was then incubated in a solution of (PQQ)GDH–aptamer complex. Electrochemical measurements were then carried out in a solution containing the reagent (glucose) spiked with a charge carrier, 1-methoxyphenazine methosulphate (m-PMS), to facilitate the movement of charges between the enzyme and the electrode. (**b**) Concentration-dependent response of sensor with thrombin (*circles*) and negative control, BSA (*triangles*). *Inset*: magnified plot between 0 and 60 nM thrombin. Figures were adapted with modifications from (58)

were then carried out in the same manner, i.e., in a solution containing the enzymatic reagent (glucose) spiked with the charge carrier, m-PMS. They found that this new enzyme improved the LoD of their sensor by two orders of magnitude (10 nM). This sensor shows a linear response to thrombin in the concentration range of 40–100 nM (58) (Fig. 4b). Nevertheless, the developed thrombin sensor is a one-time use sensor because of the strong affinity of thrombin to its aptamer, making it difficult to regenerate the electrode surface for reuse.

Alternatively, if the target protein only has one aptamer-binding site, it is also possible to use a labeled antibody as the signaling beacon if it binds at a different site. This was the detection scheme implemented by Centi et al. in their C-reactive protein (CRP) sensor (59). CRP is a known serum biomarker for systematic inflammation and has been implicated in many chronic illnesses, such as diabetes, cancers, Alzheimer's, and cardiovascular diseases (60). Due to the instability of RNA in biological fluids, Centi et al. modified the RNA-based CRP-binding aptamer with F-pyrimidines in order to increase its stability and immobilized it onto a streptavidin-coated magnetic bead (59). Upon binding to CRP, the magnetic beads were washed and incubated in a solution containing biotinylated IgG-anti-CRP antibodies. The signaling enzyme, alkaline phosphatase (ALP), was cross-linked to streptavidin and then immobilized onto the magnetic bead assembly. The sensor was exposed to different concentrations of CRP in the presence of the enzymatic substrate (α-naphthyl phosphate which is hydrolyzed into 1-naphthol). Target binding was detected by differential pulse voltammetry (DPV) on a screen-printed electrode placed on a magnetic block which attracts the magnetic beads onto the electrode surface (Fig. 5). The LoD for this sensor is 5.4×10^{-2} mg/L (~2.15 nM); the dynamic range of this sensor is 0.1–50 mg/L. A major advantage of immobilizing the aptamer onto a magnetic bead versus onto a solid electrode support is that a bead can be suspended in the biological sample and is thus less dependent on the diffusion of the target to reach the aptamer for a binding event to occur (i.e., reduce the binding equilibration time).

One method to potentially improve the LoD is to use nanostructures as the signaling agent in a dual binding site-based sensing scheme. Yang et al. created a dual binding site-based thrombin electrochemical sensor (53), for which a CdSe quantum dot (QD) was attached to the signaling aptamer in place of an enzyme. After the signaling aptamer was bound onto the electrode surface, the CdSe QDs were dissolved with the addition of a 0.1 M nitric acid solution. Thus generated cadmium ions were then detected via square-wave stripping voltammetry. The LoD for this sensor (1 pM) is four orders of magnitude better than that of Ikebukuro's ((PQQ)GDH)-tagged sensor (59).

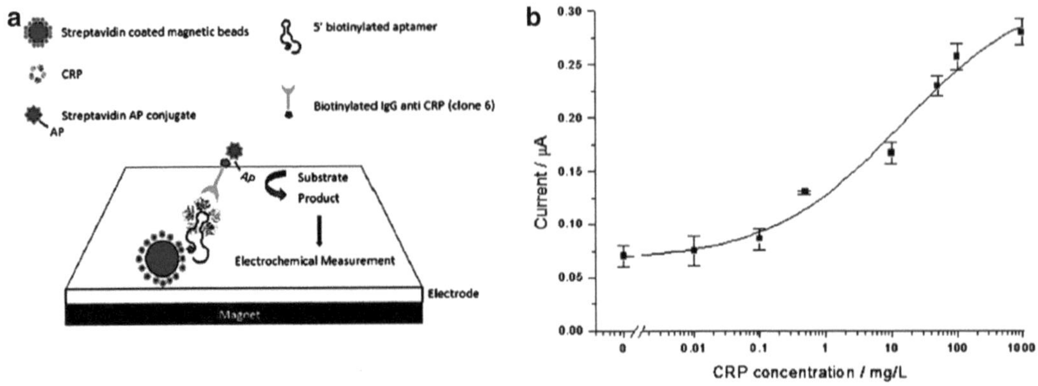

Fig. 5 (**a**) Schematic of C-reactive protein sensor using streptavidin-coated magnetic beads and anti-CRP antibody. Biotinylated aptamers were bound onto the magnetic beads via streptavidin–biotin interaction; unbound streptavidin sites were then blocked by free biotin. CRP was introduced into a solution of the aptamer-modified magnetic beads at varying concentrations. Biotinylated IgG anti-CRP antibodies were then allowed to interact with the target and the subsequent addition of streptavidin–alkaline phosphatase conjugates allowed for the enzymatic reaction to take place. Prior to the detection of the enzymatic product, the magnetic beads were concentrated onto a magnet-backed electrode. (**b**) Dose–response curve of sensor for CRP in 10× diluted serum. *Solid line*: theoretical sigmoidal data interpolation. Copyright Wiley-VCH Verlag GmbH & Co. KGaA. Reproduced with permission from (59)

Hansen et al. demonstrated the ability of this type of dual binding site-based electrochemical sensor to detect multiple targets with great selectivity and sensitivity in a mixed target sample (54). In their design, a mixed monolayer of thrombin- and lysozyme-binding aptamers was immobilized onto a gold electrode. CdS and PbS QDs were attached onto thrombin and lysozyme respectively via coupling reactions; the QD-modified proteins were then allowed to bind to their respective aptamers on the electrode surface (Fig. 6). Upon immersion in a solution containing unlabeled thrombin or lysozyme, competitive displacement of the QD-labeled signaling proteins on the sensor surface occurs. The QD labels left on the electrode are dissolved in 0.1 M nitric acid. The Cd^{2+} and/or Pb^{2+} ions in the wash solution are then detected by square-wave stripping voltammetry. The LoD of this sensor is very impressive (~0.5 pM), which was attributed to the presence of multiple QDs on the signaling proteins. This technique can be used for the creation of a sensor that is capable of simultaneously detecting a host of biomarkers and can decrease the amount of time spent in pursuit of a medical diagnosis.

However, it is not always necessary to dissolve the signaling QDs in a nanomaterial-based biosensor in order to detect the binding. Polsky et al. have devised a method using platinum nanoparticles as the signaling catalyst in anti-thrombin aptamer sandwich sensor (52). The anti-thrombin aptamer is immobilized onto a gold electrode; after the thrombin binds onto the electrode

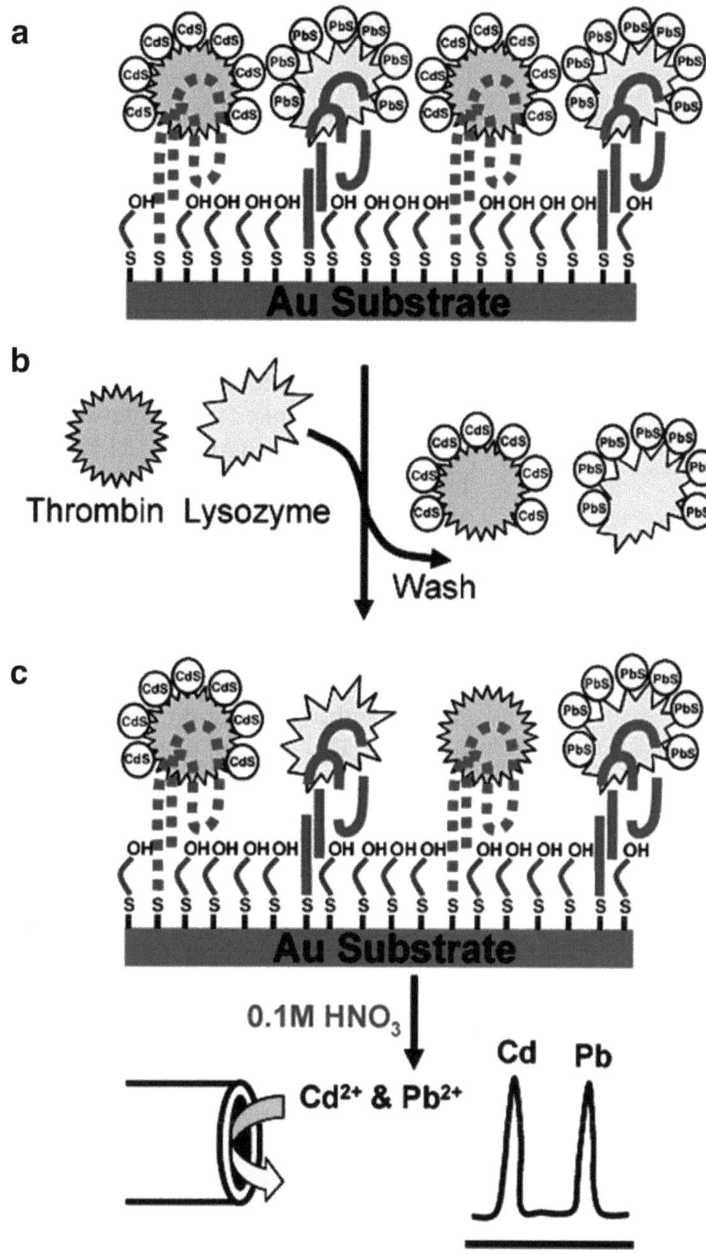

Fig. 6 Detection scheme of a sandwich-type aptamer sensor using the competitive displacement of the target. (**a**) The mixed monolayer of thrombin and lysozyme aptamers on the gold electrode surface bound to QD-labeled targets. (**b**) Upon competitive binding with unlabeled targets in a sample, loss of QD-labeled targets on the electrode surface occurs. The remaining QDs are then dissolved in nitric acid (**c**) and detected via square-wave stripping voltammetry. Figures were adapted with modifications from (54)

surface, a signaling aptamer is bound onto another binding site on the thrombin. For this sensor design, a platinum nanoparticle is attached to the end of the signaling aptamer, which is used to catalyze the reduction of hydrogen peroxide to water. The LoD of this sensor is close to 1 nM.

Although the majority of aptamer-modified electrodes tend to use gold as the electrode material, it is also possible to use other electrode materials. As an alternative to gold, Xiang et al. designed a thrombin sensor using an aptamer-modified glassy carbon electrode (2). The amine-modified anti-thrombin aptamer was immobilized onto the glassy carbon electrode following the electrochemical polymerization of *p*-amino-benzoic acid on the electrode surface. A second amine-modified thrombin-binding aptamer was covalently linked to a single-walled carbon nanotube (SWCNT) that had been treated with ALP. The ability for the SWCNT to bind to multiple enzymes essentially allows it to act as a signal amplifier. During cyclic voltammetry in the presence of the substrate, *p*-aminophenylphosphate (*p*-APP), it was shown that this sensor has a LoD of 10 ng/mL (2), corresponding to a molar concentration of about 277 pM. Xiang et al. then introduced redox recycling to this sensor design with the addition of a secondary enzyme, diaphorase (DI), into the system. The product of the *p*-APP dephosphorylation by ALP, *p*-aminophenol (*p*-AP), is electrochemically oxidized to *p*-quinoneimine (*p*-QI) on the surface of the SWCNT. In the presence of DI, *p*-QI is then reduced back into *p*-AP which can then be cycled back into *p*-QI. The DI is reduced by the addition of NADH in the system (Fig. 7). The LoD of the sensor utilizing redox recycling is 0.3 pg/mL, corresponding to 8.3 fM. This is clearly a very significant improvement over single-enzyme labels and shows a better sensitivity when compared to the nanoparticle-labeled detection systems.

2.2 Conformational Change-Based Detection Schemes

Although dual binding site-based electrochemical aptamer biosensors can be designed to have very low limits of detection, their fabrication and their detection schemes can get complicated and the need for multiple binding and washing steps makes them unsuitable for a one-step, user-friendly biosensor device. Another common detection scheme is based on the change in the three-dimensional conformation of the aptamer strand upon target binding; this in turn results in a variation in the electrochemical response of the sensor. In these types of sensors, the aptamers are typically end labeled with a redox marker and upon target binding either the aptamer can fold into a more compact structure, thus bringing the redox marker closer to the electrode surface and increasing the rate of electron transfer, or the aptamer can unfold, thus moving the redox marker away from the electrode surface and decreasing the electrochemical signal (Fig. 8). It is important to first fold the aptamer into the appropriate secondary structure prior

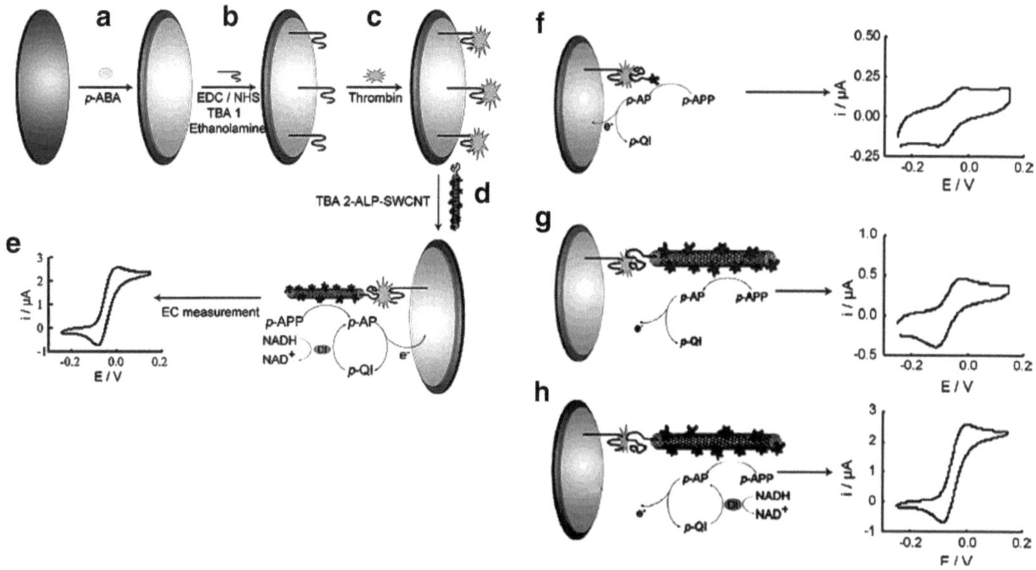

Fig. 7 Schematic for the fabrication of an ultrasensitive detection of thrombin with the use of alkaline phosphatase-modified single-walled carbon nanotube (SWCNT) and the detection of thrombin with this sensor. (**a**) *p*-amino benzoic acid (*p*-ABA) is electrically polymerized onto a glassy carbon electrode. (**b**) *N*-(3-dimethylamminopropyl)-*N*-ethylcarbodiimide hydrochloride–*N*-hydroxysulfosuccinimid coupling (EDC-NHS) of the 15 nucleotides thrombin-binding aptamer (TBA1) and the temporary blocking of the active sites by ethanolamine. (**c**) Displacement of ethanolamine by the thrombin target binding onto the electrode surface. (**d**) Binding of the 29 nucleotides thrombin-binding aptamer-modified alkaline phosphatase–SWCNT conjugate (TBA2-ALP–SWCNT) onto the electrode-bound thrombin. (**e**) Electrochemical measurement of the current generated by the catalytic reaction in the presence of *p*-aminophneylphosphate (*p*-APP), nicotinamide adenine dinucleotide (NADH, NAD^+), and disphorase (DI). (**f**) Diagram and cyclic voltammogram for the single-enzyme detection scheme with 1 μg mL^{-1} thrombin with 1 mM *p*-APP substrate. (**g**) Diagram and cyclic voltammogram for the enzymatic detection of 10 ng mL^{-1} thrombin with 1 mM *p*-APP substrate using TBA2-ALP–SWCNT conjugates without DI in solution. (**h**) Diagram and cyclic voltammogram for the enzymatic detection of 1 ng mL^{-1} thrombin with 1 mM *p*-APP substrate using TBA2-ALP–SWCNT conjugates with 2 μM DI and 4 mM NADH in solution. The current signal increases by about three magnitudes in this ultrasensitive enzymatic sandwich-type design (**h**) compared the typical enzymatic sandwich-type design (**f**). Figures were adapted with modifications from (3)

to immobilizing onto the electrode. This is usually done via the heating and slow cooling of the aptamer into its optimal three-dimensional structure in the immobilization buffer which normally contains appropriate salts and adjusted to physiological pH.

As mentioned previously, the thrombin-binding aptamer is known to fold into a compact G-quartet structure upon binding to thrombin (18). Radi et al. took advantage of this feature by designing an electrochemical sensor for thrombin using the 15-nucleotide thrombin-binding aptamer that has been modified on the 5′ and 3′ end with a thiol and a ferrocene group, respectively (33). The aptamer was immobilized via the 5′-thiol onto a gold electrode surface that was then passivated with 2-mercaptoethanol forming a mixed monolayer. Because of the short length of the aptamer,

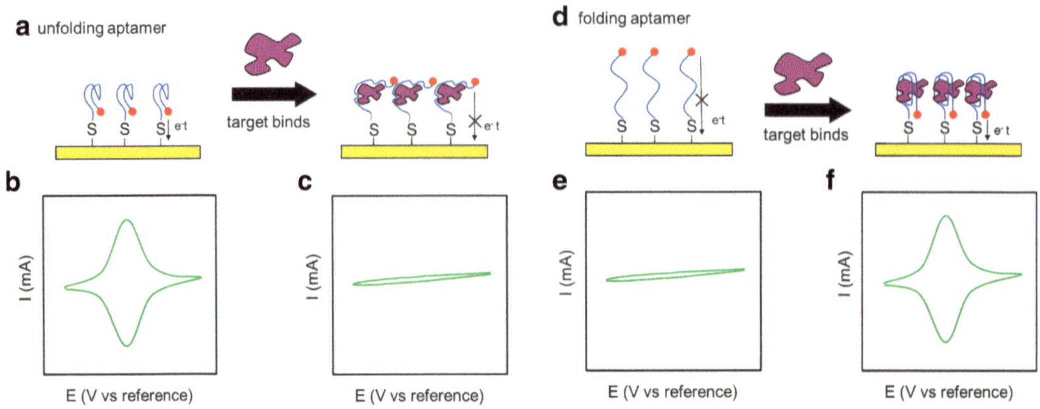

Fig. 8 Diagrams of an electrochemical aptamer sensor using the conformational change-based detection scheme. There are two possible modes of conformational change. Part (**a**) illustrates the aptamer unfolding mode. Upon binding to the target, the aptamer sensor unfolds causing a disruption to the electron transfer between the end-labeled redox marker and the electrode surface. Depictions of cyclic voltammograms before (**b**) and after (**c**) target binding. For the conformational change-based detection scheme, the theoretical cyclic voltammograms will be symmetrical because of the surface-bound redox label. Part (**d**) illustrates the aptamer folding mode. Upon binding to the target, the aptamer sensor folds up, bringing the redox marker close to the electrode surface and allowing electron transfer to occur. Depictions of cyclic voltammograms before (**e**) and after (**f**) target binding in this second scenario

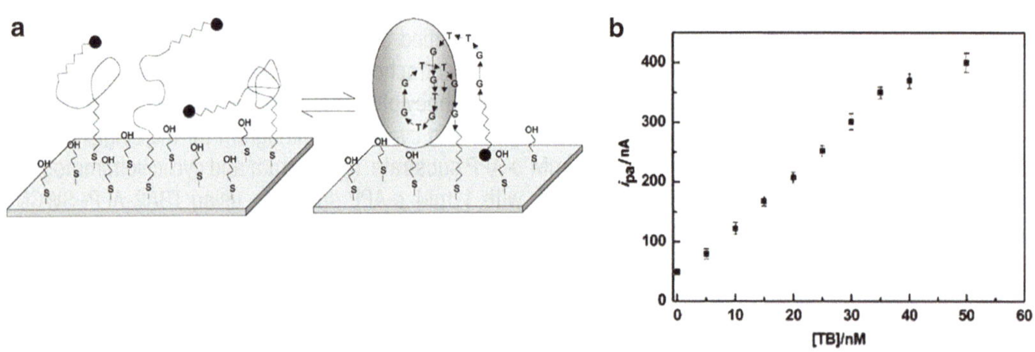

Fig. 9 (**a**) Schematic of an electrochemical thrombin sensor using the conformational change-based detection scheme. Target binding folds the aptamer into a G-quartet structure, bringing the ferrocene marker close to the electrode surface. *Closed black circles* represent the ferrocene moiety. (**b**) Calibration curve of the sensor clearly illustrating the linear range between 5.0 nM and 35 nM. The limit of detection is 0.5 nM. Figures were adapted with modifications from (33)

Radi et al. suggest that the folding of the aptamer will bring the ferrocene moiety to the electrode surface, thus decreasing the distance and increasing the electron transfer rate (Fig. 9). Since the thrombin-binding aptamer also spontaneously folds into the G-quartet structure in the presence of high salt concentrations (61), Radi et al. examined the effect of different monovalent cations on the folding of the immobilized thrombin-binding aptamer with fluorescence spectroscopy. While the effect of K^+ ions was far more prominent than that of the Li^+ or Na^+, it was determined that the binding experiments in the absence of all salts showed the highest sensitivity. The sensor was then characterized by Radi et al. (33), using cyclic voltammetry (CV), differential pulse voltammetry (DPV), and electrochemical impedance spectroscopy (EIS), and by Sánchez et al. (62), using chronoamperometry (CA), chronopotentiometry (CP), and square-wave voltammetry (SWV). It was demonstrated that this sensor has a LoD of 0.1 nM using SWV as the electrochemical method of detection; it was possible to regenerate the sensor up to 25 times by placing it in a stirred solution of 0.1 M hydrochloric acid for 60 s (62).

Xiao et al. have also created a thrombin sensor with the 32-nucleotide thrombin-binding aptamer that was 3′-end labeled with the redox marker methylene blue (MB) (63). Their sensor has been demonstrated to be able to detect the presence of thrombin even in a complex biological matrix such as fetal calf serum. In contrast to the thrombin sensor of Radi et al. this sensor was created with a longer thrombin-binding aptamer and so the electrochemical signal does not increase but rather decrease with increasing concentrations of thrombin. Xiao et al. proposed that this is due to the fact that the longer aptamer (a single-stranded DNA) is more flexible which allows it to move toward the electrode dynamically (Fig. 10a). This increases the probability of the redox marker colliding with the electrode prior to binding to thrombin; after binding the aptamer is no longer as flexible and thus the electrochemical signal decreases. Because the ionic strength of the environment correlated to the degree to which the thrombin-binding aptamer folds into a G-quartet in the absence

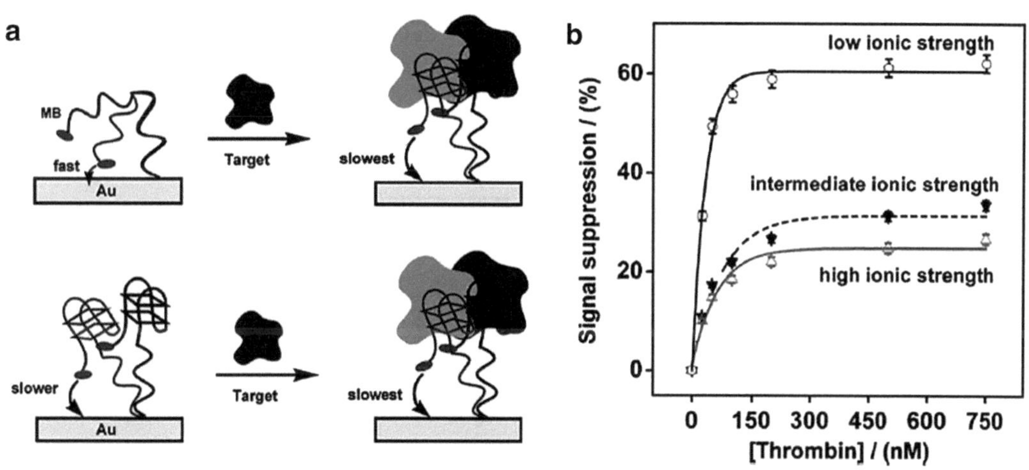

Fig. 10 (**a**) The two possible structures of a conformational changed-based sensor using the 32-base long thrombin-binding aptamer. *Top*: the sensor in low ionic strength environment; *bottom*: in high ionic strength environment. (**b**) Concentration dependence plots of the sensor under different ionic strengths (64). Copyright Wiley-VCH Verlag GmbH & Co. KGaA. Reproduced with permission

of its target, Xiao et al. tested the ability of their sensor to operate in buffers of different ionic strength (64). As expected, they found that in high and medium ionic strength buffers, their sensor was less responsive to increased concentrations of thrombin, as a result of increased prevalence of G-quartets on the initial sensor surface (Fig. 10b). The results clearly show that in addition to optimizing the aptamer sensor architecture, it is also necessary to optimize the operating conditions in order to achieve optimal sensor sensitivity.

Like the thrombin-binding aptamer, the cocaine-binding aptamer is suited for incorporation into a conformational change-based architecture due to its binding-dependent secondary structure. The cocaine-binding aptamer was first identified by Stojanovik et al. as a three-way junction that bound cocaine at the center (65) (Fig. 11a). Despite being partially folded prior to the binding of cocaine, the aptamer is still flexible at one end; this flexibility allowed Baker et al. to create an electrochemical sensor to detect cocaine in a complex biological matrix (67). Methylene blue (MB) was attached to the 3′ end of the aptamer as a redox tag; the aptamer is then partially folded and immobilized onto a gold electrode surface at the 5′ end (Fig. 11b). Upon binding to cocaine, the aptamer folds itself into a rigid three-way junction and brings the MB tag close to the electrode surface and increasing the electrochemical signal. This sensor was found to be inactive to common cocaine cutting agents such as flour and sugar while simultaneously being responsive to cocaine despite the presence of the masking agent cobaltous thiocyanate. Swensen et al. then went on to create a real-time microfluidic electrochemical sensor for cocaine using the same electrode design (68). This real-time sensor had a response time less than one minute. After the sensor was saturated with cocaine, it was possible to regenerate the sensor after flushing the system with cocaine-free solution.

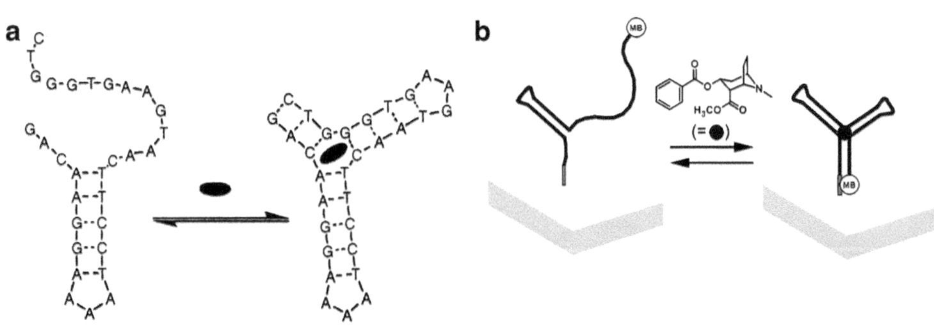

Fig. 11 (**a**) Structure of the cocaine-binding aptamer before and after binding its target. Figure adapted with modifications from (66). (**b**) The detection schematic of the conformational change-based electrochemical cocaine sensor. *MB* redox marker methylene blue. Figure adapted with modifications from (67)

Although using target binding-induced aptamer conformational change to detect analyte is an elegant approach to the signal transduction task, it is not easily generalized because not all aptamers will fold into or unfold from a more compact form upon target binding. Another option would be to engineer aptamer-based nucleic acid assemblies that will mimic these types of binding-activated structural changes.

The simplest design for these assemblies is to use a complementary nucleic acid to hybridize to a labeled aptamer immobilized onto the electrode surface. This produces a double-stranded nucleic acid that is much less flexible than a single-stranded aptamer and will protrude away from the electrode as a result, limiting electrode transfer from the end-tethered redox marker to the electrode. Upon adding target analyte, competitive displacement of the complementary strand will cause the aptamer assembly to dehybridize leaving only a single-stranded aptamer on the electrode surface. The single-stranded aptamer is flexible and will be able to "bend" toward the electrode which allows direct electron transfer to occur. Zuo et al. created an aptamer-based adenosine triphosphate (ATP) electrochemical sensor based on this design (Fig. 12a) (69). Their sensor had a LoD of 10 nM and was highly selective to ATP (i.e., does not respond to similar molecules such as cytidine triphosphate, uridine triphosphate, and guanidine triphosphate). This sensor was then used to detect ATP in vitro in a freshly lysed cell culture to demonstrate its selectivity in a complex sample matrix.

Alternatively, it is feasible to immobilize the labeled complementary strand on the electrode instead; such a strand should be designed so that it will fold into a hairpin structure after the displacement of the aptamer strand (as this will improve electrochemical signal at the sensor). Lu et al. created an ATP sensor with the anti-ATP-binding aptamer, which had complementary bases on its

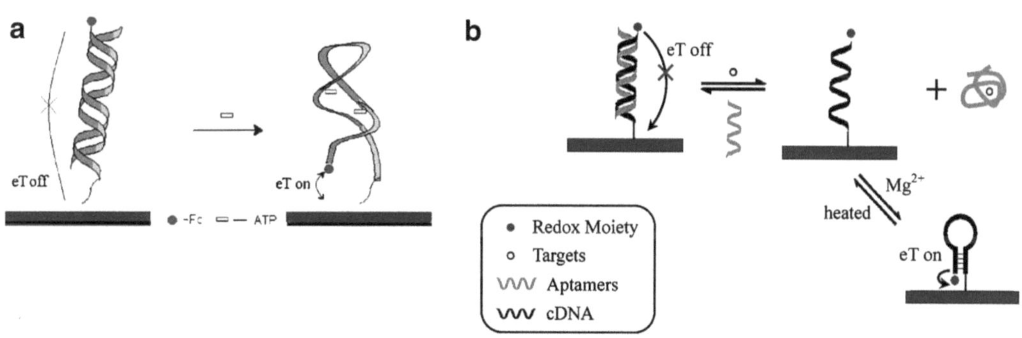

Fig. 12 (**a**) A simple aptamer-based assembly with complementary DNA that is displaced by the target for ATP detection (69). Schematic depicts the electron transfer (eT) state of ferrocene (Fc) before (off) and after (on) ATP binding. (**b**) A variation of (**a**) where the complementary strand is immobilized onto the electrode and folds into a compact structure upon displacement of the aptamer due to target binding. Figures were adapted with modifications from (70)

5′ and 3′ ends; they also created a thrombin sensor with the anti-thrombin aptamer which does not have this feature (Fig. 12b) (70). The introduction of additional bases on the 3′ end of the thrombin-binding aptamer created complementary bases to form a hairpin. The addition of Mg^{2+} to the target binding buffer was required to induce the Watson–Crick base pairing that forms the hairpin after removal of the aptamer strand upon target binding. One of the advantages of Lu et al.'s detection scheme was that it was possible to regenerate the sensor under milder conditions in comparison to sensors where the aptamer was directly immobilized onto the electrode. This is due to the high affinity of aptamers for their targets which makes it difficult to regenerate the aptamer sensor without the use of high concentrations of chaotropic salts.

2.3 Conductivity Change-Based Detection Schemes

Aside from forming binding-dependent sensor architecture, it is also possible to design an aptamer assembly that can modulate charge transfer through the base pairing of a double-stranded DNA to depend on the binding state of the aptamer (71). In a perfectly matched DNA duplex, the aromatic rings of the nucleotide are stacked on top of one another. As suggested, this allows charge transfer to occur through the DNA base-stacking; if the base-stacking is interrupted due to base mismatches then that charge transfer cannot occur (72). Beyond the scope of this chapter, several reviews on the topic of DNA charge transfer have been published previously (72–75).

Although the design of DNA conductivity charge-based detection schemes is most easily utilized in aptamers that have specific binding-induced structural changes, e.g., binding sites that form bulges in the double-stranded adenosine aptamer, helical aptamer-based electrochemical sensors that detect binding through charge transfer can be universally applied, as demonstrated by Huang et al. in their thrombin sensor (76). This thrombin sensor was constructed from a three-way junction between the thrombin-binding aptamer and a double helical DNA "conduction path." It was proposed that the binding of thrombin to the aptamer domain will cause the aptamer stem to fold tightly into a G-quartet (17, 18, 20), allowing the three-way junction to close up (76). The otherwise interrupted DNA conduction path will be repaired and allow charge transfer between the redox marker and the electrode to take place (Fig. 14a). Five different three-way junctions, each composed of a different combination of nucleotides, were examined for their charge transfer properties by using polyacrylamide gel electrophoresis assays. Once the best three-way junction was selected, a ferrocene marker was used to label the aptamer assembly, which was then immobilized onto the gold electrode (Fig. 14b). Square-wave voltammetry was performed on the electrodes both before and after thrombin binding. The LoD of this sensor was found to be 5 pM with a linear range between 5 and 150 pM. This sensor does not respond to comparable concentrations of BSA, avidin, IgA,

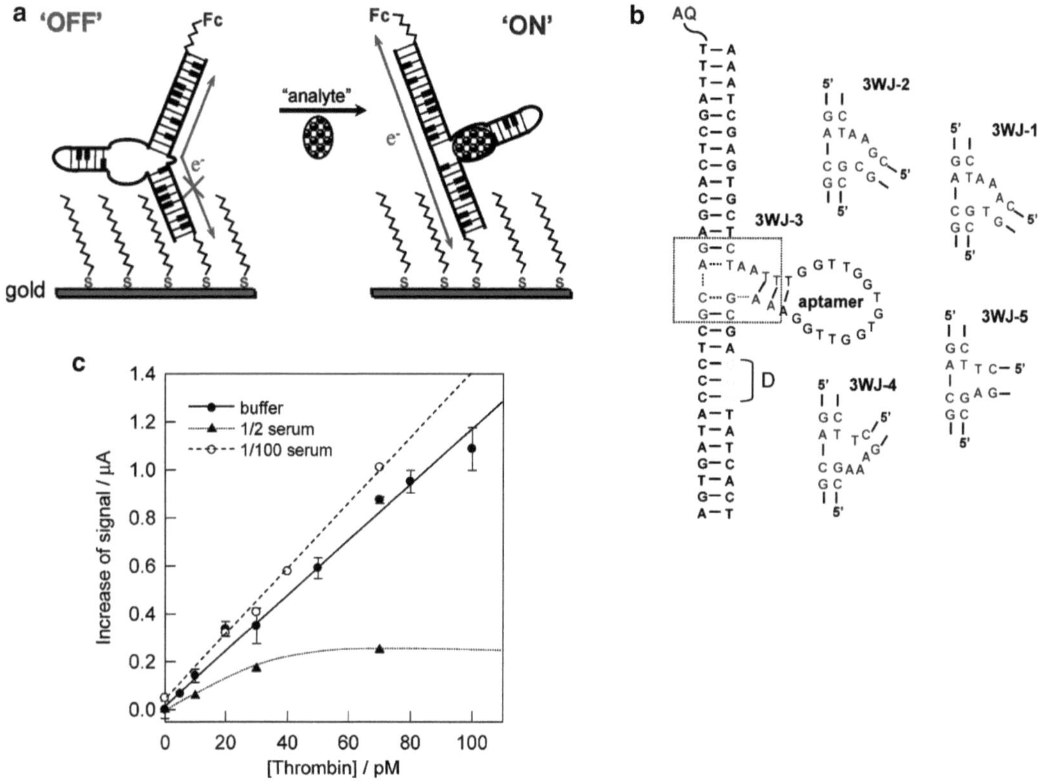

Fig. 13 (**a**) Upon binding to the analyte thrombin the base pairing of the signaling strand is reconstructed and electron transfer between the ferrocene redox marker and the electrode can occur. (**b**) Rational design of the three-way junction between the thrombin-binding aptamer and reporter stem. Anthraquinone (AQ) was used as the electron donor in gel-based assays for charge transfer to the triple G (D) marked in *blue*. Electron flows from AQ under UV irradiation, causing oxidative damage to guanine nucleotides. Five different junctions (3WJ-1–5) were examined. 3WJ-3 was shown to have the most guanine damage at D and thus the best electron transfer properties. (**c**) Sensor response to thrombin in buffer and buffer-diluted serum. Figures were adapted with modifications from (76)

and IgG; however, this sensor showed a marked decrease in the linear range within a serum matrix (Fig. 13c).

2.4 Surface Property Change-Based Detection Schemes

Aside from the above-mentioned detection schemes, it is also possible to detect the binding of a target to an aptamer-based electrochemical sensor via a change of the surface property of the aptamer-modified electrode. While aptamer-modified electrodes are negatively charged due to the phosphate backbone, the overall charge on the electrode surface can be altered upon the binding of a positively charged target. This, of course, will affect the diffusion of a solution-based redox marker toward the electrode and will lead to different redox responses of the sensing electrode.

Rodriguez et al. first demonstrated this principle in their lysozyme sensor (78). Lysozyme is a common antimicrobial

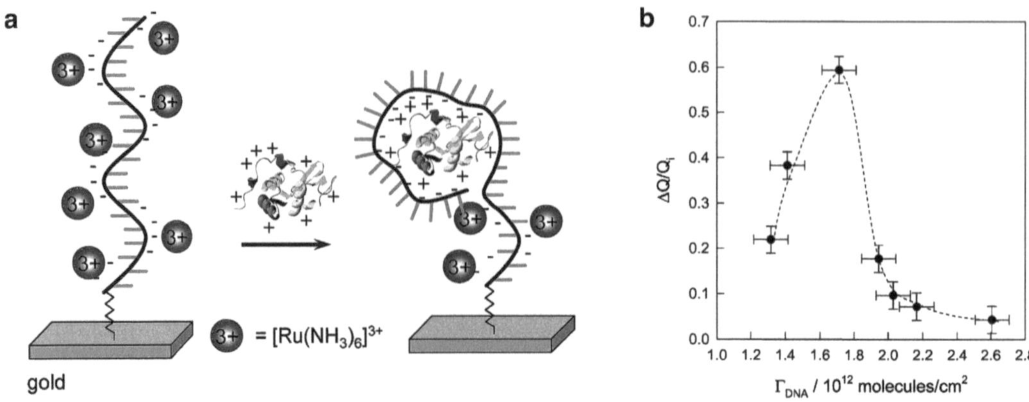

Fig. 14 (**a**) Change in charge density of the aptamer construct upon binding the positively charged lysozyme target causes a dissociation of the charged $(Ru(NH_3)_6)^{3+}$ redox marker to the phosphate backbone. (**b**) A plot of the sensor response vs. surface coverage of aptamer at 10 μg/mL lysozyme. The optimal surface coverage of aptamer on the surface of the electrode was found to be 1.7×10^{12} molecules/cm^2. Figures were adapted with modifications from (77)

enzyme that is positively charged at physiological pH. The single-stranded lysozyme-binding aptamer was immobilized onto an ITO electrode. Initially, the negatively charged oligonucleotide repelled the negatively charged $Fe(CN)_6^{3-/4-}$ redox markers in the physiological buffer solution and showed high charge transfer resistance during electrochemical impedance spectroscopy (EIS) measurements; however, upon binding to the positively charged lysozyme to the sensor, the resistance showed a marked decrease as a result of the decreased overall negative charges at the electrode surface. The LoD of this sensor was 0.2 μg/mL.

Cheng et al. have utilized the same phenomenon to create a lysozyme sensor with the positively charged redox marker, $Ru(NH_3)_6^{3+}$ (Fig. 14a) (77). However, by using such a positively charged redox marker, the number of electrostatically bound redox markers to the electrode surface decreases upon the binding of lysozyme. This is a classic example of a "signal-off" sensor where increasing target binding results in a decreased electrochemical signal. The LoD of this sensor was found to be 0.5 μg/mL, which is comparable to the sensor of Rodriguez et al. (78). Cheng et al. showed that there is an optimal surface density of aptamers on the sensor surface that will give the best electrochemical response (77). If the surface density is too high, the electrode response diminishes rapidly (Fig. 14b); this is most likely due to the tightly packed aptamer strands on the surface causing steric hindrance and preventing the lysozyme from binding. However, if the surface density is low, there is not enough aptamer strands on the surface to bind the target and the response is weaker. It is important to note that this method of detection only works if the binding buffer is not above the p*I* of the protein target; otherwise, the protein will

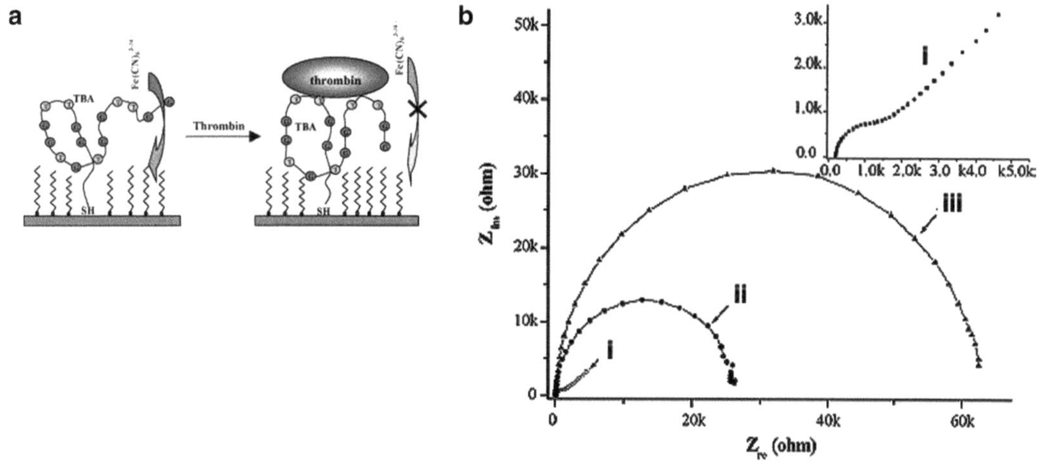

Fig. 15 (**a**) Detection scheme of surface property change-based thrombin electrochemical sensor. Upon binding of thrombin, the electroactive species $Fe(CN)_6^{3-/4-}$ is prevented from accessing the electrode surface. (*TBA* thrombin-binding aptamer.) (**b**) Nyquist plots of (i) the bare gold electrode, (ii) TBA-modified gold electrode after surface passivation with mercaptohexanol, and (iii) after 20 min incubation with a 1.0 μM thrombin containing buffer. *Inset*: magnified version of (i). Figures modified from (79)

be neutral and will not affect the overall charge density of the electrode surface.

Another important effect of protein target binding to the surface of an electrode is that the surface itself is being physically blocked off from the bulk of the buffer. As a result, electron transfer from the bulk of the buffer will be limited; this effect will hold even if the protein is not particularly large. Cai et al. (79) and Radi et al. (32) reported the design of a thrombin sensor that took advantage of this effect independently. Cai et al. immobilized the 15-mer thrombin-binding aptamer on a gold electrode that was then further modified with mercapto-1-hexanol to ensure full passivation of the gold surface from the ferricyanide redox markers in the buffer (Fig. 15) (79). Despite the surface passivation, the ferricyanide in the buffer can still be oxidized by the electrode. However, upon the binding of thrombin, the ferricyanide is blocked from accessing the electrode and charge transfer does not occur as readily. As a result, the charge transfer resistance, as measured by EIS, at the sensor surface increased; the LoD for this sensor was 0.1 nM. Radi et al. also immobilized the 15-mer thrombin-binding aptamer onto gold but protected the gold electrode surface with 2-mercaptoethanol instead (32). While the LoD for this sensor was 2 nM, which is less sensitive compared to the one by Cai et al., this sensor could be regenerated and reused up to 15 times by treating the electrode surface with 2.0 M sodium chloride.

Zhang et al. used the same detection scheme in the design of a thrombin sensor with a polyamidoamine dendrimer to improve sensitivity (80). A G_4-NH_2 polyamidoamine dendrimer was covalently

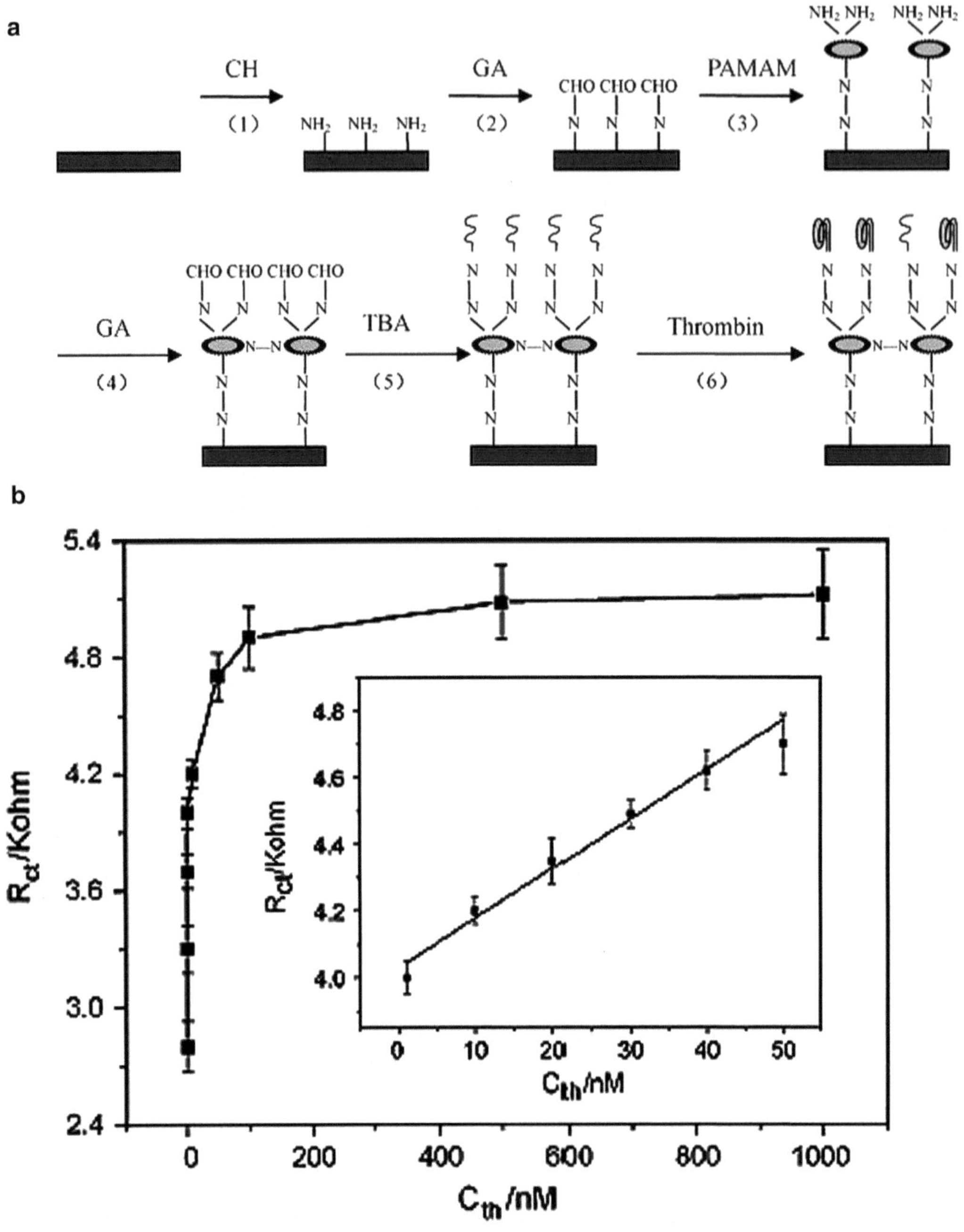

Fig. 16 (**a**) Schematic of the fabrication of a dendrimeric electrochemical thrombin sensor that uses the surface property change-based detection scheme. Upon binding to the thrombin target, the electrode surface is passivated and the conductivity of the sensor decreases. (*CH* cysteamine hydrochloride, *GA* glutaraldehyde, *PAMAM* G_4-amino-terminated polyamindoamine dendrimer, *TBA* thrombin-binding aptamer.) (**b**) Electrochemical impedance spectroscopy (EIS) response (R_{ct} resistance of charge transfer) of the dendrimeric electrochemical sensor at different concentrations of thrombin. *Inset*: Calibration curve of the EIS response. Figures were adapted with modifications from (80)

linked to a gold electrode surface and then coupled to the thrombin-binding aptamer to form a dendrimeric aptamer electrode which will allow a higher surface concentration of the aptamer strands (Fig. 16). Upon thrombin binding, the impedance at the electrode surface increases. The LoD of this sensor was 0.01 nM, a significant improvement from the sensors prepared on a "normal" electrode (32).

Aside from increasing surface density of the aptamer, there are other approaches to increasing target binding-induced surface property changes. Protein target denaturation can significantly increase the surface area that is passivated by unfolding the protein. For example, Xu et al. created a modified target detection protocol by including a denaturing step with 6 M guanidine hydrochloride just prior to electrochemical measurements (9). Their non-dendrimeric thrombin sensor had a LoD of 10 fM, which is several orders of magnitude better than the dendrimeric version of the same sensor. The combination of dendrimeric aptamers and target denaturation is a powerful design that could potentially produce sensors with significantly improved sensitivity.

3 Potential Clinical Applications as Implantable Biosensors

As mentioned above, one of the greatest advantages of using aptamers as the detection elements in biosensors is that they can be selected for any target of interest. The selection of aptamers and their applications have been very popular research topics since the 1990s. Aptamers targeting neurologically active small molecules such as dopamine (81, 82) and cocaine (65), as well as large proteins such as prions (15, 83, 84), α-amino-3-hydroxy-5-methyl-4-isoxazolepropionic acid (AMPA) receptors (85), and α-synuclein (86), have been reported in the literature.

The implantable aptamer-based sensors for real-time detection of biologically active molecules seem to be feasible in the near future, but have not yet been realized. Many important details must be kept in mind in creating such a sensor. It must be noted that an implantable sensor needs to be placed where the concentration of the target is meaningful for the clinical diagnosis. In most cases this will likely be within the circulatory system, where it will be possible to monitor the concentration of the biomarker in the bloodstream, or within the ventricles, in order to monitor the concentrations within the cerebrospinal fluid. Regardless, the sensor must be easily accessible, preferably near the surface of the skin, in order to allow for the leads to connect to an external microcontroller. Alternatively, an internal microcontroller can also be used but will require the use of wireless technology in order to be able to report the results to a healthcare provider.

Due to these considerations, the choice of aptamer targets is mostly limited to biomolecules that are present in these two locations in sufficient concentrations to be detected. While most aptamers have target binding affinities in the range of a few micromolars to a few nanomolars, for the purposes of creating an in vivo sensor it would be prudent to select for an aptamer that has an even better affinity to improve the LoD, as well as the linear working range, of the sensor. Aside from binding affinity, the binding kinetics should also be taken into account as it will ultimately determine the response time of the sensor. Depending on the application, a response time in the order of a few minutes to an hour may be acceptable, but in other situations a real-time response may be desired. However, the aptamer-target association/dissociation kinetics is correlated to the binding affinity (36), so that a balance will need to be achieved in order to create a real-time sensor for implantation. All of these factors will affect the SELEX protocol during the aptamer selection process.

Once the aptamer has been selected, the design of sensor architecture must also be considered. An implantable aptamer-based electrochemical sensor cannot rely on the use of solution-based redox markers that are not native to the human body and it cannot rely on detection protocols that cannot be done in vivo; therefore any detection protocol that requires the sample manipulation after target binding, such as the dual binding site-based detection scheme, cannot be used. A labeled detection method, such as the conformational change-based or conductivity change-based detection scheme, may be better for implantation in the body as it removes the need for high concentrations of a physiological redox marker for signal detection. Depending on the aptamer's secondary structure, it may be possible to exploit structural changes caused by target binding and eliminate the need to create an aptamer-nucleic acid assembly to do the same. The sensor must also be highly selective for its target in the complex biological fluid that it will be implanted into. Finally, as biofouling and nonspecific adsorption will always occur on sensor surfaces, the sensor response will eventually diminish over time. An implantable aptamer-based sensor will need to be responsive within the implantation time frame, whether this is in the order of a few hours to monitor a patient in the operating room or a few days to monitor a patient in a critical care setting.

4 Concluding Remarks

Biosensors have the ability to significantly help on the time and labor required for standard medical diagnostic testing and as a result will improve the delivery of timely treatment. Point-of-care biosensors can even eliminate the need for patients to visit medical laboratory facilities and allow patients to perform the tests themselves in the privacy and comfort of their homes. While this will require a visit to the physicians to confirm the diagnosis, it can save a vast amount of cost for the healthcare system. Due to the versatility of aptamers, they are the ideal detection element for biosensors. Because electrochemical sensors are the most easily adaptable for mass production, the engineering of aptamer-based electrochemical sensors will likely lead to the creation of an implantable aptamer-based sensor that provides real-time monitoring of the levels of biomarkers for the patient.

Acknowledgements

We are unable to provide more details of the relevant research of many colleagues due to space limitations. We hope all those who have contributed to make "aptamer-based electrochemical biosensors" exciting will accept a collective acknowledgment. This research has been financially supported by the Natural Science and Engineering Research Council (NSERC) and the Canadian Institute for Health Research (CIHR).

References

1. DropSens SL (2010) Dropsens: Screen-printed electrodes. http://www.dropsens.com/
2. Xiang Y, Zhang YY, Qian XQ, Chai YQ, Wang J, Yuan R (2010) Ultrasensitive aptamer-based protein detection via a dual amplified biocatalytic strategy. Biosens Bioelectron 25: 2539–2542
3. Wang J (2008) Electrochemical glucose biosensors. Chem Rev 108:814–825
4. Ellington AD, Szostak JW (1990) In vitro selection of RNA molecules that bind specific ligands. Nature 346:818–822
5. Tuerk C, Gold L (1990) Systematic evolution of ligands by exponential enrichment: RNA ligands to bacteriophage T4 DNA polymerase. Science 249:505–510
6. Tombelli S, Minunni M, Mascini M (2007) Aptamers-based assays for diagnostics, environmental and food analysis. Biomol Eng 24:191–200
7. Homann M, Goringer HU (1999) Combinatorial selection of high affinity RNA ligands to live african trypanosomes. Nucleic Acids Res 27:2006–2014
8. O'Sullivan C (2002) Aptasensors - the future of biosensing? Anal Bioanal Chem 372:44–48
9. Xu Y, Yang L, Ye XY, He PA, Fang YZ (2006) An aptamer-based protein biosensor by detecting the amplified impedance signal. Electroanalysis 18:1449–1456
10. Mayer G (2009) The chemical biology of aptamers. Angew Chem Int Ed 48:2672–2689
11. Lin JC, Thirumalai D (2008) Relative stability of helices determines the folding landscape of adenine riboswitch aptamers. J Am Chem Soc 130:14080–14081
12. Ellington AD, Szostak JW (1992) Selection in vitro of single-stranded DNA molecules that fold into specific ligand-binding structures. Nature 355:850–852
13. Pileur F, Andreola M-L, Dausse E, Michel J, Moreau S, Yamada H, Gaidamakov SA, Crouch RJ, Toulme J-J, Cazenave C (2003) Selective inhibitory DNA aptamers of the human RNase H1. Nucleic Acids Res 31:5776–5788
14. Okazawa A, Maeda H, Fukusaki E, Katakura Y, Kobayashi A (2000) In vitro selection of hematoporphyrin binding DNA aptamers. Bioorg Med Chem Lett 10:2653–2656
15. Weiss S, Proske D, Neumann M, Groschup MH, Kretzschmar HA, Famulok M, Winnacker EL (1997) RNA aptamers specifically interact with the prion protein PrP. J Virol 71: 8790–8797
16. Bock LC, Griffin LC, Latham JA, Vermaas EH, Toole JJ (1992) Selection of single-stranded DNA molecules that bind and inhibit human thrombin. Nature 355:564–566
17. Kelly JA, Feigon J, Yeates TO (1996) Reconciliation of the X-ray and nmr structures

of the thrombin-binding aptamer d(GGTTGGTGTGGTTGG). J Mol Biol 256:417–422
18. Wang KY, McCurdy S, Shea RG, Swaminathan S, Bolton PH (1993) A DNA aptamer which binds to and inhibits thrombin exhibits a new structural motif for DNA. Biochemistry (Mosc) 32:1899–1904
19. Tasset DM, Kubik MF, Steiner W (1997) Oligonucleotide inhibitors of human thrombin that bind distinct epitopes. J Mol Biol 272:688–698
20. Macaya RF, Waldron JA, Beutel BA, Gao HT, Joesten ME, Yang MH, Patel R, Bertelsen AH, Cook AF (1995) Structural and functional characterization of potent antithrombotic oligonucleotides possessing both quadruplex and duplex motifs. Biochemistry (Mosc) 34:4478–4492
21. Cerchia L, Hamm J, Libri D, Tavitian B, de Franciscis V (2002) Nucleic acid aptamers in cancer medicine. FEBS Lett 528:12–16
22. Brody EN, Willis MC, Smith JD, Jayasena S, Zichi D, Gold L (1999) The use of aptamers in large arrays for molecular diagnostics. Mol Diagn 4:381–388
23. Dausse E, Gomes SD, Toulme JJ (2009) Aptamers: a new class of oligonucleotides in the drug discovery pipeline? Curr Opin Pharmacol 9:602–607
24. Herne TM, Tarlov MJ (1997) Characterization of DNA probes immobilized on gold surfaces. J Am Chem Soc 119:8916–8920
25. Steel AB, Herne TM, Tarlov MJ (1998) Electrochemical quantitation of DNA immobilized on gold. Anal Chem 70:4670–4677
26. Anne A, Bouchardon A, Moiroux J (2003) 3′-Ferrocene-labeled oligonucleotide chains end-tethered to gold electrode surfaces: novel model systems for exploring flexibility of short DNA using cyclic voltammetry. J Am Chem Soc 125:1112–1113
27. Campuzano S, Pedrero M, Montemayor C, Fatás E, Pingarrón JM (2006) Characterization of alkanethiol-self-assembled monolayers-modified gold electrodes by electrochemical impedance spectroscopy. J Electroanal Chem 586:112–121
28. Ge B, Huang Y-C, Sen D, Yu H-Z (2007) Electrochemical investigation of DNA-modified surfaces: from quantitation methods to experimental conditions. J Electroanal Chem 602:156–162
29. Anne A, Bonnaudat C, Demaille C, Wang K (2007) Enzymatic redox 3′-end-labeling of DNA oligonucleotide monolayers on gold surfaces using terminal deoxynucleotidyl transferase (tdt)-mediated single base extension. J Am Chem Soc 129:2734–2735
30. McEwen GD, Chen F, Zhou A (2009) Immobilization, hybridization, and oxidation of synthetic DNA on gold surface: electron transfer investigated by electrochemistry and scanning tunneling microscopy. Anal Chim Acta 643:26–37
31. Warsinke A, Stöcklein W, Leupold E, Micheel E, Scheller FW (2005) In: Paleček E, Wang J, Scheller F (eds) Electrochemistry of nucleic acids: towards electrochemical sensors for genomics and proteomics, vol 1. Elsevier B. V, Amsterdam, pp 451–483
32. Radi A-E, Sánchez JLA, Baldrich E, O'Sullivan CK (2005) Reusable impedimetric aptasensor. Anal Chem 77:6320–6323
33. Radi A-E, Acero Sánchez JL, Baldrich E, O'Sullivan CK (2005) Reagentless, reusable, ultrasensitive electrochemical molecular beacon aptasensor. J Am Chem Soc 128:117–124
34. Ravelet C, Grosset C, Peyrin E (2006) Liquid chromatography, electrochromatography and capillary electrophoresis applications of DNA and RNA aptamers. J Chromatogr A 1117:1–10
35. Balamurugan S, Obubuafo A, Soper SA, Spivak DA (2008) Surface immobilization methods for aptamer diagnostic applications. Anal Bioanal Chem 390:1009–1021
36. Peyrin E (2009) Nucleic acid aptamer molecular recognition principles and application in liquid chromatography and capillary electrophoresis. J Sep Sci 32:1531–1536
37. Okanda FM, El Rassi Z (2007) Biospecific interaction (affinity) CEC and affinity nano-LC. Electrophoresis 28:89–98
38. Stratis-Cullum DN, McMasters S, Pellegrino PM (2009) Evaluation of relative aptamer binding to *Campylobacter jejuni* bacteria using affinity probe capillary electrophoresis. Anal Lett 42:2389–2402
39. Miyachi Y, Shimizu N, Ogino C, Kondo A (2010) Selection of DNA aptamers using atomic force microscopy. Nucleic Acids Res 38:e21–e29
40. Cox JC, Hayhurst A, Hesselberth J, Bayer TS, Georgiou G, Ellington AD (2002) Automated selection of aptamers against protein targets translated in vitro: from gene to aptamer. Nucleic Acids Res 30:e108
41. Hamula CLA, Guthrie JW, Zhang HQ, Li XF, Le XC (2006) Selection and analytical applications of aptamers. Trends Anal Chem 25: 681–691
42. Osborne SE, Ellington AD (1997) Nucleic acid selection and the challenge of combinatorial chemistry. Chem Rev 97:349–370
43. Yang Y, Yang DL, Schluesener HJ, Zhang ZR (2007) Advances in selex and application of aptamers in the central nervous system. Biomol Eng 24:583–592
44. Ruff KM, Snyder TM, Liu DR (2010) Enhanced functional potential of nucleic acid aptamer libraries patterned to increase secondary structure. J Am Chem Soc 132:9453–9464

45. Torres-Chavolla E, Alocilja EC (2009) Aptasensors for detection of microbial and viral pathogens. Biosens Bioelectron 24: 3175–3182
46. Metzler DE (1977) Biochemistry: the chemical reactions of living cells. Academic, New York
47. Famulok M, Mayer G, Blind M (2000) Nucleic acid aptamers: from selection in vitro to applications in vivo. Acc Chem Res 33:591–599
48. Pieken WA, Olsen DB, Benseler F, Aurup H, Eckstein F (1991) Kinetic characterization of ribonuclease-resistant 2′-modified hammerhead ribozymes. Science 253:314–317
49. Integrated DNA Technologies, Inc. (2011) Integrated DNA technologies. http://www.idtdna.com
50. Eurofins MWG Operon (2011) Eurofins MWG Operon. http://www.operon.com
51. Pale ek E, Fojta M, Jelen F, Vetterl V (2002) In: Wilson GS (ed) Bioelectrochemistry, vol 9. Wiley-VCH, Weinheim, pp 365–429
52. Polsky R, Gill R, Kaganovsky L, Willner I (2006) Nucleic acid-functionalized pt nanoparticles: catalytic labels for the amplified electrochemical detection of biomolecules. Anal Chem 78:2268–2271
53. Yang H, Ji J, Liu Y, Kong JL, Liu BH (2009) An aptamer-based biosensor for sensitive thrombin detection. Electrochem Commun 11:38–40
54. Hansen JA, Wang J, Kawde A-N, Xiang Y, Gothelf KV, Collins G (2006) Quantum-dot/aptamer-based ultrasensitive multi-analyte electrochemical biosensor. J Am Chem Soc 128:2228–2229
55. Ikebukuro K, Kiyohara C, Sode K (2004) Electrochemical detection of protein using a double aptamer sandwich. Anal Lett 37:2901–2909
56. Harrison DC (1931) Glucose dehydrogenase: a new oxidising enzyme from animal tissues. Biochem J 25:1016–1027
57. Sode K, Kojima K (1997) Improved substrate specificity and dynamic range for glucose measurement of *Escherichia coli* PQQ glucose dehydrogenase by site directed mutagenesis. Biotechnol Lett 19:1073–1077
58. Ikebukuro K, Kiyohara C, Sode K (2005) Novel electrochemical sensor system for protein using the aptamers in sandwich manner. Biosens Bioelectron 20:2168–2172
59. Centi S, Bonel Sanmartin L, Tombelli S, Palchetti I, Mascini M (2009) Detection of c reactive protein (crp) in serum by an electrochemical aptamer-based sandwich assay. Electroanalysis 21:1309–1315
60. Lavie CJ, Milani RV, Verma A, O'Keefe JH (2009) C-reactive protein and cardiovascular diseases—is it ready for primetime? Am J Med Sci 338:486–492
61. Baldrich E, Restrepo A, O'Sullivan CK (2004) Aptasensor development: elucidation of critical parameters for optimal aptamer performance. Anal Chem 76:7053–7063
62. Sánchez J, Baldrich E, Radi AG, Dondapati S, Sánchez P, Katakis I, O'Sullivan C (2006) Electronic 'off-on' molecular switch for rapid detection of thrombin. Electroanalysis 18:1957–1962
63. Xiao Y, Lubin AA, Heeger AJ, Plaxco KW (2005) Label-free electronic detection of thrombin in blood serum by using an aptamer-based sensor. Angew Chem Int Ed 44:5456–5459
64. Xiao Y, Uzawa T, White RJ, DeMartini D, Plaxco KW (2009) On the signaling of electrochemical aptamer-based sensors: collision- and folding-based mechanisms. Electroanalysis 21:1267–1271
65. Stojanovic MN, de Prada P, Landry DW (2000) Fluorescent sensors based on aptamer self-assembly. J Am Chem Soc 122: 11547–11548
66. Stojanovic MN, de Prada P, Landry DW (2001) Aptamer-based folding fluorescent sensor for cocaine. J Am Chem Soc 123:4928–4931
67. Baker BR, Lai RY, Wood MS, Doctor EH, Heeger AJ, Plaxco KW (2006) An electronic, aptamer-based small-molecule sensor for the rapid, label-free detection of cocaine in adulterated samples and biological fluids. J Am Chem Soc 128:3138–3139
68. Swensen JS, Xiao Y, Ferguson BS, Lubin AA, Lai RY, Heeger AJ, Plaxco KW, Soh HT (2009) Continuous, real-time monitoring of cocaine in undiluted blood serum via a microfluidic, electrochemical aptamer-based sensor. J Am Chem Soc 131:4262–4266
69. Zuo X, Song S, Zhang J, Pan D, Wang L, Fan C (2007) A target-responsive electrochemical aptamer switch (TREAS) for reagentless detection of nanomolar ATP. J Am Chem Soc 129:1042–1043
70. Lu Y, Li X, Zhang L, Yu P, Su L, Mao L (2008) Aptamer-based electrochemical sensors with aptamer: complementary DNA oligonucleotides as probe. Anal Chem 80:1883–1890
71. Sankar CG, Sen D (2004) DNA helix-stack switching as the basis for the design of versatile deoxyribosensors. J Mol Biol 340:459–467
72. Genereux JC, Barton JK (2010) Mechanisms for DNA charge transport. Chem Rev 110:1642–1662
73. Priyadarshy S, Risser SM, Beratan DN (1996) DNA is not a molecular wire: protein-like electron-transfer predicted for an extended pi-electron system. J Phys Chem 100:17678–17682

74. Giese B (2002) Long-distance electron transfer through DNA. Annu Rev Biochem 71:51–70
75. Wagenknecht H-A (2006) Electron transfer processes in DNA: mechanisms, biological relevance and application in DNA analytics. Nat Prod Rep 23:973–1006
76. Huang YC, Ge B, Sen D, Yu H-Z (2008) Immobilized DNA switches as electronic sensors for picomolar detection of plasma proteins. J Am Chem Soc 130:8023–8029
77. Cheng AKH, Ge B, Yu H-Z (2007) Aptamer-based biosensors for label-free voltammetric detection of lysozyme. Anal Chem 79: 5158–5164
78. Rodriguez MC, Kawde A-N, Wang J (2005) Aptamer biosensor for label-free impedance spectroscopy detection of proteins based on recognition-induced switching of the surface charge. Chem Commun 4267–4269
79. Cai H, Lee TM-H, Hsing IM (2006) Label-free protein recognition using an aptamer-based impedance measurement assay. Sens Actuators B 114:433–437
80. Zhang Z, Yang W, Wang J, Yang C, Yang F, Yang X (2009) A sensitive impedimetric thrombin aptasensor based on polyamidoamine dendrimer. Talanta 78:1240–1245
81. Mannironi C, Di Nardo A, Fruscoloni P, Tocchini-Valentini GP (1997) In vitro selection of dopamine RNA ligands. Biochemistry (Mosc) 36:9726–9734
82. Walsh R, DeRosa MC (2009) Retention of function in the DNA homolog of the RNA dopamine aptamer. Biochem Biophys Res Commun 388:732–735
83. Gilch S, Schatzl HM (2009) Aptamers against prion proteins and prions. Cell Mol Life Sci 66:2445–2455
84. Hossain MT, Shibata T, Kabashima T, Kai M (2010) Aptamer-mediated chemiluminescence detection of prion protein on a membrane using trimethoxyphenylglyoxal. Anal Sci 26:645–647
85. Du M, Ulrich HN, Zhao XR, Aronowski JS, Jayaraman V (2007) Water soluble RNA based antagonist of ampa receptors. Neuropharmacology 53:242–251
86. Tsukakoshi K, Harada R, Sode K, Ikebukuro K (2010) Screening of DNA aptamer which binds to α-synuclein. Biotechnol Lett 32:643–648

Chapter 16

Electrochemiluminescent Biosensors: Neuroscience Applications

Robert J. Forster and Tia E. Keyes

Abstract

The convergence of photonics, electrochemistry, materials, and biomedical sciences at the nanoscale opens up significant new opportunities. For example, *ElectroChemiLuminescence* (*ECL*), in which an electronically excited state is electrochemically created that then goes on to emit light, opens up new possibilities for the sensitive and selective detection of biomolecules. However, key biomarkers are often present at low concentration demanding the development of highly sensitive, even single molecule, assays. Moreover, in traditional systems, the ECL generated often overlaps significantly with blood and tissue making in vivo measurements challenging. In this contribution, we present key examples of materials, light generation mechanisms, and detection strategies that can significantly impact on these challenges to create new sensors capable of elucidating key biological processes and providing early detection of disease biomarkers.

Key words Electrochemiluminescence, Biosensor, Diagnostics, Biomarker detection, Ultrasensitive analysis, In-vivo sensing, Microelectrode, Microsensor

1 Background

The successful treatment of disease depends on early detection and appropriate therapy, but many human diseases, especially neurodegenerative diseases, do not exhibit clinical symptoms until they are well established. The key to improving patient prognosis is to quantify appropriate biomarkers of disease (proteins, antibodies, nucleic acids, and cells) that are found in abnormal quantities in body fluids or tissue when disease is present. For example, microRNAs (miRNA) are short nucleotide sequences (typically 20–25 bases) and are a promising class of biomarkers that are present in the tissue and blood, plasma, or serum of patients facilitating early detection, prognostic assessment, and monitoring of therapeutic efficacy (1–4). However, *the central challenge is that the concentration of miRNA is typically ultralow*, i.e., picomolar or lower. For example, panels of miRNA (5), antibodies, and proteins have been identified that

Stéphane Marinesco and Nicholas Dale (eds.), *Microelectrode Biosensors*, Neuromethods, vol. 80,
DOI 10.1007/978-1-62703-370-1_16,

diagnose and predict overall survival rates of neurological diseases, but a 1 μl sample of a patient's blood may contain less than 1,000 copies of one of these biomarkers, i.e., in the absence of clinical symptoms the concentration of biomarkers is typically pico- to femto-molar. Moreover, these disease biomarkers must be detected in the presence of tens of thousands of proteins and other nucleic acids that have the potential to interfere with the analysis. These extreme demands of sensitivity and selectivity mean that complex analytical strategies, such as Polymerase Chain Reaction, PCR, and microarray approaches, are common. However, amplification approaches, such as PCR, and Nucleic acid sequence-based amplification, NASBA, can be affected by handling errors (6) and are not applicable to protein biomarkers.

Luminescent detection approaches can have the sensitivity required to directly detect these biomarkers without amplification. However, the need for an optical excitation source complicates the instrumentation required. Electrochemiluminescence represents an attractive alternative detection methodology since an optical excitation source is not required and the response can be turned on and off simply by changing the applied potential. Moreover, the response is only generated close to the electrode surface which offers enhanced discrimination over solution phase interferences. As illustrated in Fig. 1, in *Electrogenerated ChemiLuminescence or ElectroChemiLuminescence* (*ECL*), electrochemically generated intermediates undergo a highly exergonic reaction to produce an electronically excited state that then goes on to emit light. These electron transfer reactions are sufficiently

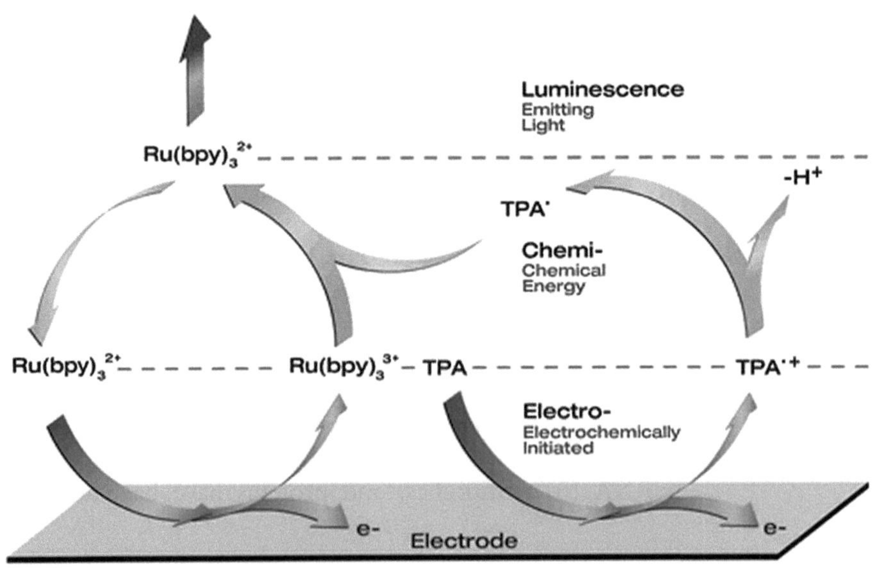

Fig. 1 Generation of electrochemiluminescence from $[Ru(bpy)_3]^{2+}$ using tripropyl amine, TPA, as a co-reactant

exergonic to allow the excited states of luminophores including polycyclic aromatic hydrocarbons and metal complexes to be created *without* photoexcitation. For example, oxidation of $[Ru(bpy)_3]^{2+}$ in the presence of guanine (7, 8), or a range of amino acids (9) results in light emission, which is analogous to that produced by photoexcitation (bpy is 2,2′-bipyridine).

Electrochemiluminescence (10–18) offers an attractive transduction strategy for the direct (no molecular amplification) detection of ultralow concentrations of disease biomarkers in microlitres of blood within an analysis time of less than tens of seconds. In this chapter, we introduce the concept and principles of electrochemiluminescence and discuss its use for in vitro detection of neurotransmitters and consider potential applications for implantable probes for in vivo detection.

2 Experimental Setup

As discussed below, the ECL reagent can be in solution, as a monolayer of labeled antibody immobilized on an electrode surface or as a thicker polymer film, e.g., an ECL reagent with a polymeric nafion film. The experimental setup for measuring ECL differs significantly depending on whether the target analyte is in a sample that will be measured in vitro or in vivo. First, the experimental setup typically used for in vitro measurement is illustrated in Fig. 2. In its simplest incarnation the potential is controlled using a potentiostat, and the emitted light is detected with a photomultiplier tube (PMT) or a CCD detector if multianalyte detection (where each ECL reagents emits at a different wavelength) is needed. Since the light is generated at the electrode interface, the collection efficiency can be enhanced by reproducibly positioning a fiber optic bundle directly opposite the electrode and coupling the other end to a PMT. The entire electrode assembly was contained inside a light-tight box. Many potentiostats feature an auxiliary input channel that can be used to capture the ECL intensity dependent voltage generated by the PMT. Alternatively, a data acquisition card can be used to capture the ECL, which can then be correlated with the current flow. Beyond this flexible setup for research, there are commercial options that are capable of performing close to one hundred different assays, including those for thyroid diseases, tumor and cardiac markers, cancer research, cell signaling pathways, nucleic acids, fertility therapies, immunogenicity assays, and analytes relevant to infectious diseases (19, 20).

To translate these assays to execute in vivo measurements, e.g., in cells, explanted tissue, or living subjects, what is needed are single, integrated, pen like devices that can address small sample volumes. Optrodes are fiber optic devices with an integrated biochemical transducer system, which changes its optical properties upon

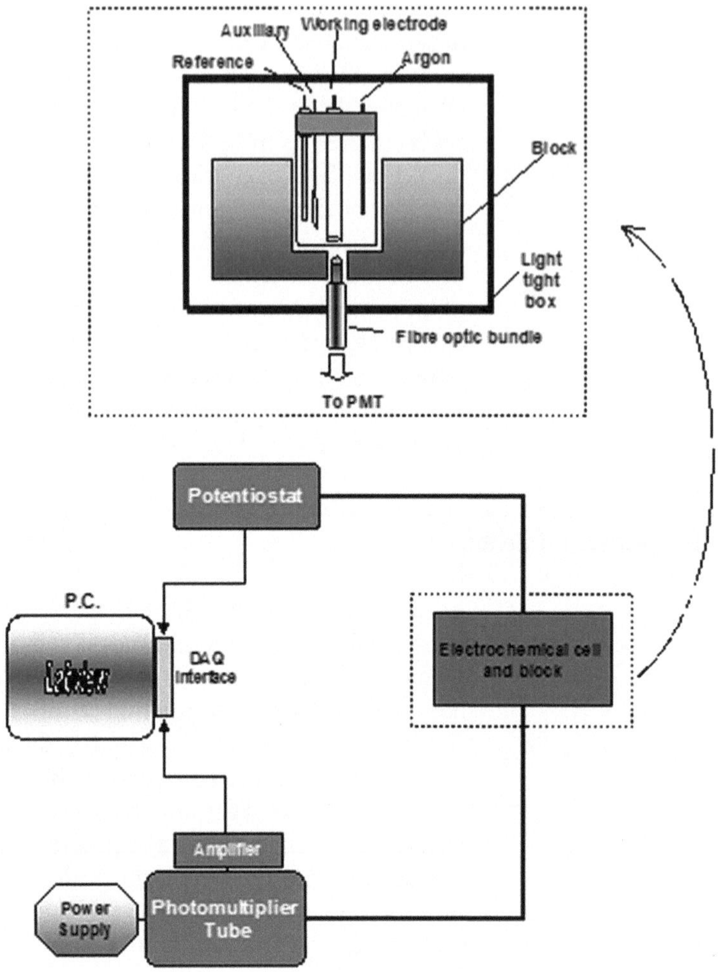

Fig. 2 Typical experimental setup for electrochemiluminescence measurements involving simultaneous detection of light and current

interaction with the analyte (21). These sensors can be constructed using single, dual, or multiple fiber arrangements in which the same optical fiber is used to deliver the exciting radiation to the sample and collect the optical response. Optrodes are commercially available, e.g., for photometric end point recognition at several different wavelengths, but somewhat surprisingly have not been extensively developed for in vivo electrochemiluminescence. In ECL, the fiber would only be used to collect the emission, but the device needs to be capable of applying a potential so as to drive the emission. For example, it ought to be possible to deposit a metal coating around the outside of the fiber that would reflect the collected ECL while allowing a potential to be applied to the ECL reagent immobilized at the sensor tip. Given the ease with which micron dimensioned fibers can currently be produced, such an integrated device could represent a powerful platform for advance in neurosciences.

3 ECL Generation Mechanisms

ECL has several attractive features including the absence of a background optical signal, i.e., unlike optically driven fluorescence measurements, the ECL response is measured against a dark background. Also, the reaction kinetics can be precisely controlled through the applied potential. It is also compatible with solution phase and thin film formats and significant opportunities exist to enhance the ECL intensity with nanomaterials such as metallic nanoparticles and nanotubes. These features combine to make ECL a highly sensitive and selective analytical method. For example, ECL generating species have been used extensively as labels on biological molecules allowing many clinically relevant analytes to be determined at sub-picomolar concentrations using conventional antibody or nucleic acid assay approaches (22, 23).

There are two dominant pathways to the production of ECL, i.e., the annihilation and co-reactant pathways. In the annihilation pathway, two species are electrochemically generated, e.g., the oxidized and reduced forms of the luminophore are produced close to an electrode surface by a potential step or sweep. These species then interact to produce both a ground state and an electronically excited state which then relaxes by emission. For example, ruthenium tris-bipyridyl type systems, $[Ru(bpy)_3]^{2+}$ (Fig. 3), can be oxidized at approximately +1.2 V and reduced at approximately −1.6 V to produce the oxidized, 3+, and reduced, 1+, forms.

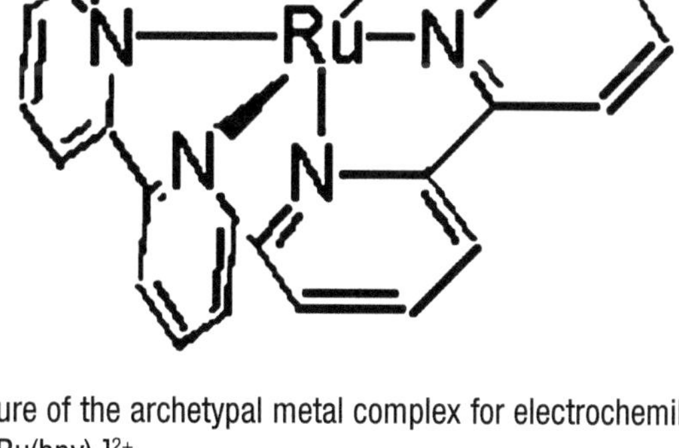

Fig. 3 Structure of the archetypal metal complex for electrochemiluminescence generation, $[Ru(bpy)_3]^{2+}$

Electron transfer between these species produces one ground state and one excited state metal complex which then emits at 610 nm. This pathway is not attractive for biomolecule detection since the reactive precursors typically cannot be produced in water due to electrolysis of the solvent, making it necessary to use organic solvents.

The co-reactant pathway represents the dominant pathway for the generation of ECL. In this approach, a species such as tripropylamine (24) is present in the solution and is either oxidized or reduced in the same potential step as the ECL species. Through electron transfers or chemical reactions that follow electron transfer, the co-reactant generates a product that reacts with the ECL luminophore to generate an excited state. In the case of Ru-bpy-based ECL, amines, amino acids, and certain nucleic acids, such as guanine, generate intense ECL. A key advantage of the co-reactant approach is that it facilitates ECL generation in aqueous solution opening up a wide range of assays for molecules of diagnostic or biological relevance.

4 Electrochemiluminescent Materials

4.1 Monomeric Metal Complexes

The spectroscopic and electrochemical properties of the ruthenium tris bpy complex illustrated in Fig. 3 make it attractive for the development of ECL-based sensors. However, other complexes have been developed to tune the wavelength of maximum emission, e.g., so that it is outside the major blood and tissue absorbances, to enhance its emission intensity by increasing its luminescent lifetime or by introducing reactive groups such as carboxylic acids or amines that can couple to biomolecules. For example, the development of novel transition metal complexes that incorporate ligands that control their redox and photophysical properties and can be immobilized on surfaces, or act as labels for biomolecules such as nucleic acids and antibodies that can bind a neurologically relevant molecule, continues to attract significant interest. For example, the hydrophobicity, excited state location within the complex, luminescence lifetime, wavelength of maximum emission, and both ground and excited state redox potentials can be optimized for particular applications by changing the identity of the peripheral ligands (25–27). The ability to control these features directly impacts on the overall performance of assays, e.g., longer excited state lifetimes may lead to a brighter ECL response making the assay more sensitive, while reducing the redox potentials may eliminate parasitic Faradaic reactions thus making the assay more selective. However, a key finding has been that due to the complexity of the ECL generation process, the photoluminescence efficiency does not directly predict the ECL intensity. Moreover, there is often no obvious relationship between the photoluminescence quantum yield and the ECL intensity. For example, the photoemission efficiency of $[Ru(bpy\text{-carboxylic acid ethyl ester})_3]^{2+}$ exceeds that of $[Ru(bpy)_3]^{2+}$, yet it does not generate intense ECL. In contrast, luminophores with lower photoluminescence quantum yields show higher ECL than $[Ru(bpy)_3]^{2+}$. Complexes of other metals, such as iridium(III) diimine complexes, may offer brighter emission further decreasing the limit of detection (28).

4.2 Multicenter Luminophores

A key objective is to create high sensitivity ECL sensors capable of detecting low concentrations of biomarkers so that disease can be detected earlier; more accurate diagnosis can be made or relationships between lifestyle, e.g., diet and exercise, and disease development and progression understood. An attractive possibility is to combine several luminophores into a single entity so as to create a brighter electrochemiluminescence. Significant effort is being devoted to the development of dimers and dendrimers with the objective of having a brighter overall emission, although this may not correspond to higher emission intensity per emitting center. For example, 1,4-bis(4′-methyl-2,2′-bipyridin-4-yl)benzene, bphb, acts as a bridge between two ruthenium centers within the complex $[(bpy)_2Ru(bphb)Ru(bpy)_2]^{4+}$ (29). This dimeric species produced two- to threefold more intense emission than $[Ru(bpy)_3]^{2+}$ in aqueous and nonaqueous solution using annihilation and co-reactant approaches. Significantly, these and other investigations reveal that in order to achieve a brighter emission, there must be weak electronic coupling between the metal centers. Other weakly coupling bridges have been explored notably, amino acids and peptides, e.g., lysine (30). However, even in systems where the centers interact weakly, multimetallic systems typically gave an increase in ECL intensity of only about 30% with respect to mononuclear compounds. This behavior arises from the fact that the upper limit of achievable intensity is determined by the quantum yield of the luminophore, but in the case of ECL, steps such as the rate of diffusion or heterogeneous electron transfer may limit the actual intensity obtained.

Figure 4 shows an approach in which a number of luminophores are linked to a single carrier, e.g., a dendrimer, so that multiple labels are immobilized at a single site within the biomolecule (31). This multiple labeling strategy may offer an enhanced ECL response. For example, in antibody or nucleic acid assays, the overall brightness of the "label" dictates the limit of detection which can be achieved using this multiple labeling strategy even if the emitters are individually less efficient when part of the larger structure. For example, eight $[Ru(bpy)_3]^{2+}$ units have been immobilized at the periphery of a carbosilane dendrimer platform (32). ECL of the functionalized dendrimer was five times that of a control

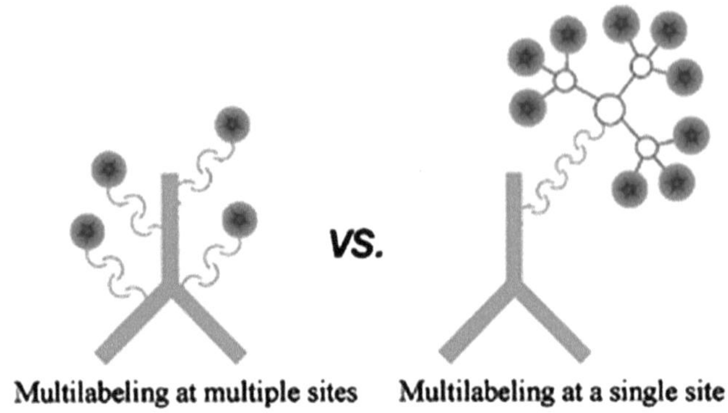

Fig. 4 Labeling of a biomolecule at multiple and single sites using dendrimers

monometallic species. However, beyond sensitivity it is important to note that labeling at multiple sites, especially with large dyes, can result in the loss of biological activity of the molecules as well as problems with precipitation. Moreover, where a wide dynamic range over a higher analyte concentration is needed, single site labeling is preferred.

4.3 Bioconjugation

Ruthenium and osmium polypyridine type complexes continue to play a pivotal role in ECL-based assay development. This dominance is due to the close to ideal reversible voltammetry and attractive photophysical properties and because of the synthetic versatility of the polypyridine and related ligands. For example, Fig. 5 illustrates the attachment of *n*-hydroxysuccinimide ester, phosphoramadites, or amine functional groups to the bipyridine moieties that facilitates covalent attachment to biological molecules such as antibodies and DNA (13). This strategy has opened up many analytical applications that have sensitivities which rival those achieved using photoluminescence or radiolabeling methods. Also, these reactive groups can be used to create monolayers of the luminophores on a wide variety of electrode surfaces opening up the possibility of detecting antibodies labeled with a suitable co-reactant. Alternatively, certain amino acids or nucleic acids can themselves act as the co-reactants.

One of the difficulties frequently encountered in photoluminescence detection using ruthenium tris-bpy is quenching by dissolved oxygen. Significantly, this problem can be avoided in ECL generation. For example, in the cross-reaction pathway, when the concentration of the ECL luminophore is greater than that of the co-reactant, the emission intensity becomes independent of the O_2 concentration. For in vivo applications, the ECL reagent must be immobilized on an optrode surface, and the concentration within the thin films is typically a significant fraction of 1 M. Therefore, it ought to be possible to engineer the system so as to minimize the sensitivity to oxygen concentration.

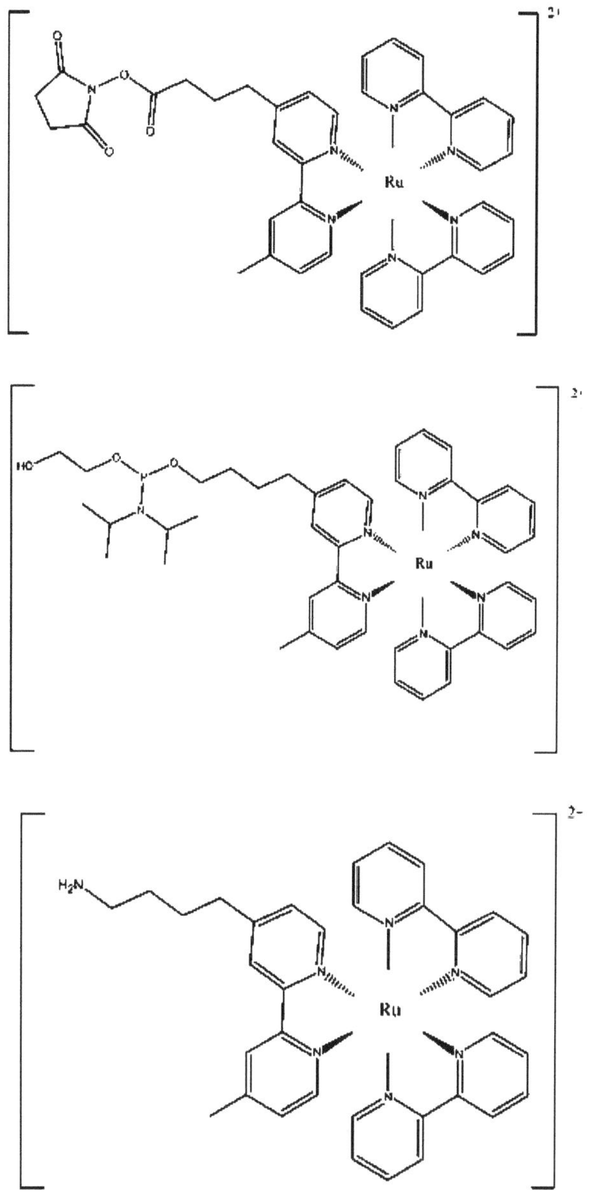

Fig. 5 $[Ru(bpy)_3]^{2+}$–NHS ester, –phosphoramadite, and –amine for ECL labeling of biological molecules

5 Neurologically Relevant Analytes

Amperometric, and more recently electrochemiluminescent, electrochemical sensors have contributed significantly to understanding underlying mechanisms, detection, and on-going monitoring of disease (33, 34). In this section, we present some examples of the electrochemiluminescent detection of neurologically relevant analytes in vitro. Adrenaline [1-(3,4-dihydroxyphenyl)-2-methyloaminoethanol] is one of several catecholamines that are produced within the adrenal medulla, and it plays an important role as a neurotransmitter. Pharmaceutically, it is widely used for the treatment of neurological disorders (35). Significantly, adrenaline strongly enhances the ECL intensity of lucigenin using a potential sweep to positive potentials (36). Under optimum conditions, the ECL intensity increases linearly with increasing adrenaline concentration from 40 to 200 nM. A detection limit of 24 nM and a relative standard deviation of 2.7% were obtained for 100 nM adrenaline ($n = 11$). This robust approach was successfully applied to the determination of adrenaline in pharmaceutical samples. An exciting possibility would be to develop a micro- or nanoscale optrode modified with a suitable thin film, e.g., containing $[Ru(bpy)_3]^{2+}$ centers whose ECL can be quenched by adrenaline, so as to measure catechols in vivo.

Adenosine is one of the most important modulators in the brain (37, 38). It is an endogenous nucleoside, not a neurotransmitter, in addition to its direct pre- and postsynaptic actions on neurones; adenosine is rich in nuances of priming, triggering, and inhibiting the action of several neurotransmitters and neuromodulators. For example, caffeine's stimulatory effects arise primarily from its binding to the same receptors as adenosine in the CNS. Section 16.4.3 provides more complete information on aptamers, but in general terms they are single-stranded DNA or RNA molecules that are ideal candidates as molecular recognition elements in sensors (39, 40). For example, the highly enhanced electrochemiluminescence from a nitrogen doped TiO_2 nanotube array has been exploited to develop an ultrasensitive ECL aptasensor for the detection of adenosine in cancer cells (41).

Fang and coworkers (42) have developed an aptamer-based electrochemiluminescent sensor for the detection of adenosine. Figure 6 illustrates the preparation of a solid-state ECL sensing platform for detecting adenosine based on a ferrocene-labeled structure-switching signaling aptamer. This approach relies on energy transfer from an excited state ruthenium center to the ferrocene unit to detect binding. Depending on the specific redox and optical properties of the ruthenium and ferrocene units, energy and electron transfer can contribute significantly to the quenching mechanism (43). However, from the perspective of sensing, the key feature is that the quenching efficiency depends strongly on the separation of the ruthenium and ferrocene centers and targets that change this separation can be detected with high sensitivity.

The luminescent substrate of $[Ru(bpy)_3]^{2+}$–AuNPs on a gold electrode was prepared first (Step a), to obtain an ECL intensity switch of immobilized Ferrocene-aptamer-$[Ru(bpy)_3]^{2+}$–AuNPs (Step b), the Ferrocene-aptamer was attached onto the electrode via gold-thiol bond formation, and the film was then back filled

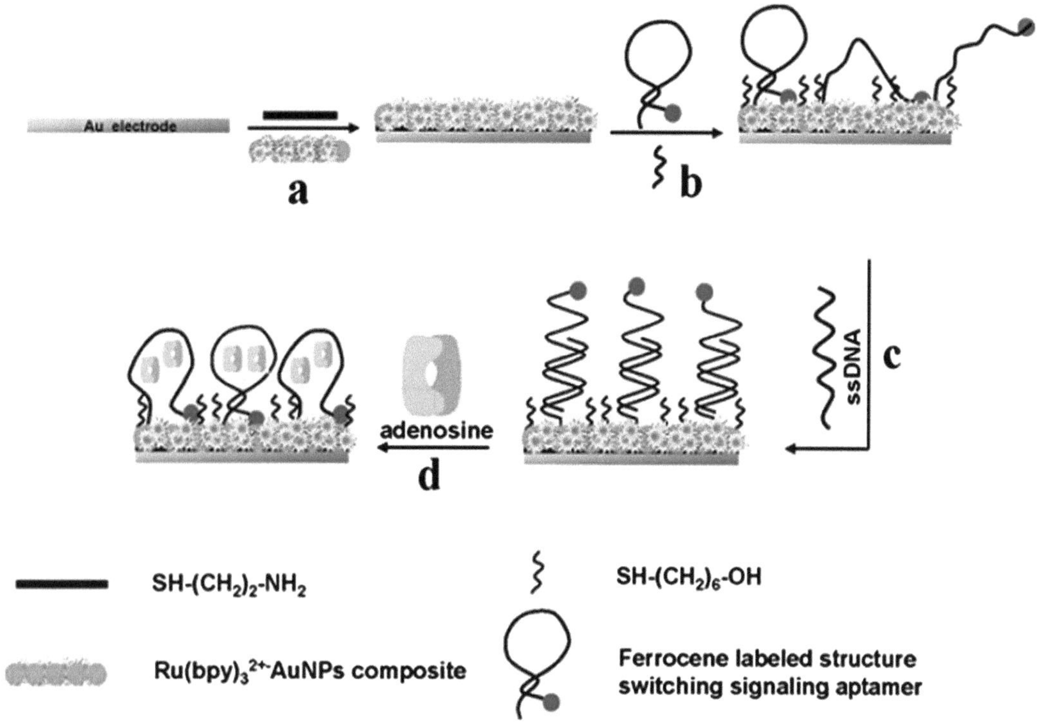

Fig. 6 Schematic representation of the preparation of solid-state electrochemiluminescence sensing platform for the detection of adenosine

with SH–$(CH_2)_6$–OH so as to control the Ferrocene-aptamer distribution on the electrode surface. In Step c, the immobilized Ferrocene-aptamer was hybridized with a short complementary ssDNA for the aptamer to form dsDNA on the electrode surface, and the first ECL signal detected using tripropylamine as the co-reactant. Finally, the dsDNA modified electrode was incubated with adenosine (Step d). The Fc-aptamer-adenosine (Fc-aptamer-A) modified electrode was used to detect the second ECL signal. The ECL intensity decreases significantly when the short complementary ssDNA to anti-adenosine aptamers was replaced by adenosine. The difference in ECL intensity was then used to quantify adenosine.

Significantly, as illustrated in Fig. 7, control experiments using an antiadenosine aptamer, a random aptamer, and a random molecular beacon aptamer demonstrate that only the anti-adenosine aptamer showed a significant ECL response while the other two aptamers gave an almost negligible response. These results suggest that the unique three-dimensional DNA/target complex leading to a change in the separation of the emissive ruthenium center and ferrocene quenching unit can only be formed by the anti-adenosine aptamer and adenosine. The observation that adenosine binding

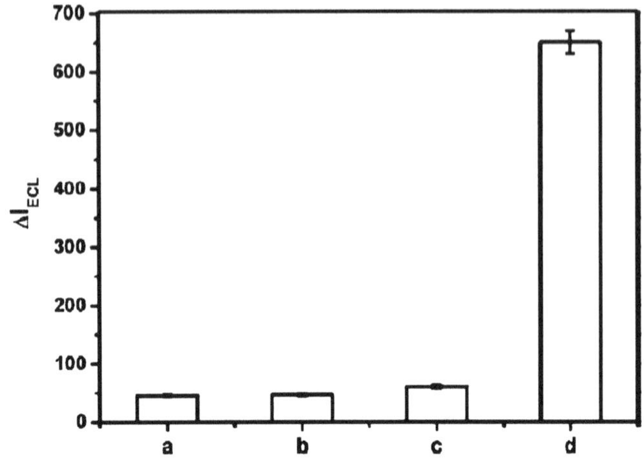

Fig. 7 Comparison of the difference in electrochemiluminescence intensity, ΔI_{ECL}, of the adenosine binding to specific aptamers, for (**a**) control, (**b**) random aptamer, (**c**) random molecular beacon aptamer, (**d**) antiadenosine aptamer

triggers structural switching of the aptamer to give a significant change in ECL intensity has potentially significant implications for in vivo sensing. For example, it ought to be possible to use this structural switching strategy to develop an optrode for where an exogenous co-reactant does not need to be added. For example, if the ferrocene unit was replaced by a bound co-reactant, e.g., a tertiary amine or poly-guanine sequence, adenosine binding ought to lead to the turning on of ECL. However, the challenge with strategies of this type is that the co-reactant is typically consumed in the ECL generating reaction and unless provision is made for its regeneration, the light response to target binding is transient and irreversible. This structural switching approach has recently been extended to the ultratrace detection of thrombin (44).

In common with optically driven fluorescence detection strategies, depending on its optical and photophysical properties, the presence of an analyte can either produce/enhance, or it can decrease, the ECL intensity. ECL quenching assays (45) rely on the analyte's optical or redox properties being such that it can either accept energy from the electronically excited ECL reagent (e.g., if there is significant spectral overlap between the emission spectrum of the ECL reagent and the absorbance spectrum of the analyte) or it oxidises or reduces the electronically excited ECL reagent. Either process is useful analytically since the ECL intensity will *decrease* with *increasing* analyte concentration. However, as with all quenching assays one must be careful to ensure that other processes such as chemical degradation or loss of the ECL reagent that also cause the ECL intensity to decrease are not incorrectly associated with the presence of the analyte. ECL quenching assays represent a useful approach to the detection of

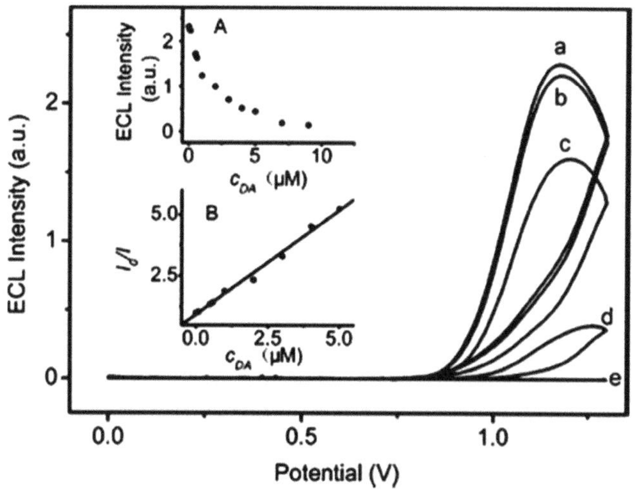

Fig. 8 ECL curves of CdTe QDs dissolved in air-saturated phosphate buffer (PBS), pH 9.3 containing 0 (a), 0.05 (b), 0.5 (c), 5 (d), and 50 μM (e) of DA at an ITO electrode, scan rate 100 mV s^{-1}. *Inset A*: relationship between ECL intensity and quencher concentration. *Inset B*: linear calibration plot for DA

phenols and benzoquinone compounds, such as adrenaline, noradrenaline, and dopamine, since these undergo electron transfer reactions with the excited states of common ECL reagents (46). Direct detection of several different analytes within a single complex sample poses significant challenges, but when combined with techniques, such as liquid chromatography or electrophoresis, excellent analytical performance can be obtained. For example, Wang has achieved linear ranges of 0.1–10 mM for both dopamine and adrenaline with detection limits of 10 nM and 30 nM for dopamine and adrenaline, respectively, by first separating the two analytes using capillary electrophoresis (47).

Semiconductor nanocrystals (NCs) or quantum dots (QDs) have received considerable interest due to their excellent luminescent properties and applications in many areas of fundamental importance (48, 49). To date, there have been relatively few reports on electrochemiluminescence from QDs because their solubility and stability can be relatively low. The electrochemiluminescence detection of catecols using mercaptopropionic acid-capped CdTe NCs has been recently reported (50). As shown in Fig. 8, a stable and intense ECL was obtained at about 1.2 V vs. Ag/AgCl in phosphate buffer (pH 9.3).

The ECL emission involved the generation of superoxide ions at the ITO electrode surface, which then injected an electron into CdTe NPs to form a CdTe anion species. Then, collisions between the CdTe anion species and the oxidation products lead to the formation of species in an excited state that emitted light at around 580 nm. Interestingly, common interferences such as uric acid and ascorbic acid were found to only weakly quench the ECL emission of the CdTe QDs. It is perhaps important to note that because this ECL mechanism relies on the production of superoxide and involves a relatively high pH, it is unlikely to be directly applicable to the detection of catechols in tissue or living cells. However, it is likely that quantum dots can be engineered to circumvent these barriers to application in biological systems in the future.

Alzheimer's disease is the most common form of dementia, and amyloid beta (Aβ) deposits are widely considered to be the fundamental cause of the disease (51, 52). Levels of amyloid beta in plasma could be useful as a noninvasive biomarker in diagnosing the disease. Electrochemiluminescence detection may provide significant advantages over conventional ELISA methods for plasma amyloid beta quantification since the approach lends itself to multiplexed detection, shorter processing times, and it can be applied to smaller sample volumes.

Axonal damage is intimately connected with the loss of neurological function. Electrochemiluminescence-based assays for neurofilament subunits have emerged as promising biomarkers for neurodegeneration. A key advantage of these assays is their superior sensitivity and ability to address smaller sample volumes than conventional ELISAs. For example, solid-phase sandwich immunoassays have been developed to determine the neurofilament heavy chain protein (NfHSMI35) in CSF (53). These studies used commercial 96-well plates (Multi-Array® plates, Meso Scale Discovery, Gaithersburg, MD) that have integrated screen-printed carbon ink electrodes on the bottom of the wells. Capture antibodies were immobilized onto the wells and nonspecific binding sites blocked with bovine serum albumin. The sample or control was then allowed to bind to the capture antibody before the secondary antibody labeled with sulfonated $[Ru(bpy)_3]^{2+}$ was bound. The ECL intensity was then measured using a commercial MSD Sector Imager 2400 plate reader. A limit of detection of 2.4 pg ml^{-1}, a mean intra-assay coefficient of variation (CV) of 4.8% and an inter-assay CV of 8.4% were achieved. The median NfHSMI35 value in CSF for the reference cohort, "normals" was 27.1 pg ml^{-1}, while patients with multiple sclerosis (MS) had a median of 46.2 pg ml^{-1}, $n=95$, amyotrophic lateral sclerosis (ALS) patients showed values of 160.1 pg ml^{-1}, $n=50$, mild cognitive impairment/Alzheimer's disease (MCI/AD) 65.6 pg ml^{-1}, $n=20$, Guillain–Barre syndrome (GBS) 91.0 pg ml^{-1}, $n=20$, or subarachnoid hemorrhage (SAH) 345.0 pg ml^{-1}, $n=20$, i.e., all of these conditions have higher NfHSMI35 values than the reference population. The overall analytical performance in terms of sample volume, sensitivity, precision, and accuracy is superior to other approaches, notably ELISA. The improved performance and small sample volume requirement qualify this method in experimental settings and clinical trials designed to perform a number of tests on limited amounts of material.

Beyond the detection of neurotransmitters and biomarkers of neurodegenerative diseases, ECL-based sensors show significant promise for the monitoring of therapeutics. For example, Chen and coworkers fabricated a $Ru(bpy)_3^{2+}$-zirconia-Nafion-based composite modified electrode for ECL detection of different pharmaceuticals (lidocaine, tramadol, ofloxacin) coupled with an electrophoretic microchip suggesting potential applications in clinical diagnostics for neurology (54, 55).

Dehydrogenase enzymes are considered to play important roles in neurological disease. For example, the age-induced apoptotic death of cerebellar neurons in culture is associated with over-expression of a 38-kDa particulate protein identified as glyceraldehyde-3-phosphate dehydrogenase (GAPDH). Electrochemiluminescence has been repeatedly demonstrated to be a highly efficient approach for the detection of a wide range of dehydrogenases that could be easily applied to neurologically relevant examples. For example, thin films of alcohol dehydrogenase, ADH, and $[Ru(bpy)_3]^{2+}$ functionalized Au NPs have been fabricated using a layer-by-layer method (Zhang et al., 2007) to produce an ECL biosensor for the detection of ADH (56). Specifically, the positively charged $[Ru(bpy)_3]^{2+}$ species was immobilized on ITO electrodes via electrostatic interaction with negatively charged Au NPs. Au NPs strongly bind to the amine and cysteine residues of ADH enzyme facilitating the electron transfer and allowing the ECL detection of dehydrogenases. These approaches ought to find increasing application in neurological sciences.

5.1 Multianalyte Detection

Walt and coworkers have developed multiplexed sandwich immunoassays using electrochemiluminescence imaging resolved at the single bead level (57). The sensing microarray uses ECL to detect multiple antigens simultaneously. As illustrated in Fig. 9, the multiplexed ECL platform consists of 3.1 μm polystyrene microspheres loaded into the wells of an electrode, prepared from etched fiber optic bundles coated with gold. The approach is widely applicable since the surfaces of the microspheres are modified with appropriate capture antibodies. Individual microspheres are internally encoded with different concentrations of a europium (Eu^{3+}) luminophore (so as to control their emission intensity) and their surfaces modified with different antibodies to allow them to be distinguished from one another in the array. The approach could be further multiplexed by using ECL labels with different emission maxima. This powerful approach has been demonstrated using three bead types by incubating the array first in a sample containing antigen and then in a solution of the biotinylated detection antibody. The final step involves attaching the $[Ru(bpy)_3]^{2+}$ ECL label by exposing the array to a solution containing a streptavidin-modified Ru^{2+} complex (SA-Ru). The $[Ru(bpy)_3]^{2+}$ label should bind only to beads where antigen is present, and ECL from the

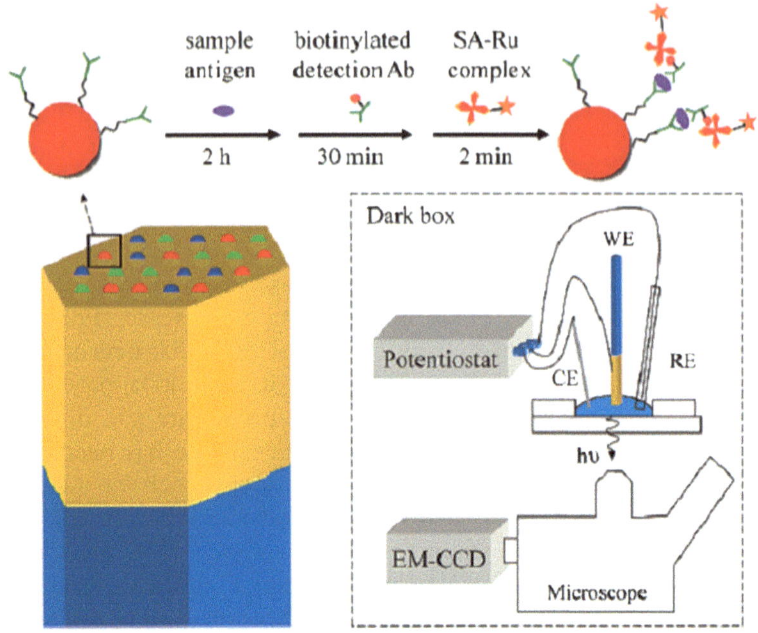

Fig. 9 A sandwich immunoassay is performed by exposing antibody-functionalized microbeads to three solutions (1) an antigen containing sample, (2) biotinylated detection antibodies, and (3) streptavidin modified with a $[Ru(bpy)_3]^{2+}$ complex (SA-Ru). The beads are housed in microwells created from an etched gold-coated fiber-optic bundle. The gold-coated fiber bundle acts as the working electrode (WE) for ECL. CE and RE refer to counter electrode and reference electrode

array can be generated by applying a suitable voltage step or sweep and visualizing the array with a microscope. The method enables multiplexed assays because all the individual sensing beads in the array are simultaneously imaged. Duplex and triplex assays have been demonstrated as well as cross-reactivity effects investigated. Beyond its ability to detect multiple analytes, it is important to note that this approach is likely to result in shorter analysis times since the sample and beads can be agitated before the analysis. This mixing step allows the analyte to bind much more rapidly than if it has to diffuse to a capture surface. When the analyte concentration is low, i.e., nanomoles to picomoles, it takes tens or even hundreds of seconds for a detectable fraction of a sensor surface to become modified with the analyte where transport occurs simply by diffusion even when the sample volume is a few microlitres.

6 Challenges, Opportunities, and Prospects for Neuroscience Applications

In common with many other chronic diseases, the treatment of neurological diseases in the future will involve the early detection of transformational processes during routine screening using minimally invasive methods such as proteomic pattern analysis from blood samples. Electrochemiluminescence has achieved many of the analytical performance characteristics required to deliver this vision including wide dynamic range, high reproducibility, and ultralow limits of detection.

6.1 Selectivity

In vivo measurements demand high levels of selectivity due to the complex nature of the samples, and interferences are likely to pose significant challenges. ECL assays can suffer from the same interferences as optically driven emission assays, e.g., nonspecific binding of matrix components (especially proteins), ionic strength effects, and redox active interferences that can quench the emission by electron transfer to or from the electronically excited state. Also, when ECL reagents with long lived excited states, e.g., ruthenium tris bipyridyl complexes, are used, quenching of the excited state by molecular oxygen can occur, but this problem can be minimized by using a high concentration of the ECL reagent. However, it is important to note that ECL has a key advantage in this regard in that the signal is generated close to the electrode exclusively and the bulk sample typically makes a negligible contribution. Therefore, even if the bulk sample contains interferences capable of generating or quenching ECL, their contribution can be minimized by depositing a highly selective capture layer. Moreover, an enhanced discrimination between individual analytes in a complex sample and between analytes and interferences can be achieved by introducing a separation component, e.g., by using a dialysis membrane or perhaps an integrated microfluidic sampling device.

6.2 Co-reactants

Detecting targets in *solution* is simplified by the ability to add an excess of co-reactant such as oxalate or tripropyl amine so that the ECL intensity generated depends on the quantity of the ECL reagent immobilized that is directly controlled by the target concentration. For example, in an antibody sandwich assay the ECL intensity depends directly on the quantity of the labeled antibody bound which depends on the target concentration. However, adding a large concentration of an exogenous co-reactant is often not feasible in vivo. Therefore, direct detection is restricted to those molecules that can themselves act as co-reactants. These include oxalate and peroxydisulfate as well as amines that are common in numerous biologically and pharmacologically important compounds including alkylamines, antibiotics, antihistamines, opiates, nicotinamide, and the reduced form of NADH (i.e., adenine dinucleotide). Many other analytes can be determined *indirectly* if they undergo energy or electron transfer reactions with the excited state ECL reagent.

6.3 Time Resolution

An important performance parameter in neuroscience is the response time of the sensor. For example, second-by-second neurotransmitter detection would be highly desirable for some applications. Once created, the exited state lifetime of the ECL reagent is identical to that observed for optical excitation, i.e., of the order of nanoseconds. However, unlike optical excitation that is effectively instantaneous, the creation of the excited state electrochemically involves several additional steps including electron transfer across the electrode/film interface and, for 3D capture films, charge transport through the layer so as to (re)generate the precursor to the excited state, e.g., oxidizing a Ru^{2+} tag bound to a secondary antibody to create the Ru^{3+} species that can interact with the co-reactant/target analyte to create the emissive species, Ru^{2+*}. The time constant for heterogeneous electron transfer typically ranges from microseconds to seconds so that systems involving only a monolayer of capture agent can achieve fast response times. However, due to the low total quantity of the ECL reagent immobilized, the ECL intensity and more importantly the signal-to-noise ratio may be low. Increasing the film thickness, e.g., to micron thicknesses using a polymeric film, increases the ECL emission intensity. However, the rate of heterogeneous electron transfer decreases exponentially with increasing distance and homogeneous charge transport, e.g., by electron self-exchange between adjacent ECL centers, is required. This process is rather slow and typical homogeneous charge transport diffusion coefficients range from 10^{-12} to 10^{-8} cm^2 s^{-1}, i.e., 3 to 6 orders of magnitude slower than diffusion in aqueous solution. Therefore, it will take between 1 second and 10,000 s to fully regenerate all of the ECL reagents immobilized within a 1 μm thick film. Thus, by carefully engineering the redox properties of the film, it is possible to achieve useful response times while immobilizing sufficient ECL reagent to give a high signal-to-noise ratio but only if materials with good electronic conductivity are used.

6.4 Spatial Resolution

The identification and treatment of some diseases requires in vivo measurements at high spatial resolution. While conventional optically driven luminescent approaches can provide information about the concentration of key disease biomarkers in vivo, its central problem is the limited penetration depth and scattering of light in tissue. When used with micrometer dimensioned optrodes, electrochemiluminescence opens up the possibility of excellent spatial resolution similar to that achieved using amperometric biosensors based on microelectrodes (58, 59).

Electrochemiluminescent optrodes are based on optical fibers functionalized so as to make selected regions electrically conducting which are then modified with biological recognition molecules (60). Multiple analytes can be detected by incorporating several different antibody/DNA—luminophore combinations making them very useful for investigating biosystems. Microscale sensors can be created that allow concentrations of biomolecules to be mapped over wide concentration ranges with good spatial resolution, even inside single living cells. The field is now perfectly positioned to exploit the latest advances in metal-coated optical fibers developed primarily for surface plasmon resonance (SPR) fiber sensors and photonic crystal fiber (PCF) sensors (61). These fibers include tapered fibers, side-polished fibers (or D-shape fiber) and one-side metal-coated fibers and structures with modified fiber tips with flat or angled structures. These structures have also been further modified with several types of gratings, heterogeneous core fibers, and nano-holes, which promise to further enhance the electrochemiluminescence intensity, e.g., through plasmonic effects, or the ability to discriminate individual luminophores. Moreover, the development of these high sensitivity sensors will also benefit greatly from the latest advances in surface modification for bioimplantation which has seen sensors, e.g., for glucose, function for over a year after implantation in living animals (62).

References

1. Calin GA, Croce CM (2006) MicroRNA signatures in human cancers. Nat Rev Cancer 6:857
2. Bartel DP (2004) MicroRNAs: genomics, biogenesis, mechanism, and function. Cell 116:281
3. Tavazoie SF, Alarco C, Oskarsson T, Padua D, Wang Q, Bos PD, Gerald WL, Massague J (2008) Endogenous human microRNAs that suppress breast cancer metastasis. Nature 451:147
4. Sidransky D (2002) Emerging molecular markers of cancer. Nat Rev Cancer 2:210
5. Chen Y, Stallings RL (2007) Differential patterns of microRNA expression in neuroblastoma are correlated with prognosis, differentiation, and apoptosis. Cancer Res 67:976
6. Compton J (1991) Nucleic acid sequence-based amplification. Nature 350:91–92
7. Dennany L, Forster RJ, Rusling JF (2003) Simultaneous direct electrochemiluminescence and catalytic voltammetry detection of DNA in ultrathin films. J Am Chem Soc 125:5213
8. Dennany L, Forster RJ, White B, Smyth M, Rusling JF (2004) Direct electrochemiluminescence detection of oxidized DNA in ultrathin films containing [Os(bpy)(2)(PVP)(10)](2+). J Am Chem Soc 126:8835
9. Dennany L, O'Reilly E, Forster RJ (2006) Electrochemiluminescent monolayers on metal oxide electrodes: detection of amino acids. Electrochem Commun 8:1588
10. Pyati R, Richter M (2007) ECL-electrochemical luminescence. Annu Rep Prog Chem Sect C Phys Chem 103:12
11. Miao W (2008) Electrogenerated chemiluminescence and its biorelated applications. Chem Rev 108:2506
12. Richter M (2004) Electrochemiluminescence (ECL). Chem Rev 104:3003
13. Bard A, Debad J, Leland J, Sigal G, Wilbur J, Wohlstadter J (2000) Chemiluminescence, electrogenerated. In: Meyers RA (ed) Encyclopedia of analytical chemistry. Wiley, Chichester, p 9842
14. Aboul-Enein H, Stefan R, van Staden J, Zhang X, Garcia-Campana A, Baeyens W (2000) Recent developments and applications of chemiluminescence sensors. Crit Rev Anal Chem 30:271
15. Knight A (2001) Electrogenerated chemiluminescence. In: Garcia-Campana A, Baeyens W (eds) Chemiluminescence in analytical chemistry. Marcel Dekker, New York, pp 211–247
16. Richter M (2002) Surface plasmon resonance biosensors. In: Ligler FS, Rowe-Taitt CA (eds) Optical biosensors present and future. Elsevier, New York, pp 173–205
17. Bard A (2004) In: Bard AJ (ed) Electrogenerated chemiluminescence. Marcel Dekker, New York
18. Gorman B, Francis P, Barnett N (2006) Tris(2,2′-bipyridyl)ruthenium(II) chemiluminescence. Analyst 131:616
19. Debad JD, Glezer EM, Leland JK, Sigal GB, Wohlstadter J (2004) Clinical and biological applications in ECL. In: Bard AJ (ed) Electrogenerated chemiluminescence, chap. 8. Marcel Dekker, New York, p 359

20. Meso Scale Discovery (Gaithersburg, MD, USA; http://www.meso-scale.com/); BioVeris Corporation (Gaithersburg, MD, USA; http://www.bioveris.com/)
21. Kuswandi B, Andres R, Narayanaswamy R (2001) Optical fibre biosensors based on immobilised enzymes. Analyst 126:1469
22. Forster RJ, Bertoncello P, Keyes TE (2009) Electrogenerated chemiluminescence. Annu Rev Anal Chem 2:359
23. Bertoncello P, Forster RJ (2009) Nanostructured materials for electrochemiluminescence (ECL)-based detection methods: recent advances and future perspectives. Biosens Bioelectron 24:3191
24. Noffsinger J, Danielson N (1987) Generation of chemiluminescence upon reaction of aliphatic-amines with Tris(2,2′-Bipyridine) Ruthenium(III). Anal Chem 59:865
25. Kanoufi F, Bard A (1999) Electrogenerated chemiluminescence. 65. An investigation of the oxidation of oxalate by tris(polypyridine) ruthenium complexes and the effect of the electrochemical steps on the emission intensity. J Phys Chem B 103:10469
26. Zhou M, Robertson G, Roovers J (2005) Comparative study of ruthenium(II) tris(bipyridine) derivatives for electrochemiluminescence application. Inorg Chem 44:8317
27. Brooks S, Vinyard D, Richter M (2006) Electrogenerated chemiluminescence of (bis-bipyridyl)ruthenium(II) acetylacetonate complexes. Inorg Chim Acta 359:4635
28. Li MJ, Jiao P, Lin M, He W, Chen GN, Chen X (2011) High electrochemiluminescence of a new water-soluble iridium(III) complex for determination of antibiotics. Analyst 136:205
29. Richter M, Bard A, Kim W, Schmehl R (1998) Electrogenerated chemiluminescence. 62. Enhanced ECL in bimetallic assemblies with ligands that bridge isolated chromophores. Anal Chem 70:310
30. Staffilani M, Hoss E, Giesen U, Schneider E, Harti F, Josel H, De Cola L (2003) Multimetallic ruthenium(II) complexes as electrochemiluminescent labels. Inorg Chem 42:7789
31. Zhou M, Roovers J, Robertson G, Grover C (2003) Multilabeling biomolecules at a single site. 1. Synthesis and characterization of a dendritic label for electrochemiluminescence assays. Anal Chem 75:6708
32. Zhou M, Roovers J (2001) Dendritic supramolecular assembly with multiple Ru(II) tris(bipyridine) units at the periphery: synthesis, spectroscopic, and electrochemical study. Macromolecules 34:244
33. Koudelka-Hep M, van der Wal PD (2000) Microelectrode sensors for biomedical and environmental applications. Electrochim Acta 45:2437
34. Wang J (2006) Electrochemical biosensors towards point-of-core cancer diagnostics. Biosens Bioelectron 21:1887
35. Deftereos TN, Calokerinos AC, Efstathiou CE (1993) Flow-injection chemiluminometric determination of epinephrine, norepinephrine, dopamine and L-Dopa. Analyst 118:627
36. Su Y, Wang J, Chen G (2005) Determination of epinephrine based on its enhancement for electrochemiluminescence of lucigenin. Talanta 65:531
37. Burnstock G (2007) Physiology and pathophysiology of purinergic neurotransmission. Physiol Rev 87:659
38. Sebastião AM, Ribeiro JA (2000) Fine-tuning neuromodulation by adenosine. Trends Pharmacol Sci 21:341
39. Hermann T, Patel DJ (2000) Biochemistry – adaptive recognition by nucleic acid aptamers. Science 287:820
40. Kirby R, Cho EJ, Gehrke B, Bayer T, Park YS, Neikirk DP, McDevitt JT, Ellington AD (2004) Aptamer-based sensor arrays for the detection and quantitation of proteins. Anal Chem 76:4066
41. Yuan Tian C, Xu J-J, Chen H-Y (2012) A novel aptasensor for the detection of adenosine in cancer cells by electrochemiluminescence of nitrogen doped TiO^2 nanotubes. Chem Commun 48:8234
42. Wang X, Dong P, He P, Fang Y (2010) A solid-state electrochemiluminescence sensing platform for detection of adenosine based on ferrocene-labeled structure-switching signaling aptamer. Anal Chim Acta 658:128
43. Lee EJ, Wrighton MS (1991) Competitive energy and electron-transfer quenching of excited ruthenium polypyridyls by ferrocene derivatives. J Am Chem Soc 113:8562
44. Zhang J, Chena P, Wua XY, Chen J-H, Xua L, Chen G, Fua F (2011) A signal-on electrochemiluminescence aptamer biosensor for the detection of ultratrace thrombin based on junction-probe. Biosens Bioelectron 26:2645
45. McCall J, Alexander C, Richter MM (1999) Quenching of electrogenerated chemiluminescence by phenols, hydroquinones, catechols, and benzoquinones. Anal Chem 71:2523
46. Guo Z, Dong S (2005) Electrogenerated chemiluminescence determination of dopamine and epinephrine in the presence of ascorbic acid at carbon nanotube/nafion-Ru(bpy)(2+)(3) composite film modified glassy carbon electrode. Electroanalysis 17:607
47. Kang J, Yin X-B, Yang X, Wang E (2005) Electrochemiluminescence quenching as an

indirect method for detection of dopamine and epinephrine with capillary electrophoresis. Electrophoresis 26:1732
48. Chen S, Truax L, Sommers J (2000) Alkanethiolate protected PbS nanoclusters, synthesis, spectroscopic and electrochemical studies. Chem Mater 12:3864
49. Michalet X, Pinaud FF, Bentolila LA, Tsai JM, Doose S, Li JJ, Sundaresan G, Wu AM, Gambhir SS, Weiss S (2005) Quantum dots for live cells, in vivo imaging, and diagnostics. Science 307:538
50. Liu X, Jiang H, Lei J, Ju H (2007) Anodic electrochemiluminescence of CdTe quantum dots and its energy transfer for detection of catechol derivatives. Anal Chem 79:8055
51. Hardy J, Allsop D (1991) Amyloid deposition as the central event in the etiology of Alzheimer's-disease. Trends Pharmacol Sci 12:383
52. Mudher A, Lovestone S (2002) Alzheimer's disease - do Tauists and Baptists finally shake hands? Trends Neurosci 25:22
53. Kuhlea J, Regeniter A, Leppert D, Mehling M, Kappos L, Lindberga RLP, Petzold A (2010) A highly sensitive electrochemiluminescence immunoassay for the neurofilament heavy chain protein. J Neuroimmunol 220:114
54. Ding S, Xu J, Zhang W, Chen H (2006) Tris(2,2′-bipyridyl)ruthenium(II)-zirconia-nafion composite modified electrode applied as solid-state electrochemiluminescence detector on electrophoretic microchip for detection of pharmaceuticals of tramadol, lidocaine and ofloxacin. Talanta 70:572
55. Ding S, Xu J, Chen H (2005) Tris(2,2′-bipyridyl)ruthenium(II)-zirconia-nafion composite films applied as solid-state electrochemiluminescence detector for capillary electrophoresis. Electrophoresis 26:1737
56. Zhang L, Liu B, Dong S (2007) Bifunctional nanostructure of magnetic care luminescent shell and its application as solid-state electrochemiluminescence sensor material. J Phys Chem B 111:10448
57. Deiss F, LaFratta CN, Symer M, Blicharz TM, Sojic N, Walt DR (2009) Multiplexed sandwich immunoassays using electrochemiluminescence imaging resolved at the single bead level. J Am Chem Soc 131:6088
58. Forster RJ, Keyes TE (2006) Ultramicroelectrodes. In: Zoski C (ed) Handbook of electrochemistry. Elsevier, Amsterdam, pp 155–260
59. Forster RJ (1994) Microelectrodes: new dimensions in electrochemistry. Chem Soc Rev 23:289–297
60. Brogan KL, Walt DR (2005) Optical fibre-based sensors: application to chemical biology. Curr Opin Chem Biol 9:494
61. Lee B, Roha S, Park J (2009) Current status of micro- and nano-structured optical fiber sensors. Opt Fiber Technol 15:209
62. Gough DA, Kumosa LS, Routh TL, Lin JT, Lucisano JY (2010) Function of an implanted tissue glucose sensor for more than 1 year in animals. Sci Transl Med 2:42

Chapter 17

Molecularly Imprinted Polymers: Promising Advanced Materials for In Vivo Sensing

Yi Ge, Samir Akhtar, Farhan Mirza, Sergey Piletsky, Shenqi Wang, and Dan Fei

Abstract

Molecularly imprinted polymers are now well known as synthetic polymeric receptors or robust artificial antibodies ("plastibodies") and have attracted considerable attention from the scientific and industrial community due to their inherent simplicity, reusability, robust polymer network, and cost-effectiveness. In this chapter, the concept and principles of molecularly imprinted polymers are introduced and illustrated. Their fascinating properties are further described and summarized with a focus on the medical/biological aspects followed by a comprehensive review of published research results on the sensor applications of molecularly imprinted polymers as biomaterials and/or biocompatible materials. In addition, a future outlook is presented demonstrating the great potentials of molecularly imprinted polymers for in vivo sensing.

Key words Molecularly imprinted polymer, MIP, In vivo, Sensor, Diagnostic

1 What Are Molecularly Imprinted Polymers?

Molecularly imprinted polymers (MIPs) are known as synthetic polymeric receptors or robust artificial antibodies (or "plastibodies") that have gained a surging scientific interest within the last decade (1–3). An MIP is a three-dimensional polymer network formed with complimentary binding sites to a specific target molecule, which can be referred to as the template. Once the template is removed, by washing and/or extraction, cavities that are specific to the molecule are left behind and are functionally similar to enzymes or antibodies. Therefore the template, which could be either a biomolecule or synthetic compound, would require at least one functional group and be resistant to the polymerization process for this procedure to be successful. The procedure to form an MIP is termed "molecular imprinting," and a summary diagram of the process is shown in Fig. 1.

The key requirement of an MIP is its ability to recognize specific molecules, an essential process for all living organisms

Stéphane Marinesco and Nicholas Dale (eds.), *Microelectrode Biosensors*, Neuromethods, vol. 80,
DOI 10.1007/978-1-62703-370-1_17, © Springer Science+Business Media, LLC 2013

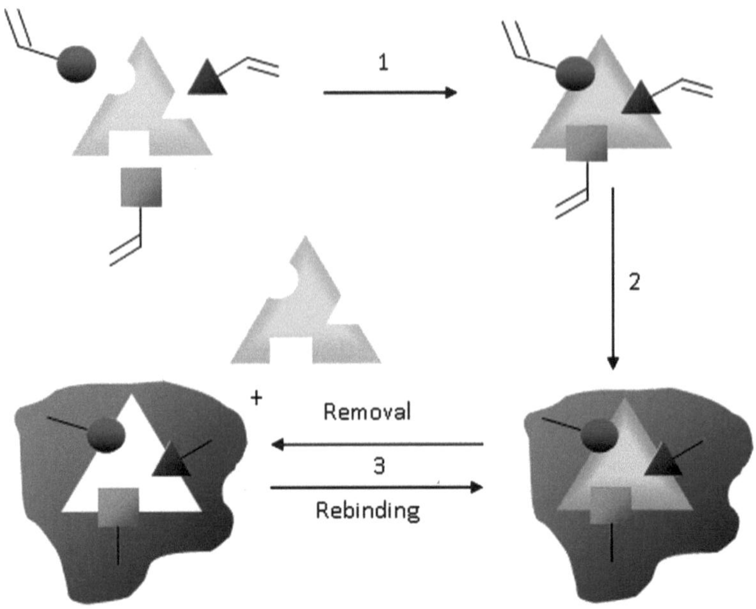

Fig. 1 Molecular imprinting (**1**) Formation of a pre-polymerizable complex (*large triangle* with *small circle*, *triangle*, and *square*) from the template (*large triangle* without *small circle*, *triangle*, and *square*) and functional monomers (*small circle*, *square*, and *triangle*) by interactions (non-covalent, covalent, and/or metal coordination interactions) that occur between complementary functionalities in the template molecule and functional monomer units; (**2**) Polymerization with excess cross-linking agent to produce the MIP; (**3**) Template removal by washing and/or extraction, leaving specific recognition sites that are complementary to the template in terms of size, shape, and chemical functionality orientations, thus enabling subsequent recognition of the template during the rebinding process. Adapted from (4)

within nature. If the processes are considered at a cellular level, then biological molecule recognition plays a pivotal role in such procedures as respiration, immune system functionality, nutrition, and cellular recognition. Theoretically MIPs have the potential to mimic these natural biological actions. Additionally both MIPs and natural molecule recognition have the same overall aim: which is to form a cavity that is able to favorably form chemical interactions with the specific template/target molecule.

Molecularly imprinted polymers hold many advantages to biological receptors and other biomolecules, including being resistant to upper and lower pH, pressure changes, and temperature (<180 °C). Although MIPs tend to be larger, more rigid, and insoluble when compared to biological receptors, they can be manipulated and constructed around a target molecule, even for that which a natural binder does not exist. They can also be cheaper than conventional biological antibodies and as they can be tailor synthesized for the target molecule required, depending on size, this can mean the possibility for thousands and even millions of binding sites (3–5). As with naturally occurring biological receptors,

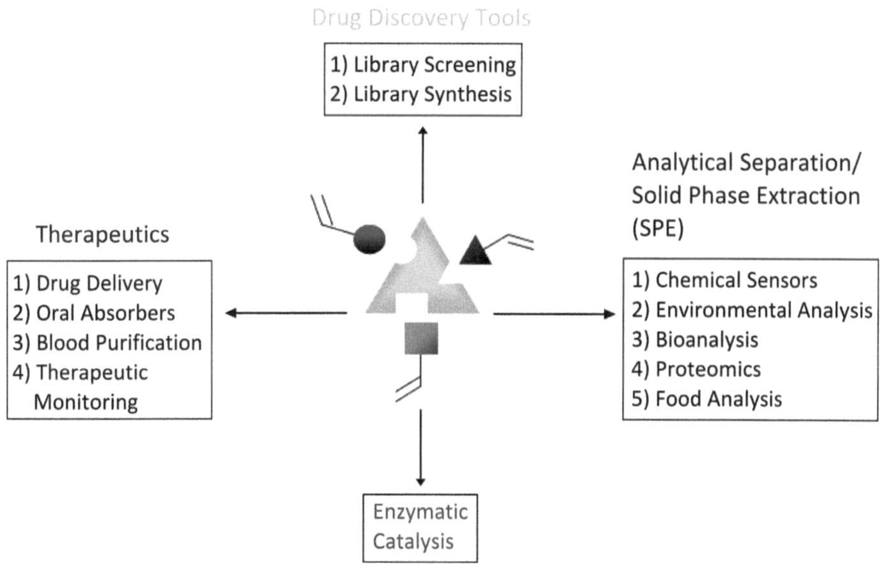

Fig. 2 Some areas of application envisaged for MIPs in the near future

such as proteins, MIPs are reusable and very robust which also makes them very cost-effective to utilize.

MIPs are envisaged to play a function within a wide variety of sectors such as sensing, drug discovery, and therapeutics, among others. The potential uses of MIPs are summarized in Fig. 2.

2 A Brief History of MIPs

Molecular imprinting technology is not a recently discovered concept. Early work was conducted by Polyakov in this area within the 1930s (6, 7) where several silica gels were prepared, and it was noted that in the presence of a solvent, the gels demonstrated preferential binding for specific solvents. Further work was published by Dickey in 1949 (8) using silica gels prepared with dyes. After removal of the dye, the silica gels demonstrated the strongest re-adsorption for the dye used in preparation and decreasing adsorption for dyes of decreasing similarity in molecular structure. The work conducted by Dickey on silica can be thought of as the first imprinted materials. Later in the early 1970s, Wulff et al. (and independently Takagishi and Klotz) began to introduce organic polymer molecular printing (9). Wulff et al. used sugar derivatives as the template and prepared synthetic polymers with appropriate receptors by using a covalent approach (9, 10). Later, a non-covalent approach to prepare MIPs was pioneered by Arshady and Mosbach (11), and Sellergren and Andersson (12) demonstrated the first true MIP using semi-covalent interactions. Around these key developments, MIPs have gained significant scientific interest, and the dramatic increase within the last 20 years in the number of literature regarding MIPs focusing on design, development, and applications reflect the broad interest this area is gaining within the scientific community (6). There is great potential for research science in this area, and it is not difficult to imagine the possibility of custom creating a specific receptor for a target molecule, and as seen in this brief history, MIP techniques have developed a long way since the 1930s.

3 Methods of MIPs Preparation

Preparing MIPs broadly consists of three general stages (1) firstly the selection of components beginning with the target molecule as the template then leading onto the individual monomers. Additionally the appropriate cross-linker, solvent, and initiator must be selected; (2) the second stage of production is the formation of the monomer–template complex within the solution (13). High levels of cross-linking agent are required to form stable well-defined recognition sites where the functional groups can be arranged into a stable configuration (14); (3) the final stage is the polymerization after which the template is removed by washing or a combination of washing with chemical treatment.

The approaches to produce MIPs can be generally divided into a few categories, based on the type of interactions between the template and the functional monomer, which are covalent molecular, non-covalent, and metal-coordinating molecular imprinting. Although the categories seem distinct, it is useful to note that one or more of the interactions mentioned could be involved within the process of MIP-template formation. A detailed diagram of the process to produce MIPs detailing the interactions can be seen in Fig. 3.

As an example MIP preparation recipe, Matsui et al. (15) developed an MIP that was selective for the herbicide Atrazine. They dissolved the template Atrazine in chloroform, added the functional monomer Methacrylic Acid (MAA), the cross-linking monomer Ethylene Glycol Dimethacrylate (EDMA), followed by the initiator 2,2′-azobisisobutyronitrile (AIBN) and initiated polymerization with UV light. Removal of the template yielded binding sites complimentary to Atrazine.

3.1 Covalent Imprinting

Covalent imprinting is where the template is bound to one or several polymerization groups covalently. Once the template–MIP complex is formed (after synthesis), the covalent bonds must be cleaved. After the template is removed, the target molecule can rebind due to covalent bonds being formed in the same locations, with the main advantages being that the functional groups are associated with the template only, and the stability of covalent

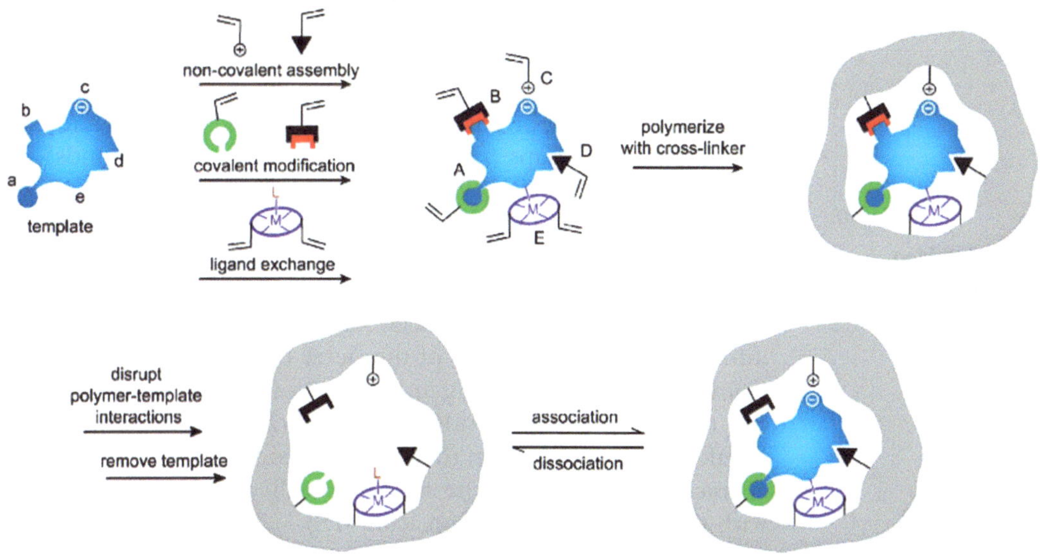

Fig. 3 Detailed outline of process to produce an MIP. The complimentary interactions between the template (shape at beginning) and the functional monomers can vary in nature, as seen where: A represents reversible covalent bonds, B represents covalently attached polymerizable binding groups that are activated for non-covalent interactions by template cleavage, C represents electrostatic interactions, D represents hydrophobic or van der Waals interactions and E represents co-ordination with a metal center/. Once the functional monomers have formed interactions with the template molecule, the next step is the polymerization with a cross-linker(s), cross-linking reaction, or other process (*gray*). The template molecule is then removed and can be rebound. The diagram represents the process used in vinyl polymerization; however, the general process is similar for other methods such as in sol–gel, etc. Adapted from (6)

bonding yields more homogeneous binding sites. Only certain compounds can be imprinted in this way, including diols, aldehydes, ketones, amines, and carboxylic acids (6). Covalent imprinting methods have historically been the more prevalent (until recently) where it was first developed by Wulff et al. (9).

Semi-covalent imprinting is also possible and refers to the method which utilizes a covalent or partly covalent template structure in the polymerization step but is distinct from covalent imprinting in that the rebinding step is non-covalent (16), and this was pioneered by Sellergren and Andersson (12).

3.2 Non-Covalent Imprinting

Non-covalent imprinting adducts, which are formed between the template and functional monomer, are based on non-covalent interactions, such as hydrogen bonding, ion pairing, dipole–dipole interactions, and Van der Waals forces. Unlike covalent imprinting these adducts are unstable and rearrange in relation to the imprinting process (6, 16). After production the template is removed by washing with a solvent or a mixture of solvents. The non-covalent approach is the most popular method of MIP synthesis (17) and was introduced in the early 1980s (11).

Non-covalent imprinting is usually based on commodity acrylics or methacrylic monomers (e.g., methacrylic acid (MAA)) which are then cross-linked with ethylene glycol dimethacrylate (EDMA). Hydrogen bonding or acid–base interactions between the template and functional groups (of low molecular weight compounds) are desirable (14).

3.3 Metal Ion Imprinting

Metal ions can take part in imprinting either as a template or can directly be involved in the template–polymer interactions. With the latter binding is possible as metal ions are able to accept electrons into their "unfilled orbitals of the outer coordination sphere" (6) from the heteroatoms of ligands. Therefore, the binding strength can range from weak bonds to strong bonds (which can behave covalently), depending on the metal, its oxidation state, and ligand characteristics (6, 16). The first approach was reported in 1985 with the imprinting of amino acids (18).

4 Sizes of MIP Templates

Imprinting of small molecules such as sugars, steroids, pesticides, certain pharmaceuticals, amino acid, peptides, and nucleotide bases is at the moment a very common area of research. The small size of the target molecules allows free movement into and out of the binding site within the MIP (3, 5).

Although MIP size is generally measured on the microscale, recently there has been a great scientific interest of regularly shaped molecularly imprinted polymers on the nanoscale (MIP nanoparticles), due to their properties. The binding kinetics are better due to the higher surface-to-volume ratios of MIP nanoparticles (providing easy accessibility for the templates). The most common synthesis strategies are "precipitation polymerization, mini-, and micro-emulsion polymerization core-shell approaches (with subsequent grafting) and living radical polymerization processes" (19).

Although creating MIPs for smaller molecules can be accomplished with some effort, molecular imprinting for larger molecules, such as proteins (which can range in molecular weight from 6,000 Da to several million Da), provides more of a challenge. A protein is considered as a macromolecule, which is a large molecule formed from smaller structural units due to some kind of polymerization. The difficulty for the proteins can actually be reaching the actual artificial binding site, an issue which can affect other types of macromolecules. Additionally the complexity of the protein surface (which can contain a large number of functional groups on its surface) can also cause difficulty—leading to nonspecific interactions occurring. Proteins can be incompatible with the organic solvents used in the synthesis of some MIPs; their sensitivity to pH, temperature, and ionic strength leading to denaturing can all cause issues when attempting to form related MIPs (3).

There are three main approaches for protein imprinting: bulk imprinting where the template is imprinted in the bulk of the polymer matrix and functional monomers are polymerized together into an MIP; surface imprinting is where the recognition and binding site are close or on the surface of the MIP and the template is only partially imprinted; and epitope imprinting where only a small section of the protein is used as the template but the MIP can identify the whole protein (3, 5).

Bulk imprinting on proteins has been successfully conducted on hemoglobin (20) and lysozyme (21) and polyacrylamide is common to use as it shows preferential selectivity for the templates. However, as polyacrylamide reduces density and therefore stability of the MIPs, alternative materials which can be used are hydrogels (22) and sol–gels (3, 21).

5 Rebinding of MIP Templates

The rebinding of the target molecule to the MIP depends highly on the medium, and the optimum medium can be predicted based on the MIP structure and morphology as well as the template structure (14). Thc ability of MIPs to selectively rebind with their templates (which can be a very wide variety of molecules, including those for which no natural/biological receptor exists) encourages their use for the purposes of acting as separating agents for solutions and also for sensing. Therefore MIP research has lead to a wide variety of sectors, including environmental monitoring (23). While there is the potential and much research for direct monitoring on human health, indirect health environment monitoring can be important, such as the use of MIPs for monitoring toxins produced by bacteria within water supplies (24) or other contaminants (25).

6 MIPs for Biosensing and Diagnostics

Accurate disease diagnosis has been a crucial part of medicine throughout history, with emphasis on earlier diagnosis giving clinicians a better chance to successfully treat patients. Recent advances at the cellular level have afforded scientists and researchers better knowledge regarding disease symptoms and biomarkers, leading to improved detection and novel research challenges.

Specific molecule recognition is an essential requirement of a successful biological diagnostic tool. MIPs currently form an intense area of scientific research, and their potential use within the medical diagnostics field has been well recognized, as several attempts have currently focused on producing complements to biological molecules. Since MIPs are capable of recognizing certain biological molecules, the next step from a medical perspective would be to detect those molecules which can provide evidence for the presence or progression of a disease.

Biosensors are devices which have a biological recognition element to detect the presence and concentration of analyte (through biochemical interactions) and provide a meaningful analytical signal through a transducer. The two key components of a biosensor therefore are the transducer, responsible for converting the signal received into meaningful information, and the biological recognition element. This biological recognition element can involve a wide variety of structures including enzymes, immune cells/proteins, DNA, organelles, and even whole cells. Since the function of the biological recognition element is to recognize complimentary molecules, MIPs are capable of reproducing this function. Due to their high selectivity and robustness, MIPs would be useful for detection in a wide range of environments and changes in conditions such as temperature or pressure changes.

Biosensors can be classified due to the type of biorecognition element or the transducer, and some examples of different transduction methods are electrochemical, optical, mass-sensitive, thermal, and magnetic. It has been reported that electrochemical sensors (where the signal is converted from the electrochemical interactions between the analyte and electrode) are the most common due to their low cost, good limits of detection, and ability to be miniaturized and automated (10, 26). While selectivity is important for biosensors, stability provides one of the bigger challenges that MIPs can solve (27). Within a biosensor, an MIP could artificially reproduce the functions performed by the biorecognition element, not only to detect a specific analyte, but to detect that analyte to a good degree of selectivity. A generalized diagram representing biosensors is illustrated in Fig. 4.

MIP research to produce electrochemical sensors has increased over the 1990–2000 decade (Fig. 5), while the first paper on this area was published in the early 1990s (in 1993), several other publications have subsequently arisen, with the largest number of papers from that decade being published in 2000 (29). And now with some 470 papers being published in 2011, it indicates the progression of research on this ever growing topic.

MIPs produced for a target analyte have been shown to be selective and produce a superior conductivity signal when compared to nontarget specific MIP, indicating their suitability for use within biosensors. However, it has also been seen that when the MIPs are saturated with the analyte, the difference between the targeted and unspecific MIPs were constant, indicating that the binding sites had become saturated and therefore no further uptake was possible

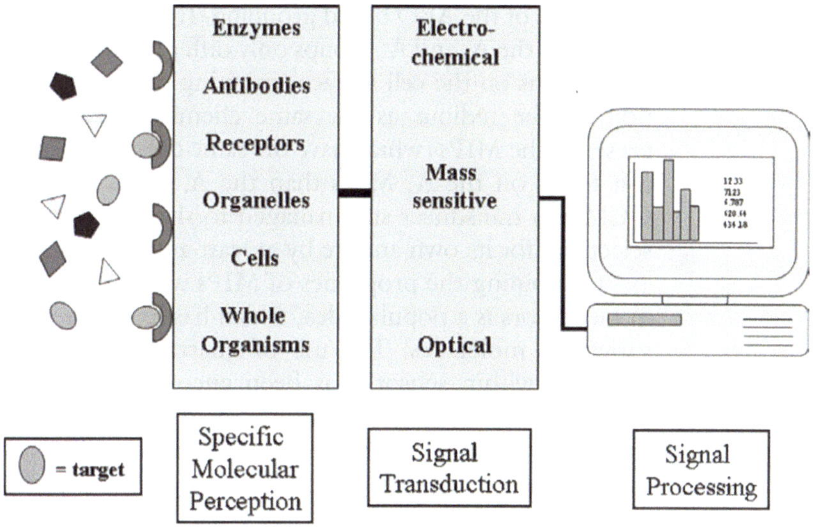

Fig. 4 The general structure and components of a biosensor, where it is envisaged that MIPs could artificially reproduce the function (performed by biomolecules) labeled "specific molecular perception" (the biorecognition element). Adapted from (28)

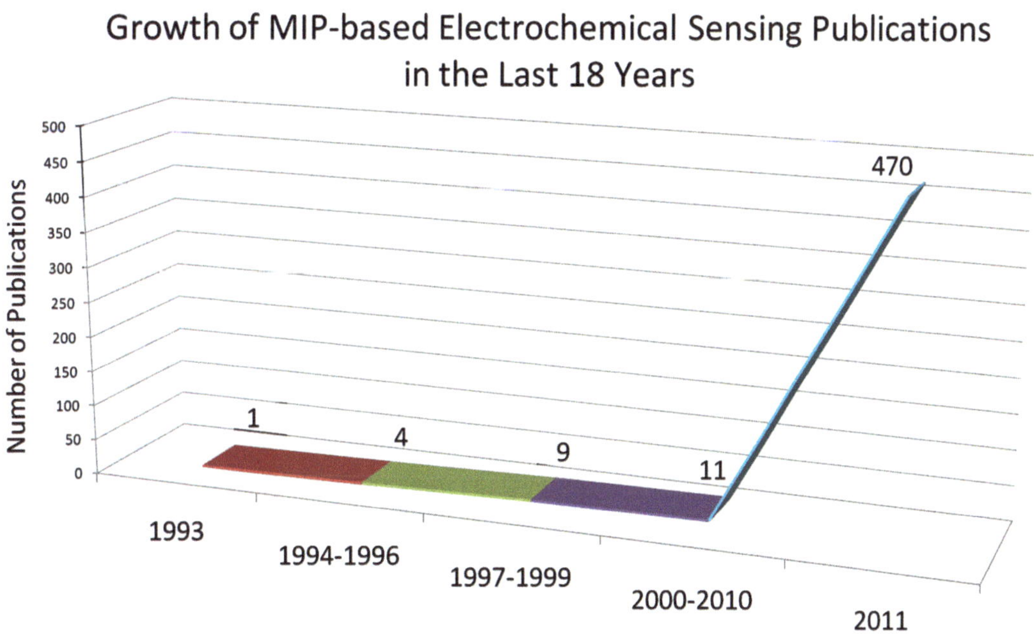

Fig. 5 Publication distribution for electrochemical sensors involving MIPs arranged by year. Modified from (29)

(27). Larger MIPs can cause a longer duration to response, due to diffusion times, however, the binding kinetics and therefore faster response times can be achieved by utilizing smaller MIP particles (27). However, the selectivity of MIPs has been shown to be very useful, even being able to differentiate between the A_1 and A_2

subgroups of the ABO blood grouping. It is important to note here that while the A_1 and A_2 groups only differ by the number of identical antigens on the cell surface (meaning that ordinary differentiation can be tedious as the same chemical recognition sites are present), the MIPs (which have the same chemical recognition sites, just fewer on the A_2 MIPs than the A_1 MIPs) combined with a QCM as a transducer still managed to show at least a preferential selectivity for its own analyte by at least a threefold factor (30).

Combining the properties of MIPs with piezoelectric material in biosensors is a popular idea, which has led to the detection of a variety of molecules. The use of quartz crystal (a piezoelectric material) within sensors has been encouraged because they are simple to construct and provide a rapid response. All but one of the classes of crystals which are lacking a center of symmetry are piezoelectric, and with the addition of external pressures or an electric field, the crystals become polarized. When an alternating electrical current is applied, alternating stress is produced and so various vibration modes can be produced, including those which are sensitive to mass examples including quartz crystal microbalance (QCM) and surface acoustic wave (SAW) devices (31). However, the surface of the piezoelectric crystal is nonselective and cannot effectively identify a specific molecule; therefore, the surface has to be modified with biomaterials (e.g., antibodies, enzymes, etc.), other chemicals or even molecularly imprinted polymers for selective targeting.

The QCM device in conjunction with an MIP has been shown to selectively detect albumin (32), a major blood protein, as well as molecules such as peptides which are related to viruses (33). In terms of accuracy of these devices as compared to conventional methods, when the QCM–albumin–MIP was tested on a sample of human serum the result was found to have only a 4% difference to the result obtained by clinical assay, indicating a good degree selectivity compared to clinical assay (32).

MIP biosensors have been shown to work on low-molecular weight molecules, such as l-nicotine, where they were able to distinguish between another similar low-molecular weight molecule, l-cotinine, which only differs by one oxygen atom (34). Figure 6 shows some MIP nanoparticles on a transducer. Using a conjugated thin film to immobilize the MIPs and with impedimetric detection allows the possibility to detect 10 nM with response increase of 45% as compared to a bare Pt electrode. With Impedance measurements the interactions between the MIPs and the analyte affects the formation of a double layer at the electrode surface, which produces a signal in the form of capacitive changes at a frequency range and allows charge transfer from the electrolyte to the electrode (34) *(for more information, please see Chap. 4.1 on electrochemical impedance spectroscopy).*

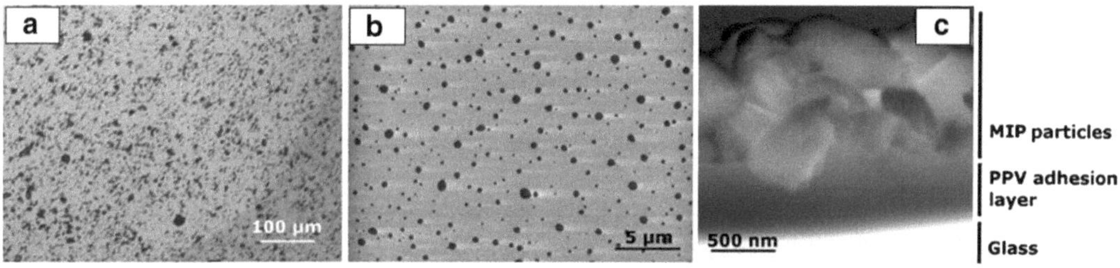

Fig. 6 (**a**) MIP particles on a transducer under an optical microscope, (**b**) scanning electron microscope (SEM) image of MIPs, and (**c**) is the cross-sectional image (with an accelerating voltage 15 keV and a large-field detector). Adapted from (34)

Diabetes biosensors are very common devices, as the monitoring of blood glucose levels in patients is crucial to effective management. Yoshimi et al. utilized the gating effect of MIPs (with glucose as the template, but without the utilization of an enzyme) grafted with an electrode and found that the anodic current intensity could relate to glucose concentration. The anodic current intensity (affected by glucose) was analyzed after cyclic voltammetry, and the response time to stepwise changes was found to be about 3–5 min (35).

MIPs have even been exploited to identify DNA. In the experiment the MIP gel acted as an electrophoretic matrix and hindered the target DNA (due to the interactions between the DNA and the MIP), and the distance travelled is compared to that in polyacrylamide gel. The solution is simple and cost-effective, but the MIP may not be able to distinguish T.A base pairs from A.T (base pair that was detected), which could cause errors with mutation detection (36).

However, molecularly imprinted polymer-based biosensors that have an optical component have been researched recently. Wu et al. utilized MIPs designed for formaldehyde, and when the target molecule diffuses into the MIP film it could cause changes in film density and refractive index which could then cause optical reflective changes which are then measured (by illuminating the MIP with a laser). As the reflective intensity shifts were proportional to the amount of formaldehyde present, exposing the MIP to compounds similar to formaldehyde, specifically benzaldehyde, formic acid, methanol, and ethylaldehyde, the specificity of the MIP can be identified. While exposing the MIP to similar compounds, there was a response; however, the measured response to formaldehyde was found to be much higher than any other substances, proving its specificity (37). An optical-based sensor was again used to produce the first research on an MIP membrane for digoxin (glycosylated steroid like drug, which can be prescribed for certain cardiac issues but can be toxic outside a strict concentration) (38).

7 MIPs and Implantable Biosensors

Miniaturized implantable biosensors form highly desirable diagnostic and monitoring tools for the practice of medicine. If these systems could be automated so that regardless of the patient's condition (i.e., conscious, sleeping, exercising, etc.) and additionally provide the ability to continuously monitor changes within the subject, then the advantages could be immense especially in conditions such as diabetes. However, complications can arise and for implantable biosensors, as well as other implantable devices, to be successful there are many challenges to overcome, including the body's natural responses to foreign objects (39).

Additionally implantable biosensors, such as those to monitor glucose, are only reliable within the short term partly due to tissue reactions including inflammation and fibrosis. The accumulation of inflamed cells around the sensor then can interrupt the effective monitoring of glucose levels, and therefore if inflammation is suppressed, then the implanted sensor has an increased chance to perform effectively for a longer period of time (40). To overcome this problem, very recently Wang et al. published an encouraging research work where a pH-sensitive MIP coated with nanosphere/hydrogel composite (Fig. 7) was developed as a protective coating for an implantable biosensor (40). Dexamethasone (DX) is a drug that is able to control inflammation, and so the MIP nanospheres/hydrogel was designed to allow a controlled 6-week release of dexamethasone-21 phosphate disodium (DXP).

The molecularly imprinted pH-sensitive nanospheres were fabricated by UV-initiated precipitation polymerization using DXP as the template molecule. The DXP loading and release experiments showed that the MIP nanospheres exhibited a higher loading level and slower release rate than non-imprinted polymer (NIP) nanospheres due to the interaction of DXP with the DXP-imprinted cavities within the MIP nanospheres. Moreover, the MIP nanospheres showed a faster DXP release rate at a lower pH value within the pH range tested (i.e., 6.0–7.4), which is desirable for suppressing inflammation, a response that induces an acidic microenvironment. In contrast, the NIP nanospheres did not show a notable pH-responsive DXP release behavior. The hydrogel poly(2-hydroxyethyl methacrylate (HEMA)-*N*-vinyl-2-pyrrolidinone (NVP)-2-methacryloyloxyethyl phosphorylcholine (MPC)) was prepared by UV polymerization. The MIP nanospheres were then successfully incorporated into the hydrogel. The equilibrium water content and swelling kinetics of the MIP nanospheres/hydrogel composite were similar to those of pure hydrogel. The MIP nanospheres/hydrogel composite demonstrated a much better controlled DXP release profile than the

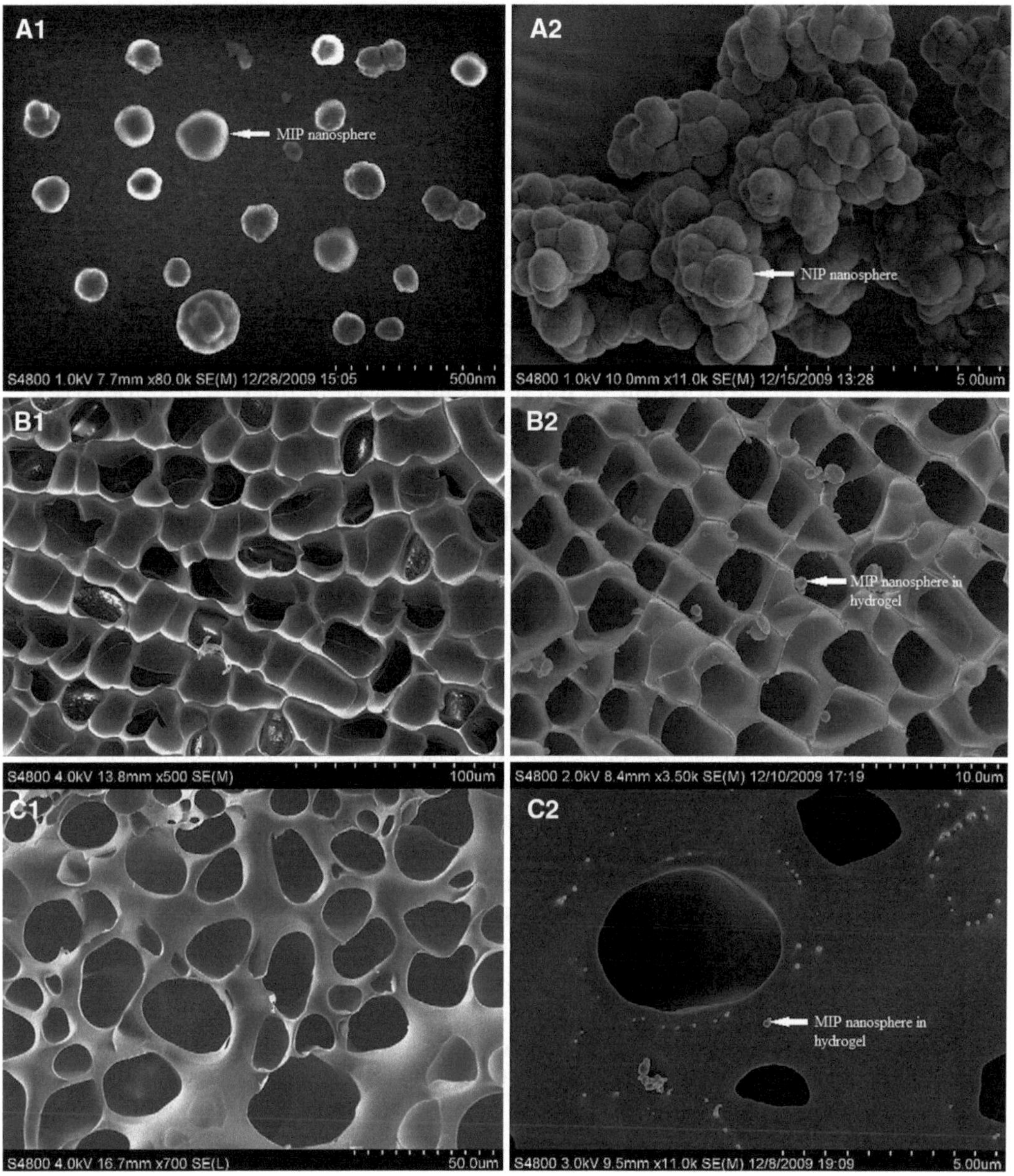

Fig. 7 SEM images of the nanospheres, hydrogel, and composite: (**A1**) MIP nanospheres; (**A2**) NIP nanospheres; (**B1**) cross section of the hydrogel; (**B2**) cross section of the MIP nanospheres/hydrogel composite; (**C1**) surface of the hydrogel; and (**C2**) surface of the MIP nanospheres/hydrogel composite. The above images show that the only discernable difference between MIPs and NIPs is the non-tendency of MIPs to form agglomerates. Adapted from (40)

pure hydrogel. This pH-sensitive MIP nanospheres/hydrogel composite designed as a coating for implantable biosensors could potentially suppress the inflammatory response of the implanted biosensors efficiently thereby effectively improving their lifetime.

8 Conclusions and Future Outlook

In conclusion, clearly as MIPs can be considered as an emerging field, the future potential for further research and practical application of molecularly imprinted polymers is immense. Within the last few years there has been a vast increase in the number of publications on this topic, indicating a shift in focus onto further research into MIPs.

As our understanding of microorganisms, diseases, and molecules progress, MIPs are likely to become more and more useful in creating artificial receptors for targeted templates. There is a good possibility that the medical, environmental protection, and security sectors will be interested in exploiting the potential for MIPs. As molecularly imprinted polymers are more resistant to changes in pH and temperature than biological receptor molecules (e.g., enzymes) they have advantages for use in extreme or unusual environments. Additionally the possibility of tailoring MIPs to custom targets provides the ability to selectively target the detection of specific pathogens and molecules. Furthermore, the preparation of MIPs is relatively easy and cost-effective.

Since the initial function of MIPs has been specific to molecule recognition, it follows that such polymers are "born" for detection/sensing of molecules. MIPs have been successfully developed as various biosensors for a large range of biomolecules such as glucose, DNA, formaldehyde, and digoxin. However, more investigations are still required in order to further improve the solubility and biocompatibility of MIPs, especially in biomedical applications such as medical devices, diagnostics, drug delivery, and biological assays. Additionally due to their construction property, MIPs tend to be more rigid than their biological equivalents.

It is very impressive that since the 1930s, MIPs have progressed a great deal and with the explosion of interest within the last two decades, the progression and manipulation of molecularly imprinted polymers have taken leaps forward. Further research continues and as more sectors begin to see the potentials, the list of functionality of MIPs continues to grow. For example, the great potentials of MIPs as advanced materials for in vivo sensing system seem to have already been realized. Some excellent efforts and progress have been made, but it may be some time before a clinical sample of in vivo biosensor utilizing molecular imprinting technique will be achieved.

References

1. Piletsky SA, Turner APF (2006) Molecular imprinting of polymers. Landes Bioscience, Georgetown, TX
2. Yan M, Ramström O (2005) Molecularly imprinted materials-science and technology, 1st edn. Marcel Dekker, New York, NY
3. Ge Y, Turner APF (2008) Too large to fit? Recent developments in macromolecular imprinting. Trends Biotechnol 26:218–224
4. Ge Y, Turner APF (2009) Molecularly imprinted sorbent assays: recent developments and applications. Chem Eur J 15:8100–8107

5. Haupt K (2003) Molecularly imprinted polymers: the next generation. Anal Chem 75:376A–383A
6. Alexander C, Andersson HS, Andersson LI, Ansell RJ, Kirsch N, Nicholls IA, O'Mahony J, Whitcombe MJ (2006) Molecular imprinting science and technology: a survey of the literature for the years up to and including 2003. J Mol Recognit 19:106–180
7. Polyakov MV (1931) Adsorption properties and structure of silica gel. Zh Fiz Khim 2:799–805
8. Dickey F (1949) The preparation of specific adsorbents. Proc Natl Acad Sci USA 35:227–229
9. Wulff G, Sarhan A, Zabrocki K (1973) Enzyme-analogue built polymers and their use for the resolution of racemates. Tetrahedron Lett 44:4329–4332
10. Suryanarayanan V, Wu CT, Ho KC (2010) Molecularly imprinted electrochemical sensors. Electroanalysis 22:1795–1811
11. Arshady R, Mosbach K (1981) Synthesis of substrate-selective polymers by host-guest polymerization. Makromol Chem 182:687–692
12. Sellergren B, Andersson L (1990) Molecular recognition in macroporous polymers prepared by a substrate analog imprinting strategy. J Org Chem 55:3381–3383
13. Chianella I, Lotierzo M, Piletsky SA, Tothill IE, Chen B, Karim K, Turner APF (2002) Rational design of a polymer specific for microcystin-LR using a computational approach. Anal Chem 74:1288–1293
14. Sellergren B (1999) Polymer- and template-related factors influencing the efficiency in molecularly imprinted solid-phase extractions. Trends Anal Chem 18:164–174
15. Matsui J, Miyoshi Y, Doblhoff-Dier O, Takeuchi T (1995) A molecularly imprinted synthetic polymer receptor selective for Atrazine. Anal Chem 67:4404–4408
16. Mayes AG, Whitcombe MJ (2005) Synthetic strategies for the generation of the molecularly imprinted organic polymers. Adv Drug Deliv Rev 57:1742–1778
17. Sellergren B, Allender CJ (2005) Molecularly imprinting polymers: a bridge to advanced drug delivery. Adv Drug Deliv Rev 57:1733–1741
18. Fujii Y, Matsutani K, Kikuchi K (1985) Formation of a specific co-ordination cavity for a chiral amino acid by template synthesis of a polymer Schiff base cobalt(III) complex. J Chem Soc Chem Commun 7:415–417
19. Poma A, Turner APF, Piletsky SA (2010) Advances in the manufacture of MIP nanoparticles. Trends Biotechnol 28:629–637
20. Guo TY, Xia YQ, Hao GJ, Zhang BH, Fu GQ, Yuan Z, He BL, Kennedy JF (2005) Chemically modified chitosan beads as matrices for adsorptive separation of proteins by molecularly imprinted polymer. Carbohydr Polym 62:214–221
21. Ou SH, Wu MC, Chou TC, Liu CC (2004) Polyacrylamide gels with electrostatic functional groups for the molecular imprinting of lysozyme. Anal Chim Acta 504:163–166
22. Odabaşi M, Say R, Denizli A (2007) Molecular imprinted particles for lysozyme purification. Mater Sci Eng 27:90–99
23. Breton F, Euzet P, Piletsky SA, Giardi MT, Rouillon R (2006) Integration of photosynthetic biosensor with molecularly imprinted polymer-based solid phase extraction cartridge. Anal Chim Acta 569:50–57
24. Chianella I, Piletsky SA, Tothill IE, Chen B, Turner APF (2003) MIP-based solid phase extraction cartridges combined with MIP-based sensors for the detection of microcystin-LR. Biosens Bioelectron 18:119–127
25. Valero-Navarro A, Salinas-Castillo A, Fernández-Sánchez JF, Segura-Carretero A, Mallavia R, Fernández-Gutiérrez A (2009) The development of a MIP-optosensor for the detection of monoamine naphthalenes in drinking water. Biosens Bioelectron 24:2305–2311
26. Farré M, Kantiani L, Pérez S, Barceló D (2009) Sensors and biosensors in support of EU directives. Trends Anal Chem 28:170–185
27. Kriz D, Kempe M, Mosbach K (1996) Introduction of molecularly imprinted polymers as recognised elements in conductometric chemical sensors. Sens Actuators B Chem 33:178–181
28. Nakamura H, Karube I (2003) Current research activity in biosensors. Anal Bioanal Chem 377:446–468
29. Blanco-López MC, Lobo-Castañón MJ, Miranda-Ordieres AJ, Tuñón-Blanco P (2004) Electrochemical sensors based on molecularly imprinted polymers. Trends Anal Chem 23:36–48
30. Jenik M, Seifner A, Krassnig S, Seidler K, Lieberzeit PA, Dickert FL, Jungbauer C (2009) Sensors for bioanalytes by imprinting—polymers mimicking both biological receptors and the corresponding bioparticles. Biosens Bioelectron 25:9–14
31. Ramsden JJ (1994) Experimental methods for investigating protein adsorption kinetics at surfaces. Q Rev Biophys 27:41–105
32. Lin TY, Hu CH, Chou TC (2004) Determination of albumin concentration by MIP-QCM sensor. Biosens Bioelectron 20:75–81
33. Tai DF, Lin CY, Wu TZ, Chen LK (2005) Recognition of dengue virus protein using epitope-mediated molecularly imprinted film. Anal Chem 77:5140–5143

34. Thoelen R, Vansweevelt R, Duchateau J, Horemans F, D'Haen J, Lutsen L, Vanderzande D, Ameloot M, VandeVen M, Cleij TJ, Wagner P (2008) A MIP-based impedimetric sensor for the detection of low-MW molecules. Biosens Bioelectron 23:913–918
35. Yoshimi Y, Narimatsu A, Nakayama K, Sekine S, Hattori K, Sakai K (2009) Development of an enzyme-free glucose sensor using the gate effect of a molecularly imprinted polymer. J Artif Organs 12:264–270
36. Ogiso M, Minoura N, Shinbo T, Shimizu T (2006) Detection of a specific DNA sequence by electrophoresis through a molecularly imprinted polymer. Biomaterials 27:4177–4182
37. Wu N, Feng L, Tan Y, Hu J (2009) An optical reflected device using a molecularly imprinted polymer film sensor. Anal Chim Acta 653:103–108
38. González GP, Hernando PF, Alegría JSD (2009) An optical sensor for the determination of digoxin in serum samples based on a molecularly imprinted polymer membrane. Anal Chim Acta 638:209–212
39. Vaddiraju S, Tomazos I, Burgess DJ, Jain FC, Papadimitrakopoulos F (2010) Emerging synergy between nanotechnology and implantable biosensors: a review. Biosens Bioelectron 25:1553–1565
40. Wang C, Javadi A, Ghaffari M, Gong S (2010) A pH-sensitive molecularly imprinted nanospheres/hydrogel composite as a coating for implantable biosensor. Biomaterials 31:4944–4951

INDEX

Stéphane Marinesco and Nicholas Dale (eds.), *Microelectrode Biosensors*, Neuromethods, vol. 80,
DOI 10.1007/978-1-62703-370-1, © Springer Science+Business Media, LLC 2013

T

V

W

MIX
Papier aus verantwortungsvollen Quellen
Paper from responsible sources
FSC® C105338

If you have any concerns about our products,
you can contact us on
ProductSafety@springernature.com

In case Publisher is established outside the EU,
the EU authorized representative is:
Springer Nature Customer Service Center GmbH
Europaplatz 3, 69115 Heidelberg, Germany

Printed by Libri Plureos GmbH
in Hamburg, Germany